Springer Texts in Statistics

Springer Texts in Statistics (STS) includes advanced textbooks from 3rd- to 4th-year undergraduate courses to 1st- to 2nd-year graduate courses. Exercise sets should be included. The series editors are currently Genevera I. Allen, Richard D. De Veaux, and Rebecca Nugent. Stephen Fienberg, George Casella, and Ingram Olkin were editors of the series for many years.

More information about this series at http://www.springer.com/series/417

Bing Li · G. Jogesh Babu

A Graduate Course on Statistical Inference

Springer

Bing Li
Department of Statistics
Penn State University
University Park, PA, USA

G. Jogesh Babu
Department of Statistics
Penn State University
University Park, PA, USA

ISSN 1431-875X ISSN 2197-4136 (electronic)
Springer Texts in Statistics
ISBN 978-1-4939-9759-6 ISBN 978-1-4939-9761-9 (eBook)
https://doi.org/10.1007/978-1-4939-9761-9

This Springer imprint is published by the registered company Springer Science+Business Media, LLC part of Springer Nature.
The registered company address is: 233 Spring Street, New York, NY 10013, U.S.A.

To our parents:
Jianmin Li and Liji Du
Nagarathnam and Mallayya

and to our families:
Yanling, Ann, and Terrence
Sudha, Vinay, and Vijay

Preface

It is our goal to write a compact, rigorous, self-contained, and accessible graduate textbook on statistical estimation and inference that reflects the current trends in statistical research.

The book contains three main themes: the finite-sample theory, the asymptotic theory, and Bayesian statistics. Chapters 2 through 4 are devoted to the finite-sample theory, which includes the classical theory of optimal estimation and hypothesis test, sufficiency, completeness, ancillarity, and exponential families. Chapters 5 to 6 are devoted to Bayesian statistics, covering prior and posterior distributions, Bayesian decision theory for estimation, hypothesis testing, and classification, empirical Bayes, shrinkage estimates. Chapters 8 through 11 are devoted to asymptotic theory, covering consistency and asymptotic normality of maximum likelihood estimation and estimating equations, the Le Cam-Hajek convolution theorem for regular estimates, and the asymptotic analysis of a wide variety of hypothesis testing procedures. Two chapters on preliminaries are included to make the book self-contained: Chapter 1 contains preliminaries for the finite-sample theory and Bayesian statistics; Chapter 7 for the asymptotic theory.

The topics and treatment of some material are different from a typical textbook on statistical inference, which we regard as a special feature of this book. For example, we devoted a chapter on estimating equations and used it as a unifying mechanism to cover some useful methodologies such as the generalized linear models, generalized estimation equations, quasi likelihood estimation, and conditional inference. We include a systematic exposition of the theory of regular estimates, from regularity, contiguity, the convolution theory, to asymptotic efficiency. This theory was then used in conjunction with the Local Asymptotic Normal (LAN) assumption to develop asymptotic local alternative distributions and the optimal properties for a wide variety of hypothesis testing procedures that can be written as quadratic forms in the limit.

One of the features of the book is the systematic use of a parsimonious set of assumptions and mathematical tools to streamline some recurring regularity conditions, and theoretical results that are fundamentally similar. This makes the development of the methodology more transparent and interconnected, and the book a coherent whole. For example, the conditions "differentiable under the integral sign (DUI)", and "stochastic equicontinuity" are repeatedly used throughout many chapters of the book; the geometric projection and the multivariate Cauchy-Schwarz inequality are used to unify different types of optimal theories; the structures of asymptotic estimation and hypothesis testing echo their counterparts in the finite-sample theory.

This book can be used either as a one-semester or a two-semester textbook on statistical inference. For the two-semester courses, the first six chapters can be used for the first semester to cover finite-sample estimation and Bayesian statistics, and the last five for the second semester to cover asymptotic statistics. For a one-semester course, there are several pathways depending on the instructor's emphasis. For example, one possibility is to use Chapters 1, 3, 4, 7, 10, 11 for an advanced course on hypothesis testing; another possibility is to use Chapters 1, 2, 5, part of 6, 7, 8, 9 as an advanced course on point estimation and Bayesian statistics.

The book grew out of the lecture notes for two graduate-level courses that we have taught for more than two decades at the Pennsylvania State University. Over this period we have revamped the courses several times to adapt to the evolving trends, emphases, and demands in theoretical and methodological research. The authors are grateful to the Department of Statistics of the Pennsylvania State University for its constant support and the stimulating research and education environment it provides. The authors also gratefully acknowledge the support from the National Science Foundation grants.

State College
April 2019

Bing Li
G. Jogesh Babu

Contents

1

Probability and Random Variables

A brief outline of the important ideas and results from classical theory of measure and probability are presented in this chapter. This is not intended for the first reading of the subject, but rather as a review and a reference. Occasionally some proofs are presented, but in most cases they are omitted, and such omissions are indicated by saying "it is true ...". These proofs are easily found in standard texts on measure theory and probability, such as Billingsley (1995), Rudin (1987), and Vestrup (2003). In the last section we lay out some basic notations that will be repeatedly used throughout the book.

1.1 Sample space, events, and probability

Probability theory has three basic elements: outcomes, events, and probability. An outcome is a result of an experiment. An experiment here means any action that can have a number of possible results, but which result will actually occur cannot be predicted with certainty before the experiment is performed. For example, tossing a coin is an experiment, and the coin turning up heads is an outcome; rolling a die is an experiment, and the die turns up 6 is an outcome. The set of all outcomes of an experiment is the sample space, which is denoted by Ω. For example, in the experiment of tossing a pair of dice, the sample space is the set of 36 possible combinations of (i, j), $i, j = 1, \ldots, 6$. A set of outcomes, or a subset of Ω, is an event. Of course a single outcome is itself an event, but when we think of it as an event, we think of it as a subset, rather than an element, of the sample space.

1.2 σ-field and measure

In probability we are concerned with a class of events, which is formulated as an algebraic structure called the σ-field. Intuitively, a σ-field describes the world in which events occur, whose likelihood we would like to assess.

B. Li and G. J. Babu, *A Graduate Course on Statistical Inference*,
Springer Texts in Statistics, https://doi.org/10.1007/978-1-4939-9761-9_1

Definition 1.1 *A σ-field (or σ-algebra) over a nonempty set Ω is any collection of subsets in Ω that satisfies the following three conditions:*

1. *$\Omega \in \mathcal{F}$,*
2. *If $A \in \mathcal{F}$, then $A^c \in \mathcal{F}$,*
3. *If $A_1, A_2, \ldots$ is a sequence of sets in $\mathcal{F}$, then $\bigcup_{n=1}^{\infty} A_n \in \mathcal{F}$.*

A set A in a σ-field $\mathcal{F}$ is called a $\mathcal{F}$-measurable set, a measurable set, or an $\mathcal{F}$-set. It is easy to show that the collection of all subsets of Ω is a σ-field, and the collection $\{\varnothing, A, A^c, \Omega\}$, where $A \subset \Omega$, is a σ-field. Another simple example is $\{\varnothing, \Omega\}$, which is, in fact, the smallest σ-field.

It is true that the intersection of any collection of σ-fields is itself a σ-field. Let $\mathcal{A}$ be a collection of subsets of Ω. Then the σ-field generated by $\mathcal{A}$ is defined as the intersection of all σ-fields that contain $\mathcal{A}$. This is well defined because the collection of all subsets of Ω, which must contain $\mathcal{A}$, is a σ-field. If $\Omega = \mathbb{R}^k$, the k-dimensional Euclidean space, and if $\mathcal{A}$ is the collection of all open sets in Ω, then the σ-field generated by $\mathcal{A}$ is written as $\mathcal{R}^k$. Members of $\mathcal{R}^k$ are called the Borel sets. (Therefore $\mathcal{R}^k$ is the collection of all Borel sets).

A set Ω, together with a σ-field $\mathcal{F}$ of its subsets, is called a measurable space, and is written as $(\Omega, \mathcal{F})$.

A measure is a mechanism that enable us to assign probability to each event in a σ-field. It can also be understood as the length, the area, or the volume, and so on, of a set.

Definition 1.2 *A measure μ defined on a measurable space $(\Omega, \mathcal{F})$ is a mapping $\mu : \mathcal{F} \to [0, \infty]$ such that*

1. *$\mu(\varnothing) = 0$,*
2. *If $A_1, A_2, \ldots \in \mathcal{F}$ and $A_i \cap A_j = \emptyset$ whenever $i \neq j$, then*

$$\mu\Big(\bigcup_{n=1}^{\infty} A_n\Big) = \sum_{n=1}^{\infty} \mu(A_n).$$

A measure μ is called a σ-finite measure if there is a sequence of $\mathcal{F}$-sets $\{A_n\}$ such that $\Omega = \cup_{n=1}^{\infty} A_n$ and $\mu(A_n) < \infty$. It is called a finite measure if $\mu(\Omega) < \infty$. It is called a probability measure if $\mu(\Omega) = 1$. There is no essential difference between a finite measure and a probability measure; the latter is introduced simply to conform to our daily convention that the largest probability is 100%.

A set Ω, together with a σ-field $\mathcal{F}$ of its subsets, and a measure μ defined on $(\Omega, \mathcal{F})$, is called a measure space. In the special case where μ is a probability measure on $\mathcal{F}$, $(\Omega, \mathcal{F}, \mu)$ is called a probability space. Often P is used, instead of μ, to represent a probability measure.

Example 1.1 It is true that there exists a unique measure λ on $(\mathbb{R}^k, \mathcal{R}^k)$ such that for each open rectangle A in $\mathbb{R}^k$, $\lambda(A)$ is the volume of the rectangle. This measure is called the Lebesgue measure.

Example 1.2 Consider the measure space $(\mathbb{R}, \mathcal{R}, \lambda)$, where λ is the Lebesgue measure. Then λ is σ-finite: let $A_n = (-n, n)$, then $\bigcup_{n=1}^{\infty} A_n = \mathbb{R}$, $\lambda(A_n) < \infty$.

Example 1.3 Let $\Omega = \{\omega_1, \omega_2, \ldots\}$ be a countable set. Let $\mathcal{F}$ be the class of all subsets of Ω. Then $\mathcal{F}$ is a σ-field. Let $\mu : \mathcal{F} \to \mathbb{R}$ be defined as follows: for any subset A of Ω, $\mu(A) =$ the number of elements in A. Then it is true that μ is a measure on $\mathcal{F}$. This measure is called the counting measure.

1.3 Measurable function and random variable

Let $(\Omega, \mathcal{F})$ and $(\Omega', \mathcal{F}')$ be two measurable spaces. Let $f : \Omega \to \Omega'$ be a mapping from Ω to Ω'. For any set $A' \subseteq \Omega'$, let $f^{-1}(A')$ denotes the set $\{\omega \in \Omega : f(\omega) \in A'\}$. Suppose we have a measure μ on $(\Omega, \mathcal{F})$. Then we could measure any set A in $\mathcal{F}$ by $\mu(A)$. But can we somehow use μ to measure any set A' in $\mathcal{F}'$? One possibility is to use the measure μ of the set in Ω that maps to A'. To do so we need this set, $f^{-1}(A')$, to be in $\mathcal{F}$. This motivates the following definition of the measurability of f.

Definition 1.3 *The mapping f is measurable $\mathcal{F}/\mathcal{F}'$ if, whenever $A' \in \mathcal{F}'$, $f^{-1}(A') \in \mathcal{F}$.*

Suppose $(\Omega', \mathcal{F}') = (\mathbb{R}^k, \mathcal{R}^k)$. If $f : \Omega \to \mathbb{R}^k$ is measurable $\mathcal{F}/\mathcal{R}^k$, then we will say f is measurable with respect to $\mathcal{F}$, or measurable $\mathcal{F}$, or, simply, measurable.

As mentioned at the beginning of this section, if μ is a measure on $(\Omega, \mathcal{F})$, then any mapping $f : \Omega \to \Omega'$ that is measurable with respect to $\mathcal{F}/\mathcal{F}'$ induces a measure on $\mathcal{F}'$, in the following way. For any $A' \in \mathcal{F}'$, define the set function

$$\nu(A') = \mu(f^{-1}(A')).$$

It is true that ν is a measure in $(\Omega', \mathcal{F}')$. This measure is written as $\mu \circ f^{-1}$.

Suppose that $(\Omega, \mathcal{F}, \mu)$ is a measure space, and $(\Omega', \mathcal{F}')$ is a measurable space. Suppose that the function $f : \Omega \to \Omega'$ is measurable $\mathcal{F}/\mathcal{F}'$. Let $B \in \mathcal{F}'$. We say that $f \in B$ almost everywhere μ (a.e. μ) if

$$\mu(\{\omega \in \Omega : f(\omega) \notin B)\}) = (\mu \circ f^{-1})(B^c) = 0.$$

If P is a probability measure, then $f \in B$ almost everywhere P is also called $f \in B$ almost everywhere P. Note that in this case, $P(\{\omega : f(\omega) \notin B\}) = 0$ is equivalent to $P(\{\omega : f(\omega) \in B\}) = 1$.

For convenience, we abbreviate sets such as

$$\{\omega : f(\omega) \notin B\}, \quad \{\omega : f(\omega) \in B\}$$

as $\{f \in B\}$ and $\{f \notin B\}$.

In the context of probability theory, a mapping $X : \Omega \to \mathbb{R}^k$ that is measurable with respect to $\mathcal{F}/\mathcal{R}^k$ is also called a random vector. If $k = 1$, then the mapping is called a random variable. For general measurable spaces $(\Omega, \mathcal{F})$ and $(\Omega', \mathcal{F}')$, a mapping $X : \Omega \to \Omega'$ that is $\mathcal{F}/\mathcal{F}'$ measurable is called a random element. The notation $X(\omega)$ represents an evaluation, or a realization, of an element $\omega \in \Omega$.

Suppose that $X = (X_1, \ldots, X_k)^T$ is a k-dimensional random vector on $(\Omega, \mathcal{F}, P)$. Then X induces a probability measure, $P \circ X^{-1}$ on $\mathcal{R}^k$. We call this probability measure the distribution of X, and write it as P_X. Let $a_1, \ldots, a_k$ be numbers in $\mathbb{R}$ and let $a = (a_1, \ldots, a_k)^T$. The function $F_X : \mathbb{R}^k \to [0, 1]$ defined by

$$\begin{aligned} F_X(a) =& P(\{\omega : X_1(\omega) \le a_1, \ldots, X_k(\omega) \le a_k\}) \\ =& P_X(\{x : x_1 \le a_1, \ldots, x_k \le a_k\}) \end{aligned}$$

is called the cumulative distribution function of X. Evidently, the measure P_X uniquely determines the function F_X through the above relation. In fact, F_X also uniquely determines P_X. For this reason F_X is also called the distribution of X.

Let $(\Omega, \mathcal{F})$ and $(\Omega', \mathcal{F}')$ be measurable spaces, and $X : \Omega \to \Omega'$ be a random element. Let Ω_X be the range of X, $\{X(\omega) : \omega \in \Omega\}$. Then, regardless of whether $\Omega_X \in \mathcal{F}'$, the class of sets $\{\Omega_X \cap A : A \in \mathcal{F}'\}$ is a σ-field. We denote this σ-field as $\mathcal{F}_X$, and say that X is defined on $(\Omega, \mathcal{F})$ and takes values in $(\Omega_X, \mathcal{F}_X)$.

The σ-field generated by a random vector X, written as $\sigma(X)$, is the intersection of all σ-fields with respect to which X is measurable. Two random vectors X and Y on $(\Omega, \mathcal{F})$ of dimensions k and ℓ are independent if, for any $A \in \sigma(X)$ and $B \in \sigma(Y)$, we have $P(A \cap B) = P(A)P(B)$. It is true that $\sigma(X)$ is equal to the collection of sets $\{X^{-1}(A_1) : A_1 \in \mathcal{R}^k\}$. It follows that X and Y are independent if and only if, for any $A_1 \in \mathcal{R}^k$ and $B_1 \in \mathcal{R}^\ell$, one has $P(X \in A_1, Y \in B_1) = P(X \in A_1)P(Y \in B_1)$. The independence of several random vectors, or a sequence of vectors, are defined similarly.

1.4 Integral and its properties

Let $(\Omega, \mathcal{F}, \mu)$ be a measure space. Let $\{A_1, \ldots, A_k\}$ be a finite and measurable partition of Ω; that is, $A_1, \ldots, A_k$ are disjoint and $\cup_{i=1}^k A_i = \Omega$. Let $f : \Omega \to \mathbb{R}$ be a nonnegative measurable function (with respect to $\mathcal{F}$). Consider the following sum

$$\sum_i \left[\inf_{\omega \in A_i} f(\omega) \right] \mu(A_i) \tag{1.1}$$

In this summation, $\mu(A_i)$ or $\inf_{\omega \in A_i} f(\omega)$ are allowed to be ∞, and we adopt the following convention:

$$\begin{cases} 0 \cdot \infty = \infty \cdot 0 = 0, \\ x \cdot \infty = \infty \cdot x = \infty, & \text{if } 0 < x < \infty \\ \infty \cdot \infty = \infty \end{cases} \tag{1.2}$$

The supremum of this sum (1.1) over all finite measurable partitions is defined to be the integral $\int f d\mu$. That is

$$\int f d\mu = \sup \sum_i \left[\inf_{\omega \in A_i} f(\omega) \right] \mu(A_i).$$

If this number is finite, we say that f is integrable. If it is ∞, then we say that f is not integrable, but has integral ∞.

For an arbitrary measurable function $f : \Omega \to \mathbb{R}$, let f^+ be defined by

$$f^+(\omega) = \begin{cases} f(\omega) & \text{if } f(\omega) \geq 0 \\ 0 & \text{otherwise} \end{cases}$$

and let $f^- = (-f)^+$. Then f^+ and f^- are nonnegative measurable functions, and $f = f^+ - f^-$. The integral of f is defined as

$$\int f d\mu = \int f^+ d\mu - \int f^- d\mu,$$

if at least one of the terms on the right is finite. f is said to be integrable if both terms on the right are finite. If $\int f^+ d\mu = \infty$ and $\int f^- d\mu < \infty$, then f is not integrable, but has definite integral ∞. If $\int f^- d\mu = \infty$ and $\int f^+ d\mu < \infty$, then f is not integrable but its definite integral is $-\infty$. If both terms are ∞, then integral of f is not defined.

Let A be a measurable set. Let I_A be the indicator function of A; that is $I_A : \Omega \to \{0, 1\}$ is defined by

$$I_A(\omega) = \begin{cases} 1 & \text{if } \omega \in A \\ 0 & \text{if } \omega \notin A \end{cases}.$$

It is true that $f I_A$ is a measurable function if f is measurable. Then $\int f I_A d\mu$ will be written as $\int_A f d\mu$, and is called the integral of f with respect to measure μ over the set A, whenever it exists.

An integral has the following properties.

Theorem 1.1

1. *Let $\{A_1, \ldots, A_k\}$ be a finite measurable partition of Ω, and $f : \Omega \to \mathbb{R}$ be a nonnegative simple function; that is, $f(\omega) = \sum_i x_i I_{A_i}(\omega)$ for some nonnegative numbers $x_1, \ldots, x_k$. Then $\int f d\mu = \sum_i x_i \mu(A_i)$.*
2. *If f and g are integrable and $f \leq g$ almost everywhere, then $\int f d\mu \leq \int g d\mu$.*

3. *If f and g are integrable and α and β are real numbers, then $\alpha f + \beta g$ is integrable and*

$$\int \alpha f d\mu + \beta g d\mu = \alpha \int f d\mu + \beta \int g d\mu.$$

A property is said to hold almost everywhere with respect to a measure μ (a.e. μ), if it holds for ω outside a measurable set of μ-measure zero.

Note that, part 2 implies that if $f = g$ a.e. μ, then $\int f d\mu = \int g d\mu$. An important consequence of part 1 is that, for any $A \in \mathcal{F}$,

$$\mu(A) = \int I_A d\mu = \int_A d\mu.$$

That is, the measure of a set A is the integral of the measure over that set.

Theorem 1.2 *Suppose that $f : \Omega \to \mathbb{R}$ is measurable and nonnegative. Then*

1. *$f = 0$ a.e. μ if and only if $\int f d\mu = 0$.*
2. *If $\int f d\mu < \infty$ then $f < \infty$ a.e. μ.*

Corollary 1.1 *If f and g are measurable $\mathcal{F}$ and integrable μ and if $\int_A f d\mu = \int_A g d\mu$ for all $A \in \mathcal{F}$, then $f = g$ a.e. μ.*

Proof. Note that

$$\int |f - g| d\mu = \int_{\{f>g\}} (f - g)\, d\mu + \int_{\{f<g\}} (g - f)\, d\mu$$

Because $\{f > g\} \in \mathcal{F}$ and $\{f < g\} \in \mathcal{F}$, the right hand side is zero. Hence $\int |f - g| d\mu = 0$ which, by part 1 of Theorem 1.2, implies that $f = g$ a.e. μ. □

Corollary 1.2 *If $f : \Omega \to \mathcal{F}$ is measurable $\mathcal{F}$ and integrable μ, and $\int_A f d\mu \leq 0$ for all $A \in \mathcal{F}$, then $f \leq 0$ almost everywhere μ.*

Proof. Since $\int_A f d\mu \leq 0$ for every $A \in \mathcal{F}$, we have $\int_{f>0} f d\mu \leq 0$. Hence $\int I(f > 0) f d\mu \leq 0$. But we know that $I(f > 0) f \geq 0$. So $\int I(f > 0) f d\mu \geq 0$. Therefore this integral must be 0. By part 1 of Theorem 1.2, $I(f > 0) f = 0$ a.e. μ. However, we know that

$$\{\omega : I(f > 0) f = 0\} = \{\omega : f = 0\} \cup \{\omega : f \leq 0\} = \{\omega : f \leq 0\}.$$

Therefore, $f \leq 0$ a.e. μ. □

Suppose that $(\Omega, \mathcal{F}, P)$ is a probability space, and $X : \Omega \to \mathbb{R}$ is a random variable. Suppose that $f : \mathbb{R} \to \mathbb{R}$ is measurable $\mathcal{R}/\mathcal{R}$ and that $f \circ X$ is integrable with respect to P. Then

$$\int f \circ X dP = \int f[X(\omega)] P(d\omega)$$

is called the expectation of $f(X)$, and is written as $E[f(X)]$. If X^2 is integrable, then $E[X - E(X)]^2$ is called the variance of X, and is written as $\text{var}(X)$.

1.5 Some inequalities

A set A in a vector space is convex if, for any $a_1, a_2 \in A$, the line segment

$$\{(1-\lambda)a_1 + \lambda a_2 : \lambda \in [0,1]\}$$

is in A. A real-valued function f defined on a convex set A is convex if, for any $a_1, a_2 \in A$ and $\lambda \in (0,1)$,

$$f((1-\lambda)a_1 + \lambda a_2) \leq (1-\lambda)f(a_1) + \lambda f(a_2).$$

Such a function is strictly convex if the above inequality is strict whenever $a_1 \neq a_2$. Let $(\Omega, \mathcal{F}, P)$ be a probability space. We say that a probability measure is degenerate if it is concentrated on a single point. The next theorem is taken from Perlman (1974).

Theorem 1.3 (Jensen's inequality) *Let f be a convex function defined on a convex subset C of $\mathbb{R}^p$, and let $X \in \mathbb{R}^p$ be an integrable random vector such that $P(X \in C) = 1$. Then*

1. *$E(X) \in C$;*
2. *$Ef(X)$ exists and $f(E(X)) \leq E(f(X))$;*
3. *if f is strictly convex and the distribution of X is not degenerate then $f(E(X)) < E(f(X))$.*

Two positive integers, p and q, are called a conjugate pair if $1/p + 1/q = 1$. If $p = 1$, then $(1, \infty)$ is also defined as a conjugate pair. Let $(\Omega, \mathcal{F}, \mu)$ be a measure space.

Theorem 1.4 *Suppose (p, q) is a conjugate pair, and f and g are measurable functions on Ω. Then the following inequalities hold:*

$$\int |fg| d\mu \leq \left(\int |f|^p d\mu \right)^{1/p} \left(\int |g|^q d\mu \right)^{1/q}, \tag{1.3}$$

and

$$\left(\int |f+g|^p d\mu \right)^{1/p} \leq \left(\int |f|^p d\mu \right)^{1/p} + \left(\int |g|^p d\mu \right)^{1/p}. \tag{1.4}$$

The first inequality is the Hölder's inequality; the second is the Minkowski's inequality.

1.6 Logical statements modulo a measure

Recall that a statement holds almost everywhere μ if it holds everywhere outside a measurable set of μ measure zero. This convention induces a logical deduction system modulo a measure, which is now developed further for later use. Let $(\Omega, \mathcal{F}, \mu)$ be a measure space. Let S be a logical statement. We say S is a measurable statement if $\{\omega : S\} \in \mathcal{F}$. Here, we should always understand $\{\omega : S\}$ as $\{\omega \in \Omega : S\}$. We say that a statement holds for ω' if $\omega' \in \{\omega : S\}$. We say that S holds on a subset A of Ω if S holds for every $\omega \in A$; that is, $A \subseteq \{\omega : S\}$. We say that a statement S holds if it holds for every $\omega \in \Omega$; that is $\{\omega : S\} = \Omega$. If a statement holds for ω, then we write $S(\omega)$. As an example, let $f : \Omega \to \mathbb{R}$ be a measurable function. Let S be the statement that $f > 0$. Because f is measurable $\mathcal{F}$, $f > 0$ is a measurable statement. The sentence "$f > 0$ holds for ω" means $f(\omega) > 0$. So the symbol, $(f > 0)(\omega)$ means $f(\omega) > 0$.

In measure theory, many statements hold not for Ω but for a subset A of Ω with $\mu(A^c) = 0$ for a measure μ on Ω. For example, later on we will learn that if $E(X^2) = 0$, then all we can conclude is $P(X \neq 0) = 0$. We cannot conclude $X(\omega) = 0$ for every $\omega \in \Omega$. If a statement S only holds for ω on a set A with $\mu(A^c) = 0$, then we say S holds modulo μ, and write

$$S \quad [\mu].$$

For example, $f > 0 \quad [\mu]$ means $\mu(f \leq 0) = 0$. What is interesting — and extremely convenient — is that modulo μ statements obey the usual logical laws, in the sense of the following proposition.

Proposition 1.1 *Let $S_1, \ldots, S_k$ be k logical statements. If, for every $\omega \in \Omega$,*

$$S_1(\omega), \ldots, S_k(\omega) \implies S(\omega), \tag{1.5}$$

then $S_1 \ [\mu], \ldots, S_k \ [\mu] \implies S \ [\mu]$.

Proof. The expression (1.5) means that

$$\{\omega : S_1\} \cap \cdots \cap \{\omega : S_k\} \subseteq \{\omega : S\},$$

which is equivalent to $\{\omega : S\}^c \subseteq \{\omega : S_1\}^c \cup \cdots \cup \{\omega : S_k\}^c$. Consequently,

$$\mu(\{\omega : S\}^c) \leq \mu(\{\omega : S_1\}^c) + \cdots + \mu(\{\omega : S_k\}^c).$$

So if each term on the right is 0 then the term on the left is also 0. □

A practical implication of this proposition is that, when we are dealing with a finite set of statements each of which holds modulo μ, we can make logical deductions ignoring μ, and at the end state the conclusion modulo μ. The premise of this simplification is that the modulus measure to be the same for every statement involved. For different measures, the following proposition is helpful.

Let μ and ν be measures on $(\Omega, \mathcal{F})$. We say that ν is absolutely continuous with respect to μ if, for every $A \in \mathcal{F}$, $\mu(A) = 0 \implies \nu(A) = 0$. In this case we write $\nu \ll \mu$. We say ν and μ are equivalent if $\nu \ll \mu$ and $\mu \ll \nu$. We write $\mu \equiv \nu$.

Proposition 1.2 *Suppose $\nu \ll \mu$ and S is a logical statement. Then $S\ [\mu] \Rightarrow S\ [\nu]$. In particular, if $\nu \equiv \mu$, then $S\ [\mu] \Leftrightarrow S\ [\nu]$.*

The next example illustrates a typical logical deduction modulo a measure — the type in which we will often be engaged.

Example 1.4 Let ν and μ be two measures defined on a measurable space $(\Omega, \mathcal{F})$ such that $\nu \ll \mu$. Let f_1, f_2, f_3 be measurable functions from $\Omega \to \mathbb{R}$. Suppose we know $f_1 f_3 = f_2 f_3\ [\nu]$ and $f_3 \neq 0\ [\mu]$. Then

$$f_1 f_3 = f_2 f_3\ [\nu], \qquad f_3 \neq 0\ [\nu],$$

which implies $f_1 = f_2\ [\nu]$. □

1.7 Integration to the limit

Suppose that $(\Omega, \mathcal{F}, \mu)$ is a measure space and f and $\{f_n : n = 1, 2, \ldots\}$ are real-valued measurable functions on Ω. If f_n converges (a.e. μ) to a function f, will the integral $\int f_n d\mu$ also converge to the integral $\int f d\mu$? This is not always true. Here is a counter example.

Example 1.5 Consider the measure space $(\mathbb{R}, \mathcal{R}, \lambda)$, where λ is the Lebesgue measure. Let

$$f_n(x) = n I_{(0,1/n)}(x).$$

Then $f_n(x) \to 0$ for all $x \in \mathbb{R}$ but

$$\int f_n d\lambda = 1$$

for all n. Thus $\lim_n \int f_n d\lambda = 1$ and $\int \lim_n f_n d\lambda = 0$. □

We see that the point-wise convergence does not always imply the convergence of the integral. Nevertheless, under reasonable conditions the above situation can be ruled out. We now give several sufficient conditions under which integration to the limit is valid.

Theorem 1.5 (Monotone Convergence Theorem) *Suppose that $\{f_n\}$ and f are measurable $\mathcal{F}$. If $0 \leq f_n \uparrow f$ a.e. μ, then $\int f_n d\mu \uparrow \int f d\mu$.*

This is the basic result for integration to the limit, from which other sufficient conditions can be reasonably easily deduced.

Theorem 1.6 (Fatou's Lemma) *Suppose that* $\{f_n\}$ *are measurable* $\mathcal{F}$ *and* $f_n \geq 0$. *Then*

$$\int \liminf_n f_n d\mu \leq \liminf_n \int f_n d\mu. \tag{1.6}$$

Proof. Let $g_n = \inf_{k \geq n} f_k$. Then $0 \leq g_n \uparrow \liminf_n f_n \equiv g$. Hence $\int g_n d\mu \uparrow \int g d\mu$. But $g_n \leq f_n$. So $\int g_n d\mu \leq \int f_n d\mu$. Hence

$$\liminf_n \int f_n d\mu \geq \liminf_n \int g_n d\mu = \int g d\mu \equiv \int \liminf_n f_n d\mu,$$

as desired. □

Note that, under the conditions of Fatou's lemma alone, it is not true that

$$\int \limsup_n f_n d\mu \geq \limsup_n \int f_n d\mu. \tag{1.7}$$

However, if $f_n \leq g$ for some integrable g, then $g - f_n$ satisfies the conditions of Fatou's lemma, and we have

$$\int \liminf_n (g - f_n) d\mu \leq \liminf_n \int (g - f_n) d\mu,$$

which implies

$$\int g d\mu - \int \limsup_n f_n d\mu \leq \int g d\mu - \limsup_n \int f_n d\mu.$$

Since $\int g d\mu$ is finite, we can cancel it out from both sides of the equality, which then reduces to (1.7). The directions of the inequalities (1.6) and (1.7) can be easily memorized if we notice that bringing limit (lim sup or lim inf) inside an integral makes the integral more extreme (bearing in mind that the lim sup case requires a dominating functions).

Note that, if $\liminf_n f_n = \limsup_n f_n = \lim_n f_n$, then (1.6) and (1.7) imply

$$\liminf_n \int f_n d\mu = \limsup_n \int f_n d\mu = \int \lim_n f_n d\mu.$$

That is, $\lim_n \int f_n d\mu$ exists and coincides with $\int \lim_n f_n d\mu$. Essentially, this is the argument of the Lebesgue's Dominated Convergence Theorem, though a more careful treatment than outlined above would allow us to remove the requirement $f_n \geq 0$. See, for example, Billingsley (1995, page 209).

Theorem 1.7 (Lebesgue's Dominated Convergence Theorem) *Let* $\{f_n\}$ *be a sequence of measurable functions such that* $|f_n| \leq g$ *a.e.* μ, *where* g *measurable* $\mathcal{F}$ *and is integrable* μ. *If* $f_n \to f$ *a.e.* μ, *then* f *and* f_n *are integrable* μ *and* $\int f_n d\mu \to \int f d\mu$.

The following theorem is an immediate consequence of Lebesgues' Dominated Convergence Theorem.

Theorem 1.8 (Bounded Convergence Theorem) *Suppose that $\mu(\Omega) < \infty$ and that $\{f_n\}$ is a uniformly bounded sequence of measurable functions; that is, $|f_n| \le C$ for some $C > 0$. Then $f_n \to f$ a.e. μ implies that $\int f_n d\mu \to \int f d\mu$.*

1.8 Differentiation under integral

Closely related to passing the limit inside an integral is passing a derivative inside an integral. After all, derivative is a form of limit. So, inevitably, the verification of the validity of this operation relies on the Dominated Convergence Theorem (Theorem 1.7), whose sufficient condition in this case is the Lipschitz condition. Differentiation under the integral sign will be used heavily in the rest of the book. Instead of having to state the complicated sufficient conditions involved every time we use this, we devote this section to streamlining the condition of passing a derivative through an integral.

Let $(\Omega, \mathcal{F}, \mu)$ a measure space. Let Θ be an open subset of $\mathbb{R}^k$. Let $g : \Omega \times \Theta \to \mathbb{R}$, and $g(\cdot, \theta)$ is measurable for each $\theta \in \Theta$. Let B be a measurable set in Ω.

Definition 1.4 *Suppose*

1. *for each $x \in \Omega$, $g(\theta, x)$ is differentiable with respect to θ, and $[\partial g(\theta, x)/\partial\theta] I_B(x)$ is integrable with respect to μ;*
2. *for each $\theta \in \Theta$, $g(\theta, x) I_B(x)$ is integrable with respect to μ the function $\theta \mapsto \int_B g(\theta, x) d\mu(x)$ is differentiable;*
3. $$\frac{\partial}{\partial\theta} \int_B g(\theta, x) d\mu(x) = \int_B \frac{\partial g(\theta, x)}{\partial\theta} d\mu(x). \tag{1.8}$$

Then we say that g is differentiable with respect to θ under the integral over B with respect to μ, and state this as "g satisfies DUI(θ, B, μ)".

The following theorem gives sufficient conditions for DUI(θ, B, μ). It is essentially the dominated convergence theorem applied to quotient. In this case dominating function is related to the L_1-Lipschitz condition. Let e_i denote the p-dimensional vector whose ith component is 1 and the rest of the components are 0.

Theorem 1.9 *Let B be a measurable set. Suppose*

1. *$g(\theta, x)$ is differentiable with respect to θ, and $[\partial g(\theta, x)/\partial\theta] I_B(x)$ is integrable μ;*

2. there is a function $g_0(x)$ integrable μ such that, for each θ_1, θ_2 in Θ

$$|g(\theta_2, x) - g(\theta_1, x)| \leq g_0(x)\|\theta_2 - \theta_1\|$$

for each $x \in B$.

Then g satisfies DUI(θ, B, μ).

The second condition is simply the L_1-Lipschitz condition for the variable θ.

Proof. Let $f(\theta) = \int_B g(\theta, x) d\mu(x)$. Recall that a p-variate function, say $f(\theta)$, is differentiable at θ_0 if and only if, for every θ in a neighborhood of θ_0, the function $f((1-t)\theta_0 + t\theta)$ is differentiable with respect to t at $t = 0$. Now for any $t \in \mathbb{R}$,

$$\frac{f((1-t)\theta_0 + t\theta) - f(\theta_0)}{t} = \int_B \frac{g((1-t)\theta_0 + t\theta, x) - g(\theta_0, x)}{t} d\mu(x).$$

Since

$$\left| \frac{g((1-t)\theta_0 + t\theta, x) - g(\theta_0, x)}{t} \right| \leq g_0(x)\|\theta - \theta_0\|$$

for all $x \in B$, by the dominated convergence theorem (Theorem 1.7),

$$[f_t'((1-t)\theta_0 + t\theta)]_{t=0} = \int_B [g_t'((1-t)\theta_0 + t\theta, x)]_{t=0} d\mu(x)$$

This shows that $f(\theta)$ is differentiable at θ_0. Now take θ to be e_i, $i = 1, \ldots, p$, to prove the equality (1.8), where e_i is the k-dimensional vector with its i component being 1 and other components being 0. □

1.9 Change of variables

Suppose $(\Omega, \mathcal{F})$ and $(\Omega', \mathcal{F}')$ are measurable spaces, and $T : \Omega \to \Omega'$ is a mapping measurable $\mathcal{F}/\mathcal{F}'$.

Theorem 1.10 *A function $f : \Omega' \to \mathbb{R}$ is measurable $\mathcal{F}'$ and integrable $\mu \circ T^{-1}$ if and only if $f \circ T$ is measurable $\mathcal{F}$ and integrable μ, in which case*

$$\int_{A'} f(\omega')(\mu \circ T^{-1})(d\omega') = \int_{T^{-1}A'} (f \circ T)(\omega)\mu(d\omega). \tag{1.9}$$

So, the expectation of $f(T)$ can be represented in various ways:

$$E[f(T)] = \int f(T(\omega))P(d\omega)$$
$$= \int f(x)(P \circ T^{-1})(dx) \tag{1.10}$$
$$= \int y(P_T \circ f^{-1})(dy),$$

where $P_T = P \circ T^{-1}$

Example 1.6 Let $(\Omega, \mathcal{F}, P)$ be a probability space and let $X : \Omega \to [0,1]$ be a random variable. Suppose that the distribution $P_X = P \circ X^{-1}$ has the following probability density with respect to the Lebesgue measure

$$f(x) = \begin{cases} 1 & 0 \leq x \leq 1 \\ 0 & \text{otherwise} \end{cases}$$

We can find the expectation $E(e^X)$ by definition or using the above change of variable theorem. By definition,

$$E(e^X) = \int_0^1 e^x f(x)dx = \int_0^1 e^x dx = e - 1.$$

Alternatively, let $Y = e^X$. The range of Y is $\Omega_Y = [1, e]$. For each $a \in \Omega_Y$,

$$F_Y(a) = P(Y \leq a) = P(e^X \leq a) = P(X \leq \log a) = \int_0^{\log a} dx = \log a.$$

Thus applying the third line of (1.10) we have

$$E(Y) = \int_1^e y(1/y)dy = e - 1.$$

We see that both methods give the same answer. □

1.10 The Radon-Nikodym Theorem

Suppose $(\Omega, \mathcal{F}, \mu)$ is a measure space and $\delta \geq 0$ is a measurable function. Then it can be shown that the set function

$$\nu(A) = \textstyle\int_A \delta d\mu, \quad \text{for all } A \in \mathcal{F} \tag{1.11}$$

defines a measure on $(\Omega, \mathcal{F})$. The above equation is often abbreviated as

$$d\nu = \delta d\mu.$$

Note that (1.11) implies that if $\mu(A) = 0$, then $\nu(A) = 0$. This leads to the following definition.

Definition 1.5 *Let μ and ν be two measures on a measurable space $(\Omega, \mathcal{F})$. Then ν is said to be absolutely continuous with respect to μ if, whenever $\mu(A) = 0$, $A \in \mathcal{F}$, we have $\nu(A) = 0$. In this case we write $\nu \ll \mu$.*

The next theorem is the Radon-Nikodym theorem, which says that not only (1.11) implies $\nu \ll \mu$, but the converse implication is also true provided that μ and ν are σ-finite.

Theorem 1.11 (Radon-Nikodym Theorem) *Suppose that μ and ν are σ-finite measures. Then the following two statements are equivalent:*

1. *$\nu \ll \mu$,*
2. *there exists a measurable $\delta \geq 0$ such that $\nu(A) = \int_A \delta d\mu$ for all $A \in \mathcal{F}$.*

By Corollary 1.1, if $\delta \geq 0$ and $\delta' \geq 0$ both satisfy *2* then $\delta = \delta'$ a.e. μ. The function δ is called the Radon-Nikodym derivative of ν with respect to μ, and is written as $d\nu/d\mu$. The Radon-Nikodym derivative $d\nu/d\mu$ is also known as the density of ν with respect to μ. Radon-Nikodym theorem is the key to many important probability concepts, such as conditional probability and conditional expectation. The next theorem generalizes the equality $\int I_A d\nu = \int I_A \delta d\mu$ to arbitrary measurable functions.

Theorem 1.12 *Suppose δ is the density of ν with respect to μ and f is measurable $\mathcal{F}$. Then, f is integrable with respect to ν if and only if $f\delta$ is integrable with respect to μ, in which case*

$$\int_A f d\nu = \int_A f \delta d\mu.$$

for all $A \in \mathcal{F}$.

The next theorem connects three probability measures.

Theorem 1.13 *Let P, Q, μ be probability measures on a measurable space $(\Omega, \mathcal{F})$. If $P \ll Q \ll \mu \ll P$, then*

$$\mu\{p = 0\} + \mu\{q = 0\} = 0 \tag{1.12}$$

$$P\left(\frac{q}{p}\frac{dP}{dQ} = 1\right) = 1, \tag{1.13}$$

where $p = \frac{dP}{d\mu}, q = \frac{dQ}{d\mu}$.

Proof. By Theorems 1.11, and 1.12, we have (1.12), and for all $A \in \mathcal{F}$,

$$P(B) = \int_B \frac{dP}{dQ} dQ = \int_B \frac{dP}{dQ} q \, d\mu.$$

By (1.12), the right-hand side can be rewritten as

$$\int_{B,p>0} \frac{dP}{dQ} q \, d\mu = \int_{B,p>0} \frac{dP}{dQ} \frac{q}{p} p \, d\mu = \int_{B,p>0} \frac{dP}{dQ} \frac{q}{p} dP = \int_B \frac{dP}{dQ} \frac{q}{p} dP.$$

The result now follows by another application of Theorem 1.11. □

1.11 Fubini's theorem

Let $(\Omega_1, \mathcal{F}_1, \mu_1)$ and $(\Omega_2, \mathcal{F}_2, \mu_2)$ be two measure spaces. Let $\Omega = \Omega_1 \times \Omega_2$ be the Cartesian product of Ω_1 and Ω_2. We now describe a σ-field (say $\mathcal{F}$) on Ω, and a measure μ on $(\Omega, \mathcal{F})$.

Let $\mathcal{F}$ be the σ-field generated by the set of class

$$\{A \times B : A \in \mathcal{F}_1, B \in \mathcal{F}_2\}.$$

A member of the above class is called a measurable rectangle. So $\mathcal{F}$ is the σ-field generated by measurable rectangles. Sometimes the σ-field $\mathcal{F}$ is also written as $\mathcal{F}_1 \times \mathcal{F}_2$. But note that this is *not* a Cartesian product in the usual sense, which would have been $\{(A, B) : A \in \mathcal{F}_1, B \in \mathcal{F}_2\}$.

Let $E \in \mathcal{F}$. Consider the sections of E:

$$E^{\omega_2} = \{\omega_1 : (\omega_1, \omega_2) \in E\}, \quad E_{\omega_1} = \{\omega_2 : (\omega_1, \omega_2) \in E\}.$$

Theorem 1.14 *If $E \in \mathcal{F}$, then, for each $\omega_1 \in \Omega_1$, $E_{\omega_1} \in \mathcal{F}_2$, and for each $\omega_2 \in \Omega_2$, $E^{\omega_2} \in \mathcal{F}_1$. Moreover, if $f : \Omega \to \mathbb{R}$ is measurable $\mathcal{F}$. Then, for each $\omega_1 \in \Omega_1$, $f(\omega_1, \cdot)$ is measurable $\mathcal{F}_2$ and for each $\omega_2 \in \Omega_2$, $f(\cdot, \omega_2)$ is measurable $\mathcal{F}_1$.*

It is true that the function $\omega_2 \mapsto \mu_1(E^{\omega_2})$ is $\mathcal{F}_2$-measurable, and the function $\omega_1 \mapsto \mu_2(E_{\omega_1})$ is $\mathcal{F}_1$-measurable. Thus the following set functions on $\mathcal{F}$ are well defined:

$$\pi'(E) = \int_{\Omega_1} \mu_2(E_{\omega_1})\mu_1(d\omega_1), \quad \pi''(E) = \int_{\Omega_2} \mu_1(E^{\omega_2})\mu_2(d\omega_2), \quad E \in \mathcal{F}.$$

It turns out that if μ_1 and μ_2 are σ-finite, these two set function are in fact the same function and they define a measure uniquely associated with μ_1 and μ_2.

Theorem 1.15 *If $(\Omega_1, \mathcal{F}_1, \mu_1)$ and $(\Omega_2, \mathcal{F}_2, \mu_2)$ are σ-finite measure spaces, then*

1. *$\pi'(E) = \pi''(E)$ for all $E \in \mathcal{F}$;*
2. *Let π be this common set function. Then π is a σ-finite measure on $\mathcal{F}$;*
3. *π is the only set function on $\mathcal{F}$ such that $\pi(A \times B) = \mu_1(A)\mu_2(B)$ for measurable rectangles.*

The measure π is called the product measure of μ_1 and μ_2, and will be written as $\mu_1 \times \mu_2$.

According to this definition and the discussion at the end of Section 1.3, two random vectors X and Y are independent if and only if their joint distribution is the product measure of their marginal distributions. That is,

$$P \circ (X, Y)^{-1} = (P \circ X^{-1}) \times (P \circ Y^{-1}).$$

Fubini's and Tonelli's Theorems are concerned with the integration of a measurable function f defined on Ω with respect to the product measure π. We first state Tonelli's Theorem.

Theorem 1.16 (Tonelli's Theorem) *Suppose that $(\Omega_1, \mathcal{F}_1, \mu_1)$ and $(\Omega_2, \mathcal{F}_2, \mu_2)$ are σ-finite measure spaces and $f : \Omega \to \mathbb{R}$ is a nonnegative and measurable $(\mathcal{F}/\mathcal{R})$ function. Then the functions*

$$\int_{\Omega_2} f(\omega_1, \omega_2)\mu_2(d\omega_2), \quad \int_{\Omega_1} f(\omega_1, \omega_2)\mu_1(d\omega_1) \tag{1.14}$$

are measurable with respect to $\mathcal{F}_1$ and $\mathcal{F}_2$, respectively, and

$$\begin{aligned}\int_{\Omega_1} \left[\int_{\Omega_2} f(\omega_1, \omega_2)\mu_2(d\omega_2)\right] \mu_1(d\omega_1) &= \int_{\Omega_2} \left[\int_{\Omega_1} f(\omega_1, \omega_2)\mu_1(d\omega_1)\right] \mu_2(d\omega_2) \\ &= \int_{\Omega} f(\omega_1, \omega_2)\pi(d(\omega_1, \omega_2)). \end{aligned} \tag{1.15}$$

Thus, when f is nonnegative and measurable, its integration with respect to the product measure can always be computed iteratively with respect to one measure at a time, and the order of the iterative integration does not matter. If f is not nonnegative, this is still true but requires the additional condition that f be integrable with respect to π. This is called Fubini's Theorem.

Theorem 1.17 (Fubini's Theorem) *Suppose $(\Omega_1, \mathcal{F}_1, \mu_1)$ and $(\Omega_2, \mathcal{F}_2, \mu_2)$ are σ-finite measure spaces and $f : \Omega \to \mathbb{R}$ is measurable $(\mathcal{F}/\mathcal{R})$ and integrable with respect to π. Then the functions defined in (1.14) are finite and measurable on A_1 and A_2, respectively, with $\mu_1(\Omega_1 \setminus A_1) = 0$ and $\mu_2(\Omega_2 \setminus A_2) = 0$. Moreover, equality (1.15) still holds.*

1.12 Conditional probability

Let $(\Omega, \mathcal{F}, P)$ be a probability space, and let A be a member of $\mathcal{F}$. From the elementary probability theory we know that if B is a member of $\mathcal{F}$ such that $P(B) \neq 0$, then the conditional probability of A given B is

$$P(A|B) = \frac{P(A \cap B)}{P(B)}.$$

We now define the conditional probability of A given a σ-field, which is a generalization of the conditional probability given a set.

Let $\mathcal{G} \subset \mathcal{F}$ be a σ-field. Let ν be the set function on $\mathcal{G}$ given by

$$\nu(G) = P(A \cap G) \quad \text{for} \quad G \in \mathcal{G}$$

It is true that ν is a measure on $\mathcal{G}$. The measure P, being a measure on $\mathcal{F}$, is also a measure on $\mathcal{G}$. Moreover, if $P(G) = 0$, then

$$\nu(G) = P(G \cap A) = 0.$$

Hence $\nu \ll P$. By the Radon-Nikodym Theorem there is a nonnegative function f that is measurable with respect to $\mathcal{G}$ such that for all $G \in \mathcal{G}$, $\nu(G) = \int_G f dP$. That is

$$P(G \cap A) = \int_G f dP \quad \text{for all} \quad G \in \mathcal{G}.$$

Furthermore, by Corollary 1.1, if there is a nonnegative function g that is measurable with respect to $\mathcal{G}$ satisfying the above relation, then $g = f$ a.e. P. This function f is a version of conditional probability of A given $\mathcal{G}$. More generally, we have the following definition.

Definition 1.6 *Let $A \in \mathcal{F}$, and $\mathcal{G}$ be a sub-σ-field of $\mathcal{F}$. Then any function $f : \Omega \to \mathbb{R}$ that satisfies the following conditions*

1. *f is measurable $\mathcal{G}$ and integrable P,*
2. *for each $G \in \mathcal{G}$,*

$$\int_G f dP = P(A \cap G),$$

is called the conditional probability of A given $\mathcal{G}$, and is written as $P(A|\mathcal{G})$.

We emphasize that the conditional probability $P(A|\mathcal{G})$ is defined as a $\mathcal{G}$-measurable function rather than a number. Following Billingsley (1995, Section 33) we use $P(A|\mathcal{G})_\omega$ to evaluation of this function at ω. The next theorem shows that this function *resembles* a probability even thought it is not a probability.

Theorem 1.18 *The function $P(A|\mathcal{G})$ has the following properties almost everywhere P:*

1. $0 \leq P(A|\mathcal{G}) \leq 1$,
2. $P(\varnothing|\mathcal{G}) = 0$,
3. *If A_n is a sequence of disjoint $\mathcal{F}$-sets, then $P(\cup_n A_n|\mathcal{G}) = \sum_n P(A_n|\mathcal{G})$.*

Proof. 1. We know, for any $G \in \mathcal{G}$, $\int_G P(A|\mathcal{G}) dP = P(A \cap G)$. Hence, for any $G \in \mathcal{G}$,

$$0 \leq \int_G P(A|\mathcal{G}) dP \leq P(G).$$

By the first inequality and Corollary 1.2, $P(A|\mathcal{G}) \geq 0$ a.e. P. The second inequality implies that $\int_G (P(A|\mathcal{G}) - 1) dP \leq 0$ for all $G \in \mathcal{G}$. By Corollary 1.2 again, $P(A|\mathcal{G}) - 1 \leq 0$ a.e. P. This proves part 1.

2. Since $f = 0$ satisfies

$$\int_G 0 dP = P(\varnothing \cap G)$$

for all $G \in \mathcal{G}$, f is a version of $P(\varnothing|G)$.

3. Let $f = \sum_n P(A_n|\mathcal{G})$. Then

$$\int_G f dP = \int_G \sum_n P(A_n|\mathcal{G}) dP = \sum_n \int_G P(A_n|\mathcal{G}) dP$$
$$= \sum_n P(A_n \cap G) = P(A \cap G).$$

Hence $\sum_n P(A_n|\mathcal{G})$ is a version of $P(A|\mathcal{G})$. □

Now let $X : \Omega \to \mathbb{R}$ be a random variable. Let $B \in \mathcal{R}$. We write

$$P(X^{-1}(B)|\mathcal{G}) = P(\{\omega : X(\omega) \in B\}|\mathcal{G})$$

as $P(X \in B|\mathcal{G})$. So far we have defined the conditional probability as a $\mathcal{G}$-measurable function on Ω for a fixed set $B \in \mathbb{R}$. Since this construction can be carried out for each $B \in \mathcal{R}$, $P(X \in B|\mathcal{G})$ can also be viewed as a mapping from $\mathcal{R} \times \Omega$ to $[0, 1]$. By intuition, as B varies in $\mathcal{R}$, and for a fixed $\omega \in \Omega$, $P(X \in B|\mathcal{G})$ should behave like a probability measure on Ω. If fact, if it were not for the qualification "almost everywhere P", Theorem 1.18 amounts exactly to this statement. This statement is valid in the following sense.

1.13 Conditional expectation

Let $(\Omega, \mathcal{F}, P)$ be a probability space and $\mathcal{G} \subset \mathcal{F}$ be a sub σ-field. The definition of conditional expectation has similar motivation as that of conditional probability, as demonstrated in the proof of the next theorem.

Theorem 1.19 *Suppose that $f : \Omega \to \mathbb{R}$ is measurable $\mathcal{F}$ and integrable P. Suppose $\mathcal{G}$ is a sub σ-field of $\mathcal{F}$. Then there is a function $f_0 : \Omega \to \mathbb{R}$ such that*

1. *f_0 is measurable $\mathcal{G}$ and integrable P,*
2. *$\int_G f_0 dP = \int_G f dP$ for all $G \in \mathcal{G}$.*

Proof. First, suppose that $f \geq 0$. Define $\nu : \mathcal{G} \to \mathbb{R}$ as the set function

$$\nu(G) = \int_G f dP, \quad G \in \mathcal{G}.$$

Then, it can be shown that ν is a measure on $\mathcal{G}$. Moreover, if $P(G) = 0$, then $\nu(G) = \int_G f dP = 0$. Hence $\nu \ll P$. By the Radon-Nikodym Theorem, there is a nonnegative function $f_0 : \Omega \to \mathbb{R}$, measurable $\mathcal{G}$, $f_0 \geq 0$, such that

$$\nu(G) = \int_G f_0 dP \quad \text{for all } G \in \mathcal{G}. \tag{1.16}$$

More generally, suppose that f is measurable $\mathcal{F}$ and integrable P. Let $f = f^+ - f^-$. Then f^+ and f^- are also measurable $\mathcal{F}$ and integrable P. Let f_0^+ and f_0^- be constructed as above, then f_0 is measurable $\mathcal{G}$ and integrable P such that (1.16) is satisfied. □

This leads to the definition of conditional expectation.

Definition 1.7 *Let $f : \Omega \to \mathbb{R}$ be a mapping that is measurable $\mathcal{F}$ and integrable P. Suppose $f_0 : \Omega \to \mathbb{R}$ is a mapping satisfying the following conditions:*

1. *f_0 is measurable $\mathcal{G}$ and integrable P,*
2. *For all $G \in \mathcal{G}$ we have $\int_G f dP = \int_G f_0 dP$.*

Then the function f_0 is called a version of the conditional expectation of f given $\mathcal{G}$, and is written as $E(f|\mathcal{G})$.

Again, if the sub σ-field $\mathcal{G}$ is generated by some random element T, then the $E(f|\sigma(T))$ is abbreviated as $E(f|T)$. Also notice that $E(I_A|\mathcal{G})$ coincides with the definition of $P(A|\mathcal{G})$.

We now describe several useful properties of the conditional expectation. The next theorem and corollary connect the conditional expectation and the projection in $L_2(P)$-space.

Theorem 1.20 *Let $(\Omega, \mathcal{F}, P)$ be a probability space, and $\mathcal{G}$ be a sub σ-field of $\mathcal{F}$. Suppose $f : \Omega \to \mathbb{R}$ is measurable $\mathcal{F}/\mathcal{R}$ and integrable with respect to P. Then the following statements are equivalent:*

1. *$h = E(f|\mathcal{G})$ $[P]$;*
2. *for any function $g : \Omega \to \mathcal{R}$ measurable $\mathcal{G}/\mathcal{R}$ such that fg is integrable P, we have*

$$\int (f - h) g dP = 0. \tag{1.17}$$

Proof. 1 ⇒ 2. By 1, for any $G \in \mathcal{G}$, $\int I_G h dP = \int I_G f dP$. In other words, equality (1.17) holds for all indicator functions measurable $\mathcal{G}$. Use three-step argument to complete the proof of this part.
2 ⇒ 1. Any I_G is measurable $\mathcal{G}$ such that $f I_G$ is integrable. □

Let $L_2(P)$ be the collection of all functions that are measurable $\mathcal{F}/\mathcal{R}$ and square-integrable P; that is,

$$\{f : \ f \text{ measurable } \mathcal{F}/\mathcal{R}, \ \int f^2 dP < \infty\}.$$

Corollary 1.3 *Let $(\Omega, \mathcal{F}, P)$ be a probability space, and $\mathcal{G}$ be a sub σ-field of $\mathcal{F}$. Suppose $f \in L_2(P)$. Then the following statements are equivalent:*

1. *$h = E(f|\mathcal{G})$ $[P]$;*

2. for any function $g : \Omega \to \mathcal{R}$, $g \in L_2(P)$, we have

$$\int (f - h)g dP = 0. \tag{1.18}$$

Proof. We only need to show fg is integrable. This follows from Hölder's inequality (1.3). □

Corollary 1.3 can be interpreted as $E(f|\mathcal{G})$ is the projection of f on to the subspace of $L_2(P)$ consisting of members of $L_2(P)$ that are measurable with respect to $\mathcal{G}$. Define the inner product in $L_2(P)$ as

$$\langle f_1, f_2 \rangle = \int f_1 f_2 dP.$$

Then $L_2(P)$, along with this inner product, is a Hilbert space. Let $L_2(P|\mathcal{G})$ be the subset of $L_2(P)$ consisting of measurable functions of $\mathcal{G}$ that are P-square integrable. Then, it can be shown that $L_2(P|\mathcal{G})$ is a subspace of $L_2(P)$. The identity (1.18) can be rewritten in the form

$$\langle f - E(f|\mathcal{G}), h \rangle = 0$$

for all $h \in L_2(P|\mathcal{G})$. This is precisely the definition of projection of f on to the subspace $L_2(P|\mathcal{G})$. See, for example, Conway (1990, page 10). The next Proposition is another useful property of conditional expectation.

Proposition 1.3 *Suppose U, V are members of $L_2(P)$ and $\mathcal{G}$ is a sub σ-field of $\mathcal{F}$. Then*

$$E[E(U|\mathcal{G})V] = E[UE(V|\mathcal{G})]. \tag{1.19}$$

Proof. We have

$$\begin{aligned} E[E(U|\mathcal{G})V] &= E\{E[(U|\mathcal{G})V]|\mathcal{G}\} \\ &= E[E(U|\mathcal{G})E(V|\mathcal{G})] \\ &= E\{E[E(V|\mathcal{G})U|\mathcal{G}]\} \\ &= E[UE(V|\mathcal{G})]. \end{aligned}$$

This completes the proof. □

The identity (1.19) can be generalized to random vectors in an obvious way. This proposition can be interpreted as "the conditional expectation is a self-adjoint operator." We can regard $A : U \mapsto E(U|\mathcal{G})$ as a mapping from $L_2(P)$ to $L_2(P)$ that transforms any member U of $L_2(P)$ to the member $E(U|\mathcal{G})$ of $L_2(P|\mathcal{G})$. Thus, the identity (1.19) can be rewritten as

$$\langle AU, V \rangle = \langle U, AV \rangle,$$

which is the defining relation of a self-adjoint operator, see for example Conway (1990, page 33).

1.14 Conditioning on a random element

In the last two sections we considered conditional probability and expectation conditioned on a sub σ-field $\mathcal{G}$ of $\mathcal{F}$. We now consider the special case where $\mathcal{G}$ is generated by random elements. Let $(\Omega_1, \mathcal{F}_1, P)$ be a probability space and $(\Omega_2, \mathcal{F}_2)$ be a measurable space. Let $T : \Omega_1 \to \Omega_2$ be a measurable function. Let $T^{-1}(\mathcal{F}_1)$ denote the collection of sets $\{T^{-1}(A) \in \mathcal{F}_1 : A \in \mathcal{F}_2\}$. Recall that $\sigma(T)$ is the intersection of all sub σ-fields of $\mathcal{F}_1$ with respect to which T is measurable. The following fact would be useful.

Lemma 1.1 $T^{-1}(\mathcal{F}_2) = \sigma(T)$.

Proof. It suffices to show the following three statements

1. $T^{-1}(\mathcal{F}_2)$ is a σ-field;
2. T is measurable $T^{-1}(\mathcal{F}_2)/\mathcal{F}_2$;
3. for each $A \in \mathcal{F}_2$, $T^{-1}(A) \in \sigma(T)$.

The details are left as an exercise. □

For a function $f : \Omega_1 \to \mathbb{R}$ measurable $\mathcal{F}_1/\mathcal{R}$, we would like to further investigate the special conditional expectation $E(f|\sigma(T))$, where $\sigma(T)$ is the intersection of all sub σ-field of $\mathcal{F}_1$ with respect to which T is measurable. Recall that $E(f|\sigma(T))$ is abbreviate as $E(f|T)$.

Lemma 1.2 *A function $h : \Omega_1 \to \mathbb{R}$ is measurable $T^{-1}(\mathcal{F}_2)/\mathcal{R}$ if and only if there is a mapping $g : \Omega_2 \to \mathbb{R}$ that is measurable with respect to $\mathcal{F}_2/\mathcal{R}$ such that $h = g \circ T$.*

A proof can be found in Halmos and Savage (1949). Recall that $E(h|T)$ is a function from Ω_1 to $\mathbb{R}$ that is measurable with respect to $T^{-1}(\mathcal{F}_2)/\mathcal{R}$. By the above lemma it can be written as $g \circ T$ where $g : \Omega_2 \to \mathbb{R}$ is measurable $\mathcal{F}_2/\mathcal{R}$. Two functions are of interest here': one is the composite function $g \circ T$, and the other is the function g itself. We use $E(f|\circ T)$ to denote the function g. Thus $E(f|T)$ is defined on Ω_1, but $E(f|\circ T)$ is defined on Ω_2.

Halmos and Savage (1949) gave the following construction of the conditional expectation $E(f|\circ T)$. See also Kolmogorov (1933).

Theorem 1.21 *Let $h : \Omega_1 \to \mathbb{R}$ be a measurable, nonnegative function that is integrable with respect to P. Let Q be the measure on $(\Omega_1, \mathcal{F}_1)$ defined by $dQ = h dP$. Then*

1. $Q \circ T^{-1} \ll P \circ T^{-1}$;
2. $d(Q \circ T^{-1})/d(P \circ T^{-1}) = E(h|\circ T)$.

Proof. 1. Suppose $A \in \mathcal{F}_2$ and $(P \circ T^{-1})(A) = 0$. Then $P(T^{-1}(A)) = 0$. Since, by definition, $Q \ll P$, we have $Q(T^{-1}(A)) = (Q \circ T^{-1})(A) = 0$.
2. Let $g = d(Q \circ T^{-1})/d(P \circ T^{-1})$. Then g is a function from Ω_2 to $\mathbb{R}$ measurable $\mathcal{F}_2/\mathcal{R}$. Moreover, for any $A \in \mathcal{F}_2$ we have

$$\int_A d(Q\circ T^{-1}) = \int_A gd(P\circ T^{-1}).$$

By change of variable theorem we have

$$\int_A d(Q\circ T^{-1}) = \int_{T^{-1}(A)} dQ, \quad \int_A gd(P\circ T^{-1}) = \int_{T^{-1}(A)} g\circ TdP$$

So we have

$$\int_{T^{-1}(A)} dQ = \int_{T^{-1}(A)} g\circ TdP$$

But by definition of Q, $\int_{T^{-1}(A)} dQ = \int_{T^{-1}(A)} hdP$. In other words for any $B \in T^{-1}(\mathcal{F}_2)$ we have

$$\int_B hdP = \int_B g\circ TdP$$

So $g\circ T$ is a version of $E(h|T)$. That is, $E(h|\circ T) = g$. □

To familiarize ourselves with the notation $E(h|\circ T)$, the following relations are helpful:

$$\begin{aligned} &[E(h|\circ T)](T(\omega)) = [E(h|T)](\omega), \\ &E(g\circ T|T) = g\circ T, \\ &E(g\circ T|\circ T) = g, \end{aligned}$$

where, in the second and third equality, g is a function from $\mathcal{F}_2 \to \mathbb{R}$ measurable $\mathcal{F}_2/\mathcal{R}$.

1.15 Conditional distributions and densities

Let $(\Omega, \mathcal{F}, P)$ be a probability space, where $P \ll \mu$ for some σ-finite measure μ on $(\Omega, \mathcal{F})$. Suppose X and Y are random elements defined on $(\Omega, \mathcal{F}, P)$, with X taking values in $(\Omega_X, \mathcal{F}_X)$ and Y taking values in $(\Omega_Y, \mathcal{F}_Y)$. Then the mapping

$$\mathcal{F}_Y \times \Omega_X \to \mathbb{R}, \quad (A, x) \mapsto P(Y^{-1}(A)|\circ X)_x$$

is called the conditional distribution of Y given X. This mapping is written as $P_{Y|X}$. The conditional distribution $P_{X|Y}$ is defined in the same way.

Because $P \ll \mu$, we have $P\circ X^{-1} \ll \mu\circ X^{-1}$, $P\circ Y^{-1} \ll \mu\circ Y^{-1}$ (Problem 1.8). Let

$$P_X = P\circ X^{-1}, \quad P_Y = P\circ Y^{-1}, \quad \mu_X = P\circ X^{-1}, \quad \mu_Y = \mu\circ Y^{-1}.$$

Let $\Omega_{XY} = \Omega_X \times \Omega_Y$ and $\mathcal{F}_{XY} = \mathcal{F}_X \times \mathcal{F}_Y$. Let (X, Y) be the mapping

$$(X, Y) : \Omega \to \Omega_X \times \Omega_Y, \quad \omega \mapsto (X(\omega), Y(\omega)).$$

Then (X, Y) is measurable with respect to $\mathcal{F}/(\mathcal{F}_X \times \mathcal{F}_Y)$ (Problem 1.10). Let $P_{XY} = P\circ(X, Y)^{-1}$. Then $(\Omega_{XY}, \mathcal{F}_{XY}, P_{XY})$ is a probability space, μ_{XY} is a σ-finite measure on $(\Omega_{XY}, \mathcal{F}_{XY})$, and $P_{XY} \ll \mu_{XY}$. Let

$$f_{XY} = dP_{XY}/d\mu_{XY}, \quad f_X = dP_X/d\mu_X, \quad f_Y = dP_Y/d\mu_Y.$$

The function f_{XY} is called the joint density of (X, Y); f_X the marginal density of X; f_Y the marginal density of Y. Finally, let

$$f_{Y|X}(x|y) = \begin{cases} f_{XY}(x, y)/f_X(x) & \text{if } f_X(x) \neq 0 \\ 0 & \text{if } f_X(x) = 0 \end{cases}$$

This function is called the conditional density of Y given X. It can be shown (Problem 1.22) for each $A \in \mathcal{F}_Y$, the function

$$x \mapsto \textstyle\int_A f_{Y|X}(y|x) d\mu_Y(y) \tag{1.20}$$

is a version of the conditional probability $P(A|X)$. That is, the two mappings

$$(A, x) \mapsto P(Y^{-1}(A)|\circ X)_x, \quad (A, x) \mapsto \textstyle\int_A f_{Y|X}(y|x) d\mu_Y(y)$$

are the same function modulo P.

1.16 Dynkin's π-λ theorem

Let $(\Omega, \mathcal{F}, \mu)$ be a measure space and $\mathcal{G} \subseteq \mathcal{F}$ be a sub-σ-field. Let f and g be $\mathcal{G}$-measurable function and are integrable with respect to μ. We often need to prove that $f = g$ $[\mu]$. By Corollary 1.1 it suffices to show that

$$\textstyle\int_A f d\mu = \int_A g d\mu, \text{ for all } A \in \mathcal{G}. \tag{1.21}$$

Dynkin's π-λ theorem is very useful for this purpose. A class $\mathcal{P}$ of subsets of Ω is called a π-system if

$$A_1 \in \mathcal{P}, \ A_2 \in \mathcal{P} \implies A_1 \cap A_2 \subseteq \mathcal{P}.$$

A class of subsets $\mathcal{L}$ of Ω is called a λ-system if

1. $\Omega \in \mathcal{L}$;
2. $A \in \mathcal{L} \implies A^c \in \mathcal{L}$;
3. if $A_1, A_2, \cdots$ are disjoint members of $\mathcal{L}$, then $\cup_{n=1}^{\infty} A_n \in \mathcal{L}$.

Note that a λ-system is almost a σ-field except that the former requires $A_1, A_2, \ldots$ to be disjoint. So a σ-field is always a λ-system but not vice-versa. The following theorem is the π-λ theorem; its proof can be found in Billingsley (1995, page 42).

Theorem 1.22 (Dynkin's π-λ Theorem) *If $\mathcal{P}$ is a π-system, $\mathcal{L}$ is a λ-system, and $\mathcal{P} \subseteq \mathcal{L}$, then $\sigma(\mathcal{P}) \subseteq \mathcal{L}$.*

We can use the π-λ theorem to prove $f = g\ [\mu]$ in the following way. Let $\mathcal{P} \subseteq \mathcal{G}$ be a π-system that generates σ-field $\mathcal{G}$ and let

$$\mathcal{L} = \{A \in \mathcal{F} : \textstyle\int_A f d\mu = \int_A g d\mu\} \tag{1.22}$$

If we can show $\int_A f d\mu = \int_A g d\mu$ holds on $\mathcal{P}$ (that is, $\mathcal{P} \subseteq \mathcal{L}$) and $\mathcal{L}$ is a λ-system, then $\mathcal{G} \subseteq \mathcal{L}$. That is, (1.21) also holds on $\mathcal{G}$, which implies $f = g\ [\mu]$.

Corollary 1.4 *Suppose $\mathcal{G} \subseteq \mathcal{F}$ is a sub σ-field, f and g are real-valued functions on Ω that are measurable with respect to $\mathcal{G}$ and integrable with respect to μ. Suppose $\mathcal{P} \subseteq \mathcal{G}$ is a π-system generating $\mathcal{G}$ and $\Omega \in \mathcal{P}$. Then $\int_A f d\mu = \int_A g d\mu$ for all $A \in \mathcal{P}$ implies $f = g\ [\mu]$.*

Proof. It suffices to show that $\mathcal{L}$ defined in (1.22) is a λ-system. Since $\Omega \in \mathcal{P}$, $\Omega \in \mathcal{L}$. If $B \in \mathcal{L}$, then

$$\textstyle\int_{B^c} f d\mu = \int_\Omega f d\mu - \int_B f d\mu = \int_\Omega g d\mu - \int_B g d\mu = \int_{B^c} g d\mu.$$

So $B^c \in \mathcal{L}$. If $A_1, A_2, \ldots$ are disjoint members of $\mathcal{L}$, then

$$\textstyle\int_{\cup_{n=1}^\infty A_n} f d\mu = \sum_{n=1}^\infty \int_{A_n} f d\mu = \sum_{n=1}^\infty \int_{A_n} g d\mu = \int_{\cup_{n=1}^\infty A_n} g d\mu.$$

So $\cup_{n=1}^\infty A_n \in \mathcal{L}$. □

1.17 Derivatives and other notations

We will frequently need to take derivative of a vector-valued function with respect to a vector-valued variable. Let A be a subset of $\mathbb{R}^p$, B a subset of $\mathbb{R}^q$, and $f : A \to B$ a differentiable function. Denoting the argument of f by $\theta = (\theta_1, \ldots, \theta_p)$ and the components of $f(\theta)$ by $f_1(\theta), \ldots, f_q(\theta)$, we adopt the following convention:

$$\frac{\partial f(\theta)}{\partial \theta^T} = \begin{pmatrix} \partial f_1(\theta)/\partial\theta_1 & \cdots & \partial f_1(\theta)/\partial\theta_p \\ \vdots & \ddots & \vdots \\ \partial f_q(\theta)/\partial\theta_1 & \cdots & \partial f_q(\theta)/\partial\theta_p \end{pmatrix}.$$

The transpose of the above matrix will be denoted by

$$\left(\frac{\partial f(\theta)}{\partial \theta^T} \right)^T = \frac{\partial f^T(\theta)}{\partial \theta}.$$

This convention also applies to the situation where f depends on another variable, say $x \in \mathbb{R}^k$; that is, $f = f(\theta, x)$.

The 3-dimensional array of the second derivatives

$$\{\partial^2 f_i(\theta)/\partial \theta_j \partial \theta_k : \; i = 1, \ldots, p, \; j, k = 1, \ldots, p\}$$

will be denoted by $\partial f(\theta)/\partial \theta \partial \theta^T$, which is a $p \times q \times q$ array.

We will use the letter I or $I(\theta)$ to denote the Fisher information (as defined in Section 2.4). We will use I_p to represent the p by p identity matrix. The difference between I and I_p should be emphasized: the symbol with a integer subscript always denotes the identity matrix. In addition, for a set A and a variable x we use $I_A(x)$ to represent the indicator function. We should also pay attention to the difference between I_p and I_A: the former is indexed by an integer represented by a lowercase letter; the latter is the indicator of a set represented by an uppercase letter.

Problems

1.1. Let Ω be a non-empty set. Show that the classes of subsets in Ω

$$\begin{aligned} \mathcal{F}_1 =& \{\text{all subsets of } \Omega\}, \\ \mathcal{F}_2 =& \{\varnothing, \Omega\}, \\ \mathcal{F}_3 =& \{\varnothing, \Omega, A, A^c\}, \quad \text{where } A \subset \Omega \end{aligned}$$

are σ-fields.

1.2. Let Ω be a set. Let $\mathcal{F}$ be a finite class of subsets of Ω. Suppose that $\mathcal{F}$ satisfies

1. $\Omega \in \mathcal{F}$;
2. If $A \in \mathcal{F}$ then $A^c \in \mathcal{F}$;
3. If $A, B \in \mathcal{F}$ then $A \cup B \in \mathcal{F}$.

Show that $\mathcal{F}$ is a σ-field.

1.3. Let $\Omega = \mathbb{R}$. Let $a \in \mathbb{R}$. Show that the set $\{a\}$ is a Borel set. Let $a_1, \ldots, a_k$ be numbers in $\mathbb{R}$. Show that $\{a_1, \ldots, a_k\}$ is a Borel set. Show that the set of all rational numbers and the set of all irrational numbers are Borel sets. Show that any countable set in $\mathbb{R}$ is a Borel set. A set is said to be cocountable if its complement is countable. Show that any cocountable set in $\mathbb{R}$ is a Borel set.

1.4. A number a is *algebraic* if it is a solution to any equation of the form

$$m_0 + m_1 x + \cdots + m_k x^k = 0,$$

where $m_0, m_1, \ldots, m_k$ are integers with $m_k \neq 0$ and $k = 1, 2, \ldots$. (For example, a rational number is an algebraic number with $k = 1$). A number that is not algebraic is a *transcendental* number. Show that the collection of all algebraic numbers is a Borel set. Find the Lebesgue measure of this set. (Hint: use the following facts: (i) a kth order polynomial has at most k roots; (ii) a countable union of countable sets is a countable set.)

1.5. Let Ω_1 and Ω_2 be nonempty sets, and $T : \Omega_1 \to \Omega_2$ is a function. Show that

1. $T^{-1}(\varnothing) = \varnothing$;
2. For any $A \subseteq \Omega_2$, $T^{-1}(A^c) = [T^{-1}(A)]^c$;
3. For $A \subseteq \Omega_2$ and $B \subseteq \Omega_2$, we have $T^{-1}(A \cap B) = T^{-1}(A) \cap T^{-1}(B)$;
4. For $A \subseteq \Omega_2$ and $B \subseteq \Omega_2$, we have $T^{-1}(A \cup B) = T^{-1}(A) \cup T^{-1}(B)$;
5. Let C be any non-empty set. If $\{A_c : c \in C\}$ is a collection of subsets of Ω_2, then

$$T^{-1}(\cap_{c \in C} A_c) = \cap_{c \in C} T^{-1}(A_c), \quad T^{-1}(\cup_{c \in C} A_c) = \cup_{c \in C} T^{-1}(A_c).$$

1.6. Suppose $(\Omega_1, \mathcal{F}_1)$ and $(\Omega_2, \mathcal{F}_2)$ are measurable spaces, and $T : \Omega_1 \to \Omega_2$ is measurable $\mathcal{F}_1/\mathcal{F}_2$. Let $T^{-1}(\mathcal{F}_2) = \{T^{-1}(B) : B \in \mathcal{F}_2\}$. Show that $T^{-1}(\mathcal{F}_2)$ is a sub σ-field of $\mathcal{F}_1$.

1.7. Let $(\Omega, \mathcal{F})$ and $(\Omega', \mathcal{F}')$ be measurable spaces and $T : \Omega \to \Omega'$ be a measurable mapping. Suppose that μ is a measure on $(\Omega, \mathcal{F})$. Show that $\mu \circ T^{-1}$, as defined at the beginning of Section 1.2, is a measure on $(\Omega', \mathcal{F}')$.

1.8. Suppose $(\Omega, \mathcal{F})$ and $(\Omega', \mathcal{F}')$ are measurable spaces and μ and ν are two measures defined on $(\Omega, \mathcal{F})$. Let $X : \Omega \to \Omega'$ be a random element measurable with respect to $\mathcal{F}/\mathcal{F}'$. Show that, if $\nu \ll \mu$, then $\nu \circ X^{-1} \ll \mu \circ X^{-1}$.

1.9. Suppose that $(\Omega, \mathcal{F})$ is a measurable space and $f : \Omega \to \mathbb{R}$ is measurable $\mathcal{F}/\mathcal{R}$. Show that f^+ and f^-, as defined in Section 1.4, are measurable $\mathcal{F}/\mathcal{R}$.

1.10. Suppose $(\Omega, \mathcal{F})$, $(\Omega_X, \mathcal{F}_X)$, and $(\Omega_Y, \mathcal{F}_Y)$ are measurable spaces and

$$X : \Omega \to \Omega_X, \quad Y : \Omega \to \Omega_Y$$

are functions that are measurable with respect to $\mathcal{F}/\mathcal{F}_X$ and $\mathcal{F}/\mathcal{F}_Y$, respectively. Show that the function

$$(X, Y) : \Omega \to \Omega_X \times \Omega_Y$$

is measurable with respect to $\mathcal{F}/(\mathcal{F}_X \times \mathcal{F}_Y)$.

1.11. Show that any open set in $\mathbb{R}$ can be written as a countable union of open intervals, and that the class of all open intervals also generates the Borel σ-field.

1.12. Repeat the above problem with open intervals replaced by sets of the form $(-\infty, a]$, where $a \in \mathbb{R}$.

1.13. Let $(\Omega, \mathcal{F}, \mu)$ be a measure space and $\{f_n\}$ is a sequence of measurable functions from Ω to $\mathbb{R}$. Suppose that f_n is decreasing and $f_n \to f$ a.e. μ. Suppose, furthermore, that f_n is μ-integrable for some n. Use the Monotone Convergence Theorem to show that $\int f_n d\mu \to \int f d\mu$.

1.14. Let $(\Omega, \mathcal{F}, P)$ be a probability space. Let $\{A_n : n = 1, 2, \ldots\}$ be a sequence of $\mathcal{F}$-sets. The symbols $\liminf_{n\to\infty} A_n$ and $\limsup_{n\to\infty} A_n$ stand, respectively, for the sets

$$\bigcap_{n=1}^{\infty} \bigcup_{m=n}^{\infty} A_m \text{ and } \bigcup_{n=1}^{\infty} \bigcap_{m=n}^{\infty} A_m.$$

If $\liminf_{n\to\infty} A_n = \limsup_{n\to\infty} f_n$, then we call this common set the limit of $\{A_n\}$ and write it as $\lim_{n\to\infty} A_n$.

a. Show that $\liminf_{n\to\infty} A_n$ and $\limsup_{n\to\infty} f_n$ are $\mathcal{F}$-sets.
b. Suppose that $\{A_n\}$ has a limit $\lim_n A_n = A$. Show that inequalities (1.6) and (1.7) hold for f_n and f defined by

$$f_n(\omega) = I_{A_n}(\omega), \quad f(\omega) = I_A(\omega).$$

Then use these inequalities to show that $\lim_{n\to\infty} P(A_n)$ exists and

$$\lim_n P(A_n) = P(\lim_{n\to\infty} A_n).$$

Thus P, as a set function defined on $\mathcal{F}$, is continuous. The above relation is called the continuity of probability. Evidently, this continuity also applies to any finite measure μ.

1.15. Deduce the Bounded Convergence Theorem from Lebesgue's Dominated Convergence Theorem.

1.16. Let $f_n(x)$ be the probability density of $N(0, 1/n)$, and let λ be the Lebesgue measure. Show that $\lim_n \int f_n d\lambda \neq \int \lim_n f_n d\lambda$.

1.17. Suppose $(\Omega, \mathcal{F}, \mu)$ is a measure space and $\delta \geq 0$ is integrable with respect to μ. Then the set function

$$\nu(A) = \int_A \delta d\mu$$

defines a measure on $(\Omega, \mathcal{F})$.

1.18. Suppose that $(\Omega, \mathcal{F}) = (\Omega', \mathcal{F}') = (\mathbb{R}, \mathcal{R})$. Let $T : \Omega \to \Omega'$ be defined by $T(x) = x^2$. Define on $(\Omega, \mathcal{F})$ the measure

$$\mu(A) = \int_A \frac{1}{\sqrt{2\pi}} e^{-x^2/2} d\lambda(x), \quad \text{for any } A \in \mathcal{F},$$

where λ is the Lebesgue measure. How is $\mu \circ T^{-1}$ defined in this case? Express it in terms of its density with respect to the Lebesgue measure. Suppose $f : \Omega' \to \mathbb{R}$ is defined by $f(y) = \sin(y)$. What does the general formula (1.9) reduce to in this case?

1.19. Let $(\Omega, \mathcal{F}, P)$ be a probability space and X be a nonnegative random variable. Use Tonelli's theorem to show that

$$\int_0^\infty P(X \geq t)dt = E(X).$$

If X is not necessarily nonnegative but has an integral, show that

$$E(X) = \int_0^\infty P(X^+ \geq t)dt - \int_0^\infty P(X^- \geq t)dt.$$

1.20. Suppose that $(\Omega, \mathcal{F}, \mu)$ is a measure space and $\mathcal{G} \subset \mathcal{F}$ is a σ-field. Show that μ is a measure on $\mathcal{G}$.

1.21. Let $(\Omega, \mathcal{F}, P)$ be a probability space and $\mathcal{G}$ be a sub σ-field of $\mathcal{F}$. Let $A \in \mathcal{F}$. Use the definition of conditional probability to show that $E(I_A|\mathcal{G})$ is a version of $P(A|\mathcal{G})$.

1.22. Use the definition of conditional probability to show that the set function defined in (1.20) is a version of the conditional probability $P(A|X)$.

1.23. Let $(\Omega, \mathcal{F}, P)$ be a probability space and X be a random variable. Suppose that $\mathcal{G}_1$ and $\mathcal{G}_2$ are sub σ-fields of $\mathcal{F}$ such that $\mathcal{G}_1 \subset \mathcal{G}_2$. Use the definition of conditional expectation to show that

$$E[E(X|\mathcal{G}_2)|\mathcal{G}_1] = E(X|\mathcal{G}_1) \text{ a.e. } P.$$

1.24. Let P_1 and P_2 be two probability measures defined on a measurable space $(\Omega, \mathcal{F})$. Suppose that $P_2 \ll P_1$. Let $\delta = dP_2/dP_1$ be the Radon-Nikodym derivative of P_2 with respect to P_1. Suppose that $\delta > 0$ a.e. P_1. Show that $\int_\Omega \log(\delta)dP_1 \leq 0$ unless $\delta = 1$ a.e. P_1.

1.25. Let $(\Omega_1, \mathcal{F}_1)$ and $(\Omega_2, \mathcal{F}_2)$ be measurable spaces and $T : \Omega_1 \to \Omega_2$ is measurable $\mathcal{F}_1/\mathcal{F}_2$. Show that $T^{-1}(\mathcal{F}_2)$ is the smallest sub σ-field of $\mathcal{F}_1$ with respect to which T is measurable.

1.26. If $V \geq 0$ is a random variable and U is k-dimensional random vector, then $\mu(B) = E(I_B(U)V)$ defines a measure on $\mathbb{R}^k$.

References

Billingsley, P. (1995). *Probablity and Measure.* Third Edition. Wiley.

Conway, J. B. (1990). *A course in functional analysis.* Second edition. Springer, New York.

Halmos, P. R. and Savage, L. J. (1949). Application of the Radon-Nikodym Theorem to the Theory of Sufficient Statistics. *The Annals of Mathematical Statistics*, **20**, 225–241.

Kolmogorov, A. N. (1933). *Grundbegriffe der Wahrscheinlichkeitsrechnung* (in German). Berlin: Julius Springer. Translation: *Foundations of the Theory of Probability* (2nd ed.). New York: Chelsea. (1956).

Perlman, M. D. (1974). Jensen's inequality for a convex vector-valued function on an infinite-dimensional space. *Journal of Multivariate Analysis*, **4**, 52–65.

Rudin, W. (1987). *Real and Complex Analysis.* Third Edition. McGraw-Hill, Inc.

Vestrup, E. M. (2003). *The Theory of Measures and Integration.* Wiley.

2

Classical Theory of Estimation

This chapter is a compact description of the classical theory of point estimation (see, for example, Lehmann and Casella, 1998). Bias and variance are two important criteria that characterize an estimator. The classical theory seeks optimal estimator among the class of all unbiased estimators, in the sense that it has the smallest variance. The optimal problem involved is intrinsically connected the notions of sufficiency, minimal sufficiency, completeness, Fisher information, and Cramér-Rao lower bound. For example, an unbiased estimator can always be improved by taking its conditional expectation given a sufficient statistic, and becomes optimal by taking its expectation given the complete and sufficient statistic. An important class of distributions where sufficient and complete statistics are available is the exponential family, which will also be covered in this chapter.

2.1 Families of probability measures

Let $(\Omega, \mathcal{F})$ be a measurable space. In statistics we do not know the "true" probability model, or distribution, ahead of time. Instead, we would like to estimate or infer about this distribution. This can be formulated as choosing a distribution from a family of distributions. Thus in this section we explore a variety of notions of families of distributions, which will be frequently used in subsequent development.

2.1.1 Dominated and homogeneous families

Let $\mathfrak{M}$ and $\mathfrak{N}$ be two families of measures on $(\Omega, \mathcal{F})$.

Definition 2.1 *We say that $\mathfrak{M}$ is absolutely continuous with respect to $\mathfrak{N}$, or $\mathfrak{M}$ is dominated by $\mathfrak{N}$, if for any $E \in \mathcal{F}$,*

$$P(E) = 0 \quad \text{for all } P \in \mathfrak{N} \;\Rightarrow\; Q(E) = 0 \quad \text{for all } Q \in \mathfrak{M}.$$

B. Li and G. J. Babu, *A Graduate Course on Statistical Inference*,
Springer Texts in Statistics, https://doi.org/10.1007/978-1-4939-9761-9_2

We write this statement as $\mathfrak{M} \ll \mathfrak{N}$. *If* $\mathfrak{M} \ll \mathfrak{N}$ *and* $\mathfrak{N} \ll \mathfrak{M}$, *then we write* $\mathfrak{M} \equiv \mathfrak{N}$.

If $\mathfrak{M} \ll \mathfrak{N}$ and $\mathfrak{N}$ is a singleton $\{\lambda\}$, then we also say $\mathfrak{M}$ is dominated by λ. Furthermore, if both $\mathfrak{M}$ and $\mathfrak{N}$ are singletons, say $\{\lambda_1\}$ and $\{\lambda_2\}$, then we also say λ_1 is dominated by λ_2 (which agrees with our definition of $\lambda_1 \ll \lambda_2$ in Chapter 1). Here, a useful fact is that a σ-finite measure is always equivalent to a finite measure. Indeed, let λ be a σ-finite measure on $(\Omega, \mathcal{F})$, and let $\{A_1, A_2, \ldots\}$ of members of $\mathcal{F}$ such that $\cup A_n = \Omega$ and $\lambda(A_n) < \infty$. Without loss of generality, assume $\lambda(A_n) > 0$ for all $n \in \mathbb{N}$. Let c_n be a sequence of positive number such that $\sum_{n=1}^{\infty} c_n = 1$. Let λ^* be the set function defined by

$$\lambda^*(A) = \sum_{n=1}^{\infty} c_n \lambda(A \cap A_n)/\lambda(A_n),$$

It is left as an exercise to show that λ^* is a probability measure, and $\lambda^* \equiv \lambda$.

Definition 2.2 *Let* $(\Omega, \mathcal{F})$ *be a measurable space. We say that a family of measures* $\mathfrak{M}$ *on* $(\Omega, \mathcal{F})$ *is dominated if it is dominated by a* σ*-finite measure.*

A property of a dominated family is that it has an equivalent countable subfamily. See, for example, Halmos and Savage (1949) and Lin'kov (2005).

Proposition 2.1 *Let* $\mathfrak{M}$ *be a family of probability measures on a measurable space* $(\Omega, \mathcal{F})$. *Then* $\mathfrak{M}$ *is dominated if and only if it contains a countable set of probability measures* $\mathfrak{N}$ *such that* $\mathfrak{N} \equiv \mathfrak{M}$.

Proof. Since any σ-finite measure is equivalent to a finite measure, we can, without loss of generality, assume the dominating measure λ to be a finite measure. For each $\mu \in \mathfrak{M}$, let $f_\mu = d\mu/d\lambda$, and let $K_\mu = \{f_\mu > 0\}$. An $\mathcal{F}$-set K is called a kernel if there is a $\mu \in \mathfrak{M}$ such that $\mu(K) > 0$ and $K \subseteq K_\mu$. Any countable, disjoint union of kernels is called a chain.

We note that a union of countably many chains is a chain. This is because such a set is the union of countably many kernels, which need not be disjoint. Let us write this union as $\cup_{i \in \mathbb{N}} K_i$. Let

$$M_1 = K_1, \; M_2 = K_2 \setminus K_1, \; M_3 = K_3 \setminus (K_1 \cup K_2), \ldots$$

Then M_i are disjoint kernels satisfying $\cup_{i \in \mathbb{N}} M_i = \cup_{i \in \mathbb{N}} K_i$. Thus we see that $\cup K_i = \cup M_i$ is indeed a chain.

Now let $\mathcal{C}$ be the collection of all chains and let $\lambda^\circ = \sup\{\lambda(C) : C \in \mathcal{C}\}$. Then there is a sequence $\{C_n : n \in \mathbb{N}\}$ such that $\lambda(C_n) \to \lambda^\circ$. Let $C^\circ = \cup_{n \in \mathbb{N}} C_n$. Since $C_n \subseteq C^\circ$, $\lambda(C_n) \leq \lambda(C^\circ)$, and hence $\lambda(C^\circ) \geq \lambda^\circ$. Since C° is a chain, $\lambda(C^\circ) \leq \lambda^\circ$. So $\lambda(C^\circ) = \lambda^\circ$.

Since C° is a chain, it can be written as a disjoint union $\cup_{i=1}^{\infty} K_i$, where $K_i \subseteq K_{\mu_i}$ for some $\mu_i \in \mathfrak{M}$. Let $\mathfrak{N} = \{\mu_1, \mu_2, \ldots\}$. Note that μ_i are probability measures. We now show that $\mathfrak{N} \equiv \mathfrak{M}$. Since $\mathfrak{N} \subseteq \mathfrak{M}$, we have $\mathfrak{N} \ll \mathfrak{M}$.

To show $\mathfrak{M} \ll \mathfrak{N}$, let E be a member of $\mathcal{F}$ such that $\mu_i(E) = 0$ for all $i \in \mathbb{N}$. Let μ be a member of $\mathfrak{M}$. We need to show that $\mu(E) = 0$. Note that

$$\mu(E) = \mu(EK_\mu^c) + \mu(EK_\mu) = \mu(EK_\mu).$$

We will show that $\mu(EK_\mu) = 0$. This measure can be decomposed as

$$\mu(EK_\mu) = \mu(EK_\mu C^\circ) + \mu(EK_\mu (C^\circ)^c). \tag{2.1}$$

Suppose $\mu(EK_\mu(C^\circ)^c) > 0$. Then $\lambda(EK_\mu(C^\circ)^c) > 0$, and hence

$$\lambda(EK_\mu(C^\circ)^c \cup C^\circ) = \lambda(EK_\mu(C^\circ)^c) + \lambda(C^\circ) > \lambda^\circ. \tag{2.2}$$

However, because $EK_\mu \subseteq K_\mu$, and

$$\mu(EK_\mu) \geq \mu(EK_\mu(C^\circ)^c) > 0,$$

(as implied by $\lambda(EK_\mu(C^\circ)^c) > 0$), the set $EK_\mu(C^\circ)^c \cup C^\circ$ is itself a chain. Thus the inequality (2.2) is impossible, which implies $\lambda(EK_\mu(C^\circ)^c) = 0$, which implies that the second term on the right hand side of (2.1) is 0.

Next, we show that the first term on the right hand side of (2.1) is also 0. Since $\mu_i(E) = 0$, we have $\mu_i(EK_\mu K_i) = 0$. In other words

$$\int_{EK_\mu K_i} f_{\mu_i} d\lambda = 0$$

Because $EK_\mu K_i \subseteq K_{\mu_i}$, the density f_{μ_i} is positive on this set. Then the above equality implies $\lambda(EK_\mu K_i) = 0$. However, because this is true for all $i \in \mathbb{N}$, we have $\lambda(EK_\mu C^\circ) = 0$. Hence $\mu(EK_\mu C^\circ) = 0$. □

The next example illustrate the meaning of this proposition.

Example 2.1 Let $\Omega = (0, \infty)$ and $\mathcal{F} = \mathcal{R} \cap (0, \infty)$, and consider the family of distributions $\mathfrak{M} = \{P_a : a > 0\}$ defined on $(\Omega, \mathcal{F})$, where P_a is the uniform distribution $U(0, a)$. That is, for any $B \in \mathcal{F}$,

$$P_a(B) = a^{-1}\lambda(B \cap (0, a)),$$

where λ is the Lebesgue measure. Let λ_0 be the Lebesgue measure on $(0, \infty)$. Then it is easy to see that $\mathfrak{M} \ll \lambda_0$. Let $\mathfrak{N} = \{P_n : n = 1, 2, \ldots\}$. Let us show that $\mathfrak{N} \equiv \mathfrak{M}$. Since $\mathfrak{N} \subseteq \mathfrak{M}$, we have $\mathfrak{N} \ll \mathfrak{M}$. To prove the opposite direction, let B be a member of $\mathcal{F}$ such that $P_a(B) > 0$ for some a. Let n be an integer greater than a. Then

$$P_a(B) > 0 \Rightarrow \lambda_0(B \cap (0, a)) > 0 \Rightarrow \lambda_0(B \cap (0, n)) > 0 \Rightarrow P_n(B) > 0.$$

This shows $\mathfrak{M} \ll \mathfrak{N}$.

Let us create a probability measure P that is equivalent to $\mathfrak{M}$. For any $B \in \mathcal{F}$, let

$$P(B) = \sum_{n=1}^{\infty} 2^{-n} P_n(B).$$

Then clearly P is a probability measure on $(\Omega, \mathcal{F})$. Furthermore, $P(B) > 0$ if and only if $P_n(B) > 0$ for some n. Thus $P \equiv \mathfrak{N} \equiv \mathfrak{M}$. □

A special case of the dominated family is the homogeneous family. A family of distribution $\mathfrak{M}$ on $(\Omega, \mathcal{F})$ is *homogeneous* if any pair of members of $\mathfrak{M}$ are equivalent. For example, The family of distributions $\{N(\mu, 1) : \mu \in \mathbb{R}\}$ is a homogeneous family.

2.1.2 Parametric families

Let Θ be a subset of $\mathbb{R}^p$, and $\mathfrak{M}^*$ be the family of all distributions on $(\Omega, \mathcal{F})$. Let $P : \Theta \to \mathfrak{M}^*$ be a function. Then the range of P, $\mathfrak{M} = \{P_\theta : \theta \in \Theta\}$, is called a parametric family. A basic assumption for a parametric family is identifiability.

Definition 2.3 *A parametric family $\{P_\theta : \theta \in \Theta\}$ is said to be identifiable if $P : \Theta \to \mathfrak{M}^*$ is an injection. That is, whenever $\theta_1 \neq \theta_2$, we have $P_{\theta_1} \neq P_{\theta_2}$, or equivalently, there is a set $B \in \mathcal{F}$ such that $P_{\theta_1}(B) \neq P_{\theta_2}(B)$.*

We say that a family of distributions $\mathfrak{M}$ on $(\Omega, \mathcal{F})$ is a model if the true distribution belongs to the family $\mathfrak{M}$. An important property of an identifiable parametric family is the likelihood inequality, as given in the next theorem. In the following, for a random variable Y defined on $(\Omega, \mathcal{F})$, we use $E_\theta(Y)$ to denote $\int Y dP_\theta$.

Theorem 2.1 *Let $\{P_\theta : \theta \in \Theta\}$, $\Theta \subseteq \mathbb{R}^p$, be a homogeneous and identifiable parametric family dominated by a σ-finite measure λ. Let $f_\theta = dP_\theta/d\lambda$. Let $\theta_0 \in \Theta$. Suppose $\log f_\theta$ is integrable with respect to P_{θ_0} for each $\theta \in \Theta$. Then, for any $\theta \neq \theta_0$,*

$$E_{\theta_0}(\log f_\theta) < E_{\theta_0}(\log f_{\theta_0}).$$

Proof. Since $P_\theta \ll P_{\theta_0}$, $P_\theta \ll \lambda$, and $P_{\theta_0} \ll \lambda$, the Radon-Nikodym derivative dP_θ/dP_{θ_0} is defined and

$$dP_\theta/dP_{\theta_0} = f_\theta/f_{\theta_0}.$$

By Jensen's inequality,

$$E_{\theta_0} \log(f_\theta/f_{\theta_0}) \leq \log[E_{\theta_0}(f_\theta/f_{\theta_0})] = \log \int \frac{dP_\theta}{dP_{\theta_0}} dP_{\theta_0} = \log 1 = 0. \quad (2.3)$$

By identifiability, whenever $\theta \neq \theta_0$, there is a set $B \in \mathcal{F}$ such that

$$\int_B f_{\theta_0} d\lambda \neq \int_B f_\theta d\lambda.$$

This implies $\lambda(f_{\theta_0} \neq f_\theta) > 0$. If $f_\theta / f_{\theta_0} = c$ a.e. λ for some constant c, then $f_\theta c = f_{\theta_0}$ a.e. λ. Integrating both sides of this equation with respect to λ gives $c = 1$, which implies $f_\theta = f_{\theta_0}$ a.e. λ, which is impossible. Hence f_θ / f_{θ_0} is nondegenerate under λ, and hence also nondegenerate under P_{θ_0}. By Theorem 1.3, the inequality in (2.3) is strict whenever $\theta \neq \theta_0$. □

2.1.3 Exponential families

A parametric family of special interest is the exponential family, which was introduced by Darmois (1935); Pitman (1936); Koopman (1936). For more information see also Barndorff-Nielsen (1978); McCullagh and Nelder (1989), and Lehmann and Casella (1998).

Let $(\Omega, \mathcal{F})$ be a measurable space, and $(\Omega_X, \mathcal{F}_X, \mu)$ a σ-finite measure space with $\Omega_X \subseteq \mathbb{R}^m$ and $\mathcal{F}_X \subseteq \mathcal{R}^m$. Let $X : \Omega \to \Omega_X$ be a random vector. Let $t : \Omega_X \to \mathbb{R}^p$ be a function measurable with respect to $\mathcal{F}_X / \mathcal{R}^p$, such that $\int_{\Omega_X} e^{\theta^T t(x)} d\mu(x) < \infty$ for some $\theta \in \mathbb{R}^p$. We say that a measurable function $t : \Omega_X \to \mathbb{R}^p$ is of full dimension with respect to a measure μ on $\mathcal{F}_X$ if for all $a \in \mathbb{R}^p$, $a \neq 0$, and all $c \in \mathbb{R}$, we have

$$\mu(\{x : a^T t(x) \neq c\}) > 0.$$

In other words, t has full dimension with respect to μ if the range of t does not stay within a proper affine subspace of $\mathbb{R}^p$ almost everywhere μ.

Definition 2.4 *Suppose $(\Omega_X, \mathcal{F}_X, \mu)$ is a σ-finite measure space and $t : \Omega_X \to \mathbb{R}^p$ is a function of full dimension with respect to μ. Let*

$$\Theta = \left\{ \theta \in \mathbb{R}^p : \int_{\Omega_X} e^{\theta^T t(x)} d\mu(x) < \infty \right\}.$$

For each $\theta \in \Theta$, let P_θ be the probability measure on $\mathcal{F}_X$ defined by

$$P_\theta(B) = \int_B \left(\int_{\Omega_X} e^{\theta^T t(x)} d\mu(x) \right)^{-1} e^{\theta^T t(x)} d\mu(x), \quad B \in \mathcal{F}_X. \tag{2.4}$$

The family of measures $\{P_\theta : \theta \in \Theta\}$ is called an exponential family.

Since an exponential family is determined by μ, t, we denote it by $\mathfrak{E}_p(\mu, t)$, where the subscript p indicates the dimension of θ. Several properties follow immediately.

Theorem 2.2 *An exponential family is identifiable and homogenous.*

Proof. Let P_{θ_1} and P_{θ_2} be two members of $\mathfrak{E}_p(t,\mu)$, and $B \in \mathcal{F}_X$. If $P_{\theta_1}(B) = 0$, then

$$\int_B e^{\theta_1^T t(x)} d\mu(x) = 0.$$

Since $e^{\theta_1^T t(x)} > 0$, we have $\mu(B) = 0$. Since, by (2.4), $P_{\theta_2} \ll \mu$, we have $P_{\theta_2}(B) = 0$. Hence $P_{\theta_1} \ll P_{\theta_2}$. By the same argument $P_{\theta_2} \ll P_{\theta_1}$. So $\mathfrak{E}_p(t,\mu)$ is homogenous.

If $P_{\theta_1} = P_{\theta_2}$, then $P_{\theta_1}(B) = P_{\theta_2}(B)$ for all $B \in \mathcal{F}_X$. Then

$$\int_B \left(\int_{\Omega_X} e^{\theta_1^T t(x)} d\mu(x) \right)^{-1} e^{\theta_1^T t(x)} d\mu(x) = \int_B \left(\int_{\Omega_X} e^{\theta_2^T t(x)} d\mu(x) \right)^{-1} e^{\theta_2^T t(x)} d\mu(x).$$

Let $b(\theta) = \log \int_{\Omega_X} e^{\theta^T t(x)} d\mu(x)$. The above equation implies

$$e^{\theta_1^T t(x) - b(\theta_1)} = e^{\theta_2^T t(x) - b(\theta_2)} \quad [\mu].$$

This implies $(\theta_1 - \theta_2)^T t(x) = b(\theta_1) - b(\theta_2)$ $[\mu]$. But since t has full dimension with respect to μ, we have $\theta_1 = \theta_2$. □

Theorem 2.3 *The parameter space Θ for an exponential family $\mathfrak{E}_p(t,\mu)$ is convex.*

Proof. Let $\lambda \in (0,1)$. Then $((1-\lambda)^{-1}, \lambda^{-1})$ is a conjugate pair. Let $\theta_1, \theta_2 \in \Theta$. Let $f = e^{(1-\lambda)\theta_1^T t}$ and $g = e^{\lambda \theta_2^T t}$. By Hölder's inequality,

$$\begin{aligned}
\int_{\Omega_X} e^{\{(1-\lambda)\theta_1 + \lambda\theta_2\}^T t(x)} d\mu(x) &= \int_{\Omega_X} f g d\mu \\
&\le \left(\int_{\Omega_X} f^{(1-\lambda)^{-1}} d\mu \right)^{1-\lambda} \left(\int_{\Omega_X} g^{\lambda^{-1}} d\mu \right)^{\lambda} \\
&= \left(\int_{\Omega_X} e^{\theta_1^T t(x)} d\mu(x) \right)^{1-\lambda} \left(\int_{\Omega_X} e^{\theta_2^T t(x)} d\mu(x) \right)^{\lambda},
\end{aligned}$$

where both factors on the right hand side are finite because both θ_1 and θ_2 belong to Θ. Hence $(1-\lambda)\theta_1 + \lambda\theta_2 \in \Theta$. □

A slightly more general definition of an exponential family involves a bijective transformation of the parameter θ. Suppose the conditions in Definition 2.4 hold. Let Υ be another subset of $\mathbb{R}^p$, and $\psi : \Upsilon \to \Theta$ a bijection. Then, for any $\theta \in \Upsilon$,

$$\int_{\Omega_X} e^{\psi^T(\theta) t(u)} d\mu(u) < \infty.$$

Let P_θ be the probability measure on $(\Omega_X, \mathcal{F}_X)$ defined by

$$dP_\theta = c(\theta; \psi, t, \mu) e^{\psi^T(\theta) t(x)} d\mu, \tag{2.5}$$

where

$$c(\theta; \psi, t, \mu) = \left(\int_{\Omega_X} e^{\psi^T(\theta) t(x)} d\mu \right)^{-1}.$$

Then we call the family $\{P_\theta : \theta \in \Upsilon\}$ the exponential family with respect to ψ, t, μ, and write this family as

$$\mathfrak{E}_p(\psi, t, \mu).$$

Henceforth, whenever the first argument of $\mathfrak{E}_p(\cdot, \cdot, \cdot)$ is missing we always mean the special case defined by Definition 2.4, where ψ is the identity mapping. We will consistently use $c(\theta; \psi, t, \mu)$ to represent the proportional constant for $\mathfrak{E}_p(\psi, t, \mu)$. Again, if the argument ψ is missing, then $c(\theta; t, \mu)$ means the proportional constant for $\mathfrak{E}_p(t, \mu)$.

2.2 Sufficient, complete, and ancillary statistics

Let $(\Omega_X, \mathcal{F}_X)$ and $(\Omega_T, \mathcal{F}_T)$ be measurable spaces. Let $\mathfrak{M}$ be a family of probability measures on $(\Omega_X, \mathcal{F}_X)$. Let $X : \Omega \to \Omega_X$ and $T : \Omega_X \to \Omega_T$ be mappings measurable with respect to $\mathcal{F}/\mathcal{F}_X$ and $\mathcal{F}_X/\mathcal{F}_T$, respectively. Usually, Ω_X and Ω_T are Euclidean spaces and $\mathcal{F}_X$ and $\mathcal{F}_T$ are corresponding Borel σ-fields, but we do not need to make such assumptions. The random element X represents the data. The random element T represents a statistic. Formally, a statistic is any measurable mapping defined on Ω_X. Implicit in this definition is that T does not depend on the measure $P \in \mathfrak{M}$, because otherwise T would be a mapping from $\Omega_X \times \mathfrak{M}$ to Ω_T. A statistic T is sufficient with respect to the family $\mathfrak{M}$ if, for any $B \in \mathcal{F}_X$, the conditional probability $P(B|T) = P(B|T)$ is the same for all $P \in \mathfrak{M}$.

Definition 2.5 *A statistic $T = T(X)$ is sufficient for $\mathfrak{M}$ if, for each $B \in \mathcal{F}_X$, there is a function $\kappa_B : \Omega_X \to \mathbb{R}$, measurable $T^{-1}(\mathcal{F}_T)/\mathcal{R}$, such that for each $P \in \mathfrak{M}$,*

$$P(B|T) = \kappa_B \quad a.e.\ P.$$

The point of this definition is, of course, that κ_B is the same for all $P \in \mathfrak{M}$. Note that, we do not require the measure zero set under each P to be the same for all P. In other words, our sufficiency is not defined in terms of conditional distribution but in terms of conditional probability. A common way to find sufficient statistic is to use Fisher-Neyman factorization theorem, of which the following theorem is a special case.

Lemma 2.1 *Let $P \ll P_0$ be two probability measures on $(\Omega_X, \mathcal{F}_X)$ and $T : \Omega_X \to \Omega_T$ be a function measurable $\mathcal{F}_X/\mathcal{F}_T$. Then the following assertions are equivalent.*

1. $E_{P_0}\left(\frac{dP}{dP_0}\middle|T\right) = \frac{dP}{dP_0}$ $[P_0]$;
2. $P(B|T) = P_0(B|T)$ $[P]$ *for any* $B \in \mathcal{F}_X$.

Proof. 1⇒ 2. Since both $P(B|T)$ and $P_0(B|T)$ are measurable $T^{-1}(\mathcal{F}_T)$, it suffices to show that, for any $G \in T^{-1}(\mathcal{F}_T)$,

$$\int_G P(B|T)dP = \int_G P_0(B|T)dP.$$

We have

$$\int_G P(B|T)dP = \int_G I_B dP = \int I_B \left(\frac{dP}{dP_0}\right) dP_0$$

By 1, the right hand side is

$$\int_G I_B E_{P_0}\left(\frac{dP}{dP_0}\middle|T\right) S dP_0 = \int_G E_{P_0}(I_B|T)\frac{dP}{dP_0}\, dP_0 = \int_G P_0(B|T)dP.$$

2⇒ 1. It suffices to show that, for any $B \in \mathcal{F}_X$,

$$\int_B \frac{dP}{dP_0}\, dP_0 = \int_B E_{P_0}\left(\frac{dP}{dP_0}\middle|T\right) dP_0.$$

The right-hand side is

$$\begin{aligned}\int_B E_{P_0}\left(\frac{dP}{dP_0}\middle|T\right) dP_0 &= \int I_B E_{P_0}\left(\frac{dP}{dP_0}\middle|T\right) dP_0 \\ &= \int E_{P_0}(I_B|T)\frac{dP}{dP_0}\, dP_0 \\ &= \int E_{P_0}(I_B|T) dP \\ &= \int_B dP \\ &= \int_B \frac{dP}{dP_0}\, dP_0,\end{aligned}$$

as desired. □

Theorem 2.4 *Suppose* $\mathfrak{M}$ *is a family of probability measures on* $(\Omega_X, \mathcal{F}_X)$ *that is dominated by a* σ*-finite measure* λ*. Then the following statements are equivalent:*

1. T *is sufficient with respect to* $\mathfrak{M}$;
2. *There is a probability measure* P_0 *such that* $\mathfrak{M} \equiv P_0$*, and for every* $P \in \mathfrak{M}$*, there is an* $h_P : \Omega_X \to \mathbb{R}$ *measurable* $T^{-1}(\mathcal{F}_T)/\mathcal{R}$ *such that* $\frac{dP}{dP_0} = h_P$ $[P_0]$.

3. For every $P \in \mathfrak{M}$,

$$\frac{dP}{d\lambda} = (g_P \circ T)(x) u(x) \quad a.e.\ \lambda$$

for some $g_P : \Omega_T \to \mathbb{R}$ *measurable* $\mathcal{F}_T$ *and* $u : \Omega_X \to \mathbb{R}$ *measurable* $\mathcal{F}_X$ *integrable* λ.

Note that we do not need to assume $P_0 \in \mathfrak{M}$.

Proof. 2 $\Rightarrow$ 1. Since $\frac{dP}{dP_0} = h_P\ [P_0]$, we have $E_{P_0}\left(\frac{dP}{dP_0}\Big|T\right) = \frac{dP}{dP_0}\ [P_0]$. By Lemma 2.1, $P(B|T) = P_0(B|T)\ [P]$. Let $\kappa_B = P_0(B|T)$ to complete the proof of this part.
1 $\Rightarrow$ 2. Let $\mathfrak{N} = \{P_1, P_2, \ldots\}$ be a subset of $\mathfrak{M}$ such that $\mathfrak{N} \equiv \mathfrak{M}$. Let

$$P_0 = \sum_{n=1}^{\infty} 2^{-n} P_n.$$

It is left as an exercise to prove that P_0 is a probability measure and $P_0 \equiv \mathfrak{M}$. Let $B \in \mathcal{F}_X$, and let $\kappa_B : \Omega_X \to \mathbb{R}$ be a function measurable $T^{-1}(\mathcal{F}_T)/\mathcal{R}$ such that $P(B|T) = \kappa_B\ [P]$. We claim that κ_B is also the conditional probability $P_0(B|T)$; that is, $P_0(B|T) = \kappa_B\ [P_0]$. It suffices to show that, for any $G \in T^{-1}(\mathcal{F}_T)$,

$$\int_G \kappa_B dP_0 = P_0(B \cap G).$$

Note that

$$\int_G \kappa_B dP_0 = \sum_{n=1}^{\infty} 2^{-n} \int_G \kappa_B dP_n.$$

Since $P_n(B|T) = \kappa_B\ [P_n]$, we have $P_n(B \cap G) = \int_G \kappa_B dP_n$. Hence the right hand side above is

$$\sum_{n=1}^{\infty} 2^{-n} P_n(B \cap G) = \sum_{n=1}^{\infty} 2^{-n} \int I_G I_B dP_n = \int I_G I_B dP_0 = P_0(B \cap G).$$

Let P be a member of $\mathfrak{M}$. Since $P(B|T) = \kappa_B\ [P]$ and $P_0(B|T) = \kappa_B\ [P_0]$, by Lemma 2.1, part 2, we have $E_{P_0}(dP/dP_0|T) = dP/dP_0\ [P_0]$. Then $h_P = E_{P_0}(dP/dP_0|T)$ satisfies the desired condition in part 2.
2 $\Rightarrow$ 3. By 2, $dP/dP_0 = g_P \circ T\ \ [P_0]$, where $g_P : \Omega_T \to \mathbb{R}$ measurable $\mathcal{F}_T$. Since $P \ll \lambda$ and $P_0 \ll \lambda$, we have

$$dP/d\lambda = (g_P \circ T) dP_0/d\lambda\ \ [\lambda]$$

where $dP_0/d\lambda$ is integrable λ.

3 $\Rightarrow$ 2. Suppose $dP/d\lambda = (g_P \circ T)u \;\; [\lambda]$. Since $dP/d\lambda \geq 0 \;\; [\lambda]$, we have $(g_P \circ T)u = (|g_P| \circ T)|u|$. Let $dP_0 = |u| d\lambda$. Then $dP = (|g_P| \circ T) dP_0$, as desired. □

If S is another statistic such that T is measurable $\sigma(S)$ (that is, $\sigma(T) \subseteq \sigma(S)$), then we say that S refines T. It is intuitively clear that if T is sufficient and S refines T, then S is also sufficient, because T is a more concise summary of the data. This is proved in the next corollary.

Corollary 2.1 *Under the conditions in Theorem 2.4, if T is sufficient for $\mathfrak{M}$ and S refines T, then S is sufficient for $\mathfrak{M}$.*

Proof. By Theorem 2.4, $T(X)$ is sufficient for $\mathfrak{M}$ if and only if $E_{P_0}(dP/dP_0|T) = dP/dP_0$ a.e. P_0. Then

$$\begin{aligned} E_{P_0}(dP/dP_0|S) &= E[E_{P_0}(dP/dP_0|T)|S] \\ &= E_{P_0}(dP/dP_0|T) \\ &= dP/dP_0, \end{aligned}$$

which, by Theorem 2.4 again, is equivalent to $S(X)$ being sufficient. □

Sufficient statistic means we can use T as the data without losing any information about θ. This is a reduction of the original X. Naturally, if S refines T and T is sufficient we prefer T, because the latter is a greater reduction of the original data. It is then natural introduce the concept the minimal sufficient statistic.

Definition 2.6 *A statistic $T : \Omega_X \to \Omega_T$ is minimal sufficient for $\mathfrak{M}$ if*

1. *T is sufficient for $\mathfrak{M}$;*
2. *it is refined by any other sufficient statistic for $\mathfrak{M}$.*

Another useful concept is the complete statistic.

Definition 2.7 *Let $T : \Omega_X \to \Omega_T$ be measurable $\mathcal{F}_X/\mathcal{F}_T$. We say that T is complete for $\mathfrak{M}$ if, for any $g : \Omega_T \to \mathbb{R}$ measurable $\mathcal{F}_T/\mathcal{R}$ such that $g \circ T$ is P-integrable, we have*

$$\int g \circ T dP = 0 \;\; \text{for all } P \in \mathfrak{M} \Rightarrow g \circ T = 0 \; [P] \;\; \text{for all } P \in \mathfrak{M}.$$

The statistic T is bounded complete for $\mathfrak{M}$ if the above implication holds for all $\mathcal{F}_T$-measurable bounded function g.

Note that bounded completeness is a weaker assumption than completeness. These concepts are important in several ways. As we will see shortly, under mild conditions, if a sufficient statistic is complete, then it is a minimal sufficient statistic. Also, in Chapter 3, we will see that the notion of (bounded) completeness is helpful for constructing uniformly most powerful unbiased tests in the presence of nuisance parameters.

Suppose that $\mathfrak{M}$ is a dominated family of probability measures on $(\Omega_X, \mathcal{F}_X)$. Then there is a probability measure P_0 such that $P_0 \equiv \mathfrak{M}$. In the following, we say that a random vector belongs to $L_2(P_0)$ if each of its component belongs to $L_2(P_0)$.

Theorem 2.5 *Suppose there exists a minimal sufficient statistic for $\mathfrak{M}$ in $L_2(P_0)$. Then any complete and sufficient statistic for $\mathfrak{M}$ in $L_2(P_0)$ is minimal sufficient for $\mathfrak{M}$.*

Note that the statistics in the above theorem are allowed to be vectors. In the following proof, for a random variable W, $E_P(W)$ denotes the expectation of W under P; that is $E_P(W) = \int W dP$.

Proof. Let $T \in L_2(P_0)$ be a minimal sufficient statistic for $\mathfrak{M}$, and $U \in L_2(P_0)$ a complete sufficient statistic for $\mathfrak{M}$. It suffices to show that U is measurable with respect to T, for which it suffices to show $U = E_{P_0}(U|T)$. We note that

$$E_P(U - E_P(E_P(U|T)|U)) = E_P(U) - E_P(U) = 0$$

for all $P \in \mathfrak{M}$. Because both U and T are sufficient, the conditional expectations given U or T are the same for all P. In particular,

$$E_P(E_P(U|T)|U) = E_{P_0}(E_{P_0}(U|T)|U).$$

Hence

$$E_P(U - E_{P_0}(E_{P_0}(U|T)|U)) = 0$$

for all $P \in \mathfrak{M}$. By the completeness of U, we have $U = E_{P_0}(E_{P_0}(U|T)|U)$. However, because T is minimal sufficient and U is sufficient, T is measurable with respect to $\sigma(U)$. Hence $E_{P_0}(E_{P_0}(U|T)|U) = E_{P_0}(U|T)$. □

A statistic $U : \Omega_X \to \Omega_U$ is said to be an *ancillary* statistic with respect to a family $\mathfrak{M}$ of probability distributions over $(\Omega_X, \mathcal{F}_X)$ if $P \circ U^{-1}$ is the same for all $P \in \mathfrak{M}$. Basu (1955) proved the following well known result.

Theorem 2.6 (Basu's Theorem) *Suppose $T : \Omega_X \to \Omega_T$ is complete and sufficient for $\mathfrak{M}$ and $U : \Omega_X \to \Omega_U$ is ancillary for $\mathfrak{M}$. Then T and U are independent under each $P \in \mathfrak{M}$.*

Proof. We shall show that for any $B \in \mathcal{F}_U$ we have

$$P(U \in B|T) - P(U \in B) = 0 \ \ [P].$$

Since T sufficient, the first term does not depend on P, and because U is ancillary, neither does the second term. Moreover, we have

$$\int [P(U \in B|T) - P(U \in B)]dP = 0$$

for all $P \in \mathfrak{M}$. Since the integrand is a function of T independent of P, by the completeness of T we have

$$P(U \in B|T) = P(U \in B) \ \ [P]$$

This means U and T are independent. □

2.3 Complete sufficient statistics for exponential family

In this section we derive the complete and sufficient statistic for an exponential family. For this purpose it is useful to review the basic properties of analytic functions.

Definition 2.8 *A function $f : \mathbb{R} \to \mathbb{R}$ is real analytic on an open set $G \subseteq \mathbb{R}$ if for any $x \in G$, there is an open neighborhood N of x such that, for all $x' \in N$,*

$$f(x') = \sum_{n=0}^{\infty} a_n (x' - x)^n,$$

in which the coefficients $a_0, a_1, \ldots$ are real numbers.

In other words, an analytic function can be expanded as a power series locally at each point in the region in which it is analytic. Thus, equivalently, a real analytic function can be defined as a function that is infinitely differentiable and that can be expanded as a Taylor series locally at any point in G. The above definition can be generalized to functions with several variables, say $x = (x_1, \ldots, x_p)$. In that case the power series is replaced by

$$f(x') = \sum a_{i_1 \ldots i_p} (x'_1 - x_1)^{i_1} \cdots (x'_p - x_p)^{i_p},$$

where the summation is over the index set

$$\{(i_1, \ldots, i_p) : \ i_1, \ldots, i_p = 0, 1, 2, \ldots\}.$$

A complex analytic function is defined in the same way by replacing $\mathbb{R}$ by $\mathbb{C}$, the set of all complex numbers, and G by an open set in $\mathbb{C}$. It is true that any complex function from $\mathbb{C}$ to $\mathbb{C}$ that is differentiable is analytic. An analytic function is global, in the sense that the overall property of the function can be determined by its local property near a point. In this sense it behaves rather like a polynomial. We know that a kth order polynomial cannot have more than k solutions unless all the coefficients of the polynomial are 0. Similarly, if the collection of roots of an analytic function contains a limit point, then it is identically 0. This fact will be used in several places in later discussions. See, for example, Rudin (1987, page 209).

Theorem 2.7 *If f is a real analytic function on an open set G in $\mathbb{R}^p$ and if $\{x : f(x) = 0\}$ has a limit point in G, then $f(x) = 0$ on G.*

The following is an important property of an exponential family of distributions. It implies that if $g(x)$ integrable then $E_\theta g(X)$ is an analytic function of θ provided that f_θ is an exponential family.

Lemma 2.2 *Let $(\Omega_X, \mathcal{F}_X, \mu)$ be a measure space, where $\Omega_X \subseteq \mathbb{R}^p$, and Θ is a subset of $\mathbb{R}^p$ with nonempty interior. Suppose $g(x)e^{\theta^T x}$ is integrable with respect to μ for each $\theta \in \Theta$, then the integral $\int g(x)e^{\theta^T x}\mu(dx)$ is analytic function of θ in the interior of Θ. Furthermore, the derivatives of all orders with respect to θ can be taken inside the integral.*

For a proof, see, for example, Lehmann and Romano (2005, page 49).

Theorem 2.8 *Suppose $\Theta \subseteq \mathbb{R}^p$ has a nonempty interior (— that is, it has a nonempty open subset). Then the statistic $t(X)$ is sufficient and complete for the exponential family $\mathfrak{E}_p(\lambda, t)$.*

Proof. Since

$$\frac{dP_\theta}{d\lambda} = e^{\theta^T t(x)} \left(\int e^{\theta^T t(x)} d\lambda(x) \right)^{-1}$$

is measurable with respect to $t^{-1}(\mathcal{R}^p)$, by Theorem 2.4, $t(X)$ is sufficient with respect to $\mathfrak{E}_p(\lambda, t)$. Let g be an integrable function of t such that

$$E_\theta g(t(X)) = 0 \quad \text{for all} \quad \theta \in \Theta.$$

Then, by the change of variable theorem,

$$\int g(t)e^{\theta^T t} d\nu(t) = 0 \quad \text{for all} \quad \theta \in \Theta,$$

where $\nu = \lambda \circ t^{-1}$. Because this integral is an analytic function of θ, and because Θ contains an open set (and hence limit point), it is 0 over $\mathbb{R}^k$. By the uniqueness of Laplace transformation, $g(t) = 0$ $[\nu]$. By the change of variable theorem, $g(t(x)) = 0$ $[\lambda]$. Because $P_\theta \ll \lambda$, $g(X) = 0$ $[P_\theta]$ for all $\theta \in \Theta$. □

2.4 Unbiased estimator and Cramér-Rao lower bound

Let $\mathfrak{M} = \{P_\theta : \theta \in \Theta\}$, where $\Theta \subseteq \mathbb{R}^p$, be a parametric family of probability measures on $(\Omega_X, \mathcal{F}_X)$. In this section, we will always assume that $\mathfrak{M}$ is dominated by a σ-finite measure λ, and that $\mathfrak{M}$ is identifiable. For a random vector $U = (U_1, \ldots, U_k)$ defined on Ω_X, we use $E(U)$ to denote the vector $(EU_1, \ldots, EU_2)$. For two random vectors $U = (U_1, \ldots, U_k)$ and $V = (V_1, \ldots, V_\ell)$ defined on $(\Omega_X, \mathcal{F}_X)$, we use $\mathrm{cov}(U, V)$ to denote the matrix whose (i, j)th entry is $\mathrm{cov}(U_i, V_j)$. Moreover, we define $\mathrm{var}(U)$ as $\mathrm{cov}(U, U)$.

A few more words about notation for expectation. In most of this book, there is no need to make a distinction between the underlying probability space $(\Omega, \mathcal{F})$ and the induced probability space $(\Omega_X, \mathcal{F}_X)$. That is, we will simply equate these two probability spaces. In this case, a random variable X is just the identity mapping:

$$X : \Omega \to \Omega_X, \quad x \mapsto x.$$

So $E(X)$ means the integral $\int X dP = \int X(x) dP(dx) = \int x dP(dx)$. To be consistent with the previous notation, it is helpful to still regard P as a probability measure defined on $(\Omega, \mathcal{F})$.

We now introduce the unbiased estimator of a parameter. In the following, X is a random vector representing the data. Typically, X is a sample $X_1, \ldots, X_n$, where each X_i is a k-dimensional random vector. In this case X is an nk-dimensional random vector, Ω_X is $\mathbb{R}^{nk}$ and $\mathcal{F}_X$ is $\mathcal{R}^{nk}$. Intuitively, if we want to use a random vector $u(X)$ to estimate a parameter θ, then we would like the distribution of $u(X)$ to be centered at the target we want to estimate. This is formulated mathematically as the next definition. In the following we will use E_θ and var_θ to denote the mean (vector) and variance (matrix) of a random variable (vector) under P_θ.

Definition 2.9 *Let $u : \Omega_X \to \mathbb{R}^p$ be a function measurable $\mathcal{F}_X / \mathcal{R}^p$. We say that $u(X)$ is an unbiased estimator of θ if, for all $\theta \in \Theta$,*

$$E_\theta[u(X)] = \theta.$$

For two symmetric matrices A and B, we write $A \succeq B$ if $A - B$ is positive semidefinite; we write $A \succ B$ if $A - B$ is positive definite. The partial ordering represented by $\succeq$ or $\succ$ among matrices is sometimes called positive semidefinite ordering, positive definite ordering, or Louwner's ordering. The following proposition is well known. See Horn and Johnson (1985, page 465).

Proposition 2.2 *If A and B are positive definite, then*

$$A \succeq B \Leftrightarrow B^{-1} \succeq A^{-1}, \quad A \succ B \Leftrightarrow B^{-1} \succ A^{-1}.$$

The next lemma will be called the multivariate Cauchy-Schwarz inequality. It is the basis of the famous Cramér-Rao inequality. We say that a function f is square integrable with respect to a measure μ if f is measurable and f^2 is integrable μ.

Lemma 2.3 (Multivariate Cauchy-Schwarz inequality) *Let $(\Omega, \mathcal{F}, \mu)$ be a measure space, and $f : \Omega \to \mathbb{R}^p$ and $g : \Omega \to \mathbb{R}^p$ be $\mathbb{R}^p$-valued functions whose components are square-integrable with respect to μ. Suppose that the matrix $\int g g^T d\mu$ is nonsingular. Then*

$$\left(\int fg^T d\mu\right)\left(\int gg^T d\mu\right)^{-1}\left(\int gf^T d\mu\right) \leq \int ff^T d\mu. \tag{2.6}$$

The equality in (2.6) holds if and only if there is a constant matrix $A \in \mathbb{R}^{p\times p}$ such that $f = Ag \ \ [\mu]$. Moreover, if $\int ff^T d\mu$ is nonsingular then so is A.

Proof. Let

$$\delta = f - \left(\int fg^T d\mu\right)\left(\int gg^T d\mu\right)^{-1} g.$$

Then

$$\int \delta\delta^T d\mu = \int ff^T d\mu - \left(\int fg^T d\mu\right)\left(\int gg^T d\mu\right)^{-1}\left(\int gf^T d\mu\right).$$

This implies the inequality (2.6) because the left-hand side is positive semidefinite. The equality in (2.6) holds if and only if $\int \delta\delta^T d\mu = 0$, which happens if and only if $\delta = 0 \ \ [\mu]$. That is,

$$f = \left(\int fg^T d\mu\right)\left(\int gg^T d\mu\right)^{-1} g \equiv Ag \ \ [\mu].$$

Then $\int ff^T d\mu = A\int gg^T d\mu A^T$. Since $\int ff^T d\mu$ is nonsingular A must also be nonsingular. □

Let $f_\theta = dP_\theta/d\lambda$. If $\log f_\theta(x)$ is differentiable with respect to θ, then the partial derivative $\partial \log f_\theta(x)/\partial\theta$ is called the score function (as a function of θ and x). We denote the score function by $s_\theta(x)$. Let $I(\theta) = \mathrm{var}_\theta(s_\theta(X))$. This matrix is called the Fisher information. Let I_p be the $p \times p$ identity matrix.

In the rest of the book we will frequently assume that $f_\theta(x)$ has a common support, say A, for all $\theta \in \Theta$ and that, for some function $g(\theta, x)$, $g(\theta, x)f_\theta(x)$ satisfies DUI(θ, A, μ). For convenience, this lengthy statement is abbreviated in the definition below.

Definition 2.10 *We say that $g(\theta, x)f_\theta(x)$ satisfies DUI$^+(\theta, \mu)$, if the support A of $f_\theta(x)$ is independent of θ and $g(\theta, x)f_\theta(x)$ satisfies DUI(θ, A, μ).*

We usually regard the Fisher information $I(\theta)$ as appropriately defined only when both $f_\theta(x)$ and $s_\theta(x)f_\theta(x)$ satisfy DUI$^+(\theta, \mu)$. In this case, it is easy to verify (by passing first two derivatives of θ through the integral with respect to μ)

$$\begin{aligned} E_\theta[s_\theta(X)] &= 0 \\ \mathrm{var}_\theta[s_\theta(x)] &= -E[\partial s_\theta(X)/\partial\theta^T]. \end{aligned} \tag{2.7}$$

More specifically, the first identity requires $f_\theta(x)$ to satisfy DUI$^+(\theta, \mu)$; the second requires both $f_\theta(x)$ and $s_\theta(x)f_\theta(x)$ to satisfy DUI$^+(\theta, \mu)$. These identities are called the information identities. We will have a closer look at them in Chapter 8, where they play a critical role.

The next two theorems, however, only require the first equality in (2.7) to hold. So in this chapter, we define the Fisher information $I(\theta)$ as $\mathrm{var}_\theta[s_\theta(X)]$ under the first equality in (2.7) without regard of the second equality in (2.7).

We now prove the Cramér-Rao's inequality, which asserts that, under some conditions, no unbiased estimator can have variance smaller than $I^{-1}(\theta)$. This is but a special case of a very general phenomenon. Later on we will see that $I^{-1}(\theta)$ is in fact the lower bound of the asymptotic variances of all regular estimates, which cover a very wide range of estimates used in statistics.

Theorem 2.9 *Let $\mathfrak{M} = \{P_\theta : \theta \in \Theta\}$ be a dominated identifiable parametric family on a measurable space $(\Omega_X, \mathcal{F}_X)$. Let $f_\theta = dP_\theta/d\lambda$, where λ is the σ-finite dominating measure. Suppose*

1. *$U = u(X)$ is an unbiased estimator of θ such that all entries of the matrix $\mathrm{var}_\theta(U)$ are finite for each $\theta \in \Theta$;*
2. *$f_\theta(x)$ and $u(x)f_\theta(x)$ satisfy $DUI^+(\theta, \lambda)$;*
3. *$I(\theta)$ is nonsingular for each $\theta \in \Theta$.*

Then, for all $\theta \in \Theta$,

$$\mathrm{var}_\theta(U) \succeq I^{-1}(\theta). \tag{2.8}$$

Moreover, the equality holds if and only if $u(X) = \theta + I^{-1}(\theta)s_\theta(X)$.

Proof. Since f_θ satisfies $\mathrm{DUI}^+(\theta, \lambda)$, we have

$$0 = \partial(1)/\partial\theta = \int \partial f_\theta(x)/\partial\theta d\lambda(x) = E_\theta[s_\theta(X)]. \tag{2.9}$$

This implies $\mathrm{var}_\theta[s_\theta(X)] = E_\theta[s_\theta(X)s_\theta^T(X)] = I(\theta)$. Let $\delta_\theta(X) = u(X) - \theta$. By Lemma 2.3,

$$\mathrm{var}_\theta[u(X)] \succeq E_\theta[\delta_\theta(X)s_\theta^T(X)][E_\theta(s_\theta(X)s_\theta(X))]^{-1}E_\theta[s_\theta(X)\delta_\theta^T(X)]. \tag{2.10}$$

By (2.9),

$$E_\theta[s_\theta(X)\delta_\theta^T(X)] = E_\theta[s_\theta(X)u^T(X)].$$

Moreover, because $u(x)f_\theta(x)$ satisfies $\mathrm{DUI}^+(\theta, \mu)$, we have

$$E_\theta[s_\theta(X)u^T(X)] = \int u(x)(\partial f_\theta(x)/\partial\theta^T)d\mu(x) = \partial\theta/\partial\theta^T = I_p.$$

Therefore the right hand side of (2.10) is $I^{-1}(\theta)$. This proves the inequality (2.8).

By the equality part of Lemma 2.3, $\mathrm{var}_\theta[u(X)] = I^{-1}(\theta)$ if and only if

$$\delta_\theta(X) = A_\theta s_\theta(X)$$

for some matrix A_θ. It follows that

$$E_\theta[\delta_\theta(X)\delta_\theta^T(X)] = A_\theta E_\theta[s_\theta(X)\delta_\theta^T(X)] = A_\theta.$$

Hence $A_\theta = I^{-1}(\theta)$. □

We now discuss two problems related to this inequality. The first is about reaching the lower bound. The second is about the practical situations where the DUI$^+(\theta, \lambda)$ condition is violated and the lower bound is exceeded. The first problem is often referred to as the attainment problem in the literature. See, for example, Fend (1959). The next theorem gives a solution to this problem.

Theorem 2.10 *Suppose $(\Omega_X, \mathcal{F}_X)$ is a measurable space and $\mathfrak{M} = \{P_\theta : \theta \in \Theta\}$ is an identifiable parametric family dominated by a σ-finite measure λ. Let $f_\theta = dP_\theta/d\lambda$. Further, suppose that*

1. *$f_\theta(x)$ satisfies DUI$^+(\theta, \mu)$;*
2. *$I(\theta)$ is positive definite and its entries are finite.*

Then the following statements are equivalent:

1. *there is an unbiased estimator $u(X)$ of θ such that $\mathrm{var}_\theta[u(X)] = I^{-1}(\theta)$;*
2. *X has the exponential-family distribution of the form*

$$f_\theta(x) = e^{\psi^T(\theta)u(x)} \left(\int e^{\psi^T(\theta)u(x)} d\nu(x) \right)^{-1} \tag{2.11}$$

for some measure ν on Ω_X dominated by λ, and some function $\psi(\theta)$ differentiable with respect to θ satisfying

$$\frac{\partial \psi(\theta)}{\partial \theta^T} = \frac{\partial \psi^T(\theta)}{\partial \theta} = I(\theta).$$

Proof. 1 ⇒ 2. By Lemma 2.3, the equality in (2.8) holds if and only if $s_\theta(x) = A_\theta(u(x) - \theta)$ for some nonsingular matrix A_θ. This means

$$\partial \log f_\theta(x)/\partial\theta = A_\theta u(x) - A_\theta \theta$$

Hence there exist $\psi : \Theta \to \mathbb{R}^p$ and $\phi : \Theta \to \mathbb{R}$ such that

$$\partial\psi^T/\partial\theta = A_\theta, \quad \partial\phi/\partial\theta = A_\theta\theta, \quad \log f_\theta(x) = \psi^T(\theta)u(x) + \phi(\theta) + v(x).$$

Thus

$$f_\theta(x) = e^{\psi^T(\theta)u(x)} c_\theta w(x).$$

So the density of X has the form (2.11) with $d\nu = w d\lambda$.

2 ⇒ 1. Since $\psi \to \int u(x) e^{\psi^T u(x)} d\nu(x)$ and $\psi \to \int e^{\psi^T u(x)} d\nu(x)$ are analytic, $u(x) f_\theta(x)$ satisfies DUI$^+(\theta, \nu)$. Applying the fact that $u(X)$ is unbiased we find

$$E_\theta(s_\theta(X)u^T(X)) = E_\theta(s_\theta(X)(u(X) - \theta)^T) = I_p.$$

We also know that

$$s_\theta(X) = \frac{\partial \psi^T(\theta)}{\partial \theta} u(X) - \frac{\partial}{\partial \theta} \log \int e^{\psi^T(\theta)u(x)} d\nu(x) = A_\theta u(X) + b_\theta.$$

Because f_θ satisfies $\mathrm{DUI}^+(\theta, \nu)$, we have $E_\theta s_\theta(X) = 0$. Hence $b_\theta = -A_\theta \theta$; that is,

$$s_\theta(X) = A_\theta(u(X) - \theta).$$

It follows that

$$E_\theta(s_\theta(X)(u(X) - \theta)^T) = A_\theta \mathrm{var}_\theta(u(X)) = A_\theta I^{-1}(\theta).$$

Hence $A_\theta = I(\theta)$. But this implies

$$\mathrm{var}_\theta[u(X)] = I^{-1}(\theta)\mathrm{var}_\theta[s_\theta(X)]I^{-1}(\theta) = I^{-1}(\theta),$$

as desired. □

This theorem shows that the Cramér-Rao lower bound is achieved by an unbiased estimator $u(X)$ if and only if X has an exponential family distribution with $u(X)$ as its sufficient statistic. For any other distribution that satisfies the relevant $\mathrm{DUI}^+(\theta, \lambda$ assumption, this bound cannot be reached exactly.

When the $\mathrm{DUI}^+(\theta, \lambda)$ assumption is not satisfied, an unbiased estimator can have smaller variance than $I^{-1}(\theta)$, as the following example shows.

Example 2.2 Consider the case where P_θ is the uniform $U(0, \theta)$, $\theta > 0$. That is,

$$f_\theta(x) = \theta^{-1} I_{(0,\theta)}(x), \quad \theta > 0.$$

This family does not satisfy $\mathrm{DUI}^+(\theta, \mu)$ because the support of $f_\theta(x)$ depends on θ. Since

$$\partial(\theta^{-1} I_{(0,\theta)(x)})/\partial\theta = -\theta^{-2} I_{(0,\theta)(x)}$$

almost everywhere with respect to the Lebesgue measure, we have

$$\int \partial f_\theta(x)/\partial\theta \, dx = -\theta^{-1}.$$

On the other hand $\partial[\int f_\theta(x)\, dx]/\partial\theta = 0$. So

$$\frac{\partial}{\partial\theta} \int f_\theta(x) dx \neq \int \frac{\partial f_\theta(x)}{\partial\theta} dx.$$

The score function for this density is

$$s_\theta(x) = \partial \log f_\theta(x)/\partial\theta = -\partial \log(\theta)/\partial\theta + \partial \log I_{(0,\theta)}(x)/\partial\theta = -1/\theta,$$

which is defined almost everywhere with respect to the Lebesgue measure. Hence $\mathrm{var}_\theta[s_\theta(X)] = 0$. So the inequality

$$\mathrm{var}_\theta[u(X)] \geq 1/\mathrm{var}_\theta[s_\theta(X)]$$

does not hold for any unbiased estimator of θ with a finite variance.

In the above we have considered a single observation from $U(0,\theta)$ for convenience, but the conclusion for the situation where we have an i.i.d. sample from $U(0,\theta)$ is essentially the same. □

2.5 Conditioning on complete and sufficient statistics

Unbiasedness is but one way of assessing the quality of an estimator — that a good estimator should be centered around the target it intends to estimate. Another way to assess the quality of an estimator is by its variance: a good estimator should be more closely clustered around the target. In this section we study how to find estimator with small variance. We first introduced a lemma, often called the EV-VE formula.

Lemma 2.4 *Suppose that the components of U belong to $L_2(P)$. Then*

$$\mathrm{var}(U) = E[\mathrm{var}(U|\mathcal{G})] + \mathrm{var}[E(U|\mathcal{G})].$$

Proof. We have

$$\begin{aligned}\mathrm{var}(U) &= E(U-EU)(U-EU)^T \\ &= E[(U-E(U|\mathcal{G})+E(U|\mathcal{G})-EU)(U-E(U|\mathcal{G})+E(U|\mathcal{G})-EU)^T]\end{aligned}$$

Note that

$$\begin{aligned}E[(U-E(U|\mathcal{G}))(E(U|\mathcal{G})-EU)^T] &= E[(U-E(U|\mathcal{G}))E(U^T-EU^T|\mathcal{G})] \\ &= E[E(U-E(U|\mathcal{G})|\mathcal{G})(U-EU)^T] = 0.\end{aligned}$$

Hence

$$\begin{aligned}\mathrm{var}(U) &= E(U-EU)(U-EU)^T \\ &= E[(U-E(U|\mathcal{G}))(U-E(U|\mathcal{G}))^T] \\ &\qquad + E[(E(U|\mathcal{G})-EU)(E(U|\mathcal{G})-EU)^T] \\ &= \mathrm{var}(E(U|\mathcal{G})) + E(\mathrm{var}(U|\mathcal{G}))\end{aligned}$$

as desired. □

Let $(\Omega_X, \mathcal{F}_X)$ be a measurable space and $\mathfrak{M} = \{P_\theta : \theta \in \Theta\}$ be a dominated and identifiable parametric family of probability distributions on $(\Omega_X, \mathcal{F}_X)$. Let $\mathfrak{U}$ be the class of all unbiased estimators $U = u(X)$ of θ defined on Ω_X such that $\text{var}_\theta(U) < \infty$ for each $\theta \in \Theta$. The following theorem shows that, given an unbiased estimator $U \in \mathfrak{U}$ and a sufficient statistic T for $\mathfrak{M}$, one can reduce the variance of U by taking the conditional expectation on T. This is called Rao-Blackwell's inequality. See Rao (1945) and Blackwell (1947).

Theorem 2.11 *Suppose $U \in \mathfrak{U}$ and T is a sufficient statistic for $\mathfrak{M}$. Then $E(U|T) \in \mathfrak{U}$ and*

$$\text{var}_\theta(U) \succeq \text{var}_\theta[E(U|T)].$$

Proof. We have

$$\text{var}_\theta(U) = \text{var}_\theta[E(U|T)] + E_\theta[\text{var}(U|T)] \succeq \text{var}_\theta[E(U|T)].$$

But $E(U|T)$ is an unbiased estimator of θ. □

If, in addition to being sufficient, T is also complete for $\mathfrak{M}$. Then taking the conditional expectation of an unbiased estimator given T actually brings the variance to the minimum. This is called the Lehmann-Scheffe theorem (Lehmann and Scheffe, 1950, 1955).

Theorem 2.12 *Suppose $T = t(X)$ is complete and sufficient statistic for $\mathfrak{M}$, and U is a statistic in $\mathfrak{U}$ that is measurable $\sigma(t(X))$. Then for any $U' \in \mathfrak{U}$, we have $\text{var}_\theta(U) \preceq \text{var}_\theta(U')$ for each $\theta \in \Theta$.*

Proof. Since $U' \in \mathfrak{U}$, we have $E_\theta[E(U'|T)] = \theta$ for all θ. Moreover, $U = E(U|T)$ and $E_\theta[E(U|T)] = \theta$. Hence

$$E_\theta[E(U'|T) - E(U|T)] = 0 \; [P_\theta] \text{ for all } \theta \in \Theta.$$

Since T is complete,

$$E(U'|T) = E(U|T) \; [P_\theta] \text{ for all } \theta \in \Theta.$$

Hence $\text{var}_\theta(U') \succeq \text{var}_\theta[E(U'|T)] = \text{var}_\theta[E(U|T)] = \text{var}_\theta(U)$ as desired. □

So, if an unbiased estimator in $\mathfrak{U}$ is measurable with respect to a complete and sufficient statistic its variance reaches the lower bound among $\mathfrak{U}$. This leads us to introduce the following notion of the optimal estimator among unbiased estimators.

Definition 2.11 *A statistic $U \in \mathfrak{U}$ is called Uniformly Minimum Variance Unbiased Estimator (UMVUE) if for any member U' of $\mathfrak{U}$ we have $\text{var}_\theta(U) \leq \text{var}_\theta(U')$ for all $U' \in \mathfrak{U}$.*

The following result follows directly from Theorem 2.12 and Definition 2.11.

Corollary 2.2 *Suppose there is a complete and sufficient statistic T for $\mathfrak{M}$. Then for any $U_1, U_2 \in \mathfrak{U}$, we have $E(U_1|T) = E(U_2|T)$ $[P_\theta]$ for all $\theta \in \Theta$. Moreover, $E(U_1|T)$ is the UMVUE.*

The above corollary implies that there can be only one measurable function of $\sigma(T)$ that is UMVUE. In fact, the uniqueness of UMVUE need not be restricted to measurable functions of T. We now show that UMVUE is unique. We first introduce a lemma.

Lemma 2.5 *Suppose X and Y are p-dimensional random vectors defined on a probability space $(\Omega, \mathcal{F}, P)$ whose entries belong to $L_2(P)$. Then the following statements are equivalent:*

1. *For any $A \in \mathbb{R}^{p\times p}$, $\mathrm{var}(X) \preceq \mathrm{var}(X + AY)$.*
2. *$\mathrm{cov}(X, Y) = 0$.*

Proof. Assertion 1 holds if and only if, for all $t \in \mathbb{R}^p$, $t \neq 0$, we have

$$\mathrm{var}(t^T X) \leq \mathrm{var}(t^T X + t^T AY).$$

Since A can take any value in $\mathbb{R}^{p\times p}$, if we let $\tau = At$, then τ can take any value in $\mathbb{R}^p$ for any fixed $t \neq 0$. Thus, assertion 1 holds if and only if, for any fixed $t \neq 0$, $f(\tau) = \mathrm{var}(t^T X + \tau^T Y)$ is minimized at $\tau = 0$, which happens if and only if $\partial f(0)/\partial \tau = 0$ for any $t \neq 0$. Since

$$f(\tau) = \mathrm{var}(t^T X) + 2\mathrm{cov}(t^T X, Y)\tau + \tau^T \mathrm{var}(Y)\tau,$$

assertion 1 holds if and only if $\mathrm{cov}(t^T X, Y) = 0$ for any $t \neq 0$, which happens if and only if $\mathrm{cov}(X, Y) = 0$. □

Theorem 2.13 *Suppose $\mathfrak{M} = \{P_\theta : \theta \in \Theta\}$ is a dominated and identifiable parametric family. A UMVUE, if it exists, is unique $[P_\theta]$ for each $\theta \in \Theta$.*

Proof. Suppose U_1 and U_2 are both UMVUE. Then

$$\mathrm{var}_\theta(U_1 - U_2) = \mathrm{var}_\theta(U_1) - \mathrm{cov}_\theta(U_1, U_2) - \mathrm{cov}_\theta(U_2, U_1) + \mathrm{var}_\theta(U_2).$$

Since both U_1 and U_2 are UMVUE, their variances are the same. So the right-hand side of the above equality becomes

$$\begin{aligned} &2\mathrm{var}_\theta(U_1) - \mathrm{cov}_\theta(U_1, U_2) - \mathrm{cov}_\theta(U_2, U_1) \\ &\qquad = \mathrm{cov}_\theta(U_1, U_1 - U_2) + \mathrm{cov}_\theta(U_1 - U_2, U_1). \end{aligned}$$

If $A \in \mathbb{R}^{p\times p}$, then $U_1 + A(U_2 - U_1)$ is an unbiased estimate, and hence

$$\text{var}_\theta(U_1) \preceq \text{var}_\theta(U_1 + A(U_2 - U_1))$$

By Lemma 2.5, $\text{cov}_\theta(U_1, U_2 - U_1) = 0$. This also implies $\text{cov}_\theta(U_2 - U_1, U_1) = 0$. Therefore

$$\text{var}_\theta(U_1 - U_2) = 0.$$

Because $E_\theta(U_1) = E_\theta(U_2) = \theta$, the above equality implies $E_\theta\|U_1 - U_2\|^2 = 0$. Hence $U_1 = U_2$ $[P_\theta]$. □

Example 2.3 In this example we develop the UMVU estimate for θ using the Lehmann-Scheffe theorem under the setting of Example 2.2. Consider an i.i.d. sample $\{X_1, \ldots, X_n\}$ from $U(0, \theta)$. Thus $(X_1, \ldots, X_n)$ has density

$$\theta^{-n} \prod_{i=1}^{n} I_{(0,\theta)}(X_i).$$

By the factorization theorem, the sufficient statistic is $T = \max(X_1, \ldots, X_n)$. The distribution of T is $P_\theta(T < t) = \prod P_\theta(X_i < t) = (t/\theta)^n$, and its density is $n(t/\theta)^{n-1}(1/\theta)$. Hence

$$E_\theta T = \int_0^\theta tn(t/\theta)^{n-1}(1/\theta)\, dt = n\theta \int_0^1 s^n\, ds = \frac{n}{n+1}\theta.$$

Let $U = (n+1)T/n$. Then U is an unbiased estimate. We now compute its variance:

$$\text{var}_\theta U = \left(\frac{n+1}{n}\right)^2 \text{var}_\theta T = \left(\frac{n+1}{n}\right)^2 (E_\theta T^2 - (E_\theta T)^2).$$

The first two moments on the right hand side are computed to be

$$E_\theta T = \frac{n}{n+1}\theta$$

$$E_\theta(T^2) = \int_0^\theta t^2 n(t/\theta)^{n-1}(1/\theta) dt = n\theta^2 \int_0^1 (s)^{n+1} ds = \frac{n}{n+2}\theta^2$$

Therefore

$$\text{var}_\theta(U) = \left[\frac{1}{n(n+2)}\right]\theta^2.$$

By Lehmann-Scheffe's theorem, U is the UMVU estimate if T is complete. Let g be a function such that $E_\theta g(T) = 0$ for all $\theta > 0$. Then

$$\int_0^\theta g(t)t^{n-1} dt = 0$$

Take derivative with respect to θ to obtain

$$g(\theta)\theta^{n-1} = 0$$

for all $\theta > 0$. Thus $g(\theta) = 0$ for all $\theta > 0$, and T is complete. □

2.6 Fisher consistency and two classical estimators

In this section we introduce another important criterion, known as Fisher consistency, to assess an estimator, besides unbiasedness and minimal variance. Strict unbiasedness is often too strong and can exclude many useful estimators. Fisher consistency is a more practical criterion, and, more importantly, it often suggests the form of an estimator. Let $\mathfrak{M}$ be a class of all probability measures on $(\Omega_X, \mathcal{F}_X)$. A parameter can be viewed as a mapping from $\mathfrak{M}$ to $\mathbb{R}^p$. For example, the mean parameter $\theta = E_P(X)$ can be viewed as the mapping

$$\theta : \mathfrak{M} \to \mathbb{R}^p, \quad P \mapsto \int X dP.$$

Let P_0 be the true probability model. Let $X_1, \ldots, X_n$ be an independent sample from X. For $x \in \Omega_X$, let δ_x be the probability measure on $(\Omega_X, \mathcal{F}_X)$ defined by the set function $\delta_x : \mathcal{F}_X \to \mathbb{R}$ by

$$\delta_x(A) = \begin{cases} 1 & x \in A \\ 0 & x \notin A \end{cases}$$

It is left as an exercise to show that δ_x is a probability measure on $(\Omega_X, \mathcal{F}_X)$. It is called the point mass at x. The set function defined by

$$P_n = n^{-1} \sum_{i=1}^{n} \delta_{X_i}$$

is called the empirical distribution of X. It is left as an exercise to show that P_n is a probability measure and for any measurable function $f : \Omega_X \to \mathbb{R}$,

$$\int f dP_n = n^{-1} \sum_{i=1}^{n} f(X_i).$$

Definition 2.12 *Let $T = t(X_1, \ldots, X_n)$ be an $\mathbb{R}^p$-valued statistic and θ_0 be a p-dimensional vector. We say that T is a Fisher consistent estimate of θ_0 if there is a mapping $\theta : \mathfrak{M} \to \mathbb{R}^p$ such that*

$$T = \theta(P_n), \quad \theta_0 = \theta(P_0).$$

We next introduce two widely used estimators in statistics: the method of moment (Pearson, 1902) and the maximum likelihood estimator (Fisher, 1922). Let $\mathfrak{M}_0 = \{P_\theta : \theta \in \Theta\}$ be a dominated, identifiable parametric family on $(\Omega_X, \mathcal{F}_X)$. Suppose $X^1, \ldots, X^p$ are integrable with respect to P_θ. We define the solution to the following system of equations

$$\int X^k dP_\theta = \int X^k dP_n, \quad k = 1, \ldots, p \tag{2.12}$$

is the method-of-moment estimator of θ_0. Let $\mu(\theta) = (\mu_1(\theta), \ldots, \mu_p(\theta))^T$.

Theorem 2.14 *Suppose that the mapping $\theta \mapsto \mu(\theta)$ is injective and the vector*

$$\left(\int X dP_n, \ldots, \int X^p dP_n\right)^T.$$

is in the range of μ. Then the (unique) solution to (2.12) is Fisher consistent for θ_0, the parameter corresponding to the true distribution in $\mathfrak{M}_0$.

Proof. Let $\theta : \mathfrak{M} \to \mathbb{R}^p$ be a mapping that satisfies

$$\theta(P) = \mu^{-1}\left(\int X dP, \ldots, \int X^p dP\right) \text{ for any } P \in \mathfrak{M}_0 \cup \{P_n\}.$$

Then

$$\theta(P_{\theta_0}) = \mu^{-1}\left(\int X dP_{\theta_0}, \ldots, \int X^p dP_{\theta_0}\right) = \mu^{-1}\left(\mu_1(\theta_0), \ldots, \mu_p(\theta_0)\right) = \theta_0.$$

Meanwhile, if we let T be the solution to (2.12), then

$$T = \mu^{-1}\left(\int X dP_n, \ldots, \int X^p dP_n\right) = \theta(P_n).$$

Hence T is Fisher consistent. □

Let us now turn to the maximum likelihood estimate. Let

$$\ell(\theta; X_1, \ldots, X_n) = \sum_{i=1}^{n} \log f_\theta(X).$$

The random function $\theta \mapsto \ell(\theta; X_1, \ldots, X_n)$ is called the likelihood function. The maximum likelihood estimator (MLE) is defined as the maximizer of the likelihood function. That is,

$$T = \operatorname{argmax}\left\{\ell(\theta : X_1, \ldots, X_n) : \theta \in \Theta\right\}.$$

Clearly, it is equivalent to maximize the function $n^{-1}\ell(\theta, X_1, \ldots, X_n)$. Thus the maximum likelihood estimator can be defined alternatively as

$$T = \operatorname{argmax}\left\{\int \log f_\theta(X) dP_\theta : \theta \in \Theta\right\}.$$

Theorem 2.15 *Suppose $\mathfrak{M}_0$ is a dominated, identifiable parametric family and the likelihood function $\ell(\theta, X_1, \ldots, X_n)$ has a unique maximizer over Θ for each $X_1(\omega), \ldots, X_n(\omega)$. Then the maximum likelihood estimator is Fisher consistent.*

Proof. Since $\mathfrak{M}_0$ is identifiable, by Theorem 2.1, for each $\theta' \in \Theta$, the maximizer of

$$\int \log f_\theta(X) dP_{\theta'}$$

over Θ is unique and is equal to θ' itself. Let $\theta : \mathfrak{M} \to \mathbb{R}^p$ be a mapping such that

$$\theta(P) \mapsto \operatorname{argmax}\left\{\int \log f_\theta(X) dP : \theta \in \Theta\right\} \text{ for any } P \in \mathfrak{M}_0 \cup \{P_n\}.$$

Let T be the maximum likelihood estimator. Then,

$$\begin{aligned} \theta(P_{\theta_0}) &= \operatorname{argmax}\left\{\int \log f_\theta(X) dP_{\theta_0} : \theta \in \Theta\right\} = \theta_0 \\ \theta(P_n) &= \operatorname{argmax}\left\{\int \log f_\theta(X) dP_n : \theta \in \Theta\right\} = T. \end{aligned}$$

Thus T is Fisher consistent. □

Problems

2.1. Let $(\Omega, \mathcal{F})$ be a measurable space, and A be a nonempty set in $\mathcal{F}$. Show that $\{F \cap A : F \in \mathcal{F}\}$ is a σ-field of subsets of A.

2.2. Let $(\Omega, \mathcal{F}, \lambda)$ be a σ-finite measure space. Let $A_1, A_2, \ldots$ be a sequence of $\mathcal{F}$-sets such that $\lambda(A_n) < \infty$ and $\cup_n A_n = \Omega$. Let c_n be a sequence of positive numbers such that $\sum_n c_n = 1$. Define a set function $\lambda^* : \mathcal{F} \to \mathbb{R}$ as

$$\lambda^*(A) = \sum_n c_n \lambda(A \cup A_n)/\lambda(A_n),$$

where the quotients on the right are defined to be 0 if $\lambda(A_n) = 0$. Show that λ^* is a probability measure and $\lambda^* \equiv \lambda$.

2.3. Let $(\Omega, \mathcal{F})$ be a measurable space. Let $\{P_n : n \in \mathbb{N}\}$ be a sequence of probability measures on $(\Omega, \mathcal{F})$. Let $\{c_n : n \in \mathbb{N}\}$ be a sequence of positive numbers such that $\sum_{n\in\mathbb{N}} c_n = 1$. For any $A \in \mathcal{F}$, define

$$P(A) = \sum_{n\in\mathbb{N}} c_n P_n(A).$$

Prove:

1. the set function $P : \mathcal{F} \to \mathbb{R}$ is a probability measure on $(\Omega, \mathcal{F})$;
2. $\{P\} \equiv \{P_n : n \in \mathbb{N}\}$.

2.4. Let $(\Omega_1, \mathcal{F}_1)$ and $(\Omega_2, \mathcal{F}_2)$ be measurable spaces. Let $T : \Omega_1 \to \Omega_2$ be a mapping that is measurable $\mathcal{F}_1/\mathcal{F}_2$. Prove:

1. $T^{-1}(\mathcal{F}_2)$ is a σ-field;
2. if $f : \Omega_1 \to \Omega_2$ is measurable with respect to $\mathcal{F}_1/\mathcal{F}_2$, then f is measurable with respect to $T^{-1}(\mathcal{F}_2)$;
3. conclude that $T^{-1}(\mathcal{F}_2) = \sigma(T)$, where $\sigma(T)$ is the intersection of all σ-field with respect to which f is measurable.

2.5. Let $(\Omega_1, \mathcal{F}_1)$ and $(\Omega_2, \mathcal{F}_2)$ be measurable spaces, and $T : \Omega_1 \to \Omega_2$ be a mapping that is measurable with respect to $\mathcal{F}_1/\mathcal{F}_2$. Suppose $Q \ll P$ are two measures on $(\Omega_1, \mathcal{F}_1)$. Show that

$$E(dQ/dP|\circ T) = dQ\circ T^{-1}/dP\circ T^{-1}.$$

2.6. Let $(\Omega_X, \mathcal{F}_X)$ and $(\Omega_T, \mathcal{F}_T)$ be measurable spaces and $T : \Omega_X \to \Omega_T$ be measurable $\mathcal{F}_X/\mathcal{F}_T$. Let $f : \Omega_X \to \mathbb{R}$ be measurable $T^{-1}(\mathcal{F}_T)/\mathcal{R}$. Let $t_0 \in \Omega_T$, and let $x_0 \in T^{-1}(\{t_0\})$.

1. Prove that $f^{-1}(\{f(x_0)\})$ is a member of $T^{-1}(\mathcal{F}_T)$; that is $f^{-1}(\{f(x_0)\}) = T^{-1}(F)$ for some $F \in \mathcal{F}_T$;
2. From $x_0 \in f^{-1}(\{f(x_0)\}) = T^{-1}(F)$, prove that $t_0 \in F$, and hence that $T^{-1}(\{t_0\}) \subseteq T^{-1}(F)$;
3. From $T^{-1}(\{t_0\}) \subseteq f^{-1}(\{f(x_0)\})$, prove that f is constant on $T^{-1}(\{t_0\})$;
4. Conclude that, whenever $T(x) = T(x')$ for $x, x' \in \Omega_X$, we have $f(x) = f(x')$. That is, f depends on x only through $T(x)$.

2.7. Suppose $X_1, \ldots, X_n$ are i.i.d. $f_\theta(x) = \theta^{-1}e^{-\frac{x-\theta}{\theta}}$, $x > \theta$.

1. Find the MLE.
2. Find the method of moment estimator.
3. Find α and β so that $\alpha X_{(1)}$ and $\beta \bar{X}$ are unbiased.
4. Find the a linear combination of $X_{(1)}$ and $\bar{X}$ so that it is unbiased and have smallest variance among all such linear combinations.
5. Find an unbiased estimate based on X_1 and Rao-Blackwellize it. What is the improvement in the variance?

2.8. Suppose $u(X)$ is an unbiased estimate of θ. $f_\theta(x)$ is the density of X. Let $f_\theta^{(k)}(x)$ be the kth derivative of f_θ with respect to θ. Then

$$\int u f_\theta^{(k)} d\mu = \begin{cases} 1 & k = 1 \\ 0 & k = 2, \ldots, r \end{cases}$$

So for any constants $\alpha_1, \ldots, \alpha_r$, we have

$$\int u(\alpha_1 f_\theta^{(1)} + \cdots + \alpha_r f_\theta^{(r)}) d\mu = \alpha_1$$

Let $b_\theta^k = f_\theta^{(k)}/f_\theta$. These form the so called Bhattacharyya basis. (Bhattacharyya, 1946). Then the above can be rewritten as

$$\int u(\alpha_1 b_\theta^1 + \cdots + \alpha_r b_\theta^r) f_\theta d\mu = \alpha_1$$

So use the Cauchy Schwarz inequality, to get

$$\alpha_1^2 \leq \text{var}_\theta(u)\text{var}_\theta(\alpha_1 b_\theta^1 + \cdots + \alpha_r b_\theta^r)$$

Hence, for any constants $\alpha_1, \ldots, \alpha_r$ we have

$$\text{var}_\theta(u) \geq \frac{\alpha_1^2}{\text{var}_\theta(\alpha_1 b_\theta^1 + \cdots + \alpha_r b_\theta^r)}$$

2.9. Stein's Lemma: Suppose X is a random variable having exponential family density

$$e^{\theta^T t(x)} h(x) c(\theta)$$

with respect to the Lebesgue measure. Suppose x is defined on an interval (a, b) such that $\lim_{x \to x'} e^{\theta^T t(x)} h(x) = 0$ for $x' = a, b$. (a, b) is allowed to be $(-\infty, \infty)$. Then for any differentiable function g of x with $E|g'(X)| < \infty$. we have

$$E_\theta \left\{ \left[\frac{h'(X)}{h(X)} + \theta^T t(X) \right] g(X) \right\} = -E_\theta g'(X).$$

In particular, if X has a normal distribution, then

$$\text{cov}[g(X), X] = \text{var}(X) E[g'(X)].$$

A more general version of equality will be further studied in Chapter 5.

2.10. Let $\{f_\theta : \theta \in \Theta\}$ be a family of densities of X. Let $u(X)$ be an unbiased estimator of θ. Consider the function

$$\psi_\theta(x) = \frac{f_{\theta+\Delta}(x) - f_\theta(x)}{f_\theta(x)},$$

where Δ is any constant. Suppose both $u(X)$ and $\psi_\theta(X)$ have finite second moment. Show that

$$\text{var}_\theta(u(X)) \geq \frac{\Delta^2}{\text{var}_\theta(\psi_\theta(X))}.$$

This result is due to Hammersley (1950) and Chapman and Robbins (1951).

2.11. Let $\{P_\theta : \theta \in \Theta\}$ be a family of distributions of X. Let $\mathcal{N}$ be the class of all statistics $\delta(X)$ satisfying the following conditions:

1. $E_\theta \delta^2(X) < \infty$ for all $\theta \in \Theta$;
2. $E_\theta \delta(X) = 0$ for all $\theta \in \Theta$.

Use Lemma 2.5 to show that, an unbiased estimator $u(X)$ is a UMVU estimator if and only if

$$\mathrm{cov}_\theta(u(X), \delta(X)) = 0, \text{ for all } \theta \in \Theta \text{ and all } \delta \in \mathcal{N}.$$

Let $(\Omega, \mathcal{F}, \mu)$ be a measure space, and Θ be a subset of $\mathbb{R}^p$. We say that a function $g : \Theta \times \Omega \to \mathbb{R}$ is $L_k(\mu)$-Lipschitz with dominating slope $g_0 \geq 0$ if $\int g_0^k d\mu$ and

$$|g(\theta_2, x) - g(\theta_1, x)| \leq g_0(x)\|\theta_2 - \theta_1\|$$

for any $\theta_1, \theta_2 \in \Theta$. Suppose

1. $g : \Omega \times \Theta \to \mathbb{R}$ is differentiable with respect to θ modulo μ;
2. g is $L_2(\mu)$-Lipschitz with dominating slope $g_0 \in L_2(\mu)$;

Then, for any $h \in L_2(\mu)$, the function $h(x)g(\theta, x)$ is $L_1(\mu)$-Lipschitz with dominating slope hg_0.

2.12. Let $(\Omega, \mathcal{F}, \mu)$ be a measure space, and Θ be a subset of $\mathbb{R}^p$. We say that a function $g : \Theta \times \Omega \to \mathbb{R}$ is $L_k(\mu)$-Lipschitz with dominating slope $g_0 \geq 0$ if $\int g_0^k d\mu$ and

$$|g(\theta_2, x) - g(\theta_1, x)| \leq g_0(x)\|\theta_2 - \theta_1\|$$

for any $\theta_1, \theta_2 \in \Theta$. Suppose

1. $g : \Omega \times \Theta \to \mathbb{R}$ is differentiable with respect to θ modulo μ;
2. g is $L_1(\mu)$-Lipschitz with dominating slope $g_0 \in L_1(\mu)$;

Then, for any bounded function $h : \Omega \to \mathbb{R}$, the function $h(x)g(\theta, x)$ is $L_1(\mu)$-Lipschitz with dominating slope $|h|g_0$.

2.13. Suppose T is complete and sufficient with respect to $\mathfrak{M} = \{P_\theta : \theta \in \Theta\}$. Let U_1 and U_2 be two members of $\mathfrak{U}$. Show that $E(U_1|T) = E(U_2|T)$ $[P_\theta]$, for all $\theta \in \Theta$.

2.14. Show that if U be the UMVUE for θ, and V is any statistic whose components are in $L_2(P_\theta)$ and $E_\theta V = 0$ for all $\theta \in \Theta$, then $\mathrm{cov}_\theta(U, V) = 0$ for all $\theta \in \Theta$.

References

Barndorff-Nielsen, O. E. (1978). Information and Exponential Families in Statistical Theory. Wiley.

Basu, D. (1955). On statistics independence of a complete sufficient statistic. *Sankhya.* **15**, 377–380.

Bhattacharyya, A. (1946). On some analogues of the amount of information and their use in statistical estimation. *Sankhya: The Indian Journal of Statistics.* **8**, 1–14.

Blackwell, D. (1947). Conditional expectation and unbiased sequential estimation. *The Annals of Mathematical Statistics*, **18**, 105–110.

Chapman, D. G. and Robbins, H. (1951). Minimum variance estimation without regularity assumptions. *The Annals of Mathematical Statistics*, **22**, 581–586.

Darmois, G. (1935). Surles lois de probabilites a estimation exhaustive. *C. R. Acad. Sci Paris* (in French) **200**, 1265–1266.

Fend, A. V. (1959). On the attainment of Cramer-Rao and Bhattacharyya bounds for the variance of an estimate. *The Annals of Mathematical Statistics*, **30**, 381–388.

Fisher, R. A. (1922). On the mathematical foundations of theoretical statistics. *Philosophical Transactions of the Royal Society A*, **222**, 594–604.

Halmos, P. R. and Savage, L. J. (1949). Application of the Radon-Nikodym Theorem to the Theory of Sufficient Statistics. *The Annals of Mathematical Statistics*, **20**, 225–241.

Hammersley, J. M. (1950). On estimating restricted parameters. *Journal of the Royal Statistical Society, Series B*, **12**, 192–240.

Horn, R. A. and Johnson, C. R. (1985). *Matrix Analysis.* Cambridge University Press.

Koopman, B. O. (1936). On distributions admitting a sufficient statistic. *Transactions of the American Mathematical Society*, **39**, 399–409.

Lehmann, E. L. and Romano, J. P. (2005). *Testing Statistical Hypotheses.* Third edition. Springer.

Lehmann, E. L. and Casella, G. (1998). *Theory of Point Estimation.* Second edition. Springer, New York.

Lehmann, E. L. and Scheffe, H. (1950). Completeness, similar regions, and unbiased estimation, I. *Sankya*, **10**, 305–340.

Lehmann, E. L. and Scheffe, H. (1955). Completeness, similar regions, and unbiased estimation, II. *Sankya*, **15**, 219–236.

Lin'kov, Y. N. (2005). Lectures in Mathematical Statistics, Parts 1 and 2. In *Transactions of Mathematical Monographs*, **229**. American Mathematical Society.

McCullagh, P. and Nelder, J. A. (1989). *Generalized Linear Models, Second Edition*, Chapman & Hall.

Pearson, K. (1902). On the systematic fitting of curves to observations and measurements. *Biometrika*, **1**, 265–303.

Pitman, E. J. G. (1936). Sufficient statistics and intrinsic accuracy. *Proceedings of the Cambridge Philosophical Society, 32*, 567–579.

Rao, C. R. (1945). Information and the accuracy attainable in the estimation of statistical parameters. *Bulletin of the Calcutta Mathematical Society.* **37**, 81–89.

Rudin, W. (1987). *Real and Complex Analysis.* Third Edition. McGraw-Hill, Inc.

3

Testing Hypotheses for a Single Parameter

The central idea of optimal hypothesis test is the Neyman-Pearson Lemma (see Neyman and Pearson, 1933), which gives the form of the Most Powerful test for simple hypotheses. The basic idea of the Neyman-Pearson Lemma can be used to construct optimal tests for composite hypotheses, including one-sided and two-sided hypotheses. This is achieved by applying the Neyman-Pearson Lemma pointwise in the parameter spaces specified by composite hypotheses. To do so we require special assumptions on the forms of the distribution of the data, such as Monotone Likelihood Ratio and exponential family. The discussion of this chapter will be focussed on testing a scalar parameter. Vector-valued parameters will be treated in the next chapter.

3.1 Basic concepts

Scientific theories are posed as hypotheses; they uphold until refuted by sufficient evidence. While no amount of data can "prove" a scientific theory, a single instance can refute it. Suppose a hypothesis H implies an assertion A whose truth or falsehood can be determined by observation (say by experiments). If the observed facts are against A, then the hypothesis H is false. New hypotheses are then to be formulated in the hope to accommodate the observed facts that are inconsistent with the old hypothesis.

In a perfectly deterministic world, whether A is false can be determined definitely, so that we can decide whether or not to reject H with certainty — if H implies something that is false, then H itself must be false. In reality, however, the falsehood of A can in most cases only be determined with a degree of uncertainty. The need to make a decision in the face of uncertainty is the chief motivation for statistical hypothesis testing. The basic logic underlying statistical hypothesis testing is this: if H implies something unlikely, then H itself is unlikely to be true.

Let $(\Omega, \mathcal{F})$ be a measurable space, and X be a random element, which has range Ω_X, together with a σ-field $\mathcal{F}_X$. Thus X is measurable $\mathcal{F}/\mathcal{F}_X$. Let $\mathcal{P}_0$

B. Li and G. J. Babu, *A Graduate Course on Statistical Inference*,
Springer Texts in Statistics, https://doi.org/10.1007/978-1-4939-9761-9_3

and $\mathcal{P}_1$ be two disjoint families of distributions defined on $(\Omega, \mathcal{F})$. Statistical hypotheses are formulated as

$$H_0 : P \in \mathcal{P}_0 \quad \text{versus} \quad H_1 : P \in \mathcal{P}_1, \tag{3.1}$$

where H_0 is called the null hypothesis, and H_1 the alternative hypothesis. Note that the formulation (3.1) implicitly assumes that true distribution P must be in one of the two families. That is, $P \in \mathcal{P}_0 \cup \mathcal{P}_1$. If X falls into a region that has small probability under $P \in \mathcal{P}_0$ — that is, if X is unlikely whenever its distribution were from $\mathcal{P}_0$, then we can make a decision to reject H.

Our action of rejecting or not rejecting H_0 can be described by rejection region, also called critical region. A rejection region is any set $C \in \mathcal{F}_X$ such that we reject H_0 whenever X falls into C. That is, the region C describes a rule of when to reject H_0. This rule is called a nonrandomized test.

Definition 3.1 *For a nonrandomized test described by C, let*

$$\alpha = \sup_{P \in \mathcal{P}_0} P(X \in C), \quad \beta(P) = P(X \in C), \quad P \in \mathcal{P}_1.$$

Then α is called the type I error, or size, of the test, and $\beta(P)$ is called the power of the test at probability P, and $1 - \beta(P)$ is called the type II error of the test at P.

Naturally, if H_0 is true, we would like to reject H_0 with a small probability, and if H_0 is false (H_1 is true), we would like to reject H_0 with a large probability. That is, ideally, we want to choose C so that the both the type I and the type II errors are minimized. However, this is typically impossible, as the following example shows.

Example 3.1 Suppose that X has a binomial distribution

$$f_\theta(x) = \binom{n}{x} \theta^x (1-\theta)^{n-x}, \quad x = 0, 1, \ldots, n, \quad 0 \le \theta \le 1.$$

We abbreviate this statement as $X \sim b(n, \theta)$. Take $n = 2$. Then X has range $\{0, 1, 2\}$. Suppose we are interested in testing the hypothesis

$$H_0 : \theta = 1/2 \quad \text{versus} \quad H_1 : \theta = 1/6.$$

The table below gives type-I and type-II errors and power of all possible subsets of $\{0, 1, 2\}$.

C	Type I	Type II	Power
{0}	1/4	11/36	25/36
{1}	1/2	26/36	10/36
{2}	1/4	35/36	1/36
{0,1}	3/4	1/36	35/36
{0,2}	1/2	10/36	26/36
{1,2}	3/4	25/36	11/36
{0,1,2}	1	0	1
∅	0	1	0

We see that there is no critical region for which α and β are both minimized. □

From this table we also see that using a nonrandomized test we cannot control α at an arbitrary level. For example, no critical region has type-I error exactly equal to 0.05. For a technical reason it is desirable to be able to control α at an arbitrary level. This leads us to consider the following general form of test

$$\phi : \Omega_X \to [0,1], \quad \phi \text{ is measurable } \mathcal{F}_X/\mathcal{R}. \tag{3.2}$$

The evaluation of ϕ at x is the conditional probability of rejecting H_0 given the observation $X = x$; $1 - \phi(x)$ is the conditional probability of not rejecting H_0.

Definition 3.2 *A function ϕ of the form (3.2) is called a test. A test ϕ is called a randomized test if, for some $x \in \Omega_X$, $0 < \phi(x) < 1$; it is a nonrandomized test if ϕ only takes 0 or 1 as its values.*

This general definition is consistent with Definition 3.1: If ϕ only takes 0 and 1 as its values, then the set $C = \{x : \phi(x) = 1\}$ is the rejection region in Definition 3.1. If $X \in C$, we reject H_0 (with probability 1) otherwise we do not reject H_0 (or reject H_0 with probability 0). Because ϕ is assumed measurable, C is necessarily a set in $\mathcal{F}_X$.

Definition 3.3 *The test ϕ for $H_0 : P \in \mathcal{P}_0$ versus $H_1 : P \in \mathcal{P}_1$ is said to have level of significance α, if*

$$\sup_{P \in \mathcal{P}_0} \int \phi(x) dP(x) \leq \alpha. \tag{3.3}$$

The left side of (3.3) is called the size of the test.

As a special case, a nonrandomized test ϕ is of significance level α if $P(C) \leq \alpha$ for all $P \in \mathcal{P}_0$. Also note the subtle difference between the size and level of a test: the level of a test is an upper bound of the test. In other words the level of a test whose size is α' can be any number α satisfying $\alpha' \leq \alpha \leq 1$.

Definition 3.4 *The function β_ϕ on $\mathcal{P}$ given by $\beta_\phi(P) = \int \phi(x) dP(x)$ is called the power function of ϕ.*

3.2 The Neyman-Pearson Lemma

The simplest hypotheses H_0 and H_1 are ones in which $\mathcal{P}_0$ and $\mathcal{P}_1$ each contains only one distribution on $(\Omega, \mathcal{F})$. That is, $\mathcal{P}_0 = \{P_0\}$ and $\mathcal{P}_1 = \{P_1\}$. A test that contains only one distribution is called a simple hypothesis. A test that contains more than one distributions is called a composite hypothesis. Testing simple hypotheses is only of limited practical interest, but it lays the theoretical foundation for tests of composite hypotheses.

Consider the problem of testing a simple null hypothesis versus a simple alternative hypothesis

$$H_0 : P = P_0 \text{ versus } H_1 : P = P_1. \tag{3.4}$$

Let $0 \leq \alpha \leq 1$. We seek a test ϕ of size α such that its power at P_1 is greater than or equal to the power at P_1 of any other test of significance level α.

Definition 3.5 *A test ϕ for simple-versus-simple hypotheses is a Most Powerful (MP) test of size α if*

1. $\beta_\phi(P_0) = \alpha$,
2. *for any ϕ' with $\beta_{\phi'}(P_0) \leq \alpha$ we have $\beta_\phi(P_1) \geq \beta_{\phi'}(P_1)$.*

Note that $1 - \beta_\phi(P_1) \leq 1 - \beta_{\phi'}(P_1)$ and hence an MP has minimum type-II error among all tests of significance level α.

Does such a test exist? If so, is it unique? A fundamental result by Neyman and Pearson (1933) will help in answering this question. See also Lehmann and Romano (2005, Chapter 3) and Ferguson (1967, Chapter 5). Without loss of generality, we can assume that there is a common measure μ on $(\Omega, \mathcal{F})$ that dominates both P_0 and P_1 — for example, we can simply take $\mu = P_0 + P_1$.

Lemma 3.1 (Neyman-Pearson Lemma) *Let f_0 and f_1 be densities of P_0 and P_1 with respect to μ. Then any test ϕ of the form*

$$\phi(x) = \begin{cases} 1 & \text{if } f_1(x) > k f_0(x) \\ \gamma(x) & \text{if } f_1(x) = k f_0(x) \\ 0 & \text{if } f_1(x) < k f_0(x) \end{cases} \tag{3.5}$$

where $0 \leq \gamma(x) \leq 1$ and $k \geq 0$ is the MP of size $\int \phi f_0 d\mu$.

Proof. Let ϕ' be any test of level $\int \phi f_0 d\mu$. We want to show that $\int \phi dP_1 \geq \int \phi' dP_1$. Since ϕ' is a test,

$$(\phi(x) - \phi'(x))(f_1(x) - k f_0(x)) \geq 0 \text{ for all } x.$$

Thus

$$\int (\phi - \phi')(f_1 - k f_0) d\mu = \int (\phi - \phi') f_1 d\mu - k \int (\phi - \phi') f_0 d\mu \geq 0.$$

Since ϕ' is of level α, $\int \phi f_0 d\mu \geq \int \phi' f_0 d\mu$. Hence

$$\int \phi f_1 d\mu \geq \int \phi' f_1 d\mu,$$

as desired. □

The following proposition is a generalization of the intermediate value theorem.

Proposition 3.1 *Suppose $\rho : \mathbb{R} \to \mathbb{R}$ is a right continuous, nonincreasing function satisfying*

1. $\rho(0-) = 1$;
2. $\lim_{x \to \infty} \rho(x) = 0$.

Then for any $0 < \alpha \leq 1$ there exists $k \in [0, \infty)$ such that $\alpha \in [\rho(k), \rho(k-)]$.

Proof. If $\alpha = 1$ then $\alpha \in [\rho(0-), \rho(0)]$. Now suppose $1 > \alpha > 0$. Let $S = \{x : \rho(x) \leq \alpha\}$. This set is nonempty because $\rho(x) \to 0$ as $x \to \infty$. Let $k = \inf S$. It is easy to see that $k < \infty$. Since $\rho(0-) = 1$, we have $k \geq 0$. Let x_n be a sequence in S approaching k. Then, by right continuity, $\rho(x_n) \to \rho(k)$, and we have $\rho(k) \leq \alpha$. Since $\rho(x) > \alpha$ for all $x < k$, we have $\rho(k-) \geq \alpha$. □

Theorem 3.1 *For any $0 < \alpha \leq 1$, there exists a test ϕ of size α of the form (3.5) with $\gamma(x) = \gamma$, a constant, and $0 \leq k < \infty$. Furthermore, if ϕ' is MP of size $0 < \alpha \leq 1$, then it has the form (3.5) a.e. P_0 and P_1. That is,*

$$P_0(\phi \neq \phi', f_1 \neq k f_0) = 0, \quad P_1(\phi \neq \phi', f_1 \neq k f_0) = 0.$$

Note that the form (3.5) does not specify $\gamma(x)$. In other words as long as ϕ and ϕ' are the same on $\{f_1 \neq k f_0\}$, they are both of the form (3.5).

Proof. (Existence). For a test ϕ of the type (3.5) with $\gamma(x) = \gamma$, we have

$$\int \phi f_0 d\mu = P_0(f_1 > k f_0) + \gamma P_0(f_1 = k f_0).$$

Fix $0 < \alpha \leq 1$, and define

$$\rho(k) = P_0(f_1 > k f_0) = P_0(f_1/f_0 > k, f_0 > 0).$$

Then $\rho(\cdot)$ is a nonincreasing, right continuous function with left limit $\rho(k-) = P_0(f_1 \geq k f_0)$. It is then clear that

$$\rho(0-) = 1, \quad \rho(0) = P_0(f_1 > 0), \quad \lim_{k \to \infty} \rho(k) = 0.$$

It follows that $(0,1] \subseteq \cup\{[\rho(k), \rho(k-)] : 0 \leq k < \infty\}$. Hence for any $0 < \alpha \leq 1$ there is a $0 \leq k_0 < \infty$ such that

$$P_0(f_1 > k_0 f_0) \leq \alpha \leq P_0(f_1 \geq k_0 f_0).$$

If we take $k = k_0$ in (3.5) and

$$\gamma = \begin{cases} \dfrac{\alpha - P_0(f_1 > k_0 f_0)}{P_0(f_1 = k_0 f_0)} & \text{if } P_0(x : f_1 = k_0 f_0) \neq 0 \\ 0 & \text{otherwise,} \end{cases}$$

then $\int \phi f_0 d\mu = \alpha$.

(Uniqueness). Let $0 < \alpha \leq 1$, let ϕ be a test of the type (3.5) of size α, and let ϕ' be another MP of size α. Then $\int \phi f_i d\mu = \int \phi' f_i d\mu$ for $i = 0, 1$. Consequently,

$$\int (\phi - \phi')(f_1 - k f_0) d\mu = \int (\phi - \phi') f_1 d\mu - k \int (\phi - \phi') f_0 d\mu = 0.$$

Since $(\phi - \phi')(f_1 - k f_0) \geq 0$, it is 0 a.e. μ. This implies

$$\mu(\{\phi - \phi' \neq 0, f_1 - k f_0 \neq 0\}) = 0.$$

Since $P_0 \ll \mu$ and $P_1 \ll \mu$ we have the desired result. □

Existence and uniqueness of the MP test can also be established when $\alpha = 0$ if we adopt the convention (1.2). Let

$$\phi(x) = \begin{cases} 1 & \text{if } f_0(x) = 0 \\ 0 & \text{if } f_0(x) > 0. \end{cases} \tag{3.6}$$

By (1.2),

$$\{f_1 > \infty f_0\} \subset \{f_0 = 0\}, \quad \{f_1 = \infty f_0\} \subset \{f_0 = 0\}, \quad \{f_1 < \infty f_0\} \subset \{f_0 > 0\}.$$

Thus ϕ in (3.6) has the form (3.5) with $\gamma(x) = 1$ and $k = \infty$ and satisfies

$$\beta_\phi(P_0) = \int \phi f_0 d\mu = \int_{f_0 > 0} \phi f_0 d\mu = 0.$$

To prove uniqueness let ϕ' be another MP test of size $\alpha = 0$. Since $\int \phi' f_0 d\mu = 0$, $\phi' = 0$ a.e. μ on $\{f_0 > 0\}$. Since ϕ' is the most powerful

$$\begin{aligned} 0 \geq \int (\phi - \phi') f_1 d\mu &= \int_{\{f_0 = 0\}} (1 - \phi') f_1 d\mu - \int_{\{f_0 > 0\}} \phi' f_1 d\mu \\ &= \int_{\{f_0 = 0\}} (1 - \phi') f_1 d\mu \geq 0. \end{aligned}$$

Thus $\phi' = 1$ a.e. μ on $\{f_0 = 0\} \cap \{f_1 > 0\}$. Since $P_i(f_0 = f_1 = 0) = 0$ for $i = 0, 1$, we see that $\phi' = 1$ a.e. P_0, P_1 on $\{f_0 = 0\}$. Thus, for $i = 0, 1$,

$$\begin{aligned} P_i(\phi' \neq \phi) =& P_i(\phi' \neq I_{\{f_0=0\}}) \\ =& P_i(\phi' \neq 0, f_0 > 0) + P_i(\phi' \neq 1, f_0 = 0) = 0. \end{aligned}$$

In other words $\phi' = \phi$ a.e. P_0, P_1.

3.3 Uniformly Most Powerful test for one-sided hypotheses

We now begin the process of generalizing the basic idea of the Neyman-Pearson Lemma to composite hypotheses. A family of distributions indexed by a parameter in a Euclidean space is called a parametric family. Let Θ be a subset of the Euclidean space $\mathbb{R}^k$. A parametric family is a set $\mathcal{P} = \{P_\theta : \theta \in \Theta\}$, where each P_θ is a probability measure on $(\Omega, \mathcal{F})$. Let Θ_0 and Θ_1 be a partition of Θ. That is, $\Theta_0 \cap \Theta_1 = \varnothing$ and $\Theta_0 \cup \Theta_1 = \Theta$. Hypotheses (3.1) in this context reduce to

$$H_0 : \theta \in \Theta_0 \quad \text{versus} \quad H_1 : \theta \in \Theta_1.$$

For a parametric family, the power function $\beta_\phi(P_\theta)$ is a function of θ. We will write $\beta_\phi(P_\theta)$ as $\beta_\phi(\theta)$. For a measurable function f of X, we write its expectation $\int f dP_\theta$ as $E_\theta f(X)$.

In this section we consider the one-sided hypotheses. See Allen (1953), Karlin and Rubin (1956), Pfanzagl (1967), and Lehmann and Romano (2005). If Θ is an interval in $\mathbb{R}$, and $\theta_0 \in \Theta$, and if $\Theta_0 = \{\theta \in \Theta : \theta \leq \theta_0\}$ and $\Theta_1 = \{\theta \in \Theta : \theta > \theta_0\}$, then the above hypothesis takes the special form

$$H_0 : \theta \leq \theta_0 \quad \text{versus} \quad H_1 : \theta > \theta_0. \tag{3.7}$$

This is called one-sided hypotheses. In this section we develop the UMP-α test for one-sided hypotheses, and give the sufficient conditions under which such a test exists.

3.3.1 Definition and examples of UMP tests

As in the case of simple versus simple hypotheses, we would like to find a size α test that has the most power among all level α tests. When $\mathcal{P}_0$ and $\mathcal{P}_1$ are composite, however, the set of powers, $\{\beta_\phi(P) : P \in \mathcal{P}_1\}$, is no longer a number and we would like it to be large for all $P \in \mathcal{P}_1$. This is formulated rigorously as the Uniformly Most Powerful test.

Definition 3.6 *A test ϕ of size α, that is, $\sup_{P \in \mathcal{P}_0} \beta_\phi(P) = \alpha$, for testing*

$$H_0 : P \in \mathcal{P}_0 \text{ versus } H_1 : P \in \mathcal{P}_1 \tag{3.8}$$

is called a Uniformly Most Powerful test of size α *if, for any* ϕ' *of level* α*, that is,*

$$\sup_{P \in \mathcal{P}_0} \beta_{\phi'}(P) \leq \alpha,$$

we have $\beta_\phi(P) \geq \beta_{\phi'}(P)$ *for all* $P \in \mathcal{P}_1$*,*

A Uniformly Most Powerful test of size α is abbreviated as a UMP-α test.

A natural way to find a UMP-α test is to apply the MP test for simple versus simple hypotheses $H_0 : \theta = \theta_0$ versus $H_1 : \theta = \theta'$ for each fixed $\theta' > \theta_0$. However, the test ϕ obtained by such a procedure would in general depend on θ', and would therefore not be suitable for hypothesis (3.7), as it is not specific to any particular point in Θ_1.

However, in an important special case we can construct a test as described above to obtain an MP test not specific to θ'. We first illustrate this by two examples. For illustration, we first construct UMP-α test for the simple versus composite hypothesis

$$H_0 : \theta = \theta_0 \quad \text{versus} \quad H_1 : \theta > \theta_0. \tag{3.9}$$

Example 3.2 Let X be a $b(n, \theta)$ random variable and suppose we are interested in testing (3.9) for a $\theta_0 \in (0, 1)$. We first pick any fixed $\theta' > \theta_0$ and consider the simple versus simple hypotheses $H_0 : \theta = \theta_0$ versus $H_1 : \theta = \theta'$. Let

$$L(x) = \frac{f_{\theta'}(x)}{f_{\theta_0}(x)} = \left(\frac{\theta'}{\theta_0}\right)^x \left(\frac{1-\theta'}{1-\theta_0}\right)^{n-x}.$$

By the Neyman-Pearson Lemma, the MP test for θ_0 versus θ' is

$$\phi(x) = \begin{cases} 1 & \text{if } L(x) > k \\ \gamma & \text{if } L(x) = k \\ 0 & \text{if } L(x) < k \end{cases}$$

for some k and γ. Since $\theta' > \theta_0$, the likelihood ratio $L(x)$ is increasing in x. Hence ϕ is equivalent to

$$\phi(x) = \begin{cases} 1 & \text{if } x > m \\ \gamma & \text{if } x = m \\ 0 & \text{if } x < m \end{cases}$$

where γ and m are determined by $E_{\theta_0}(\phi(X)) = \alpha$; that is

$$P_{\theta_0}(X > m) + \gamma P_{\theta_0}(X = m) = \alpha.$$

To be precise, we first find m such that $P_{\theta_0}(X > m) \leq \alpha \leq P_{\theta_0}(X \geq m)$, and define $\gamma(m) = [\alpha - P_{\theta_0}(X > m)]/P_{\theta_0}(X = m)$.

Clearly, the test ϕ depends only on θ_0 and hence it is a size α MP test for $H_0 : \theta = \theta_0$ versus $H_1 : \theta = \theta'$ for every $\theta' > \theta_0$. This implies that it is a UMP-α test for $H_0 : \theta = \theta_0$ versus $H_1 : \theta > \theta_0$. □

Example 3.3 Let X denote a Gaussian random variable with density

$$f_\theta(x) = \frac{1}{\sqrt{2\pi}} e^{-\frac{1}{2}(x-\theta)^2},$$

and let $\theta' > \theta_0$. As in the previous example, the likelihood ratio,

$$\begin{aligned} L(x) = f_{\theta'}(x)/f_{\theta_0}(x) &= \exp[(x-\theta_0)^2/2 - (x-\theta')^2/2] \\ &= \exp[(\theta_0^2 - \theta'^2)/2 + x(\theta' - \theta_0)], \end{aligned}$$

is an increasing function of x. Fix a $\theta' > \theta_0$. By the Neyman-Pearson Lemma, the MP test for $H_0 : \theta = \theta_0$ versus $H_1 : \theta = \theta'$ of size α is of the form

$$\phi(x) = \begin{cases} 1 & \text{if } f_{\theta_1}(x) > k f_{\theta_0}(x) \\ \gamma(x) & \text{if } f_{\theta_1}(x) = k f_{\theta_0}(x) \\ 0 & \text{if } f_{\theta_1}(x) < k f_{\theta_0}(x) \end{cases}$$

which, because $L(x)$ is monotone increasing in x, is equivalent to

$$\phi(x) = \begin{cases} 1 & \text{if } x > k' \\ \gamma(x) & \text{if } x = k' \\ 0 & \text{if } x < k' \end{cases}.$$

Furthermore, since X is a continuous random variable, $P_{\theta_0}(X = k') = 0$. Hence we can choose $\gamma(x)$ arbitrarily without changing the size of ϕ. Choose $\gamma(x) = 0$, and the above test reduces to

$$\phi(x) = \begin{cases} 1 & \text{if } x > k' \\ 0 & \text{if } x \leq k' \end{cases},$$

where k' is determined by

$$E_{\theta_0}(\phi(X)) = \frac{1}{\sqrt{2\pi}} \int_{k'}^{\infty} e^{-\frac{1}{2}(x-\theta_0)^2} dx = \frac{1}{\sqrt{2\pi}} \int_{k'-\theta_0}^{\infty} e^{-\frac{1}{2}x^2} dx = \alpha.$$

Again, ϕ is completely determined by θ_0 and α, and is not specific to the θ' we started with. Thus ϕ is UMP-α for testing $H_0 : \theta = \theta_0$ versus $H_1 : \theta > \theta_0$. □

What makes it possible for us to construct UMP test from a collection of MP simple versus simple tests in the foregoing examples is that the likelihood ratio is monotone in X in both cases. In fact, such a construction is always possible under the condition of monotone likelihood ratio, which we now formally define. Let $\mathcal{P} = \{P_\theta : \theta \in \Theta \subset \mathbb{R}\}$ be a family of probability measures. Suppose that each P_θ in $\mathcal{P}$ is dominated by a common density μ. Let $f_\theta(x)$ denote the density of P_θ with respect to μ.

3.3.2 Monotone Likelihood Ratio

The assumption of Monotone Likelihood Ratio for constructing one-sided UMP tests, whose importance we have seen in the last two examples, is crystalized by Karlin and Rubin (1956); a more general form is given by Pfanzagl (1967). See also Ferguson (1967) and Lehmann and Romano (2005).

Definition 3.7 *Let $\mathcal{P} = \{P_\theta : \theta \in \Theta\}$ be a one-parameter family of probability measures dominated by a measure μ, and let $f_\theta = dP_\theta/d\mu$. The family of densities $\{f_\theta : \theta \in \Theta\}$ is said to have monotone likelihood ratio (MLR) in $Y(x)$ if for any $\theta_1 < \theta_2$, $\theta_1, \theta_2 \in \Theta$, the likelihood ratio $L(x) = f_{\theta_2}(x)/f_{\theta_1}(x)$ is a monotone (nondecreasing or nonincreasing) function in $Y(x)$ on a set on which $L(x)$ is defined.*

Here, we say that $L(x)$ is defined if $f_{\theta_2}(x)$ and $f_{\theta_1}(x)$ are not both 0. If $f_{\theta_1}(x) = 0$ and $f_{\theta_2}(x) > 0$, we say that $L(x)$ takes the value ∞. Also note that according to this definition, if $\mathcal{P}$ has MLR in $Y(x)$, then $Y(X)$ is sufficient for $\mathcal{P}$. MLR can also be defined more generally without using densities; see Pfanzagl (1967).

Since $L(x)$ is nondecreasing in $Y(x)$ if and only if it is nonincreasing in $-Y(x)$, for convenience we shall always assume $L(x)$ to be nondecreasing in $Y(x)$ in the rest of this chapter.

Example 3.4 Let $U(a, b)$ denote the uniform distribution on an interval (a, b), and suppose that the distribution of X belongs to the family $\{U(0, \theta) : \theta > 0\}$. This family has MLR in $Y(x) = x$. To see this, let $\theta_2 > \theta_1 > 0$. Then $f(x|\theta_i) = \theta_i^{-1} I_{(0,\theta_i)}(x)$, $i = 1, 2$. Thus $L(x)$ is θ_2/θ_1 on $(0, \theta_1)$; it is ∞ on $[\theta_1, \theta_2)$; it is not defined on $[\theta_2, \infty)$. Thus $L(x)$ is nondecreasing on the set on which it is defined.

Similarly, the family $\{U(\theta, \theta + 1); \theta > 0\}$ has MLR in x. Let $\theta_2 > \theta_1$. If $\theta_2 \geq \theta_1 + 1$, then $L(x)$ is defined on $(\theta_1, \theta_1 + 1) \cup (\theta_2, \theta_2 + 1)$ and not defined elsewhere. Note that $L(x) = 0$ on $(\theta_1, \theta_1 + 1)$ and $L(x) = \infty$ on $(\theta_2, \theta_2 + 1)$. Hence the family has MLR in x. Suppose $\theta_1 < \theta_2 < \theta_1 + 1$. Then $L(x)$ is defined on $(\theta_1, \theta_2 + 1)$ and is not defined elsewhere. Note that $L(x) = 0$ on $(\theta_1, \theta_2]$; $L(x) = 1$ on $(\theta_2, \theta_1 + 1)$; $L(x) = \infty$ on $[\theta_1 + 1, \theta_2 + 1)$. Thus the family has MLR in x. □

Example 3.5 For the double exponential

$$f_\theta(x) = \frac{1}{2\beta} \exp\left(- |x - \theta|/\beta \right),$$

where the scale parameter $\beta > 0$ is known, the likelihood ratio

$$L(x) = f_{\theta_2}(x)/f_{\theta_1}(x) = \exp\left\{ \frac{1}{\beta}\left(|x - \theta_1| - |x - \theta_2|\right) \right\},$$

is given by

$$L(x) = \begin{cases} \exp\left((\theta_1 - \theta_2)/\beta\right) & \text{if } x < \theta_1 \\ \exp\left((2x - \theta_1 - \theta_2)/\beta\right) & \text{if } \theta_1 \le x < \theta_2 \\ \exp\left((\theta_2 - \theta_1)/\beta\right) & \text{if } \theta_2 \ge x \end{cases}$$

So if $\theta_1 < \theta_2$, then L is continuous and nondecreasing in x. Hence the family has MLR in x. □

However, Cauchy family does not have MLR in x.

Example 3.6 For Cauchy distribution with density

$$f_\theta(x) = \frac{\theta}{\pi(x^2 + \theta^2)}, \quad \theta > 0,$$

the likelihood ratio is

$$L(x) = \frac{f_{\theta_2}(x)}{f_{\theta_1}(x)} = \frac{\theta_2}{\theta_1}\left(\frac{\theta_1^2 + x^2}{\theta_2^2 + x^2}\right) = \frac{\theta_2}{\theta_1}\left(\frac{\theta_1^2 - \theta_2^2}{\theta_2^2 + x^2} + 1\right).$$

Obviously $L(x)$ is symmetric in x. We know it is not a constant. Hence it is not monotone. □

The next lemma is useful in deriving UMP tests for one sided tests.

Lemma 3.2 *Suppose that the family of densities $\{f_\theta(x) : \theta \in \Theta\}$ has MLR (nondecreasing) in $Y(x)$. If ϕ is a non-decreasing and integrable function of Y, then $\beta_\phi(\theta) = \int \phi f_\theta d\mu$ is non-decreasing in θ.*

Proof. Let $\theta_1 < \theta_2$, and let

$$A = \{x : f_{\theta_2}(x) < f_{\theta_1}(x)\} \text{ and } B = \{x : f_{\theta_2}(x) > f_{\theta_1}(x)\}.$$

By definition, $L(a) < 1 < L(b)$ whenever $a \in A$, $b \in B$. This implies that $Y(b) > Y(a)$ and hence that $\phi(b) \ge \phi(a)$. Therefore

$$\begin{aligned}\beta_\phi(\theta_2) - \beta_\phi(\theta_1) &= \int \phi(f_{\theta_2} - f_{\theta_1})d\mu \\ &= \int_A \phi(f_{\theta_2} - f_{\theta_1})d\mu + \int_B \phi(f_{\theta_2} - f_{\theta_1})d\mu \\ &\geq \sup_{a\in A} \phi(a) \int_A (f_{\theta_2} - f_{\theta_1})d\mu + \inf_{b\in B} \phi(b) \int_B (f_{\theta_2} - f_{\theta_1})d\mu \geq 0,\end{aligned}$$

where the last inequality holds because

$$\inf_{b\in B} \phi(b) \geq \sup_{a\in A} \phi(a),$$

$$\int_A (f_{\theta_2} - f_{\theta_1})d\mu + \int_B (f_{\theta_2} - f_{\theta_1})d\mu = \int (f_{\theta_2} - f_{\theta_1})d\mu = 0,$$

and $f_{\theta_2} - f_{\theta_1} < 0$ on A. □

3.3.3 The general form of UMP tests

We now state the main result of this section. The theorem states, in essence, that the construction similar to those in Examples 3.2 and 3.3 always gives valid UMP test if the MLR assumption is satisfied.

Theorem 3.2 *Suppose that the family $\{f_\theta(x) : \theta \in \Theta\}$ has MLR (nondecreasing) in $Y(x)$.*

1. *If $\alpha > 0$, then the test ϕ defined by*

$$\phi(x) = \begin{cases} 1 & \text{if } Y(x) > k \\ \gamma & \text{if } Y(x) = k \\ 0 & \text{if } Y(x) < k \end{cases}, \qquad \int \phi f_{\theta_0} d\mu = \alpha \tag{3.10}$$

   is a UMP test for

$$H_0 : \theta \leq \theta_0 \text{ versus } H_1 : \theta > \theta_0. \tag{3.11}$$

2. *If $\alpha < 1$, then the test ϕ' defined by*

$$\phi'(x) = \begin{cases} 1 & \text{if } Y(x) < k' \\ \gamma' & \text{if } Y(x) = k' \\ 0 & \text{if } Y(x) > k' \end{cases}, \qquad \int \phi' f_{\theta_0} d\mu = \alpha$$

   is UMP test for

$$H_0' : \theta \geq \theta_0 \text{ versus } H_1' : \theta < \theta_0. \tag{3.12}$$

The proof follows roughly the argument used in Examples 3.2 and 3.3. Additional care must be taken, however, to extend the null hypothesis from $H_0 : \theta = \theta_0$ in the examples to $H_0 : \theta \leq \theta_0$ here, which is achieved using the monotonicity of $\beta_\phi(\theta)$, as shown in Lemma 3.2. By the MLR assumption, the likelihood ratio $L(x)$ is a function of $Y(x)$. We write $L(x)$ as $L_0(Y(x))$.

Proof of Theorem 3.2. First, we show that test (3.10) is UMP for testing $H_0 : \theta = \theta_0$ versus $H_1 : \theta > \theta_0$. To do so it suffices to show that, for any fixed $\theta_1 > \theta_0$, (3.10) is of the form

$$\phi_1(x) = \begin{cases} 1 & \text{if } f_{\theta_1}(x) > k_1 f_{\theta_0}(x) \\ \gamma(x) & \text{if } f_{\theta_1}(x) = k_1 f_{\theta_0}(x) \\ 0 & \text{if } f_{\theta_1}(x) < k_1 f_{\theta_0}(x) \end{cases}$$

for some $k_1 \geq 0$ and some function $0 \leq \gamma(x) \leq 1$. Since $\gamma(x)$ is arbitrary, any test that takes value 1 on $\{f_{\theta_1} > k_1 f_{\theta_0}\}$ and 0 on $\{f_{\theta_1} < k_1 f_{\theta_0}\}$ has the above form. Because the family $\{f_\theta : \theta \in \Theta\}$ has MLR,

$$\{x : Y(x) \leq k\} \subset \{x : L_0(Y(x)) \leq L_0(k)\}.$$

Hence

$$\{x : f_{\theta_2}(x) > L_0(k) f_{\theta_1}(x)\} \subset \{x : Y(x) > k\}.$$

Thus ϕ in (3.10) takes the value 1 on $\{f_{\theta_2} > k_1 f_{\theta_1}\}$ where $k_1 = L(k)$. For a similar reason, it takes the value 0 on $\{f_{\theta_2} < k_1 f_{\theta_1}\}$. Thus we have proved ϕ is UMP for testing $\theta = \theta_0$ versus $\theta > \theta_0$.

Next, we show that ϕ is a UMP test of size $\beta_\phi(\theta_0)$ for testing $\theta \leq \theta_0$ versus $\theta > \theta_0$. Since, by Lemma 3.2, $\beta_\phi(\cdot)$ is monotone nondecreasing, ϕ has size $\beta_\phi(\theta_0)$. Let Ψ be the class of all tests of size $\beta_\phi(\theta_0)$. That is,

$$\Psi = \{\psi : \sup_{\theta \leq \theta_0} \beta_\psi(\theta) \leq \beta_\phi(\theta_0)\}.$$

We need to show that $\beta_\phi(\theta) \geq \beta_\psi(\theta)$ for all $\theta > \theta_0$ and all $\psi \in \Psi$. Let Ψ' be the class of all tests whose power at θ_0 is no more than $\beta_\phi(\theta_0)$. That is, $\Psi' = \{\psi' : \beta_{\psi'}(\theta_0) \leq \beta_\phi(\theta_0)\}$. Clearly, $\Psi \subset \Psi'$. But we have already shown that $\beta_\phi(\theta) \geq \beta_{\psi'}(\theta)$ for all $\psi' \in \Psi'$.

The second part of the theorem follows by considering $1 - \psi$, where ψ is a size $1 - \alpha$ test of the type (3.10). □

The next theorem establishes the existence of UMP test for one-sided hypotheses.

Theorem 3.3 *Suppose that the family $\{f_\theta(x) : \theta \in \Theta\}$ has MLR in $Y(x)$. Then for any given $0 < \alpha \leq 1$ and $\theta_0 \in \Theta$, there exist $-\infty \leq k \leq \infty$ and $0 \leq \gamma \leq 1$ such that ϕ in (3.10) has size α.*

Proof. Let P_{θ_0} denote the distribution corresponding go f_{θ_0}. Choose k such that

$$P_{\theta_0}(Y(X) > k) \leq \alpha \leq P_{\theta_0}(Y(X) \geq k)$$

and define

$$\gamma = \begin{cases} \dfrac{\alpha - P_{\theta_0}(Y(X) > k)}{P_{\theta_0}(Y(X) = k)}, & \text{if } P_{\theta_0}(Y(X) = k) \neq 0 \\ 0 & \text{otherwise.} \end{cases}$$

Clearly, a test of the form (3.10) with k and γ chosen above satisfies $\beta_\phi(\theta_0) = \alpha$. By Lemma 3.2, then, the size of the test is α. □

3.3.4 Properties of the one-sided UMP test

We now further study the properties of the one-sided UMP test as given in (3.10). The next corollary shows that the UMP test (3.10) not only has the most power for $\theta > \theta_0$, but also has the *least* power for $\theta < \theta_0$.

Corollary 3.1 *Suppose* $0 < \beta_\phi(\theta_0) < 1$. *If* ϕ *is a test of the form (3.10), then, for any test* ψ *satisfying* $\beta_\psi(\theta_0) \geq \beta_\phi(\theta_0)$, *we have* $\beta_\psi(\theta) \geq \beta_\phi(\theta)$ *for all* $\theta \leq \theta_0$.

Proof. From the proof of Theorem 3.2, $1-\phi$ is the UMP test of size $1-\beta_\phi(\theta_0)$ for testing $H_0 : \theta = \theta_0$ versus $H_1 : \theta < \theta_0$. Since $\beta_{1-\psi}(\theta_0) \leq \beta_{1-\phi}(\theta_0)$, we have

$$1 - \beta_\psi(\theta) = \beta_{1-\psi}(\theta) \leq \beta_{1-\phi}(\theta) = 1 - \beta_\phi(\theta)$$

for all $\theta \leq \theta_0$. Consequently, $\beta_\psi(\theta) \geq \beta_\phi(\theta)$ for all $\theta \leq \theta_0$. □

A typical comparison between the power of the UMP test ϕ and any other test ψ of the same size is presented in Figure 3.1.

Recall that Lemma 3.2 states that if ϕ is a monotone function of $Y(x)$ then it has a nondecreasing power function. The next lemma shows that ϕ has a strictly increasing power function if the parametric family $\{P_\theta : \theta \in \Theta\}$ is identifiable and if ϕ is of the form (3.10). We say that a parametric family of probability measures $\{P_\theta : \theta \in \Theta\}$ is identifiable if, whenever $\theta_1 \neq \theta_2$, $P_{\theta_1} \neq P_{\theta_2}$, where the latter inequality means that there is a set A in $(\Omega, \mathcal{F})$ such that $P_{\theta_1}(X \in A) \neq P_{\theta_2}(X \in A)$. Thus, identifiability means different parameters correspond to different probability measures. That is, the mapping $\theta \mapsto P_\theta$ is injective.

Theorem 3.4 *Suppose that the family* $\{f_\theta(x) : \theta \in \Theta\}$ *have (nondecreasing) MLR in* $Y(x)$ *and the corresponding family of probability measures* $\{P_\theta : \theta \in \Theta\}$ *is identifiable. Let* ϕ *be the test defined in (3.10). Then* $\beta_\phi(\theta)$ *is strictly increasing over* $\{\theta : 0 < \beta_\phi(\theta) < 1\}$.

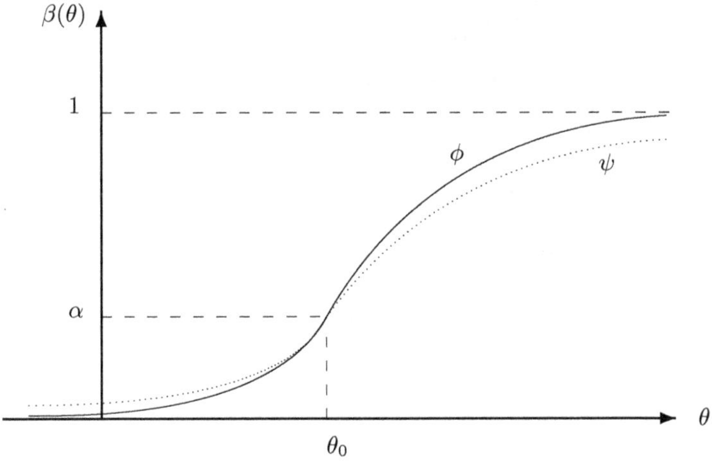

Fig. 3.1. Comparison of power functions

Proof. Let $\theta_1 < \theta_2$ be numbers in Θ. As seen earlier in the proof of Theorem 3.2, the test ϕ defined in (3.10) is equivalent to

$$\phi(x) = \begin{cases} 1 & \text{if } f_{\theta_2}(x) > k' f_{\theta_1}(x) \\ \gamma(x) & \text{if } f_{\theta_2}(x) = k' f_{\theta_1}(x) \\ 0 & \text{if } f_{\theta_2}(x) < k' f_{\theta_1}(x) \end{cases}$$

for some k' and $\gamma(x)$. By the Neyman-Pearson Lemma, ϕ is a size $\beta_\phi(\theta_1)$ MP test for $H_0 : \theta = \theta_1$ versus $H_1 : \theta = \theta_2$. If $\beta_\phi(\theta_1) = \beta_\phi(\theta_2)$, then the test $\phi^*(x) \equiv \beta_\phi(\theta_1)$ satisfies

$$(\phi - \phi^*)(f_{\theta_2} - k' f_{\theta_1}) \geq 0, \quad \text{and} \quad \int (\phi - \phi^*)(f_{\theta_2} - k' f_{\theta_1}) d\mu = 0.$$

So $\mu\{(\phi - \phi^*)(f_{\theta_2} - k' f_{\theta_1}) \neq 0\} = 0$. Because $\beta_\phi(\theta_1) \neq 0$, whenever $f_{\theta_2} \neq k' f_{\theta_1}$, we have $\phi \neq \phi^*$. And so

$$\{x : f_{\theta_2}(x) \neq f_{\theta_1}(x)\} \subset \{x : (\phi(x) - \phi^*(x))(f_{\theta_2}(x) - k' f_{\theta_1}(x)) \neq 0\}.$$

Thus we see that $\mu(f_{\theta_2} \neq k' f_{\theta_1}) = 0$. In other words $f_{\theta_2}(x) = k' f_{\theta_1}(x)$ a.e. μ, which implies $k' = 1$. This leads to a contradiction $P_{\theta_1} = P_{\theta_2}$. □

3.4 Uniformly Most Powerful Unbiased test and two-sided hypotheses

In this section we consider two-sided hypotheses, which include three types

I. $H_0 : \theta = \theta_0$ versus $H_1 : \theta \neq \theta_0$;
II. $H_0 : \theta_1 \leq \theta \leq \theta_2$ versus $H_1 : \theta < \theta_1$ or $\theta > \theta_2$;
III. $H_0 : \theta \leq \theta_1$ or $\theta \geq \theta_2$ versus $H_1 : \theta_1 < \theta < \theta_2$.

The way these tests are ordered above follows roughly according to how common they are in practice: for example hypotheses of type I are most commonly seen. Technically, however, it is easier to develop the optimal tests following the order of III→II→I, which will be the route we take.

Unlike the one-sided hypotheses, there are in general no UMP tests for two-sided hypotheses. UMP tests always exists for hypotheses III, but they do not exist for hypotheses I and II. This point is illustrated by the following example.

Example 3.7 Suppose X is distributed as $b(n, \theta), 0 < \theta < 1$, and $0 < \alpha < 1$. To test the null hypothesis $H_0 : \theta = \frac{1}{2}$ versus the two sided alternative $H_1 : \theta \neq \frac{1}{2}$ we first consider the one-sided alternative hypothesis $H_+ : \theta > \frac{1}{2}$. Then

$$\phi_+(x) = \begin{cases} 1 & \text{if } x > c_+ \\ \gamma_+(x) & \text{if } x = c_+ \\ 0 & \text{if } x < c_+, \end{cases}$$

with $E_{\frac{1}{2}}(\phi_+(X)) = \alpha$ is UMP-α test for testing H_0 versus H_+. Similarly,

$$\phi_-(x) = \begin{cases} 1 & \text{if } x < c_- \\ \gamma_-(x) & \text{if } x = c_- \\ 0 & \text{if } x > c_-, \end{cases}$$

with $E_{\frac{1}{2}}(\phi_-(X)) = \alpha$, is UMP test of size α for testing H_0 versus $H_- : \theta < \frac{1}{2}$.

Suppose ϕ_0 is a UMP-α test for H_0 versus H_1. Then it is also UMP-α for H_0 versus H_+. Consequently $E_\theta(\phi_0(X)) = E_\theta(\phi_+(X))$, for all $\theta \geq \frac{1}{2}$. Let $g(j) = \phi_0(j) - \phi_+(j)$. Then

$$0 = E_\theta[g(X)] = \sum_{j=0}^{n} g(j) \binom{n}{j} \left(\frac{\theta}{1-\theta} \right)^j (1-\theta)^n.$$

If we let $\eta = \theta/(1-\theta)$, then the above equality implies

$$\sum_{j=1}^{n} g(j) \binom{n}{j} \eta^j = 0$$

for all $\eta = \theta/(1-\theta) \geq 1$. So $g(j) = 0$ for all j. In other words $\phi_0(j) = \phi_+(j)$ for all j. Using the same argument we can show that $\phi_- \equiv \phi_0$. Thus $\phi_-(j) = \phi_+(j)$ for all j, which is impossible.

For example if $n = 4$, $\alpha = \frac{1}{16}$, then

$$\phi_+(j) = \begin{cases} 1 & \text{if } j > 3 \\ 0 & \text{if } j \le 3, \end{cases} \qquad \phi_-(j) = \begin{cases} 1 & \text{if } j < 4 \\ 0 & \text{if } j \ge 4, \end{cases}$$

but $\phi_+ \neq \phi_-$. So there is no UMP test for the 2-sided hypotheses. □

3.4.1 Uniformly Most Powerful Unbiased tests

From Example 3.7 we see that UMP tests do not in general exist for two-sided composite hypotheses, because one can always sacrifice the power for one side and make the power for the other side as large as possible. Apparently, a certain restriction should be imposed. A simple condition to impose is that the power function, for θ in the alternative, should take values greater than or equal to the size of the test. That is, the probability of rejecting the null hypothesis when false is never smaller than the probability of rejecting the null hypothesis when true. A test satisfying this condition is called an unbiased test (Neyman and Pearson, 1936). The following definition formulates the concept of unbiasedness in a more general setting than scalar parameter, which will become important for later discussions. Again, $\mathcal{P}_0$ and $\mathcal{P}_1$ denote two disjoint families of distributions.

Definition 3.8 *A size α test ϕ for testing $H_0 : P \in \mathcal{P}_0$ versus $H_1 : P \in \mathcal{P}_1$ is said to be unbiased if $\alpha \le \inf_{P \in \mathcal{P}_1} \beta_\phi(P)$.*

Definition 3.9 *A test ϕ is called Uniformly Most Powerful Unbiased test of size α (UMPU-α) if*

1. *it has size α;*
2. *for any size α unbiased test ψ we have $\beta_\phi(P) \ge \beta_\psi(P)$ for all $P \in \mathcal{P}_1$.*

If a UMP-α exists, then its power cannot fall below that of the test $\psi(x) \equiv \alpha$, for $P \in \mathcal{P}_1$. So a UMP tests are unbiased. For a large class of testing problems, where UMP tests fail to exist, there do exist UMPU tests. By a similar argument, a UMPU-α test must be itself unbiased. To see this, let ϕ be an UMPU-α test and let $\psi \equiv \alpha$. Then ψ is unbiased of size α, and hence $\beta_\phi(P) \ge \beta_\psi(P) = \alpha$ for all $P \in \mathcal{P}_1$.

A fact about size-α unbiased tests that will be useful in our discussion is that, if the power is continuous, then their power functions are constant on the boundary of the two sets of probability measures specified by the null and alternative hypotheses.

Proposition 3.2 *Suppose that on $\mathcal{P} = \mathcal{P}_0 \cup \mathcal{P}_1$ is defined a metric, say ρ, with respect to which $\beta_\phi(\cdot)$ is continuous. If ϕ is a size-α unbiased test, then $\beta_\phi(P) = \alpha$ for all $P \in \bar{\mathcal{P}}_0 \cap \bar{\mathcal{P}}_1$, where $\bar{\mathcal{P}}_0$ and $\bar{\mathcal{P}}_1$ are the closures of $\mathcal{P}_0$ and $\mathcal{P}_1$ with respect to the metric.*

Proof. Let $P \in \bar{\mathcal{P}}_0 \cap \bar{\mathcal{P}}_1$. Then there is a sequence $\{P_k'\} \subseteq \mathcal{P}_0$ and $\{P_k''\} \subseteq \mathcal{P}_1$ such that $\rho(P_k', P) \to 0$ and $\rho(P_k'', P) \to 0$. Thus

$$\alpha \geq \lim_{k\to\infty} \beta_\phi(P_k') = \beta_\phi(P) = \lim_{k\to\infty} \beta_\phi(P_k'') \geq \alpha,$$

as desired. □

The rest of this section is devoted to constructing UMPU-α tests for hypotheses II and III and the UMP-α test for hypotheses I. We first introduce some technical mechanism needed to construct optimal tests for two-sided hypotheses.

3.4.2 More properties of the exponential family

We have seen that the families with MLR property play an important role in constructing UMP test for one-sided hypotheses. Exponential families play a similar role in constructing UMPU tests for two-sided hypotheses.

Specializing to the current context, the exponential family (2.5) becomes

$$c(\theta)e^{\eta(\theta)Y(x)}, \tag{3.13}$$

where η is a monotone function of the parameter θ. Clearly, this family has the MLR property. Examples of one-parameter exponential families include normal $\{N(\mu, 1) : \mu \in \mathbb{R}\}$, and binomial $\{b(n, \theta) : 0 < \theta < 1\}$. For example, in the binomial case, the probability mass function is given by

$$f_\theta(x) = \binom{n}{x}\theta^x(1-\theta)^{n-x} = c(\theta)e^{\eta(\theta)Y(x)},$$

where $c(\theta) = (1-\theta)^n\binom{n}{x}$, $Y(x) = x$, and $\eta(\theta) = \log(\theta/(1-\theta))$. In this section we study the properties of the exponential family that will be used frequently in the subsequent exposition.

By Lemma 3.2, Theorem 2.7, and Lemma 2.2, if f_θ has the form (3.13), if ϕ is a nondecreasing function of $Y(x)$ where $Y(x)$ is as it appears in (3.13), then $\beta_\phi(\theta)$ is either strictly monotone increasing in θ or constant in θ.

For a one-parameter exponential family (3.13), since $E_\theta\psi(X) < \infty$ for $\theta \in \Theta$, the power function $\beta_\psi(\theta)$ is an analytic function for any test ψ, and the derivative with respect to θ can be brought inside the integral $\int \psi f_\theta d\mu$. (Here and in what follows, we use the dot notation to represent derivatives. For example $\dot{f}_\theta(x)$ or simply $\dot{f}_\theta$ represents $\partial f_\theta(x)/\partial\theta$, and $\dot{\beta}_\phi(\theta)$ or simply $\dot{\beta}_\phi$ represents $\partial\beta_\phi(\theta)/\partial\theta$.) Hence

$$\dot{\beta}_\psi(\theta) = \int \psi\dot{f}_\theta d\mu = \int \psi[\dot{c}_1(\theta)/c_1(\theta) + Y]f_\theta d\mu. \tag{3.14}$$

If, in the above, we take $\psi \equiv 1$, then $\beta_\psi(\theta) \equiv 1$, and $\dot{\beta}_\psi(\theta) = 0$. It follows that

$$\dot{c}_1(\theta)/c_1(\theta) = -E_\theta(Y). \tag{3.15}$$

Substitute this into (3.14) to obtain

$$\dot{\beta}_\psi(\theta) = \text{cov}_\theta(Y(X), \psi(X)). \tag{3.16}$$

In other words, the power function of a test ψ for an exponential-family distribution is the covariance between the test and the sufficient statistic.

Suppose ψ and ψ' are two tests with power functions β_ψ and $\beta_{\psi'}$. If $\beta_\psi(\theta) \geq \beta_{\psi'}(\theta)$ for all θ, and if $\beta_\psi(\theta_0) = \beta_{\psi'}(\theta_0)$ for some $\theta_0 \in \Theta_0$, then the minimum of $\beta_\psi(\theta) - \beta_{\psi'}(\theta) \geq 0$ is attained at θ_0, and so $\frac{\partial}{\partial\theta}(\beta_\psi(\theta) - \beta_{\psi'}(\theta))|_{\theta=\theta_0} = 0$. Thus, if $\psi' \equiv \alpha$, $\beta_\psi(\theta) \geq \alpha$ for all θ and $\beta_\psi(\theta_0) = \alpha$, then by (3.16), it follows that $\partial\beta_\psi(\theta_0)/\partial\theta = 0$, which in turn implies

$$E_{\theta_0}(\psi(X)) = \alpha \quad \text{and} \quad E_{\theta_0}(Y(X)\psi(X)) = \alpha E_{\theta_0}(Y(X)). \tag{3.17}$$

3.4.3 Generalized Neyman-Pearson Lemma

For the study of two-sided hypotheses the following generalized version of the Neyman-Pearson Lemma is also required (Neyman and Pearson, 1936).

Lemma 3.3 (Generalized Neyman-Pearson Lemma) *Let $f_1, f_2, \ldots, f_{m+1}$ be integrable functions with respect to a measure μ and $c_1, \ldots, c_m$ be real numbers. Let*

$$\phi^*(x) = \begin{cases} 1 & \text{if } f_{m+1}(x) > \sum_{i=1}^m c_i f_i(x) \\ \gamma(x) & \text{if } f_{m+1}(x) = \sum_{i=1}^m c_i f_i(x) \\ 0 & \text{if } f_{m+1}(x) < \sum_{i=1}^m c_i f_i(x) \end{cases}$$

Then

1. *For any test ϕ that satisfies*

$$\int \phi f_i d\mu = \int \phi^* f_i d\mu, \quad i = 1, \ldots, m, \tag{3.18}$$

   we have $\int \phi f_{m+1} d\mu \leq \int \phi^* f_{m+1} d\mu$;
2. *If, furthermore, $c_1 \geq 0, \ldots, c_m \geq 0$, then, for any ϕ that satisfies*

$$\int \phi f_i d\mu \leq \int \phi^* f_i d\mu, \quad i = 1, \ldots, m, \tag{3.19}$$

   we have $\int \phi f_{m+1} d\mu \leq \int \phi^* f_{m+1} d\mu$.

Proof. By construction,

$$(\phi^*(x) - \phi(x))\left(f_{m+1}(x) - \sum c_i f_i(x)\right) \geq 0$$

for all x. This implies

$$\int (\phi^* - \phi) f_{m+1} d\mu \geq \sum c_i \int (\phi^* - \phi) f_i d\mu.$$

If (3.18) holds then the right hand side is 0, proving assertion 1. If $c_1 \geq 0, \ldots c_m \geq 0$ and (3.19) holds, then each summand on the right hand is nonnegative. So the right hand side is nonnegative, proving assertion 2. □

The following is another variation of the Neyman-Pearson lemma that will be useful.

Lemma 3.4 *Suppose that f_1 and f_2 are two functions integrable with respect to μ. Let ϕ^* be a test satisfying*

$$\phi^*(x) = \begin{cases} 1 & \text{if } f_2(x) > k f_1(x) \\ 0 & \text{if } f_2(x) < k f_1(x) \end{cases} \tag{3.20}$$

for some $-\infty \leq k \leq \infty$. Then, for any test ϕ that satisfies $\int \phi f_1 d\mu = \int \phi^* f_1 \mu$ we have

$$\int_{f_1 \neq 0} \phi^* f_2 d\mu \geq \int_{f_1 \neq 0} \phi f_2 d\mu.$$

Proof. By construction

$$(\phi^*(x) - \phi(x))(f_2(x) - k f_1(x)) \geq 0$$

for all $x \in \Omega_X$. Hence $\int_{f_1 \neq 0} (\phi^* - \phi)(f_2 - k f_1) d\mu \geq 0$, which implies

$$\int_{f_1 \neq 0} (\phi^* - \phi) f_2 d\mu \geq k \int_{f_1 \neq 0} (\phi^* - \phi) f_1 d\mu = \int (\phi^* - \phi) f_1 d\mu = 0.$$

□

3.4.4 Quantile transformation and construction of two-sided tests

Our construction of two-sided optimal tests hinges on a type of quantile transformation, which we discuss in detail in this subsection. For a similar construction, see Ferguson (1967, Section 5.3). Let F be the distribution function of a random variable Y. For $0 < \omega < 1$, define

$$F^{-1}(\omega) = \inf\{y : F(y) \geq \omega\}.$$

This function is called the quantile function of Y. Its definition is illustrated by Figure 3.2. We will use $F(a-)$ to denote the left limit of F at a; that is, $F(a-) = P(Y < a)$. The following properties of F^{-1} will prove useful.

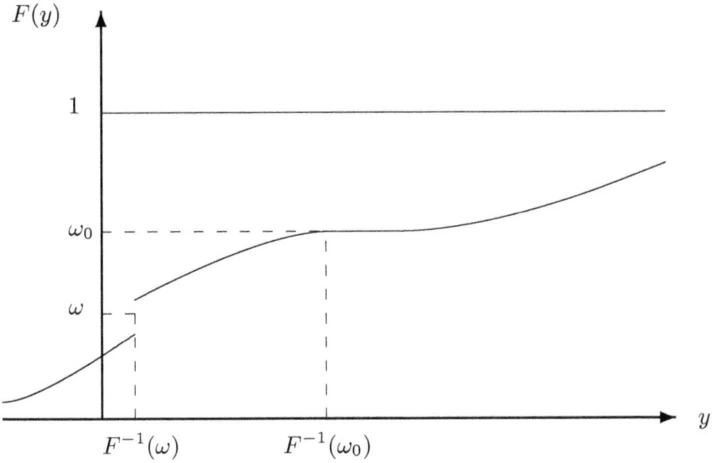

Fig. 3.2. Quantile function

Lemma 3.5 *Let* $0 < \omega < 1$. *Then:*

1. $F(y) < \omega$ *if and only if* $y < F^{-1}(\omega)$.
2. $F(y) \geq \omega$ *if and only if* $y \geq F^{-1}(\omega)$.
3. $F(F^{-1}(\omega)-) \leq \omega \leq F(F^{-1}(\omega))$.
4. *If* F *is continuous at* $F^{-1}(\omega)$, *then* $F(F^{-1}(\omega)) = \omega$.

Proof. 1. By the definition of F^{-1}, if $F(y) \geq \omega$, then $y \geq F^{-1}(\omega)$. This shows that $y < F^{-1}(\omega)$ implies $F(y) < \omega$. Now suppose $y \geq F^{-1}(\omega)$. Then $F(y) \geq F(F^{-1}(\omega))$. Let $A_\omega = \{y : F(y) \geq \omega\}$. Then there is a sequence $\{y_k\} \subseteq A_\omega$ such that $\lim_k y_k = \inf A_\omega = F^{-1}(\omega)$. Because $y_k \in A_\omega$, $y_k \geq F^{-1}(\omega)$. By right continuity of F, we have $\lim_k F(y_k) = F(F^{-1}(\omega))$. But we also know that $F(y_k) \geq \omega$ for each k. So $F(y) \geq F(F^{-1}(\omega)) \geq \omega$.

2. This statement is equivalent to statement 1.

3. That $F(F^{-1}(\omega)) \geq \omega$ has been proved in the proof of assertion 1. Also by assertion 1, whenever $y < F^{-1}(\omega)$, we have $F(y) < \omega$. So $F(F^{-1}(\omega)-) \leq \omega$.

4. This is a direct consequence of assertion 3. □

Let γ be a number in $(0, 1)$ and let $\Delta(y)$ denote the jump of F at y; that is $\Delta(y) = F(y) - F(y-)$. Let

$$G(y, \gamma) = F(y) - (1 - \gamma)\Delta(y) = F(y-) + \gamma\Delta(y),$$

Thus, $G(y, 0) = F(y-)$, $G(y, 1) = F(y)$, and

$$F(y-) \leq G(y, \gamma) \leq F(y), \quad \text{for } 0 \leq \gamma \leq 1. \tag{3.21}$$

The variable γ "compensates" any discontinuity of F. Recall that, if Y is a continuous random variable, then $F(Y) \sim U(0,1)$. This is no longer true for discrete Y. However, a variation of this can be established for general Y using $G(y,\gamma)$, which compensates for any discontinuity. For $0 < \omega < 1$, let

$$\gamma(\omega) = \begin{cases} [\omega - F(F^{-1}(\omega)-)]/\Delta[F^{-1}(\omega)] & \text{if } \Delta[F^{-1}(\omega)] > 0 \\ 0 & \text{otherwise.} \end{cases} \tag{3.22}$$

Lemma 3.6 *Let V be a random variable independent of Y, and $V \sim U(0,1)$. The random variable $W = G(Y,V)$ is distributed as $U[0,1]$.*

Proof. Let $0<\omega<1$. By (3.21), $F(Y-) \le W \le F(Y)$. By part 1 of Lemma 3.5 we have $\{Y < F^{-1}(\omega)\} = \{F(Y) < \omega\}$. Now decompose the event $\{W < \omega\}$ as

$$\{W < \omega\} = \{W < \omega, F(Y) < \omega\} \cup \{W < \omega, F(Y) \ge \omega\}. \tag{3.23}$$

The first event on the right-hand side of (3.23) can be rewritten as

$$\{W < \omega, F(Y) < \omega\} = \{F(Y) < \omega\} = \{Y < F^{-1}(\omega)\}.$$

The second event on the right-hand side of (3.23) can be rewritten as

$$\begin{aligned}\{W < \omega, F(Y) \ge \omega\} =& \{W < \omega, F(Y) \ge \omega, F(Y-) < \omega\} \\ =& \{W < \omega, Y = F^{-1}(\omega)\},\end{aligned}$$

where the first equality holds because $F(Y-) \le W$, and the second holds because $F(Y-) < \omega \le F(Y)$ implies $Y = F^{-1}(\omega)$. Moreover, $Y = F^{-1}(\omega)$, together with $W < \omega$, implies $F(Y-) < \omega \le F(Y)$. Hence

$$P(W < \omega) = P(Y < F^{-1}(\omega)) + P(W < \omega, Y = F^{-1}(\omega)). \tag{3.24}$$

If $\Delta(F^{-1}(\omega)) = 0$, then F is continuous at $F^{-1}(\omega)$ and hence the second term on the right-hand side above is 0. The first term is $F(F^{-1}(\omega)-)$, which is ω by part 4 of Lemma 3.5. Thus $P(W < \omega) = \omega$.

If $\Delta(F^{-1}(\omega)) > 0$, then, by (3.24),

$$\begin{aligned}P(W < \omega) =& P(Y < F^{-1}(\omega)) + P(Y = F^{-1}(\omega))P(W < \omega | Y = F^{-1}(\omega)) \\ =& F(F^{-1}(\omega)-) + \Delta(F^{-1}(\omega))P(V < \gamma(\omega) | Y = F^{-1}(\omega)) \\ =& F(F^{-1}(\omega)-) + \Delta(F^{-1}(\omega))P(V < \gamma(\omega)) \\ =& F(F^{-1}(\omega)-) + \Delta(F^{-1}(\omega))\gamma(\omega) = \omega,\end{aligned}$$

where the third equality follows from the independence between V and Y; the fourth follows from the assumption $V \sim U(0,1)$, and the fifth follows from the definition of $\gamma(\omega)$. Thus $P(W < \omega) = \omega$ for each $\omega \in (0,1)$, which implies $W \sim U[0,1]$. □

Now let X be a random variable (or vector) defined on $(\Omega, \mathcal{F})$ and $Y : x \mapsto Y(x)$ be a real-valued measurable function. Let F denote the distribution of Y.

Lemma 3.7 *Let $\gamma(\omega)$ be as in (3.22). For each $0 < \omega < 1$, the function defined by*

$$\phi_\omega(x) = \begin{cases} 1 & \text{if } Y(x) < F^{-1}(\omega) \\ \gamma(\omega) & \text{if } Y(x) = F^{-1}(\omega) \\ 0 & \text{if } Y(x) > F^{-1}(\omega) \end{cases} \tag{3.25}$$

satisfies

$$\phi_\omega(x) = E(I_{[0,\omega)}(G(Y(x), V))) = E(I_{[0,\omega)}(W)|X = x), \tag{3.26}$$

for all x, consequently $E(\phi_\omega(X)) = \omega$.

Proof. Let $y = Y(x)$. Then by Lemma 3.5, part 1, and (3.21), if $y < F^{-1}(\omega)$, then $G(y, \gamma) \leq F(y) < \omega$ for all $0 \leq \gamma \leq 1$. So

$$E[I_{[0,\omega)}(G(Y(x), V))] = 1.$$

If $y > F^{-1}(\omega)$, then, by Lemma 3.5, part 3, $F(y-) \geq F(F^{-1}(\omega)) \geq \omega$. So by (3.21), $G(y, \gamma) \geq \omega$ for all $0 \leq \gamma \leq 1$, which implies

$$E[I_{[0,\omega)}(G(Y(x), V))] = 0.$$

Finally, suppose $y = F^{-1}(\omega)$. When $\Delta(y) \neq 0$ we have

$$G(y, \gamma) < \omega \Leftrightarrow \gamma\Delta(y) < \omega - F(y-) = \gamma(\omega)\Delta(y) \Leftrightarrow \gamma < \gamma(\omega).$$

So $P(G(y, V) < \omega) = P(V < \gamma(\omega)) = \gamma(\omega)$. When $\Delta(y) = 0$, $G(y, \gamma) < \omega$ does not hold for any γ. So in this case $P(G(y, V) < \omega) = 0 = \gamma(\omega)$. This completes the proof. □

Lemma 3.8 *Let f denote the density of X with respect to a measure μ and h a function of x such that $\int_{f>0} |h| d\mu < \infty$. Let ϕ_ω be as defined in (3.25) with F being the distribution corresponding to $Y(X)$. Then the function $g : [0, 1] \to [0, 1]$ defined by*

$$g(\omega) = \begin{cases} 0 & \text{if } \omega = 0 \\ \int_{f>0} \phi_\omega h d\mu & \text{if } 0 < \omega < 1 \\ 1 & \text{if } \omega = 1 \end{cases}$$

is continuous on $[0, 1]$.

Proof. We have

$$\begin{aligned} g(\omega) &= \int_{f>0} \phi_\omega h d\mu = \int_{f>0} \phi_\omega \frac{h}{f} f d\mu \\ &= \int_{f>0} E[I_{[0,\omega)}(W)|X = x][h(x)/f(x)]\, f(x)\mu(dx) \\ &= E[I_{[0,\omega)}(W)(h(X)/f(X))]. \end{aligned}$$

Let $\epsilon > 0$ be such that $0 < \omega - \epsilon < \omega < \omega + \epsilon < 1$. Then

$$|g(\omega+\epsilon) - g(\omega-\epsilon)| \le E[I_{[\omega-\epsilon,\omega+\epsilon)}(W)|h(X)/f(X)|].$$

The random variable $I_{[\omega-\epsilon,\omega+\epsilon)}(W)|h(X)/f(X)|$ is dominated by $|h(X)/f(X)|$, whose expectation is finite. So by Lebesgue's Dominated Convergence Theorem,

$$\lim_{\epsilon\to 0} E[I_{[\omega-\epsilon,\omega+\epsilon)}(W)|h(X)/f(X)|] = E[I_{\{\omega\}}(W)|h(X)/f(X)|].$$

The right hand side is zero because $I_{\{\omega\}}(W) = 0$ almost everywhere. Hence g is continuous in $(0,1)$.

By a similar argument it can be shown that $\lim_{\omega\to 1} g(\omega) = 1$ and $\lim_{\omega\to 0} g(\omega) = 0$. Thus $g(\omega)$ is continuous in $[0,1]$. □

Lemma 3.9 *Let f and h be densities of two probability distributions of X with respect to μ such that $\mu\{f = 0, h > 0\} = 0$. Suppose that $h(x)/f(x)$ is a non-decreasing function of $Y(x)$. Then, for any $0 < \alpha < 1$, there exist $0 \le \gamma_1,\ \gamma_2 \le 1,\ -\infty \le t_1 < t_2 \le +\infty$ such that the test*

$$\phi(x) = \begin{cases} 1 & \text{if } t_1 < Y(x) < t_2 \\ \gamma_i & \text{if } Y(x) = t_i,\ i = 1,2 \\ 0 & \text{if } Y(x) < t_1 \text{ or } Y(x) > t_2, \end{cases} \tag{3.27}$$

satisfies $\int \phi f d\mu = \int \phi h d\mu = \alpha$.

Proof. Let ϕ_ω be as defined in (3.25) with F therein being the distribution of $Y(X)$. For $0 \le u \le 1-\alpha$, let

$$\psi_u(x) = \begin{cases} \phi_\alpha(x) & \text{if } u = 0 \\ \phi_{\alpha+u}(x) - \phi_u(x) & \text{if } 0 < u < 1-\alpha \\ 1 - \phi_{1-\alpha}(x) & \text{if } u = 1-\alpha \end{cases}$$

Clearly $0 \le \psi_u(x) \le 1$, and by Lemma 3.7, $\int \psi_u f d\mu = \alpha$ for all $u \in [0, 1-\alpha]$.

That $h(x)/f(x)$ is nondecreasing in $Y(x)$ implies that the ratio is a function of $Y(x)$. Write this ratio as $L(Y(x))$. Then

$$L(Y(x)) > L(F^{-1}(\omega)) \Rightarrow Y(x) > F^{-1}(\omega), \quad \text{and}$$
$$L(Y(x)) < L(F^{-1}(\omega)) \Rightarrow Y(x) < F^{-1}(\omega).$$

Hence $1 - \phi_\omega$ is of form (3.20) with $f_1 = f$ and $f_2 = h$. So by Lemma 3.4, taking ϕ^* and ϕ therein to be $1 - \phi_\alpha$ and $1 - \alpha$, we have

$$\int_{f>0} \phi_\alpha h d\mu \leq \alpha, \quad \text{and similarly,} \quad \int_{f>0} \phi_{1-\alpha} h d\mu \leq 1 - \alpha.$$

This implies, as $\psi_{1-\alpha} = 1 - \phi_{1-\alpha}$ and $\psi_0 = \phi_\alpha$, that

$$\int_{f>0} \psi_0 h d\mu \leq \alpha \leq \int_{f>0} \psi_{1-\alpha} h d\mu.$$

The case that $h(x)/f(x)$ is nonincreasing in $Y(x)$ can be treated similarly.

Now by construction $s(u) = \int_{f>0} \psi_u h d\mu = g(\alpha+u) - g(u)$ for $u \in [0, 1-\alpha]$. Hence by Lemma 3.8, $s(u)$ is continuous in $[0,1]$. So there exists $0 \leq u_0 \leq 1-\alpha$ such that $s(u_0) = \alpha$. But because $\mu(\{f = 0, h > 0\}) = 0$, $s(u_0) = \alpha$ implies $\int \psi_{u_0} h d\mu = \alpha$, as to be demonstrated. □

Lemma 3.10 *Suppose that f is the density of X with respect to a measure μ, that $Y(x)$ is a measurable function, and that there is an interval $(-\epsilon, \epsilon)$ such that, for each ζ in this interval, $\int e^{\zeta Y(x)} f(x) \mu(dx) < \infty$. Then, for any $0 < \alpha < 1$, there exist $0 \leq \gamma_1, \gamma_2 \leq 1$, $-\infty \leq t_1 < t_2 \leq \infty$ such that the test (3.27) satisfies*

$$\int \phi f d\mu = \alpha, \quad \int \phi Y f d\mu = \alpha \int Y f d\mu. \tag{3.28}$$

Proof. For $\zeta \in (-\epsilon, \epsilon)$, define

$$f_\zeta(x) = c(\zeta) f(x) e^{\zeta Y(x)}, \quad \text{where} \quad c(\zeta) = \left(\int f(x) e^{\zeta Y(x)} \mu(dx) \right)^{-1}.$$

Then $\{f_\zeta : \zeta \in (-\epsilon, \epsilon)\}$ is an exponential family. Let $\dot{f}(x)$ denote $\partial f_\zeta(x)/\partial \zeta|_{\zeta=0}$. Then, by (3.15),

$$\dot{f}(x)/f(x) = Y(x) - \int Y(s) f(s) \mu(ds). \tag{3.29}$$

Thus $\dot{f}(x)/f(x)$ is monotone increasing in $Y(x)$. Let $L(Y(x)) = \dot{f}(x)/f(x)$. Then

$$L(Y(x)) < L(F^{-1}(\omega)) \Rightarrow Y(x) < F^{-1}(\omega)$$
$$L(Y(x)) > L(F^{-1}(\omega)) \Rightarrow Y(x) > F^{-1}(\omega).$$

It follows that ϕ_α, as defined by (3.25), is of the form

$$\phi_\alpha(x) = \begin{cases} 1 & \dot{f}(x) < L(F^{-1})(\alpha) f(x) \\ 0 & \dot{f}(x) > L(F^{-1})(\alpha) f(x) \end{cases}$$

By Lemma 3.4, $\int_{f>0} (1-\phi_\alpha)\dot{f} d\mu \geq \int_{f>0} \phi \dot{f} d\mu$ for any test ϕ that satisfies $\int \phi f d\mu = 1 - \alpha$. In particular,

$$\int_{f>0} (1-\phi_\alpha)\dot{f} d\mu \geq (1-\alpha) \int_{f>0} \dot{f} d\mu = 0,$$

implying $\int_{f>0} \phi_\alpha \dot{f} d\mu \leq 0$. For the same reason, $\int_{f>0} \phi_{1-\alpha} \dot{f} d\mu \leq 0$. Now define $\psi_u(x)$, for $0 \leq u \leq 1-\alpha$, as in the proof of Lemma 3.9. Then

$$\int_{f>0} \psi_0 \dot{f} d\mu \leq 0 \leq \int_{f>0} \psi_{1-\alpha} \dot{f} d\mu.$$

Because f_ζ is an exponential family, $Y(X)$ has finite expectation. Consequently $\int_{f>0} |\dot{f}| d\mu < \infty$. Therefore, by Lemma 3.8, $\int \psi_u \dot{f} d\mu$ is continuous in $[0, 1-\alpha]$. So there is a $0 \leq u_0 \leq 1-\alpha$ such that $\int_{f>0} \psi_{u_0} \dot{f} d\mu = 0$ which, by (3.29), implies the second equality in (3.28). □

3.4.5 UMP test for hypothesis III

We are now ready to derive the UMP test for hypothesis III: $H_0 : \theta \leq \theta_1$ or $\theta \geq \theta_2$ versus $H_1 : \theta_1 < \theta < \theta_2$.

Theorem 3.5 *Let X be a random variable with a density given by (2.4).*

(1) For $\theta_1 < \theta_2$ and $0 < \alpha < 1$, there exist $-\infty < t_1 < t_2 < \infty$, $0 \leq \gamma_1, \gamma_2 \leq 1$, such that ϕ defined by (3.27) satisfies

$$E_{\theta_1}\phi(X) = E_{\theta_2}\phi(X) = \alpha. \tag{3.30}$$

(2) Let ϕ be a test satisfying the conditions in part 1. Then, for any ψ that satisfies $E_{\theta_1}\psi(X) \leq \alpha$, $E_{\theta_2}\psi(X) \leq \alpha$, we have $E_\theta\phi(X) \geq E_\theta\psi(X)$ for all $\theta_1 < \theta < \theta_2$.

(3) Let ϕ be a test satisfying the conditions in part 1. Then, for any test ψ satisfying $E_{\theta_1}\psi(X) = E_{\theta_2}\psi(X) = \alpha$, we have $E_\theta\psi(X) \geq E_\theta\phi(X)$ for $\theta < \theta_1$ or $\theta > \theta_2$.

(4) Any test ϕ that satisfies conditions in part 1 is a UMP-α test for testing $H_0 : \theta \leq \theta_1$ or $\theta \geq \theta_2$ versus $H_1 : \theta_1 < \theta < \theta_2$.

Proof. (1) By Lemma 3.9, there exist $0 \le \gamma_1, \gamma_2 \le 1$ and $-\infty \le t_1 < t_2 \le \infty$ such that ϕ in (3.27) satisfies (3.30). If $t_1 = -\infty$ or $t_2 = \infty$, then ϕ is a one-sided test. By Theorem 3.4, the power of ϕ is strictly monotone and hence ϕ cannot satisfy (3.30). So t_1 and t_2 are both finite.

(2) We first show that for any fixed $\theta \in (\theta_1, \theta_2)$, ϕ defined in (3.27) can be written as

$$\phi(x) \equiv \begin{cases} 1 & \text{if } f_\theta(x) > c_1 f_{\theta_1}(x) + c_2 f_{\theta_2}(x) \\ \gamma(x) & \text{if } f_\theta(x) = c_1 f_{\theta_1}(x) + c_2 f_{\theta_2}(x) \\ 0 & \text{if } f_\theta(x) < c_1 f_{\theta_1}(x) + c_2 f_{\theta_2}(x) \end{cases}$$

for some c_1, $c_2 \ge 0$ and $0 \le \gamma(x) \le 1$. By Lemma 3.3 this implies that $E_\theta(\phi(X)) \ge E_\theta(\psi(X))$ for any test ψ satisfying $E_{\theta_i}(\psi(X)) \le \alpha,\ i = 1, 2$.

Note that

$$\begin{aligned} f_\theta > c_1 f_{\theta_1} + c_2 f_{\theta_2} & \\ &\Leftrightarrow (c_1 f_{\theta_1} + c_2 f_{\theta_2})/f_\theta < 1 \\ &\Leftrightarrow [c_1 c(\theta_1)/c(\theta)] e^{(\theta_1 - \theta)Y(x)} + [c_2 c(\theta_2)/c(\theta)] e^{(\theta_2 - \theta)Y(x)} < 1. \end{aligned}$$

Let a_1, a_2 be the solution to the following system of linear equations

$$\begin{aligned} a_1 e^{(\theta_1-\theta)t_1} + a_2 e^{(\theta_2-\theta)t_1} &= 1 \\ a_1 e^{(\theta_1-\theta)t_2} + a_2 e^{(\theta_2-\theta)t_2} &= 1 \end{aligned} \tag{3.31}$$

Then

$$a_1 = (e^{(\theta_2-\theta)t_2} - e^{(\theta_2-\theta)t_1})/\det(G), \quad a_2 = (e^{(\theta_1-\theta)t_1} - e^{(\theta_1-\theta)t_2})/\det(G),$$

where

$$G = \begin{pmatrix} e^{(\theta_1-\theta)t_1} & e^{(\theta_2-\theta)t_1} \\ e^{(\theta_1-\theta)t_2} & e^{(\theta_2-\theta)t_2} \end{pmatrix}.$$

Since $\theta_1 < \theta_2$ and $t_1 < t_2$, we have

$$\det(G) = e^{\theta_2 t_2 + \theta_1 t_1 - \theta(t_1+t_2)}(1 - e^{(t_1-t_2)(\theta_2-\theta_1)}) > 0.$$

Now let $g(t) = a_1 e^{(\theta_1-\theta)t} + a_2 e^{(\theta_2-\theta)t}$. Since $\theta_1 < \theta < \theta_2$, we have $a_1 > 0$, $a_2 > 0$, and consequently

$$g''(t) = a_1(\theta_1 - \theta)^2 e^{(\theta_1-\theta)t} + a_2(\theta_2-\theta)^2 e^{(\theta_2-\theta)t} > 0$$

for all t. Thus $g(t)$ is strictly convex on $(-\infty, \infty)$. It follows that $g(t) < 1$ on $t \in (t_1, t_2)$ and $g(t) > 1$ on $(-\infty, t_1) \cup (t_2, \infty)$. Let $c_1 = a_1 c(\theta)/c(\theta_1)$ and $c_2 = a_2 c(\theta)/c(\theta_2)$. Then

$$\begin{aligned} g(Y(x)) < 1 &\Leftrightarrow f_\theta(x) > c_1 f_{\theta_1}(x) + c_2 f_{\theta_2}(x), \\ g(Y(x)) > 1 &\Leftrightarrow f_\theta(x) < c_1 f_{\theta_1}(x) + c_2 f_{\theta_2}(x). \end{aligned} \tag{3.32}$$

Thus by (3.32), $\phi(x) = 1$ if and only if $f_\theta(x) > c_1 f_{\theta_1}(x) + c_2 f_{\theta_0}(x)$ and $\phi(x) = 0$ if and only if $f_\theta(x) < c_1 f_{\theta_1}(x) + c_2 f_{\theta_2}(x)$. So by Lemma 3.3,

$$E_\theta(\psi(X)) \le E_\theta(\phi(X))$$

for any test rule ψ satisfying $E_{\theta_i}(\psi(X)) \le \alpha$.

(3) Let $\theta < \theta_1$. In this case $a_1 > 0$ and $a_2 < 0$. So

$$\lim_{t\to-\infty} g(t) = 0, \quad \lim_{t\to\infty} g(t) = -\infty.$$

Moreover, $g'(t) = 0$ has a unique solution $t = t_0$; in fact:

$$t_0 = \frac{1}{\theta_1 - \theta_2} \log\left[-\frac{a_2(\theta_2 - \theta)}{a_1(\theta_1 - \theta)}\right].$$

These facts, together with $g(t_1) = g(t_2) = 1$, imply that g is strictly increasing for $t < t_0$ and strictly decreasing for $t > t_0$ for some $t_0 \in (t_1, t_2)$. Consequently, $g(t) > 1$ if and only if $t_1 < t < t_2$. Thus, as above by (3.32), we have for some $c_1 > 0 > c_2$ that

$$\begin{aligned} 1 - \phi(x) = 1 &\Leftrightarrow g(Y(x)) < 1 \Leftrightarrow f_\theta(x) > c_1 f_{\theta_1}(x) + c_2 f_{\theta_2}(x) \\ 1 - \phi(x) = 0 &\Leftrightarrow g(Y(x)) > 1 \Leftrightarrow f_\theta(x) < c_1 f_{\theta_1}(x) + c_2 f_{\theta_2}(x). \end{aligned}$$

Hence by Lemma 3.3,

$$E_\theta(1 - \psi(X)) \le E_\theta(1 - \phi(X)),$$

whenever $E_{\theta_i}(1 - \psi(X)) = 1 - \alpha$. A similar result holds for $\theta > \theta_2$.

(4) Let ψ be any test of level α. Then $\beta_\psi(\theta_1) \le \alpha$, $\beta_\psi(\theta_2) \le \alpha$. Let ϕ be a test satisfying the conditions in part 1. Then, by part 2, $\beta_\phi(\theta) \ge \beta_\psi(\theta)$ for all $\theta \in (\theta_1, \theta_2)$. It suffices to show that ϕ is of size α. Let $\psi_1(x) \equiv \alpha$. By part 3, $\beta_\phi(\theta) \le \alpha$ for $\theta \in (-\infty, \theta_1) \cup (\theta_2, \infty)$. Hence ϕ has size α. □

Figure 3.3 illustrates the relation between the power functions of ϕ and ψ that appear in Theorem 3.5. This behavior is similar to that mentioned in Corollary 3.1.

3.4.6 UMPU tests for hypotheses I and II

Let us now turn to the UMPU tests for hypotheses I and II. First, consider hypothesis II.

Theorem 3.6 *Suppose that the density of X belong to the exponential family (3.13). Let $\theta_1 < \theta_2$ and $0 < \alpha < 1$. Then*

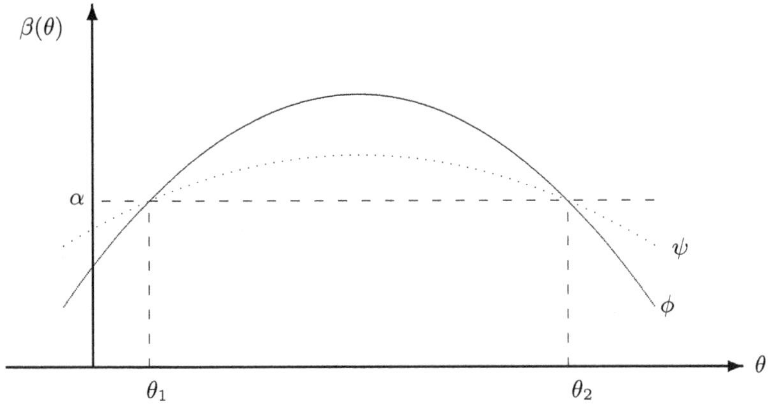

Fig. 3.3. Comparison of the power functions

(1) There exists a test of the form

$$\phi(x) = \begin{cases} 1 & \text{if } Y(x) < t_1 \text{ or } Y(x) > t_2 \\ \gamma_i & \text{if } Y(x) = t_i,\ i = 1,2 \\ 0 & \text{if } t_1 < Y(x) < t_2, \end{cases} \tag{3.33}$$

where $-\infty < t_1 < t_2 < \infty$ *that satisfies*

$$E_{\theta_1}\phi(X) = E_{\theta_2}\phi(X) = \alpha.$$

(2) Any test ϕ *that satisfies the conditions in part 1 is a UMPU-*α *for testing* $H_0 : \theta_1 \le \theta \le \theta_2$ *versus* $H_1 : \theta < \theta_1$ *or* $\theta > \theta_2$.

Proof. (1) Let ϕ^0 be a test that satisfies the conditions in part 1 of Theorem 3.5 with α in (3.30) replaced by $1-\alpha$. Then $\phi = 1 - \phi^0$ has the desired form.

(2) Let ψ be any unbiased test of size α. Because the power function $\beta_\psi(\theta)$ is continuous, we have $\beta_\psi(\theta_1) = \beta_\psi(\theta_2) = \alpha$. Let $\psi^0 = 1 - \psi$. Then $\beta_{\psi^0}(\theta_i) = 1 - \alpha$, $i = 1, 2$. Let ϕ be a test that satisfies the conditions in part 1. Then ϕ^0 is a test that satisfies the conditions in part 1 of Theorem 3.5 with α in (3.30) replaced by $1-\alpha$. By Theorem 3.5, $\beta_{\phi^0}(\theta) \le \beta_{\psi^0}(\theta)$ for all $\theta \in \Theta_1$ and $\beta_{\phi^0}(\theta) \ge \beta_{\psi^0}(\theta)$ for all $\theta \in \Theta_0$. Therefore $\beta_\phi(\theta) \ge \beta_\psi(\theta)$ for all $\theta \in \Theta_1$ and $\beta_\phi(\theta) \le \beta_\psi(\theta)$ for all $\theta \in \Theta_0$. Thus ϕ is a UMPU-α test. □

Next, consider hypothesis I.

Theorem 3.7 *Suppose that the density of* X *belongs to the exponential family (3.13). Let* $\theta_0 \in \Theta^0$ *and* $0 < \alpha < 1$. *Then*

(1) There is a test ϕ *of the form (3.33) such that*

$$E_{\theta_0}\phi(X) = \alpha \text{ and } E_{\theta_0}[Y(X)\phi(X)] = \alpha E_{\theta_0} Y(X). \tag{3.34}$$

(2) Any test ϕ that satisfies the conditions in part 1 is a UMPU-α test for testing $H_0: \theta = \theta_0$ versus $H_1: \theta \neq \theta_0$.

Proof. (1) By Lemma 3.10 there is a ϕ^0 of the form (3.27) with $-\infty \le t_1 < t_2 \le \infty$ such that (3.34) holds with α replace by $1-\alpha$. Exclude the possibilities of $t_1 = -\infty$ and $t_2 = \infty$ using the similar argument as in the proof of part 1 of Theorem 3.5. Then $\phi = 1 - \phi^0$ is the desired test.

(2) Suppose $\theta < \theta_0$. By the generalized Neyman-Pearson lemma, any test having the form

$$\phi'(x) = \begin{cases} 1 & f_\theta(x) > k_1 f_{\theta_0}(x) + k_2 \dot{f}_{\theta_0}(x) \\ 0 & f_\theta(x) < k_1 f_{\theta_0}(x) + k_2 \dot{f}_{\theta_0}(x) \end{cases} \tag{3.35}$$

that satisfies $\beta_{\phi'}(\theta_0) = \alpha$ and $\dot{\beta}_{\phi'}(\theta_0) = 0$ has maximum power at θ out of all tests ψ satisfying $\beta_\psi(\theta_0) = \alpha$ and $\dot{\beta}_\psi(\theta_0) = 0$. Thus it has maximum power at θ out of all unbiased tests of size α. We now show that any ϕ that satisfies the conditions in part 1 is of the above form for some k_1 and k_2.

Let a_1, a_2 be the solution to the system of equations

$$a_1 + a_2 t_i = e^{(\theta - \theta_0) t_i}, \ i = 1, 2.$$

Then,

$$\begin{aligned} a_1 &= \left(t_2 e^{(\theta-\theta_0)t_1} - t_1 e^{(\theta-\theta_0)t_2}\right)/(t_2 - t_1) \\ a_2 &= \left(e^{(\theta-\theta_0)t_2} - e^{(\theta-\theta_0)t_1}\right)/(t_2 - t_1). \end{aligned}$$

Clearly $a_1 > 0 > a_2$ and hence as in the proof of Theorem 3.5, it follows that

$$a_1 + a_2 t \begin{cases} < e^{(\theta-\theta_0)t} & \text{if } t < t_1 \text{ or } t > t_2 \\ > e^{(\theta-\theta_0)t} & \text{if } t_1 < t < t_2. \end{cases}$$

From the exponential family form (3.13) it can be deduced that there exist k_1, k_2 such that

$$\begin{aligned} f_\theta(x) > k_1 f_{\theta_0}(x) + k_2 \dot{f}_{\theta_0}(x) &\Leftrightarrow a_1 + a_2 Y(x) < e^{(\theta-\theta_0)Y(x)} \\ f_\theta(x) < k_1 f_{\theta_0}(x) + k_2 \dot{f}_{\theta_0}(x) &\Leftrightarrow a_1 + a_2 Y(x) > e^{(\theta-\theta_0)Y(x)}. \end{aligned}$$

It follows then that

$$\begin{aligned} f_\theta(x) > k_1 f_{\theta_0}(x) + k_2 \dot{f}_{\theta_0}(x) &\Leftrightarrow Y(x) < t_1 \text{ or } Y(x) > t_2 \\ f_\theta(x) < k_1 f_{\theta_0}(x) + k_2 \dot{f}_{\theta_0}(x) &\Leftrightarrow t_1 < Y(x) < t_2 \end{aligned}$$

Thus ϕ is of the form (3.35). A similar result holds when $\theta > \theta_0$. Thus we see that, for any unbiased test of size α, $E_\theta \phi(X) \ge E_\theta \psi(X)$ for all θ. Finally, since the null space Θ_0 is the singleton $\{\theta_0\}$, ϕ has size α. This completes the proof. □

Theorems 3.5 and 3.6 give the general forms of the UMP or UMPU tests for the two sided hypothesis. In practice, all we are left to determine are the constants $t_1, t_2, \gamma_1, \gamma_2$. For hypotheses III and II, these can be determined by

$$E_{\theta_1}\phi(X) = E_{\theta_2}\phi(X) = \alpha.$$

For hypothesis I, they can be determined by (3.34). If $Y(X)$ has a continuous distribution, then the values of γ_i are unimportant, and we usually take them to be 0 or 1. If discrete $Y(X)$, t_1 and t_2 are first determined so that the solution for γ_1 and γ_2 are between 0 and 1. Then γ_1 and γ_2 are determined. Typically numerical calculations are involved. The following example illustrate the procedure.

Example 3.8 Let $X_1, \ldots, X_n$ be i.i.d. $N(\theta, 1)$ random variables. The joint density of $X_1, \ldots, X_n$ is of the form (3.13), with $Y(X_1, \ldots, X_n) = \sum_{i=1}^n X_i$. For hypotheses III and II, t_1 and t_2 in (3.27) and (3.33) are determined by the following equations

$$P_{\theta_i}(Y < t_1) + P_{\theta_i}(Y > t_2) = \alpha, \; i = 1, 2.$$

Since, under θ, $Y \sim N(n\theta, n)$. The above equations reduce to

$$\Phi\left(\frac{t_1 - n\theta_i}{\sqrt{n}}\right) + 1 - \Phi\left(\frac{t_2 - n\theta_i}{\sqrt{n}}\right) = \alpha, \; i = 1, 2,$$

where Φ denotes the c.d.f. of a standard normal random variable. For example, if $n = 10$, $\theta_1 = 0$, $\theta_2 = 1$, and $\alpha = 0.05$. Then (t_1, t_2) is the solution to the following system of equations

$$\begin{cases} \Phi(t_1/\sqrt{10}) - \Phi(t_2/\sqrt{10}) + 0.95 = 0 \\ \Phi((t_1 - 10)/\sqrt{10}) - \Phi((t_2 - 10)/\sqrt{10}) + 0.95 = 0 \end{cases}$$

We can either solve this equation directly by, for example, the Newton-Raphson algorithm or simplify this equation using the specific symmetric structure of this problem. Note that $N(0, 10)$ is symmetric about 0 and $N(10, 10)$ is symmetric about 10, and the two distributions have the same variance. Therefore t_1 and t_2 are symmetrically placed about $t = 5$. In other words $t_1 = 5 - t_0$ and $t_2 = 5 + t_0$ for some t_0. Thus the above system of two equations reduce the following equation

$$\Phi\left(\frac{5 - t_0}{\sqrt{10}}\right) - \Phi\left(\frac{5 + t_0}{\sqrt{10}}\right) + 0.95 = 0.$$

Solving this equation numerically we find $t_0 \approx 10.18$. Thus, for testing III, we reject the H_0 if $-5.18 < Y < 15.18$, and for testing II, we reject H_0 if Y falls outside this region.

For testing I, t_1 and t_2 are determined by solving equation (3.34), which is $E_{\theta_0}(Y) = n\theta_0$ in this example. So (3.34) reduces to

$$\begin{aligned} &\Phi\left(\frac{t_1 - n\theta_0}{\sqrt{n}}\right) + 1 - \Phi\left(\frac{t_2 - n\theta_0}{\sqrt{n}}\right) = \alpha \\ &\int_{-\infty}^{t_1} t\frac{1}{\sqrt{n}}\varphi\left(\frac{t - n\theta_0}{\sqrt{n}}\right)dt + \int_{t_2}^{\infty} t\frac{1}{\sqrt{n}}\varphi\left(\frac{t - n\theta_0}{\sqrt{n}}\right)dt = \alpha n\theta_0, \end{aligned} \tag{3.36}$$

where, in the second equation, φ denotes the p.d.f. of a standard normal random variable.

Again, one can either solve these equations directly by a numerical methods or further explore the symmetric structure specific to this problem. Note that the second equation is equivalent to $\text{cov}_{\theta_0}(\phi, Y)$. Since ϕ is a function of Y we will write it as $\phi(Y)$. Since the distribution is symmetric about $n\theta_0$, this covariance is 0 if ϕ is symmetric about $n\theta_0$. Thus, if we take $t_1 = n\theta_0 - t_0$ and $t_2 = n\theta_0 + t_0$ for some $t_0 > 0$ then the second equation is automatically satisfied. Thus all we need to solve is the equation

$$\Phi\left(\frac{-t_0}{\sqrt{n}}\right) + 1 - \Phi\left(\frac{t_0}{\sqrt{n}}\right) = \alpha$$

The solution to this equation is $t_0 = \sqrt{n}\Phi^{-1}(1 - \alpha/2)$. That is, we reject the hypothesis H_0 in I if

$$Y > n\theta_0 + \sqrt{n}\Phi^{-1}(1 - \alpha/2) \quad \text{or} \quad Y > n\theta_0 - \sqrt{n}\Phi^{-1}(1 - \alpha/2).$$

For example, if $\theta_0 = 1$, $n = 10$, $\alpha = 0.05$. Then we reject the null hypothesis in I if $Y < 3.80$ or $Y > 16.20$. □

Example 3.9 Suppose $X_1, \ldots, X_n$ are i.i.d. with p.d.f.

$$f_\theta(x) = \theta x^{\theta - 1} I_{(0,1)}(x), \quad \text{where } \theta > 0,$$

and we are interested in testing hypothesis I with $\theta_0 = 1$ and $0 < \alpha < 1$. The joint p.d.f. of $X_1, \ldots, X_n$ belongs to the exponential family, and is given by

$$\theta^n e^{(\theta-1)Y(x_1,\ldots,x_n)}, \quad \text{where } Y(x_1, \ldots, x_n) = \sum_{i=1}^{n} \log x_i.$$

Let us derive the distribution of Y under $\theta_0 = 1$. We know that $Y = Y_1 + \cdots + Y_n$, where $-Y_i$ are i.i.d. Exp(1). Therefore $-Y_1 - \cdots - Y_n = -Y$ is distributed as Gamma$(n, 1)$. So Y has p.d.f.

$$(-t)^{n-1} e^t / \Gamma(n), \quad t < 0.$$

Since Y has a continuous distribution, we can ignore γ_1 and γ_2. The constants t_1, t_2 in (3.33) are determined by

$$\int_{-\infty}^{t_1} t^{n-1}e^t dt + \int_{t_2}^{0} t^{n-1}e^t dt = (-1)^{n-1}\Gamma(n)\alpha,$$
$$\int_{-\infty}^{t_1} (-t)^{n-1}te^t dt + \int_{t_2}^{0} (-t)^{n-1}te^t dt = \alpha \int_{-\infty}^{0} (-t)^{n-1}te^t dt.$$

Let Γ_n denote the cumulative distribution of a Gamma$(n, 1)$ random variable. The above equations can now be represented as

$$\Gamma_n(-t_1) + 1 - \Gamma_n(-t_2) = \alpha$$
$$\Gamma_{n+1}(-t_1) + 1 - \Gamma_{n+1}(-t_2) = \alpha.$$

where $t_1 < t_2 < 0$. One can then use a numerical method to find t_1 and t_2. For example, if $n = 10$, $\alpha = 0.05$, then $(t_1, t_2) \approx (-17.61, -4.98)$. □

Problems

3.1. Let $(\Omega, \mathcal{F}, P)$ be a probability space space and let $X : \Omega \to [a, b]$ be a random variable on $(\Omega, \mathcal{F}, P)$. Let $A \in \mathcal{F}$. Use the continuity of probability measure to show that $\rho(x) = P(\{X > x\} \cap A)$ is a right continuous function with $\rho(x-) = P(\{X \geq x\} \cap A)$. Show that for any $\alpha \in [\rho(a-), \rho(b)]$ there is an $x_0 \in [a, b]$ such that

$$P(\{X > x_0\} \cap A) \leq \alpha \leq P(\{X \geq x_0\} \cap A).$$

Here, the set A plays the role of $\{f_0 > 0\}$ in the proof of Lemma 3.1.

3.2. Let f and g be measurable functions on $(\Omega, \mathcal{F}, \mu)$. Let $A \in \mathcal{F}$. We say that $f = 0$ a.e. μ on A if $\mu(\{f \neq 0\} \cap A) = 0$. Show that $fg = 0$ a.e. μ implies $f > 0$ a.e. μ on $\{g \neq 0\}$.

3.3. Let f_0 and f_1 denote densities of uniform distribution on (0,1) and $\left(\frac{1}{2}, \frac{3}{2}\right)$ respectively. Find the most powerful test for $H_0 : f_0$ versus $H_1 : f_1$ for each $\alpha \in [0, 1]$.

3.4. Let X be a $b(2, q)$ (binomial) random variable and f_0 and f_1 denote the probability mass functions corresponding to $b(2, \frac{1}{2})$ and $b(2, \frac{2}{3})$. Find the most powerful test for $H_0 : q = \frac{1}{2}$ versus $H_1 : q = \frac{2}{3}$ for each $\alpha \in [0, 1]$.

3.5. Suppose X is a random variable defined on $(0, \infty)$. Find the most powerful test for the hypothesis

$$H_0 : f_0(x) = \frac{1}{2}e^{-x/2} \quad \text{versus} \quad H_1 : f_1(x) = e^{-x}$$

for each significance level $\alpha \in [0, 1]$.

3.6. Let $X_1, \ldots, X_n$ be an i.i.d. sample from the density $f_\theta(x)$. Construct UMPs test of size α when $f_\theta(x)$ takes the following forms.

1. $f_\theta(x) = \theta^{-1}e^{-x/\theta}, \quad x > 0, \ \theta > 0;$
2. $f_\theta(x) = \theta x^{\theta-1}, \quad 0 < x < 1, \ \theta > 0;$
3. $f_\theta(x) = \frac{1}{\sqrt{2\pi\theta}} \exp\left\{ -\frac{1}{2\theta}(x-1)^2 \right\}, \ -\infty < x < \infty, \ \theta > 0;$
4. $f_\theta(x) = 4\theta^{-4}x^3 e^{-(x/\theta)^4}, \quad x > 0, \ \theta > 0.$

3.7. Suppose X has density

$$f(x|\theta) = c(\theta)h(x)e^{\theta x}, \quad \theta \in \Theta$$

with respect to a measure μ. Here Θ is an open interval and $c(\theta) > 0$ for all $\theta \in \Theta$. Now let $\theta \in \Theta$. Show that there is an interval $(-a, a)$ on which the moment generating function $M_X(t) = E_\theta\left(e^{tX}\right)$ is finite, and, furthermore,

$$M_X(t) = c(\theta)/c(\theta + t).$$

3.8. Suppose that Y has density:

$$f_\theta(y) = c(\theta)h(y)e^{\eta(\theta)y}$$

for some one-to-one differentiable function η. Let $\phi(t)$ a function of t. Show that

$$(\partial/\partial\theta)E_\theta\phi(Y) = \dot{\eta}(\theta)\text{cov}_\theta(\phi(Y), Y),$$

where $\dot{\eta}(\theta)$ is the derivative of η with respect to θ.

3.9. Suppose X is a random variable having density f_θ where $\theta \in \Theta \subseteq \mathbb{R}$. Let π be a probability measure defined on Θ. For a $\theta_0 \in \Theta$, we are interested in testing the hypotheses $H_0 : \theta = \theta_0$ against the alternative $H_1 : \theta$ is distributed as π. Here, we have treated θ as random. Let us say that a test ϕ is best of size α if (i) $E_{\theta_0}\phi(X) = \alpha$, and (ii) for any other test ϕ' satisfying $E_{\theta_0}\phi'(X) \leq E_{\theta_0}\phi(X)$ we have

$$E_\pi E_\theta \phi(X) \geq E_\pi E_\theta \phi'(X),$$

where, for example, $E_\pi E_\theta\phi(X)$ is the integral $\int_\Theta E_\theta\{\phi(X)\}\pi(\theta)d\theta$. Show that any test of the form:

$$\phi(x) = \begin{cases} 1 & \text{if } \int_\Theta f_\theta(x)\pi(\theta)d\theta > kf_{\theta_0}(x) \\ 0 & \text{if } \int_\Theta f_\theta(x)\pi(\theta)d\theta < kf_{\theta_0}(x) \end{cases}$$

for some $k > 0$, is best of its size.

3.10. A test ϕ for testing $\theta = \theta_0$ versus $\theta > \theta_0$ is said to be local best of its size if, for any other test ϕ' with $\beta_{\phi'}(\theta_0) \leq \beta_\phi(\theta_0)$, we have

$$\dot{\beta}_\phi(\theta_0) \geq \dot{\beta}_{\phi'}(\theta_0).$$

In other words, the slope of the power at θ_0 is maximized. Now define

$$\phi_0(x) = \begin{cases} 1 & \text{if } \dot{f}_{\theta_0}(x) > k f_{\theta_0}(x) \\ \gamma(x) & \text{if } \dot{f}_{\theta_0}(x) = k f_{\theta_0}(x) \\ 0 & \text{if } \dot{f}_{\theta_0}(x) < k f_{\theta_0}(x) \end{cases}$$

where $f(x|\theta)$ denotes the density of X, $k \geq 0$, and $0 \leq \gamma(x) \leq 1$. Show that ϕ_0 is local best of its size. Here we assume that the derivative with respect to θ and the integration with respect to x can be exchanged.

3.11. Suppose that X is a random variable with pdf (with respect to μ) belonging to a parametric family $\{f_\theta : \theta \in \Theta\}$, where $\Theta \subseteq \mathbb{R}$. Consider the hypotheses $H_0 : \theta = \theta_0$ versus $H_1 : \theta \neq \theta_0$. Assume that $f_\theta(x)$ is twice differentiable with respect to θ, and that the derivative $\partial^k/\partial\theta^k$ can be moved inside the integral $\int \cdots d\mu(x)$. Let $\phi_0(x)$ be a test such that, for some k_1, k_2,

$$\phi_0(x) = \begin{cases} 1 & \text{if } \ddot{f}_{\theta_0}(x) > k_1 f_{\theta_0}(x) + k_2 \dot{f}_{\theta_0}(x) \\ \gamma(x) & \text{if } \ddot{f}_{\theta_0}(x) = k_1 f_{\theta_0}(x) + k_2 \dot{f}_{\theta_0}(x) \\ 0 & \text{if } \ddot{f}_{\theta_0}(x) < k_1 f_{\theta_0}(x) + k_2 \dot{f}_{\theta_0}(x) \end{cases}$$

and, moreover, k_1 and k_2 are so chosen that $\beta_{\phi_0}(\theta_0) = \alpha$ and $\dot{\beta}_{\phi_0}(\theta_0) = 0$. Show that, for any test ϕ that satisfies

$$\beta_\phi(\theta_0) = \alpha, \quad \dot{\beta}_\phi(\theta_0) = 0$$

we have

$$\ddot{\beta}_{\phi_0}(\theta_0) \geq \ddot{\beta}_\phi(\theta_0).$$

Comment on why such a test would be of interest.

3.12. Let (X, Y) be a bivariate random variable and, for simplicity, assume both components to be continuous. Moreover, assume that $E|Y| < \infty$. Let

$$\phi_0(x) = \begin{cases} 1 \text{ if } E(Y|x) > k \\ 0 \text{ if } E(Y|x) \leq k \end{cases}, \quad k \geq 0.$$

Show that for any function $\phi(x)$ that satisfies $0 \leq \phi(x) \leq 1$ and $E\phi(X) \leq E\phi_0(X)$, we have $E\{Y\phi(X)\} \leq E\{Y\phi_0(X)\}$.

3.13. Consider the simple versus simple hypotheses $H_0 : P$ versus $H_1 : Q$, where P and Q are probability measures with densities f and g with respect to a measure. Let ϕ_1 and ϕ_2 be two tests such that

1. $\int \phi_1 f d\mu \geq \int \phi_2 f d\mu$
2. For some $k \geq 0$, $\phi_1 \geq \phi_2$ whenever $g > kf$ and $\phi_1 \leq \phi_2$ whenever $g < kf$.

Show that $\int \phi_1 g d\mu \geq \int \phi_2 g d\mu$.

3.14. Show that the logistic distribution with location parameter θ, having density

$$f(x|\theta) = \frac{e^{x-\theta}}{(1+e^{x-\theta})^2}, \quad x \in \mathbb{R}, \quad \theta \in \mathbb{R},$$

has monotone likelihood ratio. Write down the general form of UMPU-α test for testing $H_0 : \theta < \theta_0$ against $H_1 : \theta \geq \theta_0$.

3.15. Let X_i be independent and distributed as $N(i\theta, 1)$, $i = 1, ..., n$. Show that there exists a UMP test of $H_0 : \theta \leq 0$ against $H_1 : \theta > 0$, and determine the test for a given α.

3.16. Suppose that F is a cdf and $0 < \omega < 1$. Show that

$$\inf\{y : F(y) \geq \omega\} = \sup\{y : F(y) < \omega\}.$$

So $F^{-1}(\omega)$ can be equivalently defined as either side of this equality. Give an example in which

$$\inf\{y : F(y) \geq \omega\} \neq \inf\{y : F(y) > \omega\}$$

for some ω.

3.17. Let X be distributed as Gamma$(\theta, 1)$; That is, its p.d.f. is given by

$$f_X(x;\theta) = \frac{1}{\Gamma(\theta)} x^{\theta-1} e^{-x}, \quad x > 0, \quad \theta > 0.$$

Suppose $0 < \alpha < 1$.

i. Derive the UMP-α test for $H_0 : \theta \leq 1$ versus $H_1 : \theta > 1$.
ii. Derive the UMP-α test for $H_0 : \theta \leq 1$ or $\theta \geq 2$ versus $H_1 : 1 < \theta < 2$.
iii. Derive the UMPU-α test for $H_0 : \theta = 1$ versus $H_1 : \theta \neq 1$.
iv. Derive the UMPU-α test for $H_0 : 1 \leq \theta \leq 2$ versus H_1: $\theta < 1$ or $\theta > 2$.

3.18. Suppose X has a binomial distribution $b(7, p)$. Let $\alpha = 0.20$. Construct the following tests:

i. The UMP-α test for $H_0 : p \leq 0.5$ against $H_1 : p > 0.5$;
ii. The UMPU-α test for $H_0 : 0.25 \leq p \leq 0.75$ against $H_1 : p < 0.25$ or $p > 0.75$.
iii. The UMPU-α test for $H_0 : p = 0.5$ against $H_1 : p \neq 0.5$.

3.19. Suppose that X is distributed as Exp(θ); that is,

$$f_\theta(x) = \frac{1}{\theta} e^{-x/\theta} I_{[0,\infty)}(x), \quad \theta > 0.$$

i. Find UMP-α test for testing $H_0 : \theta \leq \theta_0$ versus $H_1 : \theta > \theta_0$.
ii. Find UMPU-α test for testing $H_0 : \theta \in [\theta_0, 2\theta_0]$ versus $H_1 : \theta \notin [\theta_0, 2\theta_0]$.
iii. Find the UMPU-α test for testing $H_0 : \theta = \theta_0$ versus $H_1 : \theta \neq \theta_0$.

3.20. Mary needs to interview 5 people. Suppose each person (independently) agrees to be interviewed with probability θ, and let X be the minimal number of people she needs to ask in order to obtain the 5 interviews she needs.

i. Derive the probability mass function of X.
ii. Show that X has an exponential family distribution.
iii. Find the UMP-α test for the hypothesis $H_0 : \theta \leq 1/2$ versus $H_1 : \theta > 1/2$, where $\alpha = 0.05$.

3.21. Suppose that X is a random variable with density belonging to a parametric family $\{f_\theta : \theta \in \Theta\}$, where $\Theta \subset \mathbb{R}$. Suppose that this family has monotone likelihood ratio with respect to $Y(X)$. We are interested in testing the one-sided hypothesis $H_0 : \theta \leq \theta_0$ versus $H_1 : \theta > \theta_0$. Suppose that ϕ_1 and ϕ_2 are two tests satisfying the following conditions:

a. They are both functions of a statistic $Y(X)$ and they are monotone nondecreasing in $Y(X)$;
b. $E_{\theta_0}\phi_1(Y(X)) = E_{\theta_0}\phi_2(Y(X)) = \alpha$;
c. There is a k such that whenever $Y(x) > k$, $\phi_1(Y(x)) \geq \phi_2(Y(x))$ and whenever $Y(x) < k$, $\phi_1(Y(x)) \leq \phi_2(Y(x))$.
d. The family $\{f_\theta : \theta \in \Theta\}$ has a common support; that is, $\{x : f_\theta(x) > 0\}$ is the same set for all $\theta \in \Theta$.

Show that

i. $\beta_{\phi_1}(\theta) \geq \beta_{\phi_2}(\theta)$ for all $\theta \geq \theta_0$;
ii. $\beta_{\phi_1}(\theta) \leq \beta_{\phi_2}(\theta)$ for all $\theta \leq \theta_0$;
iii. Both tests are of size α;
iv. Both tests are unbiased.

References

Allen, S. G. (1953). A class of minimax tests for one-sided composite hypothesies. *The Annals of Mathematical Statistics.* **24**, 295–298.

Ferguson, T. S. (1967). *Mathematical Statistics: A Decision Theoretic Approach.* Academic.

Karlin, S. and Rubin, H. (1956). The theory of decisin procedures for distributions with monotone likelihood ratio. *Annals of Mathematical Statistics*, **27**, 272–299.

Lehmann, E. L. and Romano, J. P. (2005). *Testing Statistical Hypotheses*. Third edition. Springer.

Neyman, J. and Pearson, E. S. (1933). On the problem of the most efficient tests of statistical hypotheses. *Philosophy Transaction of the Royal Society of London. Series A*. **231**, 289–337.

Neyman, J. and Pearson, E. S. (1936). Contributions to the theory of testing statistical hypothesis. *Statistical Research Memois*, **1**, 1–37.

Pfanzagl, J. (1967). A techincal lemma for monotone likelihood ratio families. *The Annals of Mathematical Statistics*, **38**, 611–612.

4

Testing Hypotheses in the Presence of Nuisance Parameters

In this chapter we consider the hypothesis tests where more than one parameter is involved. Let $\mathcal{P} = \{P_\theta : \theta \in \Theta\}$ be a parametric family of distributions of a random vector X, where Θ is a subset of $\mathbb{R}^k$. The families $\mathcal{P}_0$ and $\mathcal{P}_1$ are $\{P_\theta : \theta \in \Theta_0\}$ and $\{P_\theta : \theta \in \Theta_1\}$ where $\{\Theta_0, \Theta_1\}$ is a partition of Θ. Thus the type of hypotheses we are concerned with is

$$H_0 : \theta \in \Theta_0 \quad \text{vs} \quad H_1 : \theta \in \Theta_1.$$

More specifically, we will discuss how to construct test for one of the k parameters that has optimal power for that parameter and all the rest of the parameters. Without loss of generality, we assume that component to be the first component, θ_1, of θ. For example, a typical hypothesis we will consider in this chapter is

$$H_0 : \theta_1 = a \quad \text{vs} \quad H_1 : \theta_1 \neq a. \tag{4.1}$$

where a is a specific value of θ_1. In this setting

$$\Theta_0 = \{\theta \in \Theta : \theta_1 = a\}, \quad \Theta_1 = \{\theta \in \Theta : \theta_1 \neq a\}.$$

Our goal is to find a test that maximizes the power over all Θ_1 among a reasonably wide class of tests. The component of θ to be tested is called the parameter of interest; the rest of the components are called the nuisance parameters. Tests such as (4.1) are called hypothesis tests in the presence of nuisance parameters.

Because of the uniform nature of this optimality, it is possible only under rather restrictive conditions: unlike in the single parameter case, even the one-sided hypothesis does not in general permit a UMP test. Thus in this chapter we will be dealing with UMPU tests.

In our exposition we will frequently need to divide a vector $(c_1, c_2, \ldots, c_p)$ as its first component c_1 and the rest of the components $(c_2, \ldots, c_p)$. To simplify notation we use $c_{2:p}$ to represent $(c_2, \ldots, c_p)$.

For more information on this topic, see Ferguson (1967); Lehmann and Casella (1998).

B. Li and G. J. Babu, *A Graduate Course on Statistical Inference*,
Springer Texts in Statistics, https://doi.org/10.1007/978-1-4939-9761-9_4

4.1 Unbiased and Similar tests

In Section 3.4.1 we have already noticed that, if ϕ is an unbiased test and if the function $P \mapsto \beta_\phi(P)$ is continuous, then the power function $\beta_\phi(P)$ is constant on the boundary $\bar{\mathcal{P}}_0 \cap \bar{\mathcal{P}}_1$. Under a parametric model the power function is $\beta_\phi(P_\theta)$, is abbreviated as $\beta_\phi(\theta)$. The next proposition is the parametric counterpart of Proposition 3.2.

Proposition 4.1 *If $\beta_\phi(\theta)$ is continuous in θ and ϕ is unbiased then $\beta_\phi(\theta)$ is constant on $\Theta_B = \bar{\Theta}_0 \cap \bar{\Theta}_1$.*

Thus, under the continuity of the power function, the class of all tests whose powers are constant on Θ_B contains the class of unbiased test. This means if we can find UMP test among the latter class, then we can find the UMP test among unbiased tests.

Definition 4.1 *A test ϕ satisfying $\beta_\phi(\theta) = \alpha$ on Θ_B is an α-similar test. A test ϕ is called uniformly most powerful α-similar (UMP α-similar) test for testing (4.1), if*

1. *ϕ is α-similar;*
2. *for any α-similar test ψ, $\beta_\phi(\theta) \geq \beta_\psi(\theta)$ for all $\theta \in \Theta_1$.*

If we let $\mathcal{U}_\alpha$ be the class of all unbiased tests of size α, and $\mathcal{S}_\alpha$ be the class of all α-similar tests. By Proposition 4.1, if β_ϕ is continuous for all ϕ, then $\mathcal{U}_\alpha \subseteq \mathcal{S}_\alpha$. Hence, if ϕ is UMP α-similar, then it is also UMPU test of size α as long as we can guarantee ϕ itself is in $\mathcal{U}_\alpha$. This is proved in the next theorem.

Theorem 4.1 *If the power function is continuous in θ for every test, and if ϕ is UMP α-similar test for testing (4.1) with size α, then ϕ is a UMPU-α test.*

Proof. A UMP α-similar test ϕ is unbiased because the power $\beta_\phi(\theta) \geq \beta_\psi(\theta) = \alpha$ for all $\theta \in \Theta_1$, for the α-similar test $\psi(x) \equiv \alpha$. By assumption, ϕ has size α. Therefore $\phi \in \mathcal{U}_\alpha$. □

In Section 2.2, we introduced the notions of sufficient, complete, and boundedly complete statistics. In the parametric context, it takes the following form. A statistic T is sufficient for a subset $A \subseteq \Theta$ if, for any $B \in \mathcal{F}_X$, there is a function κ_B such that

$$P_\theta(B|T) = \kappa_B \quad [P_\theta] \quad \text{for all } \theta \in A.$$

A statistic T is complete for A if, for any $\sigma(T)$-measurable function g,

$$\int g dP_\theta = 0 \quad \text{for all } \theta \in A \;\Rightarrow\; g = 0 \; [P_\theta] \text{ for all } \theta \in A. \tag{4.2}$$

A statistic T is boundedly complete for A if the above implication holds for all bounded $\sigma(T)$-measurable function g.

The basic idea underlying the construction of UMPU tests in the presence of nuisance parameters is the following. Let Λ_1 denote the parameter space of θ_1; that is,

$$\Lambda_1 = \{\theta_1 : (\theta_1, \theta_{2:p}) \in \Theta\}.$$

Suppose we want to test the hypotheses in (4.1) and suppose there is a sufficient statistic S for $\theta_{2:p}$. Then the conditional distribution $P_\theta(\cdot|S)$ depends only on θ_1. So let us write it as $P_{\theta_1}(\cdot|S)$. If this conditional distribution belongs to a one-parameter exponential family, then we can construct UMP (or UMPU) tests for the one-parameter family

$$\{P_{\theta_1}(\cdot|S) : \theta_1 \in \Lambda_1\}.$$

using the mechanism studied in Chapter 3. However, the optimal power derived in this way is in terms of the conditional distribution $P_{\theta_1}(\cdot|S)$, not the original unconditional distribution P_θ of X. To link this conditional optimality to the original unconditional optimality we assume that S is boundedly complete for $\theta_{2:p}$, and employ the notion of Neyman structure, which in some sense aligns a conditional distribution with an unconditional distribution through bounded completeness.

To do so, we first introduce the concept of Neyman structure that is closely tied to bounded completeness. Let α be a value in $(0, 1)$, and T is a statistic.

Definition 4.2 *A test $\phi(X)$ is said to have an α-Neyman structure with respect to a statistic T if*

$$E_\theta(\phi(X)|T) = \alpha$$

almost everywhere P_θ for all $\theta \in \Theta_B$.

Obviously, if a test ϕ has an α-Neyman structure with respect to S, then it is α-similar, because, for all $\theta \in \Theta_B$,

$$E_\theta \phi(X) = E_\theta E_\theta(\phi(X)|S) = \alpha.$$

The next theorem shows that, if S is sufficient and boundedly complete, then the reversed implication is true.

Theorem 4.2 *If S is sufficient and boundedly complete for $\theta \in \Theta_B$, then every test α-similar on Θ_B has an α-Neyman structure.*

Proof. Let ϕ be an α-similar test. Then $E_\theta(\phi(X)) = \alpha$ for all $\theta \in \Theta_B$, which implies

$$E_\theta\{E[\phi(X)|S]\} = \alpha$$

for all $\theta \in \Theta_B$. By sufficiency of S for $\theta \in \Theta_B$, the conditional expectation $E_\theta[\phi(X)|S]$ does not depend on $\theta \in \Theta_B$. Hence we write $E[\phi(X)|S]$ instead of $E_\theta[\phi(X)|S]$ for $\theta \in \Theta_B$. The above equality can be equivalently written as

$$E_\theta\{E[\phi(X)|S] - \alpha\} = 0$$

for all $\theta \in \Theta_B$. Since S is boundedly complete and since $E[\phi(X)|X] - \alpha$ is a bounded function of S, we have

$$E[\phi(X)|S] - \alpha = 0$$

for all $\theta \in \Theta_B$, which means that ϕ has an α-Neyman structure with respect to S. □

Sometimes we encounter the situations where Θ_B is the union of finite number of sets, and S is not complete sufficient for the whole boundary Θ_B but rather for each member of the union. More specifically, for any $a \in \Lambda_1$, let

$$\Theta(a) = \{\theta \in \Theta : \theta_1 = a\}.$$

Let $a_1, \ldots, a_s \in \Lambda_1$. Suppose Θ_B is of the form

$$\Theta_B = \cup_{r=1}^{s} \Theta(a_r) \tag{4.3}$$

and S is complete and sufficient relative to each $\Theta(a_r)$ but not necessarily on Θ_B. In this case the conclusion of Theorem 4.2 still holds, as shown in the following corollary.

Corollary 4.1 *If S is sufficient and boundedly complete for each $\Theta(a_r)$, $r = 1, \ldots, s$, then any test that is α-similar on Θ_B has an α-Neyman structure with respect to S.*

Proof. Suppose ϕ is α-similar on Θ_B. Then it is α-similar on each $\Theta(a_r)$. Because S is sufficient and complete on each $\Theta(a_r)$, by Theorem 4.2 we have

$$E_\theta[\phi(X)|S] = \alpha$$

almost everywhere P_θ all for $\theta \in \Theta(a_r)$ and for all $r = 1, \ldots, s$. This means

$$E_\theta[\phi(X)|S] = \alpha \tag{4.4}$$

almost everywhere P_θ for all $\theta \in \Theta_B$. Hence ϕ has an α-Neyman structure with respect to S. □

Note that the equality (4.4) justifies writing $E_\theta[\phi(X)|S]$ as $E[\phi(X)|S]$ in this setting as the former does not depend on θ when θ varies over Θ_B.

Thus, if we let $\mathcal{N}_\alpha$ to be the class of all tests with α-Neyman structure, then we have the following relations among the classes of unbiased tests of size α, α-similar tests, and tests with α-Neyman structures:

1. $\mathcal{U}_\alpha \subseteq \mathcal{S}_\alpha$ under condition (A),
2. $\mathcal{S}_\alpha \subseteq \mathcal{N}_\alpha$ under condition (B),
3. $\mathcal{N}_\alpha \subseteq \mathcal{S}_\alpha$,

where (A) and (B) are the conditions:

(A) the power function $\beta_\phi(\theta)$ is continuous for all tests ϕ;
(B) the set Θ_B is the union (4.3), and S is sufficient and boundedly complete for each $\Theta(a_r)$.

Thus, under conditions (A) and (B), $\mathcal{U}_\alpha \subseteq \mathcal{N}_\alpha = \mathcal{S}_\alpha$. It turns out that finding the UMPU test among $\mathcal{N}_\alpha$ is inherently a one-parameter problem. In the next section we investigate under what circumstances does there exist a sufficient and boundedly complete statistic S for $\theta_{2:p}$.

4.2 Sufficiency and completeness for a part of the parameter vector

In section 2.1.3 we introduced the exponential family for a single random variable X. We now extend it to multiple random variables. Let $X = (X_1, \dots, X_n)$, where $X_1, \dots, X_n$ are i.i.d., with each X_i having its distribution in $\mathfrak{E}_p(t_0, \mu_0)$, where $t_0 : \Omega_{X_1} \to \mathbb{R}^p$ and μ_0 is a σ-finite measure on Ω_{X_1}, a subset of $\mathbb{R}$. For brevity, we write this as

$$X \sim \mathfrak{E}_p^n(t_0, \mu_0).$$

The joint density of $(X_1, \dots, X_n)$ with respect to the product measure $\mu = \mu_0 \times \cdots \times \mu_0$ is

$$\begin{aligned}&\prod_{i=1}^n e^{\theta^T t_0(x_i)} / \int e^{\theta^T t_0(x_i)} d\mu_0(x_i)\\ &= e^{\theta^T \sum_{i=1}^n t_0(x_i)} / [\int e^{\theta^T t_0(x_i)} d\mu_0(x_i)]^n.\end{aligned}$$

Let

$$t(x_1, \dots, x_n) = \sum_{i=1}^n t_0(x_i). \tag{4.5}$$

Then the joint density of $(X_1, \dots, X_n)$ with respect to μ can be written as

$$e^{\theta^T t(x_1, \dots, x_n)} / \int e^{\theta^T t(x_1, \dots, x_n)} d\mu(x_1, \dots, x_n).$$

Using essentially the same proof as that of Theorem 2.8, we can establish the following theorem.

Theorem 4.3 *Suppose* $(X_1, \dots, X_n) \sim \mathfrak{E}_p^n(t_0, \mu_0)$, *and* Θ *has a nonempty interior. Then* $t(X_1, \dots, X_n)$ *is complete and sufficient statistic for* Θ.

For the rest of the section, the symbol X will be used to denote an i.i.d. sample $(X_1, \ldots, X_n)$.

An important property about the exponential family is that it is closed under conditioning and marginalization: that is, the conditional distributions and marginal distributions derived from the components of T also belong to the exponential family. This fact allows us to extend Theorem 2.8 the marginal and conditional distributions, so that we can speak of complete and sufficient statistics for a part of the parameter θ. This is crucial for reducing a multi-parameter problem to a one-parameter problem, so that we can use the results from Chapter 3 to tackle the new problems in this chapter.

We first introduce a lemma. Let (U, V) be a pair of random vectors defined on $(\Omega_U \times \Omega_V, \mathcal{F}_U \times \mathcal{F}_V)$, and let P and Q be the two distributions of (U, V) with $P \ll Q$. Let P_U and $P_{V|U}$ be the marginal distribution of U and conditional distribution of $V|U$ under P; let Q_U and $Q_{V|U}$ be the marginal distribution of U and conditional distribution of $V|U$ under Q.

Lemma 4.1 *Let P be the probability measure defined by*

$$dP = a(u)b(v)dQ$$

where a and b are nonnegative functions such that $\int a(u)b(v)dQ(u,v) = 1$. Then

$$\text{(a)} \qquad dP_U(u) = a(u)\left(\int b(v)dQ_{V|U}(v|u)\right)dQ_U(u),$$

$$\text{(b)} \qquad dP_{V|U}(v|u) = \frac{b(v)dQ_{V|U}(v|u)}{\int b(v')dQ_{V|U}(v'|u)}.$$

Proof. (a) By Theorem 1.21 we have

$$P{\circ}U^{-1} \ll Q{\circ}U^{-1}, \quad d(P{\circ}U^{-1})/d(Q{\circ}U^{-1}) = E[a(U)b(V)|U = u].$$

Hence

$$dP_U/dQ_U = a(u)E(b(V)|U),$$

which proves (a).

(b) Let $B \in \mathcal{F}_U \times \mathcal{F}_V$. Let P^* be the measure defined by $dP^* = I_B dP$. Then, by Theorem 1.21,

$$P^* \ll P, \quad P(B|U) = d(P^*{\circ}U^{-1})/d(P{\circ}U^{-1}).$$

Also, by Theorem 1.21,

$$\begin{aligned} d(P^*{\circ}U^{-1}) &= E_Q(I_B(U,V)a(U)b(V)|U)d(Q{\circ}U^{-1}) \\ d(P{\circ}U^{-1}) &= E_Q(a(U)b(V)|U)d(Q{\circ}U^{-1}). \end{aligned}$$

So

$$P(B|U) = \frac{E_Q(I_B(U,V)a(U)b(V)|U)}{E_Q(a(U)v(V)|U)} = \frac{E_Q(I_B(U,V)b(V)|U)}{E_Q(b(V)|U)}.$$

In particular, for any $B \in \mathcal{F}_V$ we have

$$E(I_B(V)|U) = E(I_{\Omega_U \times B}(U,V)|U) = \frac{E_Q(I_B(V)b(V)|U)}{E_Q(b(V)|U)}.$$

Another way of writing this is

$$P_{V|U}(B|u) = \int_B \frac{b(v)}{E_Q(b(V)|U)} dQ_{V|U}(v|u),$$

which is the desired equality. □

We now apply this lemma to show that the marginal and conditional distributions associated with an exponential family remain to be from an exponential family. This result is important in establishing the Neyman Structure. Suppose $t(X)$ is partitioned in $t(X) = (u(X), v(X))$ and, correspondingly, θ is partitioned into (η, ξ), so that

$$\theta^T t(X) = \eta^T u(X) + \xi^T v(X).$$

Let U, V denote the random vectors $u(X)$ and $v(X)$ and let u, v denote the specific values of U, V. Recall that μ is the n-fold product measure $\mu_0 \times \cdots \times \mu_0$.

Theorem 4.4 *Suppose $X = (X_1, \ldots, X_n) \sim \mathfrak{E}_p^n(t_0, \mu_0)$. Then:*

(a) For each fixed ξ, there is measure μ_ξ on $(\Omega_U, \mathcal{F}_U)$ such that

$$dP_U/d\mu_\xi = e^{\eta^T u} / \textstyle\int e^{\eta^T u} d\mu_\xi(u)$$

(b) For each fixed u, there is a measure μ_u on $(\Omega_V, \mathcal{F}_V)$ such that

$$dP_{V|U}(v|u)/\mu_u = e^{\xi^T v} / \textstyle\int e^{\xi^T v} dQ_{V|U}(v|u).$$

Proof. (a). Let $\nu = \mu \circ (u,v)^{-1}$. Then, the density of (U, V) with respect to ν is

$$f_\theta(u,v) = e^{\eta^T u + \xi^T v} / \textstyle\int e^{\eta^T u + \xi^T v} d\nu(u,v).$$

Let $\theta_0 = (\eta_0, \xi_0) \in \Theta$, $\theta = (\eta, \xi) \in \Theta$. Define P and Q by

$$dQ(u,v) = f_{\theta_0}(u,v) d\nu(u,v), \quad dP(u,v) = f_\theta(u,v) d\nu(u,v).$$

Let Q_U and P_U be the marginal distributions of U under Q and P, respectively, and let $Q_{V|U}$ and $P_{V|U}$ be the conditional distributions of $V|U$ under Q and P, respectively. Then,

$$\begin{aligned} dP(u,v) &= f_\theta(u,v)d\nu(u,v) \\ &= \frac{f_\theta(u,v)}{f_{\theta_0}(u,v)} f_{\theta_0}(u,v)d\nu(u,v) \\ &= c(\theta)e^{(\eta-\eta_0)^T u} e^{(\xi-\xi_0)^T v} dQ(u,v), \end{aligned}$$

where

$$c(\theta) = \frac{\int e^{\eta_0^T u + \xi_0^T v} d\nu(u,v)}{\int e^{\eta^T u + \xi^T v} d\nu(u,v)}.$$

Then, by Lemma 4.1,

$$dP_U(u) = c(\theta)e^{\eta^T u} e^{-\eta_0^T u} \left(\int e^{(\xi-\xi_0)^T v} dQ_{V|U}(v|u) \right) dQ_U(u).$$

The assertion of part (a) follows if we let $c_\xi(\eta) = c(\theta)$ and

$$d\mu_\xi(u) = e^{-\eta_0^T u} \left(\int e^{(\xi-\xi_0)^T v} dQ_{V|U}(v|u) \right) dQ_U(u).$$

(b) By Lemma 4.1 again,

$$\begin{aligned} dP_{V|U}(v|u) &= \frac{e^{\xi^T v} e^{-\xi_0^T v} dQ_{V|U}(v|u)}{\int e^{(\xi-\xi_0)^T v} dQ_{V|U}(v|u)} \\ &= \frac{e^{\xi^T v}}{\int e^{\xi^T v} dQ_{V|U}(v|u)} \frac{\int e^{\xi^T v} dQ_{V|U}(v|u) e^{-\xi_0^T v} dQ_{V|U}(v|u)}{\int e^{(\xi-\xi_0)^T v} dQ_{V|U}(v|u)}. \end{aligned}$$

Now let

$$d\mu_u(v) = \frac{\int e^{\xi^T v} dQ_{V|U}(v|u) e^{-\xi_0^T v} dQ_{V|U}(v|u)}{\int e^{(\xi-\xi_0)^T v} dQ_{V|U}(v|u)}$$

to complete the proof. □

In the next few sections, we will be concerned with the case where $\eta = \theta_1$ and $\xi = \theta_{2:p}$. The above theorem, together with Theorem 2.8, implies that, when θ_1 is fixed at a, the statistic $t_{2:p}(X)$ is sufficient and complete for $\theta_{2:p}$. For easy reference we summarize this result as a corollary. The proof follows directly from Theorem 4.4 and Theorem 2.8.

Corollary 4.2 *Suppose $X = (X_1, \ldots, X_n) \sim \mathfrak{E}_p^n(t_0, \mu_0)$. Let a be a real number such that*

1. *$(a, \theta_{2:p}) \in \Theta$;*
2. *the set $\{\theta_{2:p} : (a, \theta_{2:p}) \in \Theta\}$ has a nonempty interior in $\mathbb{R}^{p-1}$.*

Then $t_{2:p}(X)$ is sufficient and complete for $\Theta(a)$.

We will frequently need to make assumptions similar to 2 in the corollary. To simplify discussion, let Ψ be the collection of all $\theta_1 \in \mathbb{R}$ for which there exists a $\theta_{2:p} \in \mathbb{R}^{p-1}$ such that $(\theta_1, \theta_{2:p})$ is an interior point of Θ. Obviously, if the interior of Θ is nonempty, then Ψ is nonempty. Furthermore, it is easy to see that, for every $\theta_1 \in \Psi$, $\{\theta_{2:p} : (\theta_1, \theta_{2:p}) \in \Theta\}$ has a nonempty interior in $\mathbb{R}^{p-1}$.

4.3 UMPU tests in the presence of nuisance parameters

In this section we construct the UMPU tests for several hypothesis about θ_1 in the presence of the nuisance parameters $\theta_{2:p}$. First, let us show that we can restrict our attention exclusively on those tests that depend on sufficient statistics so long as the purpose is to maximize the power.

Lemma 4.2 *Suppose $T = t(X_1, \ldots, X_n)$ is sufficient for Θ. Then, for any test $\phi(X)$, there exists a $\psi \circ t(X)$ such that*

$$\beta_\phi(\theta) = \beta_{\psi \circ t}(\theta) \quad \text{for all } \theta \in \Theta. \tag{4.6}$$

Proof. The test $\psi \circ t(X) = E[\phi(X)|T]$ satisfies the asserted property. □

For the rest of this section, any test we consider will be a function of a sufficient statistic.

Theorem 4.5 *Suppose*

1. *$T = t(X_1, \ldots, X_n)$ is a sufficient statistic for Θ,*
2. *Θ_B can be written as $\cup_{r=1}^{s} \Theta(a_r)$ such that $T_{2:p}$ is sufficient and complete on each $\Theta(a_r)$,*
3. *The power function $\beta_\phi(\theta)$ is continuous in θ for all tests ϕ.*

Moreover, suppose that there is a test $\phi_0(T)$ such that

4. *ϕ_0 has α-Neyman structure with respect to $T_{2:p}$,*
5. *$E_\theta(\phi_0(T)|T_{2:p}) \le \alpha$ $[P_\theta]$ for all $\theta \in \Theta_0$,*
6. *For any test $\phi(T)$ with α-Neymen structure with respect to $T_{2:p}$, we have*

$$E_\theta(\phi_0(T)|T_{2:p}) \ge E_\theta(\phi(T)|T_{2:p}) \quad [P_\theta] \quad \text{for all } \theta \in \Theta_1. \tag{4.7}$$

Then ϕ_0 is a UMPU-α test for testing $H_0 : \theta \in \Theta_0$ versus $H_1 : \theta \in \Theta_1$.

Proof. Let $\phi(T)$ be an unbiased test of size α. By assumption 2 and Corollary 4.2, ϕ has α-Neyman structure with respect to $T_{2:p}$. Hence, by assumption 6, (4.7) holds. Then

$$E_\theta(\phi_0(T)) \ge E_\theta(\phi(T))$$

for all $\theta \in \Theta_1$. Moreover, by assumptions 4 and 5

$$\begin{aligned} E_\theta(\phi_0(T)) &\le \alpha \quad \text{for all } \theta \in \Theta_0 \\ E_\theta(\phi_0(T)) &= \alpha \quad \text{for all } \theta \in \Theta_B. \end{aligned} \tag{4.8}$$

The first inequality implies

$$\sup_{\theta \in \Theta_0} E_\theta(\phi_0(T)) \le \alpha.$$

The second equality in (4.8) and continuity of power (assumption 3) imply there is a sequence $\{\theta_k\}$ such that $\phi_{\phi_0}(\theta_k) \to \alpha$. Hence

$$\sup_{\theta \in \Theta_0} E_\theta(\phi_0(T)) = \alpha.$$

So ϕ_0 has size α. Finally, let $\psi(T) \equiv \alpha$. Then by assumption 6,

$$E_\theta(\phi_0(T)|T_{2:p}) \ge \alpha \quad \text{for all } \theta \in \Theta_1,$$

which, combined with the first line in (4.8), implies ϕ_0 is unbiased. □

This theorem provides us with a general guideline to construct UMPU tests in the presence of nuisance parameters using the techniques developed for developing UMP or UMPU tests for a single parameter. In the following we will consider four types of hypotheses, with different Θ_0, Θ_1, and Θ_B.

UMPU one-sided tests

We first consider the hypothesis

$$H_0 : \theta_1 \le a \quad \text{vs} \quad H_1 : \theta_1 > a. \tag{4.9}$$

In this case

$$\Theta_0 = \{\theta : \theta_1 \le a\}, \quad \Theta_1 = \{\theta : \theta_1 > a\}, \quad \Theta_B = \{\theta : \theta_1 = a\}.$$

Since $X \sim \mathfrak{E}_p^n(t_0, \mu_0)$, T is sufficient for Θ, $T_{2:p}$ is complete and sufficient for Θ_B, and the power function is continuous for every ϕ. Thus conditions 1 and 2 in Theorem 4.5 are satisfied.

Let $\alpha \in (0, 1)$. Let $t_{2:p}$ be an arbitrary member of $\Omega_{T_{2:p}}$. By Theorem 3.2, as applied to the conditional family $\{P_{T_1|T_{2:p}}(t_1|t_{2:p}; \theta_1) : \theta_1 \in \Psi\}$, there is a test

$$\phi_0(t) = \begin{cases} 1 & \text{if } t_1 > k(t_{2:p}) \\ \gamma(t_{2:p}) & \text{if } t_1 = k(t_{2:p}) \\ 0 & \text{if } u < k(t_{2:p}) \end{cases}, \quad E_{\theta_1 = a}(\phi_0(T)|T_{2:p} = t_{2:p}) = \alpha \tag{4.10}$$

such that, for any test $\phi(T)$ with $E_{\theta_1=a}(\phi(T)|T_{2:p} = t_{2:p}) = \alpha$, we have

$$E_{\theta_1}(\phi_0(T)|T_{2:p}=t_{2:p}) \ge E_{\theta_1}(\phi(T)|T_{2:p}=t_{2:p}) \quad \text{for all } \theta_1 > a. \tag{4.11}$$

By construction, both $\phi_0(T)$ and $\phi(T)$ have α-Neyman structure with respect to $T_{2:p}$ for the set Θ_B. Moreover, note that any test with an α-Neyman structure must be in the form of $\phi(T)$ as constructed above. Thus conditions 4 and 6 in Theorem 4.5 are satisfied. By Theorem 3.2,

$$E_{\eta}(\phi_0(T)|T_{2:p}=t_{2:p}) \le \alpha \quad \text{for all } \theta_1 \le a.$$

This implies condition 5 in Theorem 4.5. To summarize, we have the following theorem.

Theorem 4.6 *Suppose*

1. $X \sim \mathfrak{E}_p^n(t_0, \mu_0)$;
2. Θ *has a nonempty interior in* $\mathbb{R}^p$;
3. $a \in \Psi$.

Then ϕ_0 *in (4.10) is a UMPU-*α *test for hypothesis (4.9).*

The next example illustrates how to construct a UMPU-α test.

Example 4.1 Suppose that $X_1, \ldots, X_n$ are i.i.d. $N(\mu, \sigma^2)$, and we are interested in testing

$$H_0 : \mu \le 0 \quad \text{vs} \quad H_1 : \mu > 0. \tag{4.12}$$

Note that

$$\begin{aligned} f(x|\mu,\sigma^2) &= \frac{1}{\sqrt{2\pi}\sigma} \exp\{-(x-\mu)^2/(2\sigma^2)\} \\ &= \frac{1}{\sqrt{2\pi}\sigma} \exp\left(-\frac{\mu^2}{2\sigma^2}\right) \exp\left(\frac{\mu}{\sigma^2}x - \frac{1}{2\sigma^2}x^2\right) \\ &= c(\theta) e^{\theta^T t_0(x)} \end{aligned}$$

where $\theta = (\theta^1, \theta^2)$, $t_0(x) = (x, x^2)$ and

$$\theta_1 = \mu/\sigma^2, \quad \theta_2 = -1/(2\sigma^2).$$

The hypothesis (4.12) is equivalent to

$$H_0 : \theta_1 \le 0 \quad \text{vs} \quad H_1 : \theta_1 > 0.$$

Let

$$T_1 = \sum_{i=1}^n X_i, \quad T_2 = \sum_{i=1}^n X_i^2.$$

By Theorem 4.6, we should look for UMP-α one sided test for the one-parameter exponential family $\{f_{T_1|T_2}(t_1|t_2;\mu) : \mu \in \mathbb{R}\}$. That is

$$\phi_0(t_1, t_2) = \begin{cases} 1 & \text{if } t_1 > z(t_2) \\ 0 & \text{otherwise,} \end{cases} \tag{4.13}$$

where $z(t_2)$ is to be determined by $E_{\theta_1=0}(\phi(T_1)|T_2 = t_2) = \alpha$. Here and in what follows, the symbol $E_{\theta=a}$ is used to denote E_θ where θ is evaluated at a.

Later on in section 4.5 we will describe how to further simplify the test using the Basu's theorem (Theorem 2.6), but for now, let us carry through the principle provided by Theorem 4.5. For simplicity, consider the case $n = 2$. The critical point $z(t_2)$ is the solution to the equation

$$P_{\theta_1=0}(T_1 > c|T_2 = t_2) = \alpha.$$

Here and in what follows we use the standard notation $P_{\theta=a}$ to denote P_θ, where θ is evaluated at a. Under H_0, (X_1, X_2) is distributed as $N(0_2, \sigma^2 I_2)$, where 0_2 is the vector $(0, 0)$ and I_2 is the 2 by 2 identity matrix. Then, conditioning on $X_1^2 + X_2^2 = t_2$, (X_1, X_2) is uniformly distributed on the circle $\{x : \|x\| = \sqrt{t_2}\}$. The inequality $T_1 \le t_1$ corresponds to an arc on the circle if $-\sqrt{2t_2} < t_1 \le \sqrt{2t_2}$. Specifically, the distribution of $T_1|T_2$ when $\mu = 0$ is

$$F_{T_1|T_2}(t_1|t_2) = \begin{cases} 0 & \text{if } t_1 \le -\sqrt{2t_2} \\ \pi^{-1}\cos^{-1}(-t_1/\sqrt{2t_2}) & \text{if } -\sqrt{2t_2} < t_1 \le 0 \\ 1 - \pi^{-1}\cos^{-1}(t_1/\sqrt{2t_2}) & \text{if } 0 < t_1 \le \sqrt{2t_2} \\ 1 & \text{if } \sqrt{2t_2} < t_1. \end{cases}$$

Thus, if $\alpha < 1/2$, then

$$P_{\theta_1=0}(T_1 > c|T_2 = t_2) = \pi^{-1}\cos^{-1}(c/\sqrt{2t_2}) = \alpha \Rightarrow c = \sqrt{2t_2}\cos(\pi\alpha).$$

For $n \ge 2$ the development is essentially the same except that the arc length in a circle is replaced by the corresponding area in a sphere in the n-dimensional Euclidean space. However, this problem can be simplified and solved explicitly using Basu's theorem. □

The UMPU test for the hypothesis

$$H_0 : \theta_1 \ge a, \quad H_1 : \theta_1 < a$$

can be constructed using the above procedure by treating $\zeta_1 = -\theta_1$ as the parameter to be tested.

Two-sided UMPU tests

Now let us carry out the generalization for the three hypotheses I, II, III considered in section 3.2. Thus consider the following hypotheses:

I′ $H_0 : \theta_1 = a$ vs $H_1 : \theta_1 \neq a$
II′ $H_0 : a \leq \theta_1 \leq b$ vs $H_1 : \theta_1 < a$ or $\theta_1 > b$
III′ $H_0 : \theta_1 \leq a$ or $\theta_1 \geq b$ vs $H_1 : a < \theta_1 < b$,

where $a < b$. Since the procedures for developing these tests are similar, we will focus on II′. The development is parallel to the one-sided case except that the boundary set Θ_B is now the union of two sets, and the completeness and sufficiency do not apply to the union but rather to each set in the union. This aspect will be highlighted in the following development.

For hypothesis II′ we have

$$\begin{aligned}\Theta_0 &= \{\theta : a \leq \theta_1 \leq b\},\\ \Theta_1 &= \{\theta : \theta_1 < a\} \cup \{\theta : \theta_1 > b\},\\ \Theta_B &= \Theta(a) \cup \Theta(b).\end{aligned}$$

As before, T is sufficient for Θ and the power function for any test is continuous. However, in this case $T_{2:p}$ is complete and sufficient $\Theta(a)$ and $\Theta(b)$ separately, but not necessarily for their union. Nevertheless, as we have shown in Corollary 4.1, this is enough to guarantee the Neyman structure. We have then verified conditions 1, 2, 3 in Theorem 4.5.

Let Ψ and α be as defined previously, but with Θ_B replaced by the new boundary set. By Theorem 3.7, as applied to the one-parameter conditional family $\{P_{T_1|T_{2:p}}(t_1|t_{2:p};\theta_1) : \theta_1 \in \Psi\}$, there is a test

$$\phi_0(t_1, t_{2:p}) = \begin{cases} 1 & \text{if } t_1 < k_1(t_{2:p}) \text{ or } t_1 > k_2(t_{2:p}) \\ \gamma_i(t_{2:p}) & \text{if } t_1 = k_i(t_{2:p}),\ i = 1, 2 \\ 0 & \text{if } u < k_1(v) \text{ or } t_1 > k_2(t_{2:p}), \end{cases} \tag{4.14}$$

where $-\infty < k_1(t_{2:p}) < k_2(t_{2:p}) < \infty$ are determined by

$$E_{\theta_1=a}(\phi_0(T)|T_{2:p} = t_{2:p}) = E_{\theta_1=b}(\phi_0(T)|T_{2:p} = t_{2:p}) = \alpha. \tag{4.15}$$

Moreover, for any test $\phi(T)$ that satisfies the above relation, we have

$$E_\eta(\phi_0(T)|T_{2:p} = t_{2:p}) \geq E_\eta(\phi(T)|T_{2:p} = t_{2:p}), \text{ for all } \eta \notin [a, b]. \tag{4.16}$$

By construction, $\phi_0(T) \in \mathcal{N}_\alpha$ and $\phi(T)$ is an arbitrary member of $\mathcal{N}_\alpha$. Thus conditions 4 and 6 of Theorem 4.5 are satisfied. By Theorem 3.7, part 2, we also know that $\phi_0(T)$ has size α for the conditional problem. That is,

$$E_{\theta_1}(\phi_0(T)|T_{2:p} = t_{2:p}) \leq \alpha \quad \text{for all } \eta \in [a, b].$$

Thus condition 5 of Theorem 4.5 is also satisfied, leading to the next theorem.

Theorem 4.7 *Suppose*

1. $X = (X_1, \ldots, X_n) \sim \mathfrak{E}_p^n(t_0, \mu_0)$;

2. Θ *has a nonempty interior in* $\mathbb{R}^p$*;*
3. $a, b \in \Psi$*.*

Then the test specified by (4.14) and (4.15) is a UMPU-α *for testing* II$'$.

We now state the forms of UMPU tests for Hypotheses I$'$ and III$'$ without proof.

Theorem 4.8 *Suppose the three assumptions in Theorem 4.7 are satisfied. Let*

$$\phi_0(t) = \begin{cases} 0 & \textit{if } t_1 < k_1(t_{2:p}) \textit{ or } t_1 > k_2(t_{2:p}) \\ \gamma_i(t_{2:p}) & \textit{if } t_1 = k_i(t_{2:p}),\ i = 1, 2 \\ 1 & \textit{if } t_1 < k_1(t_{2:p}) \textit{ or } t_1 > k_2(t_{2:p}) \end{cases},$$

where $-\infty < k_1(t_{2:p}) < k_2(t_{2:p}) < \infty$ *that satisfies*

$$E_{\theta_1=a}(\phi_0(T_1, T_{2:p})|T_{2:p} = t_{2:p}) = E_{\theta_1=b}(\phi_0(T_1, T_{2:p})|T_{2:p} = t_{2:p}) = \alpha.$$

Then $\phi_0(T_1, T_{2:p})$ *is a UMPU-*α *test for hypothesis* III$'$.

Theorem 4.9 *Suppose* $X \sim \mathfrak{E}_p^n(t_0, \mu_0)$, $\mathrm{int}(\Theta) \neq \varnothing$, *and* $a \in \Psi$*. Let*

$$\phi_0(t_1, t_{2:p}) = \begin{cases} 1 & \textit{if } t_1 < k_1(t_{2:p}) \textit{ or } t_1 > k_2(t_{2:p}) \\ \gamma_i(t_{2:p}) & \textit{if } t_1 = k_i(t_{2:p}),\ i = 1, 2 \\ 0 & \textit{if } t_1 < k_1(t_{2:p}) \textit{ or } t_1 > k_2(t_{2:p}) \end{cases},$$

where $-\infty < k_1(t_{2:p}) < k_2(t_{2:p}) < \infty$ *satisfy*

$$\begin{aligned} &E_{\theta_1=a}(\phi_0(T_1, T_{2:p})|T_{2:p} = t_{2:p}) = \alpha, \\ &E_{\theta_1=a}(\phi_0(T_1, T_{2:p})T_1|T_{2:p} = t_{2:p}) = \alpha E_{\theta_1=a}(T_1|T_{2:p} = t_{2:p}). \end{aligned}$$

Then $\phi_0(T_1, T_{2:p})$ *is a UMPU-*α *test for hypothesis* I$'$.

The next example illustrates the construction of the UMPU test for hypothesis I' for the mean parameter of the Normal distribution.

Example 4.2 Suppose, in Example 4.1, we would like to test the hypothesis

$$H_0 : \mu = 0 \quad \text{vs} \quad H_1 : \mu \neq 0.$$

Then, by Theorem 4.9, the UMPU-α test has the form

$$\phi_0(t_1, t_2) = \begin{cases} 1 & \text{if } t_1 < k_1(t_2),\ t_1 > k_1(t_2) \\ 0 & \text{otherwise,} \end{cases}$$

where $k_1(t_2) < k_2(t_2)$ are determined by

$$P_{\theta_1=0}(T_1 < k_1 \text{ or } T_1 > k_2 | T_2 = t_2) = \alpha$$
$$E_{\theta_1=0}[(I(T_1 < k_1) + I(T_1 > k_2))T_1 | T_2 = t_2] = \alpha E_{\theta_1=0}(T_1 | T_2 = t_2).$$

We shall discuss general solution later. For now, consider the simple case $n = 2$. Since $(X_1, X_2)|T_2 = t_2$ is uniformly distributed on a circle with radius $\sqrt{t_2}$, the distribution of T_1 given T_2 is symmetric about 0. So if we take $k_2 = -k_1 = k$ then the second equation above is automatically satisfied (both sides are 0). The first equation is equivalent to

$$P_{\theta_1=0}(T_1 > k | T_2 = t_2) = \alpha/2.$$

Thus $k(t_2) = \sqrt{2t_2}\cos(\pi\alpha/2)$. □

4.4 Invariant family and ancillarity

In this and the next sections we describe a special technique that simplifies the process of finding UMPU-α test. As an illustration, consider testing the hypothesis $H_0 : \theta_1 = 0$ versus $H_1 : \theta \neq 0$. Recall that in Example 4.1 we needed to use the conditional distribution of $T_1|T_2$ to determine the critical points k_1 and k_2 of the UMPU test, which was quite complicated. Moreover, if we construct the UMPU tests from the first principle, then the critical points are in general data dependent, which cannot be derived from a single distribution such as an F or a chi-squared distribution. However, suppose we can find a transformation $g(T_1, T_2)$ such that

1. for each value of t_2, $g(t_1, t_2)$ is monotone (say, increasing) in t_1,
2. for $\theta_1 = 0$, the distribution of $g(T_1, T_2)$ does not depend on θ_2 (that is, this distribution is the same for all $\theta \in \Theta_B = \{\theta : \theta_1 = 0\}$).

Then, by Basu's theorem, $g(T_1, T_2) \perp\!\!\!\perp T_2$ under any P_θ where $\theta \in \Theta_B$. Here and in what follows, independence of U and V is denoted by the notation $U \perp\!\!\!\perp V$. Conditional probabilities such as $P_{\theta_1=0}(T_1 > k(T_2)|T_2)$ can be written as

$$P_{\theta_1=0}(T_1 > k(T_2)|T_2) = P_{\theta_1=0}(g(T_1, T_2) > g(k(T_2), T_2))$$

In other words, if we let c to be the solution of

$$P_{\theta_1=0}(g(T_1, T_2) \leq c) = \alpha$$

and let k be the (unique) solution to $g(k, T_2) = c$, then

$$P_{\theta_1=0}(T_1 > k(T_2)|T_2) = \alpha \Leftrightarrow P_{\theta_1=0}(g(T_1, T_2) > c) = \alpha.$$

and the test

$$\phi_0(T_1, T_2) = \begin{cases} 1 & \text{if } T_1 > k(T_2) \\ 0 & \text{otherwise} \end{cases}, \quad E_{\theta_1=0}(\phi_0(T_1, T_2)|T_2) = \alpha$$

is equivalent to

$$\phi_1(T_1, T_2) = \begin{cases} 1 & \text{if } g(T_1, T_2) > c \\ 0 & \text{otherwise} \end{cases}, \quad E_{\theta_1=0}(\phi_1(T_1, T_2)) = \alpha.$$

Many classical tests, such as the chi-square test, the student t-tests, and the F-tests can be shown using this mechanism to be UMPU tests. The critical step in the above procedure is to find the transformation g. For this purpose, we first introduce the notion of an invariant family of distributions.

Let $(\Omega, \mathcal{F})$ be a measurable space and $\mathcal{P}$ be a family of probability measures on $(\Omega, \mathcal{F})$. Let $\mathcal{G}$ be a set of bijections $g : \Omega \to \Omega$. Let $\circ$ denote composition of functions. Suppose $\mathcal{G}$ is a group with respect to $\circ$. That is:

1. if $g_1 \in \mathcal{G}$, $g_2 \in \mathcal{G}$, then $g_2 \circ g_1 \in \mathcal{G}$;
2. for all $g_1, g_2, g_3 \in \mathcal{G}$, $(g_1 \circ g_2) \circ g_3 = g_1 \circ (g_2 \circ g_3)$;
3. there exists $e \in \mathcal{G}$ such that $g \circ e = e \circ g$ for all $g \in \mathcal{G}$;
4. for each $g \in \mathcal{G}$ there is a $f \in \mathcal{G}$ such that $g \circ f = f \circ g = e$.

Definition 4.3 *A family of distributions $\mathcal{P}$ is said to be invariant under $\mathcal{G}$ if for every $g \in \mathcal{G}$ and $P \in \mathcal{P}$, we have $P \circ g^{-1} \in \mathcal{P}$.*

If $\mathcal{P}$ is invariant under $\mathcal{G}$ then each $g \in \mathcal{G}$ induces a function

$$\tilde{g} : \mathcal{P} \to \mathcal{P}, \quad P \mapsto P \circ g^{-1}.$$

It is left as an exercise to show that $\tilde{g}$ is bijective and the set $\tilde{\mathcal{G}} = \{\tilde{g} : g \in \mathcal{G}\}$ is itself a group. For a member P of $\mathcal{P}$, we call the set

$$\mathcal{M}(P) = \{\tilde{g}(P) : \tilde{g} \in \tilde{\mathcal{G}}\}$$

an orbit of $\tilde{\mathcal{G}}$. If $\mathcal{M}(P) = \mathcal{P}$, then we say the group $\tilde{\mathcal{G}}$ is transitive.

Theorem 4.10 *Let V be a random element defined on $(\Omega, \mathcal{F})$. Suppose:*

1. *$\mathcal{P}$ is invariant under $\mathcal{G}$;*
2. *For each $P \in \mathcal{P}$, $g \in \mathcal{G}$, $P \circ (V \circ g)^{-1} = P \circ V^{-1}$;*
3. *the group $\tilde{\mathcal{G}}$ is transitive.*

Then V is ancillary for $\mathcal{P}$.

Proof. Let $P \in \mathcal{P}$. Since $\mathcal{M}(P) = \mathcal{P}$, it suffices to show that the distribution of V is the same for all $Q \in \mathcal{M}(P)$. Let $Q_1, Q_2 \in \mathcal{M}(P)$. Then $Q_1 = P \circ g_1^{-1}$, $Q_2 = P \circ g_2^{-1}$ for some $g_1, g_2 \in \mathcal{G}$. By Exercise 4.6 we have

$$Q_1 \circ V^{-1} = (P \circ g_1^{-1}) \circ V^{-1} = P \circ (V \circ g_1)^{-1}$$
$$Q_2 \circ V^{-1} = (P \circ g_2^{-1}) \circ V^{-1} = P \circ (V \circ g_2)^{-1}.$$

By assumption 2,

$$P\circ(V\circ g_1)^{-1} = P\circ V^{-1} = P\circ(V\circ g_2)^{-1}.$$

That is, $Q_1\circ V^{-1} = Q_2\circ V^{-1}$, hence V has the same distribution under Q_1 and Q_2. □

Our treatment of this topic slightly differs from the treatment in classical texts such as Ferguson (1967); Lehmann and Romano (2005), in that

(1) we do not assume $\mathcal{P}$ to be parametric — not for generality but for greater clarity;
(2) we require $P\circ(V\circ g)^{-1} = P\circ V^{-1}$ rather than the stronger assumption $V\circ g = V$.

Example 4.3 Let $X = (X_1, \ldots, X_n)$ be an n-dimensional random vector whose density with Lebesgue measure on $(\mathbb{R}^n, \mathcal{R}^n)$ is of the form

$$f(x_1 - \mu, \ldots, x_n - \mu), \quad \mu \in \mathbb{R},$$

where f is a known density function on $\mathbb{R}^n$. Let $\mathcal{P}$ denote the class of distributions corresponding to these densities. Consider the group $\mathcal{G}$ consisting of transformations

$$g_c : \mathbb{R} \to \mathbb{R}, \quad (x_1, \ldots, x_n) \mapsto (x_1 + c, \ldots, x_n + c), \quad c \in \mathbb{R}.$$

This group is called the translation group. The random vector $Y = g_c(X)$ has density

$$f(y_1 - (\mu + c), \ldots, y_n - (\mu + c)),$$

which belongs to $\mathcal{P}$. Hence the family $\mathcal{P}$ is invariant under $\mathcal{G}$. Now let $V(x)$ be a function such that $V(x + c) = V(x)$ for all c. For example, $V(x)$ could be $x_1 - x_2$. In addition, for any $P_\mu \in \mathcal{P}$, $\mathcal{M}(P)$ corresponds to the following family of densities

$$\{f(y_1 - (\mu + c), \ldots, y_n - (\mu + c)), \quad c \in \mathbb{R}\},$$

which is $\mathcal{P}$ itself. Therefore, $V(X)$ is ancillary for μ. In the special case where $X_1, \ldots, X_n$ are i.i.d. $N(\mu, 1)$, $T = \sum_{i=1}^n X_i$ is complete and sufficient for μ. Therefore $T \perp\!\!\!\perp V(X)$. □

Example 4.4 Suppose that $(X_1, \ldots, X_n)$ has density of the form

$$\sigma^{-n} f(x_1/\sigma, \ldots, x_n/\sigma), \quad \sigma > 0.$$

Write this family of distributions as $\mathcal{P}$. Consider the group of transformations

$$g_c: (x_1, \ldots, x_n) \mapsto (cx_1, \ldots, cx_n), \quad c > 0.$$

If $Y = cX$ then the density of Y is

$$(c\sigma)^{-n} f(x_1/(c\sigma), \ldots, x_n/(c\sigma)),$$

which belongs to $\mathcal{P}$. Thus $\mathcal{P}$ is invariant under $\mathcal{G}$. Furthermore, it is easy to see that $\mathcal{M}(P) = \mathcal{P}$ for any $P \in \mathcal{P}$. Let $V(x)$ be a function that satisfies $V(x) = V(cx)$ for all $c > 0$. For example $V(x) = x_1/x_2$ satisfies this condition. Then $V(X)$ is ancillary for σ. In this special case where $X_1, \ldots, X_n$ are i.i.d. $N(0, \sigma^2)$, the statistic $T = \sum_{i=1}^n X_i^2$ is complete and sufficient for σ. Hence, by Basu's theorem $T(X) \perp\!\!\!\perp V(X)$. □

Example 4.5 Suppose that $(X_1, \ldots, X_n)$ has density of the form

$$\sigma^{-n} f((x_1 - \mu)/\sigma, \ldots, (x_n - \mu)/\sigma), \quad \sigma > 0, \ \mu \in \mathbb{R}.$$

Write this family of distributions as $\mathcal{P}$. Consider the group of transformations

$$g_c: (x_1, \ldots, x_n) \mapsto (cx_1 + d, \ldots, cx_n + d), \quad c > 0, \ d \in \mathbb{R}.$$

Similar to the previous two examples, we can show that $\mathcal{M}(P) = \mathcal{P}$ for each $P \in \mathcal{P}$ and $\mathcal{P}$ is invariant under $\mathcal{G}$. Therefore any statistic $V(X)$ satisfying $V(cX + d) = V(X)$ is ancillary for (μ, σ). If $X_1, \ldots, X_n$ are i.i.d. $N(\mu, \sigma^2)$, then such $V(X)$'s are independent of the complete and sufficient statistic $(\sum_{i=1}^n X_i, \sum_{i=1}^n X_i^2)$. For example

$$\left(\frac{X_1 - \bar{X}}{S}, \ldots, \frac{X_n - \bar{X}}{S} \right) \perp\!\!\!\perp (\bar{X}, S^2).$$

where S^2 is the sample variance $\sum_{i=1}^n (X_i - \bar{X})^2/(n-1)$ The statistic on the left is called the standardized (or studentized) residuals. □

We now give an example where the invariant family is nonparametric.

Example 4.6 Let X be a p-dimensional random vector. The distribution of X is said to be spherical if, for any orthogonal matrix A,

$$AX \stackrel{\mathcal{D}}{=} X, \text{ that is, } AX \text{ and } X \text{ have the same distribution.} \tag{4.17}$$

If X has a density with respect to the Lebesgue measure, then (4.17) means that the density of X is $cf(\|x\|)$ for some probability density function f defined on $[0, \infty)$ and some constant $c > 0$. Let τ be the transformation

$$\mathbb{R}^p \to \mathbb{R}, \quad x \mapsto \|x\|.$$

Let λ be the n-fold product Lebesgue measure. Then by the change of variable theorem

$$\int cf(\|x\|)d\lambda(x) = \int cf(s)d\lambda\circ\tau^{-1}(s)$$
$$= c\int f(s)\frac{d\lambda\circ\tau^{-1}(s)}{d\lambda(s)}\,d\lambda(s) = 1.$$

Hence

$$c = \left(\int f(s)\frac{d\lambda\circ\tau^{-1}(s)}{d\lambda(s)}\,d\lambda(s)\right)^{-1} \equiv c(f).$$

Let $\mathcal{D}$ be the class of all probability density functions defined on $[0,\infty)$. Consider the family of densities for X:

$$\{c(f)f(\|x\|) : f \in \mathcal{D}\}.$$

Denote the corresponding family of distributions of X by $\mathcal{P}$. Let $\mathcal{G}$ be the class of all functions of the form

$$x \mapsto h(\|x\|)x/\|x\|, \quad h \in \mathcal{G}_0,$$

where $\mathcal{G}_0$ is the class all bijective, positive, functions from $[0,\infty)$ to $[0,\infty)$. It is clear that $\mathcal{G}$ is a group. For example, if $y = h(\|x\|)x/\|x\|$, then $\|x\| = h^{-1}(\|y\|)$ and hence

$$x = y\|x\|/h(\|x\|) = yh^{-1}(\|y\|)/\|y\| \equiv g^{-1}(y).$$

For any $P \in \mathcal{P}$ and $g \in \mathcal{G}$, the density of $Y = g(X)$ also has a spherical distribution, because, if A is an orthogonal matrix,

$$AY = h(\|X\|)AX/\|X\| = h(\|AX\|)AX/\|AX\| \stackrel{\mathcal{D}}{=} h(\|X\|)X/\|X\|$$

where the last equality follows from $AX \stackrel{\mathcal{D}}{=} X$. Since Y is also dominated by λ, its density is also in $\mathcal{P}$. So $\mathcal{P}$ is invariant under $\mathcal{G}$. To see that and $\mathcal{M}(P) = \mathcal{P}$ for any $P \in \mathcal{P}$ (that is, $\tilde{\mathcal{G}}$ is transitive). Hence, any function $V(x)$ satisfying $V = V\circ g$ for all $g \in \mathcal{G}$ is ancillary for $\mathcal{P}$. One such function is $V(x) = x/\|x\|$.

We now show that $\|X\|$ is complete sufficient on $\mathcal{P}$. $\|X\|$ is a sufficient statistic, because the conditional distribution of X given $\|X\| = r$ is uniform on the sphere $\{x : \|x\| = r\}$. Now let u be a function of $\|X\|$, independent of f, such that

$$E_f u(\|X\|) = 0 \text{ for all } f \in \mathcal{D}.$$

Take f to be the exponential distribution te^{-st}, where $t > 0$. Then we have

$$\int_0^\infty u(s)e^{-st}tds = 0 \Rightarrow \int_0^\infty u(s)e^{-st}ds = 0$$

Consequently $v(t) = \int_0^\infty g(s)e^{-ts}ds = 0$ for all $t > 0$. Because $v(t)$ is analytic we see that $v(t) = 0$ for all $t \in \mathbb{C}$. Hence, by the uniqueness of inverse Laplace transformation, we see that $u(s) = 0$. Thus $\|X\|$ is complete. By Basu's theorem, any $V(x)$ such that $V \circ g = V$ is independent of $\|X\|$. In particular, $(X/\|X\|) \perp\!\!\!\perp \|X\|$. □

4.5 Using Basu's theorem to construct UMPU test

The development in the last section gives a general principle for constructing UMPU test using statistics that is ancillary to the nuisance parameters. The implementation is this principle is straightforward for the one-sided tests and the two-sided tests II′, III′, where $\phi_0(t_1)$ alone appears in the conditional expectations: we simply replace $\phi_0(t_1)$ by $\phi_0 \circ V(t)$, and remove conditioning. The situation is slightly more complicated for the two-sided test I′, where we encounter an additional constraint of the form

$$E_{\theta_1=a}[\phi_0(T_1)T_1|T_{2:p} = t_{2:p}] = \alpha E_{\theta_1=a}(T_1|T_{2:p} = t_{2:p}). \tag{4.18}$$

In this case, we replace $\phi_0(T_1)$ by $\phi_0(V)$ and T_1 by $t_1(V, T_{2:p})$, which is the solution in t_1 of the equation $V(t_1, t_{2:p}) = v$. Since $V \perp\!\!\!\perp T_{2:p}$, the above constraint (4.18) becomes

$$E_{\theta_1=a}[\phi_0(V)t_1(V, t_{2:p})] = \alpha E_{\theta_1=a}[t_1(V, t_{2:p})].$$

In the special case where $t_1(V, t_{2:p})$ has the linear form $a(t_{2:p})V + b(t_{2:p})$, we have

$$E_{\theta_1=a}[\phi_0(V)t_1(V, t_{2:p})] = \alpha E_{\theta_1=a}[t_1(V, t_{2:p})].$$

This is equivalent to

$$a(t_{2:p})E[\phi_0(V)V] + \alpha b(t_{2:p}) = \alpha[a(t_{2:p})E\phi_0(V) + b(t_{2:p})].$$

This happens if and only if

$$E_{\theta_1=a}[\phi(V)V] = \alpha E_{\theta_1=a}(V).$$

We now use several examples to illustrate how to use Basu's theorem to construct UMPU test.

Example 4.7 In Example 4.1, let

$$V(t_1(X), t_2(X)) = \frac{\sqrt{n-1}}{n} \frac{t_1(X)}{\sqrt{t_2(X) - t_1^2(X)/n}} = \frac{\bar{X}}{S}.$$

Write $t(x) = (t_1(x), t_2(x))$. Then $V \circ t(x) = V \circ t(cx)$ for any $c > 0$. By Example 4.4, $V \circ t(X)$ is ancillary with respect to $\Theta_B = \{(0, \sigma^2) : \sigma^2 > \infty\}$. By Basu's theorem, $V \circ t(X) \perp\!\!\!\perp t_2(X)$. In the meantime, it is easy to check by differentiation that $V(t_1, t_2)$ is an increasing function of t_1 for each fixed t_2. Therefore the UMPU test (4.13) is equivalent to

$$\phi(t_1, t_2) = \begin{cases} 1 & \text{if } V(t_1, t_2) > k(t_2) \\ 0 & \text{otherwise,} \end{cases}$$

where $k(t_2)$ is determined by the equation

$$E_{\mu=0}[\phi(T_1, T_2)|T_2 = t_2] = P_{\mu=0}(V(T_1, T_2) > k)|T_2 = t_2) = \alpha.$$

Because $V(T_1, T_2) \perp\!\!\!\perp T_2$, the above equation is equivalent to

$$P_{\mu=0}(V(T_1, T_2) > k) = \alpha.$$

Because $V(T_1, T_2) \sim t_{n-1}$. We have $k = t_{n-1}(\alpha)$, which is the one-sided t test.

The UMPU test for the two-sided hypothesis in Example 4.2 in its original form is

$$\phi(t_1, t_2) = \begin{cases} 1 & t_1 \leq k_1(t_2),\ t_1 > k_2(t_2) \\ 0 & \text{otherwise.} \end{cases}$$

We cannot directly write the two constraints in terms of V because V is not in the linear form $a(t_2)t_1 + b(t_2)$, and the second constraint

$$E_{\mu=0}(\phi(T)T_1|t_2) = \alpha E_{\mu=0}(T_1|t_2) \tag{4.19}$$

is no longer equivalent to

$$E_{\mu=0}(\phi(V)V) = \alpha E_{\mu=0}(V).$$

However, because $T_1 \perp\!\!\!\perp T_2$, the null distribution of T_1 given T_2 is the same as the marginal distribution of T_1, which is symmetric about 0. Hence, as argued earlier, the constraint (4.19) is automatically satisfied if we take $k_1(t_2) = -k_2(t_2)$. In other words the UMPU test is of the form

$$\phi(t_1, t_2) = \begin{cases} 0 & \text{if } \ -k(t_2) < t_1 < k(t_2) \\ 1 & \text{otherwise,} \end{cases}$$

where $k(t_2)$ is determined by $E_{\mu=0}(\phi(T)|t_2) = \alpha$. This can then be equivalently written in terms of V as

$$\phi(v) = \begin{cases} 0 & \text{if } V(-k(t_2), t_2) < V < V(k(t_2), t_2) \\ 0 & \text{otherwise,} \end{cases}$$

where, note that $V(-k(t_2), t_2) = -V(k(t_2), t_2)$. Because $V \perp\!\!\!\perp T_2$, $V(k(t_2), t_2)$ is a nonrandom constant determined by $E_{\mu=0}(\phi(V)) = \alpha$. This constant then must be $t_{n-1}(\alpha/2)$. □

The next example is concerned with testing the variance in the presence of the mean as the nuisance parameter.

Example 4.8 Suppose that $X_1, \ldots, X_n$ are i.i.d. $N(\mu, \sigma^2)$. We are interested in testing the hypothesis

$$H_0 : \sigma^2 \leq 1 \text{ versus } H_1 : \sigma^2 > 1. \tag{4.20}$$

In the notation of Example 4.1, we can equivalently state the hypothesis (4.20) as

$$H_0 : \theta_2 \leq -1/2 \text{ versus } H_1 : \theta_2 > -1/2.$$

The UMPU test is of the form

$$\phi(t_1, t_2) = \begin{cases} 1 & \text{if } t_2 > k(t_1) \\ 0 & \text{otherwise,} \end{cases} \tag{4.21}$$

where $k(t_1)$ is determined by

$$P_{\theta_1=-1/2}(T_2 > k(T_1)|T_1) = \alpha.$$

Let

$$V \circ t(x) = t_2(x) - t_1^2(x)/n.$$

Then $V \circ t(x+c) = V \circ t(x)$ for all $c \in \mathbb{R}$. Hence, by Example 4.3, $V \circ t(X)$ is ancillary for $\Theta_B = \{(\theta_1, -1/2) : \theta_1 \in \mathbb{R}\}$. Since T_1 is complete and sufficient for Θ_B, by Basu's theorem $V \circ t(X) \perp\!\!\!\perp t_1(X)$ for all $\theta \in \Theta_B$. As $V(t)$ is increasing in t_2 for each fixed t_1, the test (4.21) can be equivalently written as

$$\phi(t_1, t_2) = \begin{cases} 1 & \text{if } V(t_1, t_2) > k \\ 0 & \text{otherwise,} \end{cases}$$

where k is determined by the equation

$$P_{\theta_2=-1/2}(V(T_1, T_2) > k|T_1) = P_{\theta_2=-1/2}(V(T_1, T_2) > k) = \alpha$$

Since, under $\theta_2 = -1/2$, $V(T_1, T_2) \sim \chi^2_{(n-1)}$, $k = \chi^2_{(n-1)}(\alpha)$.

Now suppose we want to test the two-sided hypothesis

$$H_0 : \sigma^2 = 1 \quad \text{vs} \quad H_1 : \sigma^2 \neq 1.$$

The UMPU-α test is of the form

$$\phi(t) = \begin{cases} 1 & \text{if } t_2 < k_1(t_1), \ t_2 > k_2(t_2) \\ 0 & \text{otherwise,} \end{cases}$$

where $k_1(t_2)$, $k_2(t_2)$ are determined by

$$E_{\sigma^2=1}[\phi_0(T)|T_2] = \alpha, \quad E_{\sigma^2=1}(\phi_0(T)T_1|T_2) = \alpha E_{\sigma^2=1}(T_1|T_2).$$

Because t_2 is of the linear form $V + t_1^2/n$, by the discussion at the beginning of this section, the constraints can be replaced by

$$E_{\sigma^2=1}(\phi_0(V)) = \alpha, \quad E_{\sigma^2=1}(\phi_0(V)V) = \alpha E_{\sigma^2=1}(V) = \alpha(n-1)$$

where the last equality follows from $V \sim \chi^2_{(n-1)}$. The values of k_1, k_2 can be obtained by solving the above equation numerically. □

4.6 UMPU test for a linear function of θ

Many important statistical tests can be written as the test for a linear combination of the components of the parameter θ. Let $c_1 = (\alpha_1, \ldots, \alpha_p) \in \mathbb{R}^p$, and let $c_1 \neq 0$ be a nonzero vector. Suppose we are interested in the parameter $\eta_1 = c_1^T\theta$. Let $c_2, \ldots, c_p$ be vectors in $\mathbb{R}^p$ such that $C = (c_1, \ldots, c_p)$ is a nonsingular matrix. Then

$$\theta^T t_0 = (C^{-T}C^T\theta)^T t_0 = (C^T\theta)^T C^{-1} t_0 \equiv \eta^T u_0.$$

Let L represent the linear transformation $\theta \mapsto C^T\theta$. Then the exponential family $\mathfrak{E}_p(t_0, \mu_0)$ can be written as $\mathfrak{E}_p(u_0, \mu_0 \circ L^{-1}) = \mathfrak{E}_p(u_0, \nu_0)$. So testing hypothesis about η_1 reduces to the problem in the last section.

The choice of $c_2, \ldots, c_p$ does not affect the result. A convenient choice is as follows. Without loss of generality, assume the first component α_1 of c_1 is nonzero. Then let $c_k = e_k$, $k = 2, \ldots p$, where e_k is the p-dimensional vector whose kth entry is 1 and all the other entries are 0. In this case,

$$C^{-1} = \begin{pmatrix} 1/\alpha_1 & 0 \cdots 0 \\ -\alpha_2/\alpha_1 & 1 \quad\; 0 \\ \vdots & \ddots \\ -\alpha_p/\alpha_1 & 0 \quad\; 1 \end{pmatrix}.$$

The two examples are concerned with the well known two-sample problem for gaussian observations: the first is concerned with comparison of the variances; the second is concerned with the comparison of the means. Both examples are special cases of UMPU tests for linear combinations of θ as described above. It is somewhat surprising that, in some sense, the comparison of the means is more difficult than the comparison of the variances.

Example 4.9 Suppose that $X_1, \ldots, X_m$ are i.i.d. $N(\mu_1, \sigma_1^2)$ and that $Y_1, \ldots, Y_n$ are i.i.d. $N(\mu_2, \sigma_2^2)$. We are interested in testing

$$H_0 : \sigma_2^2/\sigma_1^2 \le \tau \quad \text{versus} \quad H_1 : \sigma_2^2/\sigma_1^2 > \tau. \tag{4.22}$$

Let $X = (X_1, \ldots, X_m)$ and $Y = (Y_1, \ldots, Y_n)$. By simple algebra we deduce the joint density of (X, Y) in the form of

$$c(\theta) e^{\theta^T t(x,y)}$$

with respect to some measure on $(\Omega_X \times \Omega_Y, \mathcal{F}_X \times \mathcal{F}_Y)$, where

$$\theta = (\theta_1, \ldots, \theta_4) = (\mu_1/\sigma_1^2, \mu_2/\sigma_2^2, -1/(2\sigma_1^2), -1/(2\sigma_2^2))$$

and

$$\begin{aligned} t(x,y) &= (t_1(x), t_2(y), t_3(x), t_4(x)) \\ &= (\sum_{i=1}^m x_i, \sum_{j=1}^n y_j, \sum_{i=1}^m x_i^2, \sum_{j=1}^n y_j^2). \end{aligned}$$

In terms of θ, the hypothesis (4.22) can be rewritten as

$$H_0 : \theta_4 \le \theta_3/\tau \quad \text{vs} \quad H_1 : \theta_4 > \theta_3/\tau.$$

Let $\eta_4 = \theta_4 - \theta_3/\tau$, $\eta_3 = \theta_3$, $\eta_2 = \theta_2$, and $\eta_1 = \theta_1$. Then, in terms of $\eta = (\eta_1, \ldots, \eta_4)$, the hypothesis further reduces to

$$H_0 : \eta_4 \le 0 \quad \text{vs} \quad H_1 : \eta_4 > 0.$$

The parameter space is

$$\Lambda = \{(\eta_1, \eta_2, \eta_3, \eta_4) : \eta_1 \in \mathbb{R},\ \eta_2 \in \mathbb{R},\ \eta_3 < 0,\ \eta_4 \in \mathbb{R}\}.$$

The boundary space is

$$\Lambda_B = \{(\eta_1, \eta_2, \eta_3, 0) : \eta_1 \in \mathbb{R},\ \eta_2 \in \mathbb{R},\ \eta_3 < 0\}.$$

Note that

$$\begin{aligned} \theta_1 t_1 + \cdots \theta_4 t_4 &= \theta_1 t_1 + \theta_2 t_2 + \theta_3 (t_3 + t_4/\tau) + (\theta_4 - \theta_3/\tau) t_4 \\ &= \eta_1 u_1 + \cdots + \eta_4 u_4. \end{aligned}$$

So the UMPU-α test is of the form

$$\phi(u) = \begin{cases} 1 & \text{if } u_4 > k(u_{1:3}) \\ 0 & \text{otherwise.} \end{cases} \tag{4.23}$$

Let u be the function $(x, y) \mapsto (u_1(x), u_2(y), u_3(x, y), u_4(y))$. Now consider the statistic

$$V\circ u(X,Y)=\frac{1}{\tau}\frac{\sum_{j=1}^{n}(Y_j-\bar{Y})^2/(n-1)}{\sum_{i=1}^{m}(X_i-\bar{X})^2/(m-1)}.$$

We write the above function as $V\circ u(x,y)$ because the right hand side is indeed such a composite function:

$$\frac{\sum_{j=1}^{n}(Y_j-\bar{Y})^2}{\sum_{i=1}^{m}(X_i-\bar{X})^2}=\frac{u_4-u_2^2/n}{u_3-u_4/\tau-u_1^2/m}. \tag{4.24}$$

Note that we have, by construction,

$$u_4-u_2^2/n\geq 0,\quad u_3-u_4/\tau-u_1^2/m\geq 0.$$

We will show

1. $V(u)$ is increasing in u_4 for each fixed (u_1,u_2,u_3);
2. $V\circ u(X,Y)$ is ancillary for Λ_B.

The validity of the first assertion is easily seen from (4.24). To show the second assertion, let $\mathcal{P}=\{P_\eta:\eta\in\Lambda_B\}$, and consider the group $\mathcal{G}$ of transformations:

$$(x_1,\ldots,x_m,y_1,\ldots,y_n)\mapsto(cx_1+d_1,\ldots,cx_m+d_1,cy_1+d_2,\ldots,cy_n+d_2),$$

where $c>0$ and $d_1,d_2\in\mathbb{R}$. Denote this transformation as g_{c,d_1,d_2}. Suppose the distribution of (X,Y) belongs to Λ_B. Then the distribution of $(\tilde{X},\tilde{Y})=g_{c,d_1,d_2}(X,Y)$ is

$$N(\tilde{\mu}_1,\tilde{\sigma}_1^2)\times\cdots\times N(\tilde{\mu}_1,\tilde{\sigma}_1^2)\times N(\tilde{\mu}_2,\tilde{\sigma}_2^2)\times\cdots\times N(\tilde{\mu}_2,\tilde{\sigma}_2^2)$$

with parameters

$$\begin{cases}\tilde{\mu}_1=c\mu_1+d_1\\ \tilde{\mu}_2=c\mu_2+d_2\\ \tilde{\sigma}_1^2=c^2\sigma_1^2\\ \tilde{\sigma}_2^2=c^2\sigma_2^2\end{cases}\Rightarrow\tilde{\eta}_4=\tilde{\theta}_4-\tilde{\theta}_3/\tau=\frac{1}{c^2}(\theta_4-\theta_3/\tau)=0.$$

In other words, the distribution of $(\tilde{X},\tilde{Y})$ stays in the family indexed by Λ_B. That is, $\mathcal{P}$ is invariant under $\mathcal{G}$. In the meantime it is easy to see that $V\circ u(\tilde{x},\tilde{y})=V\circ u(x,y)$. Finally, let $(\eta_1,\eta_2,\eta_3,0)$ be any fixed point in Λ_B. Then the distribution of $(\tilde{X},\tilde{Y})=g_{c,d_1,d_2}(X,Y)$ corresponds to the parameter

$$\tilde{\eta}_1=\eta_1/c-(2d_1/c^2)\eta_3,\quad\tilde{\eta}_2=\eta_2/c-(2d_2/c^2)\theta_3/\tau,\quad\tilde{\eta}_3=\eta_3/c^2.$$

Clearly, when (d_1,d_2,c) varies freely in $\mathbb{R}\times\mathbb{R}\times(0,\infty)$, the above parameters occupy the whole space $\mathbb{R}\times\mathbb{R}\times(-\infty,0)$. This means $\tilde{G}$ is a transitive group. Hence, by Theorem 4.10, $V\circ u(X,Y)$ is ancillary for Λ_B. By Basu's theorem, $V\circ u(X,Y)\perp\!\!\!\perp(u_1(X),u_2(Y),u_3(X,Y))$. Therefore, the test (4.23) is equivalent to

$$\phi(u) = \begin{cases} 1 & \text{if } V \circ u(X,Y) > k \\ 0 & \text{otherwise,} \end{cases}$$

where k is determined by

$$P_{\eta_4=0}(V \circ u(X,Y) > k) = \alpha.$$

Since, under any $\eta \in \Lambda_B$, $V \circ u(X,Y) \sim F_{(m-1),(n-1)}$, the critical point k is $F_{(m-1),(n-1)}(\alpha)$. □

Now let us turn to the comparison of two normal means.

Example 4.10 Consider the same setting as the above example with $\sigma_1 = \sigma_2$. Under this assumption (X, Y) follows a 3-parameter exponential family

$$\theta_1 t_1 + \theta_2 t_2 + \theta_3 (t_3 + t_4),$$

where $t_1, \ldots, t_4$ are defined as in the above example. We are now interested in testing $H_0 : \mu_1 = \mu_2$ versus $H_1 : \mu_1 \neq \mu_2$, which is equivalent to

$$H_0 : \theta_1 = \theta_2 \quad \text{vs} \quad H_1 : \theta_1 \neq \theta_2.$$

Let $\eta_1 = \theta_1 - \theta_2$, and

$$\begin{aligned} \theta_1 t_1 + \theta_2 t_2 + \theta_3 (t_3 + t_4) &= (\theta_1 - \theta_2) t_1 + \theta_2 (t_1 + t_2) + \theta_3 (t_3 + t_4) \\ &\equiv \eta_1 u_1 + \eta_2 u_2 + \eta_3 u_3. \end{aligned}$$

The UMPU test is of the form

$$\phi(u) = \begin{cases} 1 & \text{if } u_1 > k(u_2, u_3) \\ 0 & \text{otherwise.} \end{cases}$$

Now consider the statistic

$$\frac{(\bar{X} - \bar{Y})/\sqrt{m^{-1} + n^{-1}}}{\sqrt{[\sum_{i=1}^m (X_i - \bar{X})^2 + \sum_{i=1}^n (Y_i - \bar{Y})^2]/(m+n-2)}}. \tag{4.25}$$

Because

$$\begin{aligned} \bar{X} &= t_1/m = u_1/m, \\ \bar{Y} &= t_2/n = (u_2 - u_1)/n, \\ \sum_{i=1}^m (X_i - \bar{X})^2 + \sum_{i=1}^n (Y_i - \bar{Y})^2 &= u_3 - u_1^2/m - (u_2 - u_1)^2/n, \end{aligned}$$

the statistic (4.25) can be rewritten as

$$\frac{[(m^{-1}+n^{-1})u_1(X)-n^{-1}u_2(Y)]/\sqrt{m^{-1}+n^{-1}}}{\sqrt{[u_3-u_1^2/m-(u_2-u_1)^2/n]/(m+n-2)}} \equiv V\circ u(X,Y).$$

Ignoring constants, this function is

$$[(m^{-1}+n^{-1})u_1(X)-n^{-1}u_2(Y)][u_3-u_1^2/m-(u_2-u_1)^2/n]^{-1/2}.$$

To show that this is an increasing function of u_1, we differentiate it with respect to u_1 to obtain

$$\begin{aligned}&[u_3-u_1^2/m-(u_2-u_1)^2/n]^{-3/2}[(m^{-1}+n^{-1})u_1-n^{-1}u_2]^2\\&+[u_3-u_1^2/m-(u_2-u_1)^2/n]^{-1/2}(m^{-1}+n^{-1})>0.\end{aligned}$$

Use similar argument as before, the distribution of $V\circ u(X,Y)$ is symmetric about 0 given (u_2,u_3). Hence the second condition is automatically satisfied.

For this example, the full parameter space is $\Lambda=\mathbb{R}\times\mathbb{R}\times(-\infty,0)$. The boundary of the parameter space is $\Lambda_B=\{0\}\times\mathbb{R}\times(-\infty,0)$. Consider the group of transformations

$$g_{c,d}:(x,y)\mapsto(cx+d,cy+d),$$

where $c>0$ and $d\in\mathbb{R}$. Let η be a fixed point in Λ_B. For $c>0,d\in\mathbb{R}$, the random vector $(\tilde{X},\tilde{Y})$ has distribution $P_{\tilde{\eta}}$, where

$$\tilde{\eta}_1=0,\quad \tilde{\eta}_2=\eta_2/c-(2d/c^2)\eta_3,\quad \tilde{\eta}_3=\eta_3/c^2.$$

From this we can see that $\mathcal{P}=\{P_\eta:\eta\in\Lambda_B\}$ is invariant under $\mathcal{G}$ and $\tilde{\mathcal{G}}$ is a transitive group. Finally, it is easy to see that $V\circ u(g_{c,d}(x,y)=V\circ u(x,y)$ for all $c>0$ and $d\in\mathbb{R}$. Hence, by Theorem 4.10, $V(X,Y)$ is ancillary for $\mathcal{P}$. Thus the UMPU-α test has the form

$$\phi(u)=\begin{cases}1 & \text{if } V(u)>k\\ 0 & \text{otherwise}\end{cases},$$

where k is determined by

$$P_{\eta_1=0}(V\circ u(X,Y)>k)=\alpha.$$

Because $V\circ u(X,Y)\sim t_{(m+n-2)}$, the critical point k is $t_{(m+n-2)}(\alpha)$. □

4.7 UMPU test for nonregular family

So far we have been concerned with constructing the UMPU tests for exponential families. In some special cases, UMPU-tests also exist for distributions not in the exponential family. One such special case is the so called nonregular family, where the support of X_i may depend on the parameter values. Rather than discussing this family generally, we will use an example to illustrate how to construct UMPU tests for these problems. For a more general discussion about nonregular family, see Ferguson (1967, page 130).

Example 4.11 Let $X_1, \ldots, X_n$ be i.i.d. random variables uniformly distributed on (θ_1, θ_2). We are interested in testing

$$H_0 : \theta_1 \leq 0 \quad \text{vs} \quad H_1 : \theta_1 > 0.$$

Let $S = \min_{1 \leq i \leq n} X_i$ and $T = \max_{1 \leq i \leq n} X_i$. The joint density of $X_1, \ldots, X_n$ is

$$(\theta_2 - \theta_1)^n \prod_{i=1}^n I_{(\theta_1, \theta_2)}(X_i) = (\theta_2 - \theta_1)^n I_{(-\infty, \theta_2)}(T) I_{(\theta_1, \infty)}(S).$$

By the factorization theorem (Theorem 2.4), (S, T) is sufficient for $X_1, \ldots, X_n$. This tells us that any optimal test can be based on (S, T). The full parameter space is $\Theta = \{(\theta_1, \theta_2) : \theta_1 < \theta_2\}$; the boundary parameter space is $\Theta_B = \{(0, \theta_2) : \theta_2 > 0\}$. We shall now show that T is complete and sufficient for Θ_B.

Note that, for any $s < t$ satisfying $s > \theta_1$ and $t < \theta_2$ we have

$$P(s < S < T < t) = \left((t - s)/(\theta_2 - \theta_1)\right)^n.$$

From this we deduce the conditional density of $S|T$ and the marginal density of T as follows:

$$\begin{aligned} f_{S|T}(s|t) &= (n-1)(t-s)^{n-2}(t-\theta_1)^{-(n-1)} I_{(\theta_1, t)}(s), \quad \theta_1 < s < t < \theta_2 \\ f_T(t) &= n(t - \theta_1)^{n-1}/(\theta_2 - \theta_1)^n, \quad \theta_1 < t < \theta_2. \end{aligned}$$

From the first expression we see that, for any fixed θ_1, T is sufficient for θ_2. Now let $g(t)$ be a function of t such that $E_\theta g(T) = 0$. Then

$$\begin{aligned} &\int_{\theta_1}^{\theta_2} g(t) n (t - \theta_1)^{n-1}/(\theta_2 - \theta_1)^n dt = 0 \\ \Rightarrow &\int_{\theta_1}^{\theta_2} g(t)(t - \theta_1)^{n-1} dt = 0. \end{aligned}$$

Since this is true for all $\theta_2 > \theta_1$, we have

$$(\partial/\partial\theta_2) \int_{\theta_1}^{\theta_2} g(t)(t - \theta_1)^{n-1} dt = g(\theta_2)(\theta_2 - \theta_1)^{n-1} = 0 \Rightarrow g(\theta_2) = 0.$$

This means $g(t) = 0$ for all $t > \theta_1$. That is, for each fixed θ_1, T is complete for θ_2.

The UMPU-α test for this problem is

$$\phi(s, t) = \begin{cases} 1 & \text{if } s > k(t) \\ 0 & \text{otherwise,} \end{cases}$$

where

$$\alpha = E_{\theta_1 = 0}(\phi(S, T)|T = t) = (1 - (k(t)/t))^{n-1}.$$

Thus $k(t) = t(1 - \alpha^{1/(n-1)})$. □

4.8 Confidence sets

In this section we show that there is a direct association between the confidence sets and the tests of hypotheses, and use it to convert an optimal testing procedure to an optimal confidence set. We first consider the case where θ is a scalar, with no nuisance parameters involved. We will confine our discussion to nonrandomized tests. Suppose the distribution of X belongs to a parametric family $\mathcal{P} = \{P_\theta : \theta \in \Theta\}$, where $\Theta \subseteq \mathbb{R}$ is a Borel set. Let $\mathcal{F}_\Theta$ be the Borel σ-field on Θ.

Definition 4.4 *A mapping $C : \Omega_X \to \mathcal{F}_\Theta$ is called a confident set at confidence level $1 - \alpha$, if*

$$P_\theta(\theta \in C(X)) = 1 - \alpha \textit{ for all } \theta \in \Theta.$$

In the above definition, the probability $P_\theta(\theta \in C(X))$ is to be understood as

$$P_\theta(X \in \{x : \theta \in C(x)\}),$$

where X is random, but θ is non-random. To emphasize this point we will say "the probability of $C(X)$ covering θ" rather than "the probability of θ belonging to $C(X)$", though these two statements are logically equivalent. For each $a \in \Theta$, let $\phi_a : \Omega_X \to \{0, 1\}$ be a size α nonrandomized test for the simple versus composite hypothesis

$$H_0 : \theta = a \quad \text{vs} \quad H_1 : \theta \neq a. \tag{4.26}$$

Assume that, for each $x \in \Omega_X$, $a \mapsto \phi_a(x)$ is measurable $\mathcal{F}_\Theta$. Consider the following mappings

$$\begin{aligned} A &: \Theta \to \mathcal{F}_X, \quad A(a) = \{x \in \Omega_X : \phi_a(x) = 0\}, \\ C &: \Omega_X \to \mathcal{F}_\Theta, \quad C(x) = \{a \in \Theta : \phi_a(x) = 0\}. \end{aligned}$$

That is, $A(a)$ is the 'acceptance region' of the test ϕ_a, and $C(X)$ is the collection of a such that $H_0 : \theta = a$ is not rejected based on the observation X. The following relation follows from construction.

Proposition 4.2 *For each $\theta \in \Theta$,*

$$P_\theta(\theta \in C(X)) = P_\theta(X \in A(\theta)).$$

This proposition implies that $A(\theta)$ is an 'acceptance region' of a size-α test for (4.26) if and only if C is a confidence set for θ with level $1 - \alpha$.

Obviously there are infinitely many confidence sets of level $1 - \alpha$. Then, how to evaluate the performances of two confidence sets? Intuitively, a higher performing confidence set should have smaller probability of covering a wrong parameter value — if there were a confidence set that never covers a wrong parameter value, then it could tell us the correct value of θ perfectly. Thus it is reasonable to require $P_{\theta'}(\theta \in C(X))$ to be small when $\theta' \neq \theta$, so that $C(X)$ has stronger discriminating power against incorrect parameter values. This discriminating power is directly related to the power of a test, and the UMPU property can be passed on to confidence set through the relation between a test and a confidence set.

Definition 4.5 *A $(1-\alpha)$-level confidence set $C : \Omega_X \to \mathcal{F}_\Theta$ is unbiased if, for any $\theta' \neq \theta$, $P_{\theta'}(\theta \in C(X)) \leq 1-\alpha$.*

The next theorem shows that a confidence set constructed from a UMPU test inherits its optimal property.

Theorem 4.11 *Let $A(\theta)$ be the acceptance region of UMPU test of size α of the hypothesis (4.26). Let $C(x) = \{\theta : x \in A(\theta)\}$. Then*

1. *C is a $(1-\alpha)$-level confidence set;*
2. *If D is another $(1-\alpha)$-level confidence set, then*

$$P_{\theta_1}(\theta \in C(X)) \leq P_{\theta'}(\theta \in D(X))$$

for all $\theta' \neq \theta$.

Proof. 1. That $C(X)$ has level $1-\alpha$ follows directly from the definition of C. Let $\theta' \neq \theta$. Because $A(\theta)$ is unbiased, we have

$$P_{\theta'}(\theta \in C(X)) = P_{\theta'}(X \in A(\theta)) \leq P_\theta(X \in A(\theta)) = P_\theta(\theta \in C(X)).$$

Hence C is unbiased.
2. Let $D : \Omega_X \to \mathcal{F}_\Theta$ be another level $1-\alpha$ confidence set and let

$$B(\theta) = \{x \in \Omega_X : \theta \in D(x)\}.$$

Then $\phi(x) = 1 - I_{B(\theta)}(x)$ is a size α unbiased test. Hence

$$P_{\theta'}(\theta \in C(X)) = P_{\theta'}(X \in A(\theta)) \leq P_{\theta'}(X \in B(\theta)) = P_{\theta'}(X \in B(\theta)),$$

as desired. □

Example 4.12 Consider the situation in Example 3.9, where $X_1, \ldots, X_n$ are i.i.d. $N(\theta, 1)$. As we showed in that example, the UMPU-α test for $H_0 : \theta = a$ vs $H_1 : \theta \neq a$ has acceptance region

$$A(a) = \{(x_1, \ldots, x_n) : a - n^{-1/2}\Phi^{-1}(1-\alpha/2) \leq \bar{x} \leq a + n^{-1/2}\Phi^{-1}(1-\alpha/2)\}$$

So corresponding confidence set is

$$C(x) = \{a : \bar{x} - n^{-1/2}\Phi^{-1}(1-\alpha/2) \leq \theta \leq \bar{x} + n^{-1/2}\Phi^{-1}(1-\alpha/2)\}.$$

This is a $(1-\alpha)$-level unbiased confidence interval, and is optimal among all $(1-\alpha)$-level unbiased confidence intervals. □

Now let us consider the problem of obtaining confidence sets for one parameter in the presence of nuisance parameters. Let $\theta = (\theta_1, \ldots, \theta_p)$. Without loss of generality, we assume θ_1 is the parameter of interest. In this section, we assume that all tests have continuous power functions, which holds for exponential families.

Definition 4.6 *A $(1-\alpha)$-level confidence set for θ_1 is any mapping $C : \Omega_X \to \mathcal{F}_{\Lambda_1}$ such that*

$$P_\theta(a \in C(X)) = 1 - \alpha$$

for all $a \in \Lambda_1$, $\theta \in \Theta(a)$. A $(1-\alpha)$-level confidence set C for θ_1 is said to be unbiased if

$$P_\theta(a \in C(X)) \leq 1 - \alpha$$

for all $a \in \Lambda_1$ and $\theta \notin \Theta(a)$.

Let $A(a)$ be the acceptance region of a size α test for

$$H_0 : \theta_1 = a \quad \text{vs} \quad H_1 : \theta_1 \neq a \tag{4.27}$$

The following result is similar to Theorem 4.11; its proof is omitted.

Theorem 4.12 *Let $A(a)$ be the acceptance region of the UMPU test of size α for hypothesis (4.27). Let $C(X) = \{a : x \in A(a)\}$. Then*

1. *C is a $(1-\alpha)$-level confidence set for θ_1;*
2. *if D is another $(1-\alpha)$-level confidence set for θ_1, then*

$$P_\theta(a \in C(X)) \leq P_\theta(a \in D(X))$$

for all $a \in \Lambda_1$ and $\theta \notin \Theta(a)$.

The type of optimality in Theorems 4.11 and 4.12 are called Uniformly Most Accurate (UMA) in the literature, see for example, Ferguson (1967, page 260). In particular, the optimal confidence sets in the two theorems are called the UMA unbiased confidence sets (or UMAU confidence sets). We now discuss specifically how to construct construct UMAU confidence sets for exponential family $X \sim \mathfrak{E}_p^n(t_0, \mu_0)$, where μ_0 is dominated by the Lebesgue measure. By Theorem 4.9, the acceptance region is in the form of an interval

$$A(\theta_1) = \{(t_1, \ldots, t_p) : k_1(\theta_1, t_{2:p}) < t_1 < k_2(\theta_1, t_{2:p})\},$$

where we have added θ_1 as an argument of k_1, k_2 because we are considering the family of all hypothesis of the form (4.27). Suppose $k_1(\theta_1, t_{2:p})$ and $k_2(\theta_2, t_{2:p})$ are increasing functions of θ_1 for each fixed $t_{2:p}$. Let $k_1^{-1}(t_1, t_{2:p})$ and $k_2^{-1}(t_1, t_{2:p})$ be the inverses of $k_1(\theta_1, t_{2:p})$ and $k_2(\theta_1, t_{2:p})$ for each fixed

$t_{2:p}$. Then, by Theorem 4.12, the UMAU confidence set of level $1-\alpha$ is the interval

$$S(t_1,\ldots,t_n) = [k_1^{-1}(t_1,t_{2:p}),\ k_2^{-1}(t_1,t_{2:p})].$$

The next proposition shows that under some conditions (that are satisfied by exponential families) $k_1(\theta_1,t_{2:p})$ and $k_2(\theta_1,t_{2:p})$ are increasing functions of θ_1 for fixed $t_{2:p}$. Since the setting we have in mind is $X \sim \mathfrak{E}_p^n(t_0,\mu_0)$, where $T_1|T_{2:p} = t_{2:p}$ has a one-parameter exponential family distribution, we only state the result for the one-parameter case.

Proposition 4.3 *Suppose:*

1. *for each $a \in \Theta$,*

$$A(a) = \{x : k_1(a) < T < k_2(a)\} \tag{4.28}$$

   defines a size α UMPU test, $\phi_a = 1 - I_{A(a)}$, for the hypothesis (4.26);
2. *ϕ_a is a strictly unbiased test in the sense that $E_b\phi_a > E_a\phi_a = \alpha$ whenever $b \neq a$;*
3. *the family of distribution of T, say $\{P_\theta : \theta \in \Theta\}$, has monotone (nondecreasing) likelihood ratio.*

Then $k_1(a)$ and $k_2(a)$ are strictly increasing in a.

Proof. Let $a < b$, and $a, b \in \Theta_1$. Let ϕ_a and ϕ_b be UMPU tests of size α for testing

$$\begin{aligned} H_0 &: \theta = a \quad \text{vs} \quad H_1 : \theta \neq a \\ H_0 &: \theta = b \quad \text{vs} \quad H_1 : \theta \neq b, \end{aligned}$$

respectively. Since ϕ_a and ϕ_b are strictly unbiased, we have

$$E_a(\phi_b(T) - \phi_a(T)) > 0, \quad E_b(\phi_b(T) - \phi_a(T)) < 0.$$

These rule out the possibilities $A(a) \subseteq A(b)$ or $A(b) \subseteq A(a)$. In other words, among the 4 possibilities

$$k_1(a) < k_1(b) \begin{cases} k_2(a) < k_2(b) \\ k_2(a) \geq k_2(b) \end{cases} \quad \text{and } k_1(a) \geq k_1(b) \begin{cases} k_2(a) < k_2(b) \\ k_2(a) \geq k_2(b) \end{cases},$$

we are left with two possibilities

$$k_1(a) < k_1(b),\ k_2(a) < k_2(b) \quad \text{or} \quad k_1(a) \geq k_1(b),\ k_2(a) \geq k_2(b). \tag{4.29}$$

Let us further rule out the second possibility of the above two.

In this case the set $A(b)$ is positioned to the left of $A(a)$, which implies there is a point t_0 such that

$$\phi_b(t) - \phi_a(t) = I_{A(a)}(t) - I_{A(b)}(t) \begin{cases} \leq 0 & \text{if } t \leq t_0 \\ \geq 0 & \text{otherwise.} \end{cases}$$

Let r denote the likelihood ratio dP_b/dP_a. Then

$$E_b(\phi_b - \phi_a) = \int_{t \leq t_0} (\phi_b(t) - \phi_a(t))r(t)dP_a(t) + \int_{t > t_0} (\phi_b(t) - \phi_a(t))r(t)dP_a(t).$$

However, because $\phi_b - \phi_a \leq 0$ on $\{t \leq t_0\}$, the first term on the right is greater than or equal to $r(t_0) \int_{t \leq t_0} (\phi_b - \phi_a)dP_a$. Similarly, because $\phi_b - \phi_a \geq 0$ on $\{t > t_0\}$, the second term on the right is greater than or equal to $r(t_0) \int_{t > t_0} (\phi_b - \phi_a)dP_a$. Hence

$$E_b(\phi_b - \phi_a) \geq r(t_0)E_a(\phi_b - \phi_a) < 0,$$

which is a contradiction. □

We have learned in Chapter 3 that all the conditions except the strict unbiasedness are satisfied by exponential families. In fact, the strict unbiasedness also holds for exponential families. See Lehmann and Romano (2005, page 112) for a proof of this result.

Problems

4.1. Prove Proposition 4.1.

4.2. Let $(\Omega, \mathcal{F})$ be a measurable space. Let $\mathcal{G}$ be a group of bijections from Ω to Ω, and $\mathcal{P}$ be an invariant family of probability measures on $(\Omega, \mathcal{F})$. Let $\tilde{g}$ be the mapping $P \mapsto P \circ g^{-1}$.

1. Show that $\tilde{g} : \mathcal{P} \to \mathcal{P}$ is a bijection;
2. Show that $\tilde{\mathcal{G}} = \{\tilde{g} : g \in \mathcal{G}\}$ is a group.

4.3. Let $X_1, \ldots, X_n$ be i.i.d. $U(0, \theta)$ random variables with $\theta > 0$. Show that $X_{(n)} = \max_{1 \leq i \leq n} X_i$ is complete and sufficient for $\{\theta : \theta > 0\}$.

4.4. Let $X_1, \ldots, X_n$ be i.i.d. $U(\theta, \theta + 1)$ and $X_{(k)}$ be the kth order statistic. Show that $(X_{(1)}, X_{(n)})$ is sufficient but not complete for $\{\theta : \theta \in \mathbb{R}\}$.

4.5. Let $X_1, \ldots, X_n$ be i.i.d. $N(\theta, \theta^2)$. Let

$$T = \sum_{i=1}^{n} X_i, \; S = \sum_{i=1}^{n} X_i^2.$$

Show that (T, S) is sufficient but not complete for $\{\theta : \theta \in \mathbb{R}\}$.

4.6. Suppose $(\Omega_1, \mathcal{F}_1)$, $(\Omega_2, \mathcal{F}_2)$, and $(\Omega_3, \mathcal{F}_3)$ be measurable spaces and P_1 is a probability measure on $(\Omega_1, \mathcal{F}_1)$. Let $f_1 : \Omega_1 \to \Omega_2$ be a function measurable with respect to $\mathcal{F}_1/\mathcal{F}_2$ and $f_2 : \Omega_2 \to \Omega_3$ be a function measurable with respect to $\mathcal{F}_2/\mathcal{F}_3$. Show that

$$(P_1 \circ f_1^{-1}) \circ f_2^{-1} = P_1 \circ (f_2 \circ f_1)^{-1}.$$

4.7. Suppose $X \sim \text{Gamma}(2, \theta)$, $Y \sim \text{Exp}(\eta)$, that is

$$f_X(x;\theta) = \theta^{-1} x\, e^{-x/\theta}, \quad x > 0,\ \theta > 0,$$
$$f_Y(y;\eta) = \eta^{-1} e^{-y/\eta}, \quad y > 0,\ \eta > 0,$$

and X and Y are independent.

1. Derive the explicit form of the UMPU size α test for $H_0 : \theta \geq \eta$ versus $H_1 : \theta < \eta$.
2. Derive the explicit form of the UMPU size α test for $H_0 : \theta \geq 2\eta$ versus $H_1 : \theta < 2\eta$.
3. Derive the UMPU size α test for the hypothesis $H_0 : \theta = \eta$ versus $H_1 : \theta \neq \eta$. Express the critical point as the solution to an equation.

4.8. Suppose $X \sim b(n, p_1)$, $Y \sim b(m, p_2)$, and X and Y are independent. Derive the UMPU size α test for

$$H_0 : p_1 \leq p_2 \quad \text{vs} \quad H_1 : p_1 > p_2.$$

Derive the conditional distribution involved in the test.

4.9. Suppose $Y_1, \ldots, Y_n$ are random variables defined by

$$Y_i = X_i \beta + \varepsilon_i,$$

where $X_1, \ldots, X_n$ are numbers, β is the regression parameter, and $\varepsilon_1, \ldots \varepsilon_n$ are an i.i.d. sample from $N(0, \sigma^2)$. Here $X_1, \ldots, X_n$ are treated as fixed numbers (rather than random variables).

1. Find the canonical parameter $\theta = (\theta_1, \theta_2)$, and the complete and sufficient statistics, say (T_1, T_2), for the canonical parameter θ.
2. Write down the generic form of the UMPU-α test for the hypothesis $H_0 : \beta = 0$ versus $H_1 : \beta \neq 0$, as well as the constraint(s). State the hypotheses H_0 and H_1 in terms of the canonical parameter.
3. Write down the boundary of the parameter space Θ_B. Which statistic is sufficient and complete for $\theta \in \Theta_B$?
4. Let Y and X denote the vectors $(Y_1, \ldots, Y_n)$ and $(X_1, \ldots, X_n)$. Show that the statistic

$$u(Y) = \frac{X^T Y}{\sqrt{Y^T (I_n - X X^T / X^T X) Y}}$$

is ancillary for Θ_B. Here I_n is the $n \times n$ identity matrix. Comment on the relation between this statistic and the statistic in part 3.

5. Show that the test in part 2 is equivalent to

$$\phi(Y) = \begin{cases} 1 & \text{if } u(Y) > c_2, \ u(Y) < c_1 \\ 0 & \text{otherwise} \end{cases}$$

for some constants c_1 and c_2 that do not depend on Y.
6. Write down the generic form of the UMPU-α test for testing $H_0 : \sigma^2 = 1$ versus $H_1 : \sigma^2 \neq 1$, as well as the constraint(s). State the hypotheses in terms of the canonical parameter.
7. Write down the boundary of the parameter space Θ_B for the hypotheses in part 6. Which statistic is sufficient and complete for $\theta \in \Theta_B$.
8. Show that the

$$v(Y) = Y^T(I_n - XX^T/X^TX)Y.$$

is ancillary with respect to the boundary Θ_B in part 7. Comment on the relation between this statistic and the statistic in part 7.
9. Show that the test in part 6 is equivalent to

$$\phi(Y_1, \ldots, Y_n) = \begin{cases} 1 & \text{if } v(Y) > c_2, \ v(Y) < c_1 \\ 0 & \text{otherwise} \end{cases}$$

for some constants c_1 and c_2 that do not depend on Y.

4.10. Let $X_1, \ldots, X_n$ be an i.i.d. sample from a bivariate normal distribution

$$N\left(\begin{pmatrix} \mu_1 \\ \mu_2 \end{pmatrix}, \ \begin{pmatrix} \sigma_1^2 & \sigma_1\sigma_2\rho \\ \sigma_1\sigma_2\rho & \sigma_2^2 \end{pmatrix}\right).$$

We are interested in testing

$$H_0 : \rho = 0 \quad \text{vs} \quad H_1 : \rho \neq 0.$$

1. Express the joint distribution of X as an exponential family; express the above hypothesis in terms of a canonical parameter.
2. Give the general form of the UMPU size α test.
3. Show that the UMPU size α test can be equivalently expressed in terms of

$$V = \sqrt{(n-2)}\, R/\sqrt{1-R^2},$$

where R is the sample correlation coefficient

$$R = \frac{\sum_{i=1}^n (X_{i1} - \bar{X}_1)(X_{i2} - \bar{X}_2)}{\sqrt{\sum_{i=1}^n (X_{i1} - \bar{X}_1)^2 \sum_{i=1}^n (X_{i2} - \bar{X}_2)^2}}.$$

4. Show that, under $\rho = 0$, $V \sim t_{(n-2)}$.

4.11. Suppose $X \sim$ Poisson(λ_1), $Y \sim$ Poisson(λ_2), and X and Y are independant. Derive the UMPU size α test for

$$H_0 : \lambda_1 \leq \lambda_2 \quad \text{vs} \quad H_1 : \lambda_1 > \lambda_2.$$

References

Ferguson, T. S. (1967). *Mathematical Statistics: A Decision Theoretic Approach.* Academic.

Lehmann, E. L. and Romano, J. P. (2005). *Testing Statistical Hypotheses.* Third edition. Springer.

Lehmann, E. L. and Casella, G. (1998). *Theory of point estimation.* Second edition. Springer, New York.

5

Basic Ideas of Bayesian Methods

In the Bayesian approach to statistical inference, the parameter θ is treated as a random variable, which in this chapter will be written as Θ, and is assigned a distribution. This distribution represents our prior knowledge – or ignorance – about this parameter before observing the data. Once the data, as represented by a random vector X, is observed, we draw inference about Θ by the conditional distribution of $\Theta|X$. This conditional distribution is called the posterior distribution.

The term "Bayesian" comes from the well known Bayes theorem, which is a formula for computing the probability of several causes after a specific outcome is observed. The approach to inference considered in Chapters 2–4, where θ is a fixed number, is called the "frequentist" approach to distinguish it from the new "Bayesian" approach.

Bayesian analysis is a vast area and in this and the next chapter, we can only cover some basic ideas and machineries. For more extensive and specialized discussions, see Berger (1985), Lee (2012), and O'Hagan (1994).

5.1 Prior, posterior, and likelihood

Let

$$(\Omega_X, \mathcal{F}_X, \mu_X), \quad (\Omega_\Theta, \mathcal{F}_\Theta, \mu_\Theta)$$

be two σ-finite measure spaces, where Ω_X is a Borel set in $\mathbb{R}^n$, $\mathcal{F}_X$ is the Borel σ-field of subsets in Ω_X, Ω_Θ is a Borel set in $\mathbb{R}^p$, $\mathcal{F}_\Theta$ is the Borel σ-field of subsets in Ω_Θ, and μ_X, μ_Θ are σ-finite measures. Let

$$\Omega = \Omega_X \times \Omega_\Theta, \quad \mathcal{F} = \mathcal{F}_X \times \mathcal{F}_\Theta.$$

Let P be a probability measure on the measurable space $(\Omega, \mathcal{F})$ that is dominated by $\mu_X \times \mu_\Theta$. Although we have used indices such as X and Θ, up to this point we have not introduced the random vectors X and Θ, and the above

B. Li and G. J. Babu, *A Graduate Course on Statistical Inference*,
Springer Texts in Statistics, https://doi.org/10.1007/978-1-4939-9761-9_5

construction are completely independent of any random vectors. In fact, it is equally reasonable to use Ω_1, Ω_2 in place of Ω_X, Ω_Θ. Conceptually, it is clearer to think about measurable spaces and their product before introducing random elements.

Now let X be the random vector defined by

$$X : \Omega \to \Omega_X, \quad (x, \theta) \mapsto x,$$

and let Θ be the random vector defined by

$$\Theta : \Omega \to \Omega_\Theta, \quad (x, \theta) \mapsto \theta.$$

The random vector X represents the data, which in this book is usually a set of n i.i.d. random variables or random vectors. The random vector Θ represents a vector-valued parameter.

The joint probability measure P determines the marginal distributions $P_X = P \circ X^{-1}$, $P_\Theta = P \circ \Theta^{-1}$, and conditional distributions $P_{X|\Theta}$ and $P_{\Theta|X}$. As discussed in Chapter 1, conditional distributions such as $P_{\Theta|X}$ are to be understood as the mapping

$$P_{\Theta|X} : \mathcal{F}_\Theta \times \Omega_X \to \mathbb{R}, \quad (G, x) \mapsto P(\Theta^{-1}(G)|X)_x.$$

In the Bayesian context, P_Θ is called the prior distribution for Θ; P_X is called the marginal distribution of X; $P_{\Theta|X}$ is called the posterior distribution; $P_{X|\Theta}$ is called the likelihood. We are usually given the prior distribution P_Θ and the likelihood $P_{X|\Theta}$. Our goal is to compute the posterior distribution $P_{\Theta|X}$ and extract a variety of information about Θ from the posterior distribution.

Since $P \ll \mu_X \times \mu_\Theta$, we have

$$P_X = P \circ X^{-1} \ll (\mu_X \times \mu_\Theta) \circ X^{-1} = \mu_X,$$

and similarly $P_\Theta \ll \mu_\Theta$. Let

$$f_X = dP_X/d\mu_X, \quad \pi_\Theta = dP_\Theta/d\mu_\Theta, \quad f_{X\Theta} = dP/d(\mu_X \times \mu_\Theta).$$

Define

$$f_{X|\Theta}(x|\theta) = \begin{cases} f_{X\Theta}(x,\theta)/\pi_\Theta(\theta) & \text{if } \pi_\Theta(\theta) \neq 0 \\ 0 & \text{if } \pi_\Theta(\theta) = 0, \end{cases}$$

$$\pi_{\Theta|X}(\theta|x) = \begin{cases} f_{X\Theta}(x,\theta)/f_X(x) & \text{if } f_X(x) \neq 0 \\ 0 & \text{if } f_X(x) = 0. \end{cases}$$

It can be shown that $f_{X|\Theta}(x|\theta)$ is the density of $P_{X|\Theta}$ in the sense that, for each $A \in \mathcal{F}_X$, the mapping $\theta \mapsto \int_A f_{X|\Theta}(x|\theta) d\mu_X$ is a version of $P(A|\Theta)$. See Problem 5.8. The same can be said about $\pi_{\Theta|X}(\theta|x)$.

The functions π_Θ, $\pi_{\Theta|X}$, f_X, $f_{X|\Theta}$ are called, respectively, the prior density, the posterior density, the marginal density of X, and the likelihood function. If we denote the probability measure $P_{X|\Theta}(\cdot|\theta)$ by P_θ, then the likelihood $P_{X|\Theta}$ gives rise to a family of distributions

$$\{P_\theta : \theta \in \Omega_\Theta\}. \tag{5.1}$$

This corresponds to a parametric family of distributions of X in the frequentist setting. Similarly, the likelihood function $f_{X|\Theta}(\cdot|\theta)$ is simply the density of $f_\theta(x)$ of X in the frequentist context.

By construction, the posterior density can be expressed as

$$\pi_{\Theta|X}(\theta|x) = \frac{f_{X\Theta}(x,\theta)}{f_X(x)} = \frac{f_{X|\Theta}(x|\theta)\pi_\Theta(\theta)}{f_X(x)} \propto f_{X|\Theta}(x|\theta)\pi_\Theta(\theta).$$

In other words, the posterior density is the product of the prior density and the likelihood function. This fact will be useful in later discussions.

The well known Bayes theorem can be derived as follows. Let $G \in \mathcal{F}_\Theta$, $A \in \mathcal{F}_X$. Then

$$P(\Theta^{-1}(G) \cap X^{-1}(A)) = \int_{X^{-1}(A)} P(\Theta^{-1}(G)|X)dP$$

In the special case where $A = \Omega_X$, we have

$$P(\Theta^{-1}(G)) = \int_\Omega P(\Theta^{-1}(G)|X)dP$$

Hence

$$\frac{P(X^{-1}(A) \cap \Theta^{-1}(G))}{P(\Theta^{-1}(G))} = \frac{\int_{X^{-1}(A)} P(\Theta^{-1}(G)|X)dP}{\int_\Omega P(\Theta^{-1}(G)|X)dP}$$

Another way of writing this is

$$P(X \in A|\Theta \in G) = \frac{\int_{X \in A} P(\Theta \in G|X)dP}{\int_{X \in \Omega_X} P(\Theta \in G|X)dP}$$

which is the Bayes Theorem.

5.2 Conditional independence and Bayesian sufficiency

Conditional conditional independence plays a prominent role in Bayesian analysis as in many other areas of statistics. In this section we give a careful treatment of this subject. In connection with conditional independence we will also discuss the role played by sufficiency in the Bayesian approach.

Recall that we say X and Θ are independent, and write $X \perp\!\!\!\perp \Theta$, if, for any $A \in \mathcal{F}_X$ and $G \in \mathcal{F}_\Theta$ we have

$$P(X^{-1}(A) \cap \Theta^{-1}(G)) = P{\circ}X^{-1}(A) \times P{\circ}\Theta^{-1}(G). \tag{5.2}$$

Since

$$X^{-1}(A) = A \times \Omega_\Theta, \quad \Theta^{-1}(G) = \Omega_X \times G,$$

relation (5.2) is equivalent to

$$P{\circ}(X, \Theta)^{-1}(A \times G) = P{\circ}X^{-1}(A) \times P{\circ}\Theta^{-1}(G).$$

Now suppose $\mathcal{G}$ is a sub-σ-field of $\mathcal{F} = \mathcal{F}_X \times \mathcal{F}_\Theta$. We now define conditional independence given $\mathcal{G}$.

Definition 5.1 *We say that X and Θ are conditionally independent given $\mathcal{G}$ if, for any $A \in \mathcal{F}_X$, $G \in \mathcal{F}_\Theta$, we have*

$$P(X^{-1}(A) \cap \Theta^{-1}(G)|\mathcal{G}) = P(X^{-1}(A)|\mathcal{G}) \times P(\Theta^{-1}(G)|\mathcal{G}).$$

We write this relation as $X \perp\!\!\!\perp \Theta|\mathcal{G}$. Obviously, an equivalent definition of conditional independence is

$$P((X, \Theta)^{-1}(A \times G)|\mathcal{G}) = P(X^{-1}(A)|\mathcal{G}) \times P(\Theta^{-1}(G)|\mathcal{G}).$$

Let $(\Omega_T, \mathcal{F}_T)$ be another measurable space and let $T : \Omega_X \to \Omega_T$ be a function measurable $\mathcal{F}_X/\mathcal{F}_T$. We are interested in the conditional independence

$$X \perp\!\!\!\perp \Theta|T{\circ}X.$$

Intuitively, if we know $T{\circ}X$, then we don't need to know the original data X to understand Θ. It turns out this is closely related to the notion of sufficiency, as the following theorem reveals.

Theorem 5.1 *The following statements are equivalent:*

1. *For each $A \in \mathcal{F}_X$, $P(X^{-1}(A)|T{\circ}X, \Theta) = P(X^{-1}(A)|T{\circ}X)$ $[P]$;*
2. *For each $G \in \mathcal{F}_\Theta$, $P(\Theta^{-1}(G)|X) = P(\Theta^{-1}(G)|T{\circ}X)$ $[P]$;*
3. *$X \perp\!\!\!\perp \Theta|T{\circ}X$.*

Proof. 2 $\Rightarrow$ 3. Let $A \in \mathcal{F}_X$ and $G \in \mathcal{F}_\Theta$. Then

$$\begin{aligned}
P(X^{-1}(A) \cap \Theta^{-1}(G)|T{\circ}X) &= E[I_{X^{-1}(A)} I_{\Theta^{-1}(G)}|T{\circ}X] \\
&= E[E(I_{X^{-1}(A)} I_{\Theta^{-1}(G)}|X)|T{\circ}X] \\
&= E[I_{X^{-1}(A)} E(I_{\Theta^{-1}(G)}|X)|T{\circ}X] \\
&= E[I_{X^{-1}(A)} E(I_{\Theta^{-1}(G)}|T{\circ}X)|T{\circ}X] \\
&= E(I_{X^{-1}(A)}|T{\circ}X) E(I_{\Theta^{-1}(G)}|T{\circ}X).
\end{aligned}$$

Hence 3 holds.

$3 \Rightarrow 2$. By Corollary 1.1, it suffices to show that, for any $B \in \sigma(X) = X^{-1}(\mathcal{F}_X)$, we have

$$E[I_B E(I_{\Theta^{-1}(G)}|X)] = E[I_B E(I_{\Theta^{-1}(G)}|T\circ X)]. \tag{5.3}$$

Because $B \in X^{-1}(\mathcal{F}_X)$, there is an $A \in \mathcal{F}_X$ such that $B = X^{-1}(A)$. So

$$\begin{aligned} E[I_B E(I_{\Theta^{-1}(G)}|X)] &= E[I_{X^{-1}(A)} E(I_{\Theta^{-1}(G)}|X)] \\ &= E[E(I_{X^{-1}(A)} I_{\Theta^{-1}(G)}|X)] \\ &= E(I_{X^{-1}(A)} I_{\Theta^{-1}(G)}) \\ &= E[E(I_{X^{-1}(A)} I_{\Theta^{-1}(G)}|T\circ X)]. \end{aligned}$$

By 3, the right hand side is

$$\begin{aligned} E[E(I_{X^{-1}(A)}|T\circ X)E(I_{\Theta^{-1}(G)}|T\circ X)] &= E[E(I_{X^{-1}(A)} E(I_{\Theta^{-1}(G)}|T\circ X)|T\circ X)] \\ &= E[I_{X^{-1}(A)} E(I_{\Theta^{-1}(G)}|T\circ X)] \\ &= E[I_B E(I_{\Theta^{-1}(G)}|T\circ X)]. \end{aligned}$$

Thus (5.3) holds.
$1 \Rightarrow 3$. Let $A \in \mathcal{F}_X$ and $G \in \mathcal{F}_\Theta$. Then

$$\begin{aligned} E[I_{X^{-1}(A)} I_{\Theta^{-1}(G)}|T\circ X] &= E[E(I_{X^{-1}(A)} I_{\Theta^{-1}(G)}|T\circ X, \Theta)|T\circ X] \\ &= E[I_{\Theta^{-1}(G)} E(I_{X^{-1}(A)}|T\circ X, \Theta)|T\circ X] \\ &= E[I_{\Theta^{-1}(G)} E(I_{X^{-1}(A)}|T\circ X)|T\circ X] \\ &= E(I_{\Theta^{-1}(G)}|T\circ X)E(I_{X^{-1}(A)}|T\circ X). \end{aligned}$$

Hence 3 holds.
$3 \Rightarrow 1$. Again, by Corollary 1.1, it suffices to show that, for any $B \in \sigma(T\circ X, \Theta)$,

$$E[I_B E(X^{-1}(A)|T\circ X, \Theta)] = E[I_B E(X^{-1}(A)|T\circ X)]. \tag{5.4}$$

Define two set functions

$$\begin{aligned} Q_1(B) &= E[I_B E(I_{X^{-1}(A)}|T\circ X, \Theta)] \\ Q_2(B) &= E[I_B E(I_{X^{-1}(A)}|T\circ X)]. \end{aligned}$$

By Problem 5.6, $\sigma(T\circ X, \Theta)$ is of the form $\sigma(\mathcal{P})$, where $\mathcal{P}$ is the collection of sets

$$\{T^{-1}(C) \times D : C \in \mathcal{F}_T, D \in \mathcal{F}_\Theta\}.$$

Moreover $\mathcal{P}$ is a π-system, and $T^{-1}(\Omega_T) \times \Omega_\Theta = \Omega \in \mathcal{P}$.

By Corollary 1.4, it suffices to show that $Q_1(B) = Q_2(B)$ for all $B \in \mathcal{P}$. Let $B \in \mathcal{P}$. Then $B = T^{-1}(C) \times D$ for some $C \in \mathcal{F}_T$, $D \in \mathcal{F}_\Theta$. An alternative way of writing the set B is

$$B = (T^{-1}(C) \times \Omega_\Theta) \cap (\Omega_X \times D) = X^{-1}(T^{-1}(C)) \cap \Theta^{-1}(D).$$

Hence

$$\begin{aligned} Q_1(B) &= E[I_{X^{-1}(T^{-1}(C))} I_{\Theta^{-1}(D)} E(I_{X^{-1}(A)}|T{\circ}X, \Theta)] \\ &= E[E(I_{X^{-1}(T^{-1}(C))} I_{\Theta^{-1}(D)} I_{X^{-1}(A)}|T{\circ}X, \Theta)] \\ &= E(I_{X^{-1}(T^{-1}(C))} I_{\Theta^{-1}(D)} I_{X^{-1}(A)}). \end{aligned}$$

Because $I_{X^{-1}(T^{-1}(C))} I_{X^{-1}(A)} = I_{X^{-1}(T^{-1}(C)\cap A)}$, the right hand side above can be rewritten as

$$E(I_{\Theta^{-1}(D)} I_{X^{-1}(T^{-1}(C)\cap A)}) = E[E(I_{\Theta^{-1}(D)} I_{X^{-1}(T^{-1}(C)\cap A)}|T{\circ}X)]. \quad (5.5)$$

However, by 3,

$$E(I_{\Theta^{-1}(D)} I_{X^{-1}(T^{-1}(C)\cap A)}|T{\circ}X) = E(I_{\Theta^{-1}(D)}|T{\circ}X) E(I_{X^{-1}(T^{-1}(C)\cap A)}|T{\circ}X),$$

where, because $X^{-1}(T^{-1}(C)) \in \sigma(T{\circ}X)$, the second conditional expectation on the right is

$$\begin{aligned} E(I_{X^{-1}(T^{-1}(C)\cap A)}|T{\circ}X) &= E(I_{X^{-1}(T^{-1}(C))} I_{X^{-1}(A)}|T{\circ}X) \\ &= I_{X^{-1}(T^{-1}(C))} E(I_{X^{-1}(A)}|T{\circ}X). \end{aligned}$$

Hence we arrive at

$$\begin{aligned} &E[E(I_{\Theta^{-1}(D)} I_{X^{-1}(T^{-1}(C)\cap A)}|T{\circ}X)] \\ &\qquad = I_{X^{-1}(T^{-1}(C))} E(I_{X^{-1}(A)}|T{\circ}X) E(I_{\Theta^{-1}(D)}|T{\circ}X) \\ &\qquad = E(I_{X^{-1}(T^{-1}(C))} I_{\Theta^{-1}(D)}|T{\circ}X) E(I_{X^{-1}(A)}|T{\circ}X) \end{aligned}$$

Substitute this into (5.5) to obtain

$$\begin{aligned} E(I_{\Theta^{-1}(D)} I_{X^{-1}(T^{-1}(C)\cap A)}) &= E[E(I_B|T{\circ}X) E(I_{X^{-1}(A)}|T{\circ}X)] \\ &= E[I_B E(I_{X^{-1}(A)}|T{\circ}X)] = Q_2(B), \end{aligned}$$

as desired. □

The first statement of Theorem 5.1 is essentially the same as sufficiency in the frequentist context, as we described in Chapter 2. In fact, if the statement were made pointwise in θ rather than almost everywhere in the unconditional probability P, then it is exactly the same as the frequentist sufficiency. The second statement is what one might call "Bayesian sufficiency", which means that, if statistical inference is to be based on posterior probability, then we can replace X with $T{\circ}X$ without changing anything. Both of these statements can be interpreted through conditional independence in statement 3.

The next theorem shows rigorously that frequentist sufficiency implies Bayesian sufficiency. Let P_θ denote the probability measure $P_{X|\Theta}(\cdot|\theta)$.

Theorem 5.2 *If* $T \circ X$ *is sufficient for* $\{P_\theta : \theta \in \Theta\}$*, then*

$$X \perp\!\!\!\perp \Theta | T \circ X.$$

Proof. Note that $T \circ X$ is sufficient for $\{P_\theta : \theta \in \Theta\}$ means that, for any $A \in \mathcal{F}_X$, there is a $\sigma(T \circ X)$-measurable κ_A such that

$$P_\theta(A|T \circ X) = \kappa_A \quad [P_\theta] \text{ for all } \theta \in \Omega_\Theta.$$

Let

$$\begin{aligned} B_\theta &= \{x : P_\theta(A|T \circ X)_x \neq \kappa_A\} \\ B &= \{(x, \theta) : P_\theta(A|T \circ X)_x \neq \kappa_A\}. \end{aligned}$$

Then, by Tonelli's theorem,

$$\begin{aligned} P(B) &= \int_{\Omega_\Theta} \left(\int_{B_\theta} dP_{X|\Theta}(\cdot|\theta) \right) dP_\Theta \\ &= \int_{\Omega_\Theta} \left(\int_{B_\theta} dP_\theta \right) dP_\Theta. \end{aligned}$$

(pay attention to the difference between P_θ and P_Θ). By sufficiency, $\int_{B_\theta} dP_\theta = 0$ for each $\theta \in \Theta$. Hence $P(B) = 0$.

As shown in Problem 5.10, the mapping

$$(x, \theta) \mapsto P_\theta(A|T \circ X)_x$$

is (a version of) the conditional probability $(x, \theta) \mapsto P(X^{-1}(A)|T \circ X, \Theta)_{x,\theta}$. Hence

$$P(\{P(X^{-1}(A)|T \circ X, \Theta) \neq \kappa_A\}) = P(B) = 0.$$

That is,

$$P(X^{-1}(A)|T \circ X, \Theta) = \kappa_A \quad [P].$$

Let C be any set in $\sigma(T \circ X) \subseteq \sigma(T \circ X, \Theta)$. Then

$$\int_C \kappa_A dP = \int_C P(X^{-1}(A)|T \circ X, \Theta) dP = \int_C I_A dP$$

Thus κ_A is a version of $P(X^{-1}(A)|T \circ X)$, and consequently

$$P(X^{-1}(A)|T \circ X, \Theta) = P(X^{-1}(A)|T \circ X) \quad [P].$$

By Theorem 5.1, this is equivalent to $\Theta \perp\!\!\!\perp X | T \circ X$. □

The next example illustrates how to find posterior distribution $\pi_{\Theta|X}(\theta|x)$ and the marginal distribution $f_X(x)$ given the prior $\pi(\theta)$ and the likelihood $f(x|\theta)$.

Example 5.1 Suppose that $X \sim N(\theta, \sigma^2)$ and $\Theta \sim N(\mu, \tau^2)$, where σ^2, μ and τ^2 are treated as known constants. A convenient way of finding $\pi(\theta|x)$ is to view $f(x|\theta)\pi_\Theta(\theta)$ as a functional of θ and identify its functional form with a known density. Since $f_X(x)$ does not depend on θ we can ignore it in this process. Specifically, $f(x|\theta)\pi_\Theta(\theta)$ is proportional to (ignoring any multiplicative constant that does not depend on θ):

$$\begin{aligned}\exp&\left[-\frac{(x-\theta)^2}{2\sigma^2}-\frac{(\theta-\mu)^2}{2\tau^2}\right] \\ &\propto \exp\left[-\left(\frac{1}{2\sigma^2}+\frac{1}{2\tau^2}\right)\theta^2+\left(\frac{x}{\sigma^2}+\frac{\mu}{\tau^2}\right)\theta\right].\end{aligned} \tag{5.6}$$

The right hand side is of the form $\exp(-(a\theta^2+b\theta))$, where $a>0$. Let c and c_1 be constants such that

$$(\sqrt{a}\theta+c)^2 = a\theta^2+b\theta+c_1.$$

Then, $2\sqrt{a}c = b, \quad c_1 = c^2$. Hence the right hand side of (5.6) is proportional to

$$\begin{aligned}\exp[-(\sqrt{a}\theta+b/(2\sqrt{a}))^2] &= \exp(-(\sqrt{a})^2(\theta+b/(2a))^2] \\ &= \exp(-(\theta+b/(2a))^2/(2(1/\sqrt{2a})^2)].\end{aligned}$$

This corresponds to a normal density with mean

$$E(\theta|x) = -b/(2a) = \left(\frac{x}{\sigma^2}+\frac{\mu}{\tau^2}\right)\left(\frac{1}{\sigma^2}+\frac{1}{\tau^2}\right)^{-1} \tag{5.7}$$

and variance

$$\text{var}(\theta|x) = \frac{1}{2a} = \left(\frac{1}{\sigma^2}+\frac{1}{\tau^2}\right)^{-1}. \tag{5.8}$$

A more interpretable way of writing the posterior mean $E(\theta|x)$ is

$$E(\theta|x) = \left(\frac{\tau^2}{\sigma^2+\tau^2}\right)x + \left(\frac{\sigma^2}{\sigma^2+\tau^2}\right)\mu.$$

Thus the posterior mean is the weighted average of x and μ, the former is the maximum likelihood estimate based on $\{f(x|\theta) : \theta \in \Omega_\Theta\}$; the latter is the prior mean based on $\pi_\Theta(\theta)$. If τ^2 is large compared with σ^2, which means we have little prior information about Θ, then we give more weight to the maximum likelihood estimate; if τ^2 is small compare with x, then we have more prior information about Θ, and give more weight to μ.

To compute the marginal density $f_X(x)$, notice that

$$f_X(x) = f(x|\theta)\pi(\theta)/\pi(\theta|x).$$

Treating x as the variable and everything else as constants, the right hand side of the above is proportional to

$$\exp\left[-\frac{(x-\theta)^2}{2\sigma^2}-\frac{(\theta-\mu)^2}{2\tau^2}\right]\Big/\exp\left[-\frac{(\theta-E(\theta|x))^2}{2\text{var}(\theta|x)}\right]$$

We know that the expression must not involve θ — it is canceled out one way or another. Discarding all the terms depending on θ, and all constants, the above reduces to

$$\exp\left[-\frac{1}{2}\left(\frac{x^2}{\sigma^2}-\frac{E^2(\theta|x)}{\text{var}(\theta|x)}\right)\right].$$

By straight forward computation, we obtain that

$$\begin{aligned}\frac{x^2}{\sigma^2}-\frac{E^2(\theta|x)}{\text{var}(\theta|x)} &= \frac{1}{\sigma^2+\tau^2}\left(x^2-2x\mu\right)+\text{constant}\\ &= \frac{1}{\sigma^2+\tau^2}\left(x-\mu\right)^2+\text{constant}.\end{aligned}$$

From this we see that X is distributed as $N(\mu,\sigma^2+\tau^2)$. □

The next example shows how to use sufficient statistic to simplify the computation of posterior distribution.

Example 5.2 Suppose, conditioning on $\Theta=\theta$, $X_1,\ldots,X_n$ is an i.i.d. sample from $N(\theta,\sigma^2)$, and Θ is distributed as $N(\mu,\tau^2)$ for some $\mu\in\mathcal{R}$ and $\tau>0$. To find the posterior distribution $\pi(\theta|X_1,\ldots,X_n)$, note that $\bar{X}$ is sufficient for $\{N(\theta,\sigma^2):\theta\in\mathbb{R}\}$, and by Theorem 5.1,

$$\pi(\theta|X_1,\ldots,X_n)=\pi(\theta|\bar{X}).$$

But we know that $\bar{X}|\Theta=\theta\sim(\theta,\sigma^2/n)$ and $\Theta\sim N(\mu,\tau^2)$. So by Example 5.1, we have $\theta|\bar{X}\sim N(E(\theta|\bar{X}),\text{var}(\theta|\bar{X}))$, where

$$\begin{aligned}E(\theta|\bar{X}) &= \left(\frac{\tau^2}{\sigma^2/n+\tau^2}\right)\bar{X}+\left(\frac{\sigma^2/n}{\sigma^2/n+\tau^2}\right)\mu,\\ \text{var}(\theta|\bar{X}) &= \left(\frac{1}{\sigma^2/n}+\frac{1}{\tau^2}\right)^{-1}.\end{aligned}$$

From this we see that

$$E(\theta|\bar{X})\to\bar{X}\ \text{ and }\ \text{var}(\theta|\bar{X})=\sigma^2/n+o(1/n),\ \text{as } n\to\infty.$$

This result is a special case of a general fact — in a later chapter we will show that, as $n\to\infty$, the posterior distribution is concentrated at the maximum likelihood estimate $\hat{\theta}$ with asymptotic variance $1/I(\theta)$, where $I(\theta)$ is the Fisher information. This example also shows that, as the sample size increases, the effect of the prior distribution vanishes, and Bayesian estimate becomes approximately the same as the frequentist estimate. □

5.3 Conjugate families

The posterior density $P_{\Theta|X}$ is typically difficult to compute explicitly. But in a some special cases explicit and relatively simple solutions exist. One such special case is when the prior and posterior distributions are of the same form.

Definition 5.2 *We say that a family $\mathcal{P}$ of distributions on $(\Omega_\Theta, \mathcal{F}_\Theta)$ is conjugate to a likelihood $P_{X|\Theta}$ if*

$$P_\Theta \in \mathcal{P} \Rightarrow P_{\Theta|X}(\cdot|x) \in \mathcal{P} \text{ for each } x \in \Omega_X$$

where $P_{\Theta|X}$ is derived from $(P_\Theta, P_{X|\Theta})$.

By a posterior distribution derived from $(P_\Theta, P_{X|\Theta})$ we mean

$$dP_{\Theta|X}(\cdot|x) = [f(x|\theta)\pi(\theta)/f_X(x)]d\mu_\Theta.$$

Of course, if we let $\mathcal{P}$ to be sufficiently large, then it will always be conjugate to $P_{X|\Theta}$. So the concept is useful only when $\mathcal{P}$ is relatively small and is easily manipulated. The next example illustrates the idea.

Example 5.3 Suppose that $X_1, \ldots, X_n|\Theta = \theta$ are i.i.d. Poisson(θ) random variables. Then

$$f(X_1, \ldots, X_n|\theta) = \prod_{i=1}^n \frac{\theta^{X_i}}{X_i!}e^{-\theta} = \frac{\theta^{\sum X_i}e^{-n\theta}}{\prod(X_i!)}. \tag{5.9}$$

Suppose that Θ has a Gamma(α, β) distribution; that is

$$\pi(\theta) \propto \theta^{\alpha-1}e^{-\theta/\beta}, \quad \beta > 0, \ \alpha > 1. \tag{5.10}$$

Then the posterior density $\pi(\theta|X_1, \ldots, X_n)$ is proportional to

$$\theta^{\sum X_i + \alpha - 1}e^{-(n+1/\beta)\theta}.$$

Thus $\theta|X_1, \ldots, X_n$ has a Gamma(α^*, β^*) distribution with

$$\alpha^* = \alpha + \sum X_i \quad \beta^* = \frac{1}{n+1/\beta}.$$

Hence the Gamma family (5.10) is conjugate to the Poisson family (5.9). The advantage of conjugate prior is that (1) the posterior is easy to compute and (2) prior and posterior have the same distribution with parameters bearing the same interpretation. In other words, conditioning on X simply means updating the parameter of the prior distribution. □

The phenomenon can be generalized to all exponential family distributions. Let us first expand the notation of an exponential family to accommodate transformation of parameters. We now evoke the more general definition (2.5) of an exponential family, $\mathfrak{E}_p(\psi, t, \mu)$ that was introduced in Chapter 2.

Theorem 5.3 *If* $P_\Theta \in \mathfrak{E}_p(\zeta, \psi, \nu)$ *and* $P_{X|\Theta}(\cdot|\theta) \in \mathfrak{E}_p(\psi, t, \mu)$, *then* $P_{\Theta|X}(\cdot|x) \in \mathfrak{E}_p(\zeta_x, \psi, \gamma)$, *where*

$$\zeta_x(\alpha) = \zeta(\alpha) + t(x), \quad d\gamma(\theta) = d\nu(\theta) / \int e^{\psi^T(\theta)t(x)} d\mu(x).$$

Proof. Since $P_\Theta \in \mathfrak{E}_p(\zeta, \psi, \nu)$, $P_{X|\Theta}(\cdot|\theta) \in \mathfrak{E}_p(\psi, t, \mu)$, we have

$$\pi_\Theta(\theta) = e^{\zeta^T(\alpha)\psi(\theta)} / \int e^{\zeta^T(\alpha)\psi(\theta)} d\nu(\theta), \quad f(x|\theta) = e^{\psi^T(\theta)t(x)} / \int e^{\psi^T(\theta)t(x)} d\mu(x)$$

for some $\alpha \in \mathbb{R}^p$ and $\theta \in \Theta$. Consequently,

$$\begin{aligned} \pi(\theta|x) &\propto e^{(\zeta(\alpha)+t(x))^T \psi(\phi)} / [\int e^{\zeta^T(\alpha)\psi(\theta)} d\nu(\theta) \int e^{\psi^T(\theta)t(x)} d\mu(x)] \\ &\propto e^{(\zeta(\alpha)+t(x))^T \psi(\phi)} / \int e^{\psi^T(\theta)t(x)} d\mu(x). \end{aligned}$$

Hence

$$\pi(\theta|x) = \frac{e^{(\zeta(\alpha)+t(x))^T \psi(\phi)} / \int e^{\psi^T(\theta)t(x)} d\mu(x)}{\int e^{(\zeta(\alpha)+t(x))^T \psi(\phi)} / \int e^{\psi^T(\theta)t(x)} d\mu(x) d\nu(\theta)}$$

So if we let

$$\zeta_x(\alpha) = \zeta(\alpha) + t(x), \quad d\gamma(\theta) = d\nu(\theta) / \int e^{\psi^T(\theta)t(x)} d\mu(x) d\nu(\theta)$$

then the posterior density belongs to $\mathfrak{E}_p(\zeta_x, \psi, \gamma)$. □

The algebraic manipulation employed in Example 5.3 and Theorem 5.3 to construct conjugate families can be summarized as the following general scheme. We first inspect the functional form of $\theta \mapsto f_{\Theta|X}(\theta|x)$ and identify a (often parametric) family of functions of θ, say $\mathcal{F} = \{\theta \mapsto g_\alpha(\theta) : \alpha \in A\}$. If $\mathcal{F}$ is closed under multiplication; that is, for any $g_\alpha, g_\beta \in A$, their product $g_\alpha g_\beta = g_\gamma$ for some $\gamma \in A$, then any prior density in $\mathcal{F}$ would be conjugate to $f_{\Theta|X}$. The resulting posterior density is of the form g_γ.

Example 5.4 Suppose that $X|\theta$ is distributed as $N(\theta, \sigma^2)$, where σ^2 is treated as a known constant. We want to assign a conjugate prior for θ. Ignoring any constant (quantities not dependent on θ), the function $\theta \mapsto f_{X|\Theta}(x|\theta)$ is of the form

$$\theta \mapsto \exp\left[\left(-\frac{1}{2\sigma^2}\right)\theta^2 + \left(\frac{\mu}{\sigma^2}\right)\theta\right]$$

So the family $\mathcal{F}$ takes the form

$$\theta \mapsto \exp\left[\left(-\frac{1}{2\alpha_2}\right)\theta^2 + \left(\frac{\alpha_1}{\alpha_2}\right)\theta\right], \quad \alpha_1 \in \mathbb{R}, \ \alpha_2 > 0.$$

Note that, for any $g_\alpha, g_\beta \in \mathcal{F}$, their product has the form

$$\exp\left[\left(-\frac{1}{2\alpha_2}-\frac{1}{2\beta_2}\right)\theta^2+\left(\frac{\alpha_1}{\alpha_2}+\frac{\beta_1}{\beta_2}\right)\theta\right]$$

This function also belongs to $\mathcal{F}$, because

$$-\frac{1}{2\alpha_2}-\frac{1}{2\beta_2}=-\frac{1}{2\gamma_2},\quad \frac{\alpha_1}{\alpha_2}+\frac{\beta_1}{\beta_2}=\frac{\gamma_1}{\gamma_2},$$

where

$$\gamma_2=\left(\frac{1}{\alpha_2}+\frac{1}{\beta_2}\right)^{-1},\quad \gamma_1=\left(\frac{\alpha_1}{\alpha_2}+\frac{\beta_1}{\beta_2}\right)\left(\frac{1}{\alpha_2}+\frac{1}{\beta_2}\right)^{-1}.$$

So any prior of the form $\theta \sim N(\mu, \tau^2)$ is conjugate, and the posterior density is of the form derived in Example 5.1. □

The next theorem shows that, if a family $\mathcal{P}$ is conjugate to $\mathcal{F}$. Then the convex hull conv($\mathcal{P}$) is also conjugate to $\mathcal{F}$. Recall that, for a generic set S, the convex hull of S is the intersection of all convex sets that contains S. Alternatively, conv(S) can be equivalently defined as the set

$$\{\alpha_1 s_1+\cdots+\alpha_k s_k:\ \alpha_1+\cdots+\alpha_k=1,\\ \alpha_1\geq 0,\ldots,\alpha_k\geq 0,\ s_1,\ldots,s_k\in S\}.$$

If S is a class of probability measures, then, for any $P \in S$, $\alpha \in \mathbb{R}$, the product αP is the measure defined by $A \mapsto \alpha P(A)$, where A is a set in a relevant σ-field. A convex combination of a number of probability measures defined as such is called a mixture.

Theorem 5.4 *If $\mathcal{P}$ is conjugate to $P_{X|\Theta}$, then so is* conv($\mathcal{P}$).

Proof. Suppose $P_\Theta \in \mathcal{P}$. Then there exist

$$\alpha_1\geq 0,\ldots,\alpha_k\geq 0,\quad \sum_{i=1}^k \alpha_i=1,\quad P_\Theta^1,\ldots,P_\Theta^k\in\mathcal{P}$$

such that $P_\Theta=\sum_{i=1}^k \alpha_i P_\Theta^i$. It follows that

$$f(x|\theta)dP_\Theta=\sum_{i=1}^k \alpha_i f(x|\theta)dP_\Theta^i.$$

Integrating both sides over Ω_Θ, we find

$$f_X(x)=\sum_{i=1}^k \alpha_i m_i(x),$$

where

$$f_X(x) = \int_{\Omega_\Theta} f(x|\theta)dP_\Theta, \quad m_i(x) = \int_{\Omega_\Theta} f(x|\theta)dP_\Theta^i, \quad i = 1, \ldots, k.$$

Then

$$\begin{aligned} dP_{\Theta|X}(\cdot|x) &= [f(x|\theta)/f_X(x)]dP_\Theta \\ &= \sum_{i=1}^k \alpha_i [m_i(x)/f_X(x)][f(x|\theta)/m_i(x)]dP_\Theta^i \\ &= \sum_{i=1}^k \alpha_i(x) dP_{\Theta|X}^i(\cdot|x), \end{aligned}$$

where $\alpha_i(x) = \alpha_i m_i(x)/f_X(x)$, $i = 1, \ldots, k$. Since

$$\sum_{i=1}^k \alpha_i(x) = \sum_{i=1}^m \alpha_i m_i(x)/f_X(x) = 1, \quad \alpha_i(x) \geq 0, \ i = 1, \ldots, k,$$

$$P_{\Theta|X}^i(\cdot|x) \in \mathcal{P}, \quad i = 1, \ldots, k,$$

we have $P_{\Theta|X}(\cdot|x) \in \text{conv}(\mathcal{P})$. □

5.4 Two-parameter normal family

As with frequentist approach, the normal model is a classical and fundamental component of Bayesian analysis. It is important on its own but also provides intuitions to analyze other models and illuminates connections with frequentist methods. In this section we discuss specifically how to assign conjugate prior for the two-parameter normal likelihood $X|\theta, \sigma^2 \sim N(\theta, \sigma^2)$. Our approach is similar to the one used by Lee (2012), but with a more systematic use of notations, such as the Normal Inverse-Chi square distribution. We first introduce the inverse chi-square distribution.

Definition 5.3 *A random variable T is said to have an inverse $\chi^2_{(\kappa)}$ distribution with κ degrees of freedom if $1/T$ is distributed as χ^2_κ. In this case we will write $T \sim \chi^{-2}_\kappa$.*

If $\tau > 0$, then we write $T \sim \tau\chi^{-2}_{(\kappa)}$ if $T/\tau \sim \chi^{-2}_{(\kappa)}$. The next theorem gives the density of the $\chi^{-2}_{(\kappa)}$ distribution.

Theorem 5.5 *Suppose that T has an inverse chi-square distribution of κ degrees of freedom, then T has density*

$$f(t) = \frac{1}{\Gamma(\nu/2)2^{\nu/2}} t^{-\nu/2-1} e^{-1/(2t)}.$$

Proof. Let $U = 1/T$. Then $U \sim \chi^2_{(\kappa)}$, which has density

$$g(u) = \frac{1}{\Gamma(\nu/2)2^{\nu/2}} u^{\nu/2-1} e^{-u/2}. \tag{5.11}$$

Therefore

$$f_T(t) = g(1/t)|d(1/t)/dt| = g(1/t)/t^2. \tag{5.12}$$

Now combine (5.11) and (5.12) to obtain the desired density. □

We now use the inverse chi-square combined with normal distribution to construct a conjugate prior for the likelihood $N(\theta, \sigma^2)$. It will prove convenient to introduce the following family of distributions.

Definition 5.4 *Suppose λ and ϕ are random variables that take values in $\mathbb{R}$ and $(0, \infty)$, respectively. If $\lambda|\phi \sim N(a, \phi/m)$ and $\phi \sim \tau\chi^{-2}_{(k)}$, then the random vector (λ, ϕ) is said to have a Normal Inverse Chi-square distribution with parameters a, m, τ, k, where $a \in \mathbb{R}$, $\tau > 0$, and m, k are positive integers. In this case we write*

$$(\lambda, \phi) \sim \text{NICH}(a, m, \tau, k),$$

The parameter a is interpreted as the mean of λ, m the sample size in the prior distribution of λ, τ the scale of ϕ, and k the degrees of freedom of inverse chi-square distribution. The next proposition gives the form of the p.d.f. of an NICH random vector.

Proposition 5.1 *If $(\lambda, \phi) \sim \text{NICH}(a, m, \tau, k)$, then its p.d.f. is of the following form up to a proportional constant:*

$$\phi^{-1/2} \exp\left[\left(-\frac{1}{2(\phi/m)}\right)\lambda^2 + \left(\frac{a}{\phi/m}\right)\lambda\right] \phi^{-k/2-1} \exp\left(-\frac{\tau + ma^2}{2\phi}\right). \tag{5.13}$$

The proof is straightforward and is left as an exercise. It turns out that the NICH family is closed under multiplication, which is very convenient for deriving the conjugate prior and the posterior distribution.

Proposition 5.2 *The NICH family is closed under multiplication. That is, if $a_1, a_2 \in \mathbb{R}$, m_1, m_2, k_1, k_2 are positive integers, and τ_1, τ_2 are positive numbers, then*

$$\text{NICH}(a_1, m_1, \tau_1, k_1) \times \text{NICH}(a_2, m_2, \tau_2, k_2) \propto \text{NICH}(a_3, m_3, \tau_3, k_3), \tag{5.14}$$

where

$$\begin{aligned} m_3 &= m_1 + m_2, \\ a_3 &= (m_1 a_1 + m_2 a_2)/m_3, \\ \tau_3 &= \tau_1 + \tau_2 + m_1 a_1^2 + m_2 a_2^2 - m_3 a_3^2, \\ k_3 &= k_1 + k_2 + 3. \end{aligned}$$

Here and in what follows, the product of the type in (5.14) is interpreted as the product of the corresponding probability densities.

Proof. By Proposition 5.1,

$$
\begin{aligned}
&\mathrm{NICH}(a_1, m_1, \tau_1, k_1) \times \mathrm{NICH}(a_2, m_2, \tau_2, k_2) \\
&\propto \phi^{-1/2} \exp\left[\left(-\frac{1}{2(\phi/m_1)}\right)\lambda^2 + \left(\frac{a_1}{\phi/m_1}\right)\lambda\right] \phi^{-k_1/2-1} \exp\left(-\frac{\tau_1 + m_1 a_1^2}{2\phi}\right) \\
&\quad \phi^{-1/2} \exp\left[\left(-\frac{1}{2(\phi/m_2)}\right)\lambda^2 + \left(\frac{a_2}{\phi/m_2}\right)\lambda\right] \phi^{-k_2/2-1} \exp\left(-\frac{\tau_2 + m_2 a_2^2}{2\phi}\right) \\
&= \phi^{-1/2} \exp\left[-\left(\frac{1}{2(\phi/m_1)} + \frac{1}{2(\phi/m_2)}\right)\lambda^2 + \left(\frac{a_1}{\phi/m_1} + \frac{a_2}{\phi/m_2}\right)\lambda\right] \\
&\quad \phi^{-(k_1+k_2+3)/2-1} \exp\left(-\frac{\tau_2 + m_2 a_2^2}{2\phi} - \frac{\tau_2 + m_2 a_2^2}{2\phi}\right) \\
&= \phi^{-1/2} \exp\left[-\left(\frac{1}{2(\phi/(m_1+m_2))}\right)\lambda^2 + \left(\frac{1}{\phi/(m_1+m_2)} \frac{m_1 a_1 + m_2 a_2}{m_1 + m_2}\right)\lambda\right] \\
&\quad \phi^{-(k_1+k_2+3)/2-1} \exp\left(-\frac{\tau_1 + m_1 a_1^2 + \tau_2 + m_2 a_2^2}{2\phi}\right).
\end{aligned}
$$

If we write $m_1 + m_2$ as m_3, the weighted average $(m_1 a_1 + m_2 a_2)/(m_1 + m_2)$ as a_3, and $k_1 + k_2 + 3$ as k_3, then the above can be written as

$$
\begin{aligned}
&\phi^{-1/2} \exp\left[-\left(\frac{1}{2(\phi/m_3)}\right)\lambda^2 + \left(\frac{a_3}{\phi/m_3}\right)\lambda\right] \\
&\qquad \phi^{-k_3/2-1} \exp\left(-\frac{\tau_1 + m_1 a_1^2 + \tau_2 + m_2 a_2^2}{2\phi}\right),
\end{aligned}
\tag{5.15}
$$

which matches (5.13) in Proposition 5.1 except the last term $\exp(\cdots)$. We can then maneuver this term into the desired form by writing it as

$$
\exp\left(-\frac{\tau_1 + m_1 a_1^2 + \tau_2 + m_2 a_2^2 - m_3 a_3^2 + m_3 a_3^2}{2\phi}\right).
$$

Thus, if we write

$$
\tau_3 = \tau_1 + m_1 a_1^2 + \tau_2 + m_2 a_2^2 - m_3 a_3^2,
$$

then (5.15) becomes the form in Proposition 5.1. □

Now suppose that $X_1, \ldots, X_n | (\phi, \lambda)$ are i.i.d. $N(\lambda, \phi)$ random variables, where both λ and ϕ are treated as unknown parameters. From Chapter 4, we know that (T, S) is sufficient for $X_1, \ldots, X_n$, where

$$
T = \bar{X}, \quad S = \sum_{i=1}^{n} (X_i - \bar{X})^2.
$$

Hence, (T, S) is also sufficient in the Bayesian sense; that is, the conditional distribution of (λ, ϕ) given $(X_1, \ldots, X_n)$ is the same as the conditional distribution of (λ, ϕ) given (T, S). Moreover, we know that

$$T \perp\!\!\!\perp S|(\phi, \lambda), \quad T|(\phi, \lambda) \sim N(\lambda, \phi/n), \quad S|(\phi, \lambda) \sim \phi\chi^2_{(n-1)}. \tag{5.16}$$

The latter fact implies that $S \perp\!\!\!\perp \lambda|\phi$. So we can factorize the joint distribution of (T, S) as

$$f(t, s|\lambda, \phi) = f(t|\lambda, \phi/n)f(s|\phi), \tag{5.17}$$

where $f(t|\lambda, \phi)$ is the density of $N(\lambda, \phi)$ and $f(s|\phi)$ is the density of $\phi\chi^2_{(n-1)}$.

The next proposition shows that the function $(\lambda, \phi) \mapsto f(t, u|\lambda, \phi)$ is of the NICH form.

Proposition 5.3 *Suppose, given (λ, ϕ), $X_1, \ldots, X_n$ are an i.i.d. sample from $N(\lambda, \phi)$ with $n > 3$. Then the joint distribution of (T, S), viewed as a function of (λ, ϕ), is of the form* NICH$(T, S, n, n-3)$.

Proof. By (5.17), the joint density $f(t, s)$ of (T, S) is proportional to

$$\begin{aligned}
&(\phi/n)^{-1/2} \exp\left[\left(-\frac{1}{2(\phi/n)}\right)\lambda^2 + \left(\frac{t}{\phi/n}\right)\lambda + \left(-\frac{t^2}{2(\phi/n)}\right)\right] \\
&\qquad \left(\frac{s}{\phi}\right)^{(n-1)/2-1} \exp\left(-\frac{s}{2\phi}\right)\frac{1}{\phi} \\
&= (\phi/n)^{-1/2} \exp\left[\left(-\frac{1}{2(\phi/n)}\right)\lambda^2 + \left(\frac{t}{\phi/n}\right)\lambda\right]\left(\frac{\phi}{s}\right)^{-(n-3)/2-1} \\
&\qquad \exp\left(-\frac{nt^2 + s}{2\phi}\right).
\end{aligned}$$

Comparing this with the form in (5.13), we see that the right-hand side above is of the form NICH$(T, S, n, n-3)$. □

Thus, if we assign (λ, ϕ) a prior NICH(a, m, τ, k), then the posterior distribution of (T, S) is proportional to the product of two NICH families, which is again a NICH family. We state this result as the next theorem.

Theorem 5.6 *Suppose, conditioning on (λ, ϕ), $X_1, \ldots, X_n$ are i.i.d. $N(\lambda, \phi)$, and the prior distribution of (λ, ϕ) is* NICH(a, m, τ, k). *Then the posterior distribution of (λ, ϕ) given $X = (X_1, \ldots, X_n) = x$ is*

$$\text{NICH}\left(\mu(x), m + n, \tau(x), n + k\right),$$

where

$$\begin{aligned}
\mu(x) &= \frac{ma + nt(x)}{m + n}, \\
\tau(x) &= \tau + s(x) + ma^2 + nt^2(x) - (m + n)\mu^2(x),
\end{aligned}$$

where $s(x)$ and $t(x)$ are the observed values of S and T.

An equivalent way of writing the conclusion of the above theorem is

$$\phi|x \sim \tau(x)\chi^{-2}_{(n+k)}, \quad \lambda|\phi, x \sim N\left(\mu(x), \frac{\phi}{n+m}\right). \tag{5.18}$$

Assigning the prior for (λ, ϕ) in this way also gives very nice interpretations of the various parameters in the prior distribution. If we imagine our prior knowledge about (ϕ, λ) is in the form of a sample, then m would be the sample size, a would be the sample average, and τ/k would be the sample mean squared error.

We now develop the marginal posterior distribution of $\lambda|x$ based on the above joint posterior distribution of $(\lambda, \phi)|x$. By (5.18) we have

$$\frac{\lambda - \mu(x)}{\sqrt{\phi/(n+m)}}|(x, \phi) \sim N(0,1), \quad \frac{\tau(x)}{\phi}|x \sim \chi^2_{(n-1+\kappa)}. \tag{5.19}$$

Since the density of $N(0, 1)$ does not depend on (x, ϕ), we have from the first relation that

$$\frac{\lambda - \mu(X)}{\sqrt{\phi/(n+m)}} \perp\!\!\!\perp (\phi, X), \tag{5.20}$$

which implies

$$\frac{\lambda - \mu(X)}{\sqrt{\phi/(n+m)}} \perp\!\!\!\perp \frac{\tau(X)}{\phi}. \tag{5.21}$$

By relations (5.19) and (5.21), if we let

$$V = \frac{\sqrt{n+m}(\lambda - \mu(X))}{\sqrt{\tau(X)/(n-1+\kappa)}},$$

then $V \sim t_{(n-1+k)}$. Moreover, (5.20) also implies

$$\frac{\lambda - \mu(X)}{\sqrt{\phi/(n+m)}} \perp\!\!\!\perp \phi|X, \text{ which in turn implies } \frac{\lambda - \mu(X)}{\sqrt{\phi/(n+m)}} \perp\!\!\!\perp \frac{\tau(X)}{\phi}|X.$$

So the posterior distribution of V given X is also distributed as $t_{(n-1+\kappa)}$. In Chapter 6 we will discuss how to use this fact to draw Bayesian inference about λ.

5.5 Multivariate Normal likelihood

We now consider the more general situation where X is a p-dimensional random vector distributed as multivariate Normal $N(a, \Phi)$, where $a \in \mathbb{R}^p$, and $\Phi \in \mathbb{R}^{p \times p}$ is a positive definite matrix. When Φ is known, the situation is similar to Example 5.1, and it is relatively easy to generalize that example to derive the conjugate prior, posterior density, and the marginal density of X. We leave this as an exercise (see Problem 5.11).

When Φ is unknown and treated as a random parameter, we can develop the conjugate prior and posterior distribution in a parallel manner as Section 5.4. We first define the inverse Wishart distribution, which is a generalization of the inverse χ^2-distribution.

Definition 5.5 *Suppose $Z_1, \ldots, Z_k$ are p-dimensional random vectors that are i.i.d. $N(0, \Sigma)$, where Σ is a positive definite matrix. Then the distribution of $U = Z_1 Z_1^T + \cdots + Z_k Z_k^T$ is called the Wishart distribution with scale parameter Σ and degrees of freedom p. We write this as*

$$U \sim W_p(\Sigma, k).$$

If V is a p-dimension vector whose inverse V^{-1} is distributed as $W_p(\Sigma, k)$, then we say V has an inverse Wishart distribution and write this as

$$V \sim W_p^{-1}(\Sigma, k).$$

The density functions of the Wishart and the inverse Wishart distribution are given by the next Proposition. For more information about these distributions, see, for example, Mardia, Kent, and Bibby (1979).

Proposition 5.4 *The density of $U \sim W_p(\Sigma, k)$ is given by*

$$\frac{1}{2^{kp/2}\Gamma_p(k/2)}|\Sigma|^{-k/2}|U|^{(n-k-1)/2}\exp\left[-\frac{1}{2}tr(\Sigma^{-1}U)\right],$$

and that of $V \sim W_p^{-1}(\Sigma, k)$ is given by

$$\frac{1}{2^{kp/2}\Gamma_p(k/2)}|\Sigma|^{k/2}|V|^{-(k+p+1)/2}\exp\left[-\frac{1}{2}tr(\Sigma V^{-1})\right],$$

where $|\cdot|$ represents the determinant of a matrix, and Γ_p represents the multivariate Gamma function, defined by

$$\Gamma_p(a) = \pi^{p(p-1)/2}\textstyle\prod_{j=1}^{p}\Gamma[a + (1-j)/2].$$

As in Section 5.4, for developing the conjugate prior and posterior distribution for a multivariate Normal likelihood, it is convenient to introduce the following class of distributions.

Definition 5.6 *Suppose λ is a random vector that takes values in $\mathbb{R}^p$, and Φ is a random matrix that takes values in the set of all positive definite matrices. If $\lambda|\Phi \sim N(a, \Sigma/m)$ and $\Phi \sim W_p^{-1}(\Sigma, k)$, then the random element (λ, Φ) is said to have a Normal Inverse Wishart distribution with parameters a, m, Σ, k, where $a \in \mathbb{R}^p$, Σ is a positive definite matrix, and m, k are positive integers. We write this as*

$$(\lambda, \Phi) \sim \mathrm{NIW}(a, m, \Sigma, k).$$

The form of the density of $\mathrm{NIW}(a, m, \Sigma, k)$ can be derived from the product of $N(a, \Sigma/m)$ and $W_p^{-1}(\Sigma, k)$. This is stated as the next proposition. The proof is left as an exercise.

Proposition 5.5 *If $(\lambda, \Phi) \sim \mathrm{NIW}(a, m, \Sigma, k)$, then the p.d.f. of (λ, Φ) is proportional to*

$$|\Phi|^{-(k+p+2)/2} \exp[-m\lambda^T\Phi^{-1}\lambda/2 + ma^T\Phi^{-1}\lambda] \exp\{-tr[\Phi^{-1}(\Sigma + maa^T)/2]\}.$$

Similar to the NICH family, the NIW family is also closed under multiplication, as detailed by the next proposition. The proof is left as an exercise.

Proposition 5.6 *Suppose m_1, k_1, m_2, k_2 are positive integers, a_1, a_2 are vectors in $\mathbb{R}^p$, and Σ_1, Σ_2 are positive definite matrices, then*

$$\mathrm{NIW}(a_1, m_1, \Sigma_1, k_1) \times \mathrm{NIW}(a_2, m_2, \Sigma_2, k_2) \propto \mathrm{NIW}(a_3, m_3, \Sigma_3, k_3),$$

where

$$\begin{aligned} m_3 &= m_1 + m_2, \\ a_3 &= (m_1a_1 + m_2a_2)/m_3, \\ \Sigma_3 &= \Sigma_1 + \Sigma_2 + m_1a_1a_1^T + m_2a_2a_2^T - m_3a_3a_3^T, \\ k_3 &= k_1 + k_2 + p + 2. \end{aligned}$$

Note that, in the univariate case, $p = 1$, so $k_3 = p + 3$, agreeing with Proposition 5.2.

We now develop the conjugate prior and posterior distribution for the multivariate Normal likelihood. Suppose, conditioning on (λ, Φ), $X_1, \ldots, X_n$ are i.i.d. $N(\lambda, \Phi)$. Let

$$T = n^{-1}\sum_{i=1}^n X_i, \quad S = \sum_{i=1}^n (X_i - \bar{X})(X_i - \bar{X})^T.$$

Then the following statements about (T, S) hold true.

Proposition 5.7 *If (T, S) are as defined in the last paragraph, then*

1. *(T, S) is sufficient for (λ, Φ);*
2. *$T \perp\!\!\!\perp S|(\lambda, \Phi)$;*
3. *$S|\Phi \sim W_p(\Phi, n-1), \quad T|(\lambda, \Phi) \sim N(\lambda, \Phi/n)$.*

Again, abbreviating $(X_1, \ldots, X_n)$ as X. From part 1 of this proposition we see that the conditional distribution of $(\lambda, \Phi)|X$ is the same as the conditional distribution of $(\lambda, \Phi)|(T, S)$. So we only need to consider the likelihood $(T, S)|(\lambda, \Phi)$. By part 2 and part 3 of the proposition, this likelihood is of the form $N(\lambda, \Phi/n) \times W_p(\Sigma, n-1)$. The next proposition asserts that, as a function of (λ, Φ), this likelihood is of the NIW form.

Proposition 5.8 *Let $f(t, s|\lambda, \Phi)$ be the density of $N(\lambda, \Phi/m)$ and $f(s|\Phi)$ be the density of $W_p(\Phi, n-1)$. Then the function $(\lambda, \Phi) \mapsto f(t, s|\lambda, \Phi)f(s|\Phi)$ is proportional to the density of* $\mathrm{NIW}(T, n, S, n-p-2)$.

The proof is similar to that of Proposition 5.3 and so it is left as an exercise. When $p = 1$, the degrees of freedom in $\mathrm{NIW}(T, n, S, n-p-2)$ reduces to $p-3$, agreeing with Proposition 5.3. From Propositions 5.5 to 5.8, we can easily derive the posterior distribution of (λ, Φ).

Theorem 5.7 *Suppose, conditioning on (λ, ϕ), $X_1, \ldots, X_n$ are i.i.d. $N(\lambda, \Phi)$, and the prior distribution of (λ, Φ) is* $\mathrm{NIW}(a, m, \Sigma, k)$. *Then the posterior distribution of (λ, Φ) given $X = x$ is*

$$\mathrm{NIW}\left(\mu(x), m+n, \Sigma(x), n+k\right),$$

where

$$\begin{aligned}\mu(x) &= \frac{ma + nt(x)}{m+n},\\ \Sigma(x) &= \Sigma + s(x) + ma^2 + nt^2(x) - (m+n)\mu^2(x),\end{aligned}$$

where $t(x)$ and $s(x)$ are the observed values of T and S.

5.6 Improper prior

5.6.1 The motivation idea of improper prior

Sometimes we do not have much prior knowledge about the parameter Θ and would like to reflect this uncertainty in the Bayesian analysis. In this case it is desirable to use a prior distribution that is in some sense "flat." Unless the sample space Ω_Θ is a bounded set, a flat distribution cannot be a finite measure. This leads us to the notion improper priors.

Definition 5.7 *An improper prior is an infinite but σ-finite measure on Ω_Θ.*

One might ask if there is no prior information about the parameter, then why should we use the Bayesian method in the first place? Under some circumstances the Bayesian method has some technical advantages over the frequentist method. For example, when dealing with nuisance parameters we can simply integrate them out from the posterior distribution, rather than trying to condition on a sufficient statistic for the nuisance parameter, which may not be available.

A frequently used improper prior is the Lebesgue measure, which can be viewed as the uniform distribution on the whole line. The posterior distribution corresponding to an improper prior is typically a probability measure, and it often leads to similar estimates to those given by the frequentist method. However, the marginal distribution $f_X(x)$ corresponding to an improper prior is typically also improper, as illustrated by the following example.

Example 5.5 Suppose that $X|\theta \sim N(\theta, \phi)$ where ϕ is treated as a known constant. Let $\pi_\Theta(\theta) = 1$ for all $\theta \in \mathbb{R}$. That is, the improper prior is the Lebesgue measure. In this case, we have

$$\pi_{\Theta|X}(\theta|x) \propto 1 \times e^{-\frac{1}{2}\frac{(x-\theta)^2}{\phi}}.$$

We see that $\Theta|x \sim N(x, \sigma^2)$. The marginal distribution of X can be obtained formally from the equation

$$\pi_{\Theta|X}(\theta|x) f_X(x) = f_{X|\Theta}(x|\theta)\pi_\Theta(\theta) = f_{X|\Theta}(x|\theta).$$

Therefore, $f_X(x) = f_{X|\Theta}(x|\theta)/\pi_{\Theta|X}(\theta|x) = 1$, which is an improper distribution. □

As with proper priors, we can also assign improper priors sequentially when there are several parameters.

Example 5.6 Suppose that $X_1, \ldots, X_n|\lambda, \phi$ are i.i.d. $N(\lambda, \phi)$, where both λ and ϕ are unknown. Let (S, T) be defined as in Section 5.4. Recall that the likelihood $(\lambda, \phi) \mapsto f(s, t|\lambda, \phi)$ is of the form

$$\phi^{-1/2} \exp\left[\left(-\frac{1}{2(\phi/n)}\right)\lambda^2 + \left(\frac{t}{\phi/n}\right)\lambda\right]\phi^{-(n-3)/2-1}\exp\left(-\frac{s}{2\phi}\right)$$

If we assign (λ, ϕ) the improper prior $\pi(\phi) = 1/\phi$ and $\pi(\lambda|\phi) = 1$, the posterior density is of the form

$$\phi^{-1/2} \exp\left[\left(-\frac{1}{2(\phi/n)}\right)\lambda^2 + \left(\frac{t}{\phi/n}\right)\lambda\right]\phi^{-(n-1)/2-1}\exp\left(-\frac{s}{2\phi}\right).$$

Equivalently, the joint posterior distribution of (λ, ϕ) can be expressed as

$$\phi|t, s \sim s\chi^{-2}_{(n-1)}, \quad \lambda|\phi, t, s \sim N(t, \phi/n)$$

These imply

$$\left.\frac{\lambda - t}{\sqrt{\phi/n}}\right| x \sim N(0,1), \quad \left.\frac{s}{\phi}\right| x \sim \chi^2_{(n-1)}, \quad \left.\frac{\lambda - t}{\sqrt{\phi/n}} \perp\!\!\!\perp \frac{s}{\phi}\right| x.$$

So if we want to make posterior inference about the mean parameter λ, then we can use the fact

$$\left.\frac{\sqrt{n}(\lambda - t)}{\sqrt{s/(n-1)}}\right| x \sim t_{(n-1)}. \tag{5.22}$$

Since the right hand side does not depend on x, this relation also implies

$$\frac{\sqrt{n}(\lambda - t)}{\sqrt{s/(n-1)}} \perp\!\!\!\perp X, \quad \frac{\sqrt{n}(\lambda - t)}{\sqrt{s/(n-1)}} \sim t_{(n-1)}.$$

Interestingly, the statistic $\sqrt{n}(\lambda - t)/\sqrt{s/(n-1)}$ has the same form as the t statistic in the frequentist setting.

If we want to draw posterior inference about ϕ, then we can use the fact

$$\left.\frac{s}{\phi}\right| x \sim \chi^2_{(n-1)} \tag{5.23}$$

Again, this has the same form as the chi-square test in the frequentist setting. In a later chapter we will see that posterior inference based on (5.22) and (5.23) are exactly the same as discussed in Chapter 4. □

It is not always clear what is the real meaning of being noninformative. Note that in the above example we assigned $P_\Phi(\phi) = 1/\phi$ to the variance parameter Φ, which is not "flat". Is there any reason to use such priors? Also, suppose we assign a uniform prior to a parameter Θ, then any nonlinear monotone transformation of Θ would have a nonuniform distribution. Given that a one-to-one transformation of parameter does not change the family of distributions, it is not clear to which transformation should we assign the uniform prior.

5.6.2 Haar measures

The Haar measures are generalizations of the Lebesgue measure, or the (improper) uniform distribution over the whole real line. Lebesgue measure is the Haar measure under location transformations (or translations): $\theta \mapsto \theta + c$. The set of all translations form a group, and the Lebesgue is invariant under this group of transformations. This leads naturally to the question: can we construct improper distributions that are invariant under other group of transformation. In general, are there something like Lebesgue measure for *any* group of transformations? If so, then that would be a good candidate for improper prior in the more general setting.

It turns out that there are actually two types of improper distributions that are invariant under a group of transformations: the left and right Haar measure. Let $\mathcal{G}$ be a group of transformations from Ω_Θ on to Ω_Θ, indexed by members of Ω_Θ; that is

$$\mathcal{G} = \{g_t : t \in \Omega_\Theta\}. \tag{5.24}$$

In a statistical problem, the group $\mathcal{G}$ is induced by an invariant family of distributions, as discussed in Section 4.4. Specifically, suppose $\mathcal{F} = \{P_\theta : \theta \in \Omega_\Theta\}$ is parametric family of distributions of X. We assume that there is a group, say $\mathcal{H}$, of transformations from Ω_X to Ω_X such that, for each $h \in \mathcal{H}$, the induced measure $P_\theta \circ h^{-1}$ is a member of $\mathcal{F}$. That is, there exists a $\tilde{\theta} \in \Omega_\Theta$ such that $P_{\tilde{\theta}} = P_\theta \circ h^{-1}$. This induces a transformation from Ω_Θ to Ω_Θ that maps θ to $\tilde{\theta}$. The collections of all these mappings can be shown to be a group, and this is the group $\mathcal{G}$ in (5.24), which is our starting point for constructing the Haar measures.

Each $g_t \in \mathcal{G}$ induces two types of transformations of θ:

$$L_t(\theta) = g_t(\theta), \quad R_t(\theta) = g_\theta(t),$$

where L_t is called the left transformation, and R_t the right transformation.

Definition 5.8 *A measure Π on Ω_Θ is the left (right) Haar measure if it satisfies*

$$\Pi = \Pi \circ L_t^{-1} \quad (\Pi = \Pi \circ R_t^{-1}) \tag{5.25}$$

for all $t \in \Omega_\Theta$.

We can use the equations in (5.25) to determine the forms of the left and right Haar measures. Take the left Haar measure as an example. Suppose $\pi(\theta)$ is the density of Π, and assume that it is differentiable. Then the distribution of $\tilde{\theta} = L_t(\theta)$ is $\Pi \circ L_t^{-1}$, with density

$$\tilde{\pi}(\tilde{\theta}) = \pi(L_t^{-1}(\tilde{\theta}))|\det(\partial L_t^{-1}(\tilde{\theta})/\partial \tilde{\theta})|$$

where $\det(\cdot)$ denotes the determinant of a matrix and $|\det(\cdot)|$ its absolute value. The first equation in (5.25) implies that $\tilde{\pi}$ and π are the same density functions; that is,

$$\pi(L_t^{-1}(\theta))|\det(\partial L_t^{-1}(\theta)/\partial \theta^T)| = \pi(\theta),$$

for all $\theta, t \in \Omega_\Theta$. Fix θ at any $\theta_0 \in \Omega_\Theta$, and we have

$$\pi(L_t^{-1}(\theta_0)) = \frac{\pi(\theta_0)}{|\det(\partial L_t^{-1}(\theta_0)/\partial \theta^T)|} \equiv g(t),$$

for all $t \in \Omega_\Theta$. Let $h(t) = L_t^{-1}(\theta_0)$. Then it can be shown that $h : \Omega_\Theta \to \Omega_\Theta$ is a bijection. So we have $\pi(h(t)) = g(t)$ for all $t \in \Omega_\Theta$. Putting $\theta = h(t)$, we have

$$\pi(\theta) = g(h^{-1}(\theta)).$$

The right Haar measure can be derived in exactly the same way. In the next three examples we use this method to develop the left and right Haar measures for three groups of transformations: the location transformation group, the scale transformation group, and the location-scale transformation group.

Example 5.7 Let $\Omega_\Theta = \mathbb{R}$, and let $\mathcal{G}$ be the group of transformations

$$g_c(\theta) = \theta + c, \quad c \in \mathbb{R}.$$

Then

$$L_c(\theta) = g_c(\theta) = \theta + c, \quad R_c(\theta) = g_\theta(c) = \theta + c.$$

Let us first find the left Haar measure. Let $\tilde{\theta} = L_c(\theta) = \theta + c$. Then $\theta = L_c^{-1}(\tilde{\theta}) = \tilde{\theta} - c$. If the density of θ is π, then the density of $\tilde{\theta}$ is $\tilde{\pi} = \pi(\tilde{\theta} - c)$. So we want $\tilde{\pi}$ and π to be the same function for all c; that is,

$$\pi(\theta - c) = \pi(\theta)$$

for all $\theta \in \mathbb{R}$, $c \in \mathbb{R}$. Taking $\theta = 0$, we have $\pi(-c) = \pi(0)$ for all $c \in \mathbb{R}$, implying $\pi(\theta) = \pi(0)$ for all $\theta \in \mathbb{R}$. That is, the left Haar measure is proportional to the Lebesgue measure. Since $L_t = R_t$, the right Haar measure is also proportional to the Lebesgue measure. □

Example 5.8 Let $\Omega_\Theta = (0, \infty)$, and let $\mathcal{G}$ be the group of transformations

$$g_a(\theta) = a\theta, \quad a \in (0, \infty).$$

Then

$$L_a(\theta) = g_a(\theta) = a\theta, \quad R_a(\theta) = g_a(\theta) = a\theta.$$

To determine the left Haar measure, let $\tilde{\theta} = L_a(\theta) = a\theta$, then $\theta = L_a^{-1}(\tilde{\theta}) = \tilde{\theta}/a$. If the density of θ is π, then the density of $\tilde{\theta}$ is

$$\tilde{\pi} = \pi(\tilde{\theta}/a)\frac{\partial(\tilde{\theta}/a)}{\partial\tilde{\theta}} = \pi(\tilde{\theta}/a)/a.$$

The relation $\Pi \circ L_c^{-1} = \Pi$ implies that $\tilde{\pi}$ and π are the same density function. So we want $\tilde{\pi}$ and π to be the same function for all a; that is,

$$\pi(\theta/a)/a = \pi(\theta)$$

for all $\theta \in (0, \infty)$, $a \in (0, \infty)$. Take $\theta = 1$. Then we have $\pi(1/a) = a\pi(1)$ for all $a \in (0, \infty)$, implying $\pi(\theta) = \pi(1)/\theta$ for all $\theta \in \mathbb{R}$. That is, the left Haar measure has density proportional to $1/\theta$. Since $L_t = R_t$, the right Haar measure also has density proportional to $1/\theta$. □

Example 5.9 Let Ω_Θ be the parameter space of two parameters, a real number μ and a positive number σ. That is $\Omega_\Theta = \mathbb{R} \times (0, \infty)$. Let $\mathcal{G}$ be the group of transformations

$$g_{b,c}(\mu,\sigma) = (c\mu + b, c\sigma), \quad (b,c) \in \mathbb{R} \times (0,\infty).$$

Then

$$L_{b,c}(\mu,\sigma) = g_{b,c}(\mu,\sigma) = (c\mu + b, c\sigma),$$
$$R_{b,c}(\mu,\sigma) = g_{\mu,\sigma}(b,c) = (\sigma b + \mu, \sigma c).$$

To determine the left Haar measure, let $(\tilde{\mu},\tilde{\sigma}) = L_{b,c}(\mu,\sigma) = (c\mu + b, c\sigma)$. Then

$$(\mu,\sigma) = L_{b,c}^{-1}(\tilde{\mu},\tilde{\sigma}) = \left(\frac{\tilde{\mu} - b}{c}, \frac{\tilde{\sigma}}{c}\right).$$

If the density of (μ,σ) is $\pi(\mu,\sigma)$, then the density of $(\tilde{\mu},\tilde{\sigma})$ is

$$\tilde{\pi}(\tilde{\mu},\tilde{\sigma}) = \pi\left(\frac{\tilde{\mu} - b}{c}, \frac{\tilde{\sigma}}{c}\right) \left|\det\left(\frac{\partial((\tilde{\mu} - b)/c, \tilde{\sigma}/c)}{\partial(\tilde{\mu},\tilde{\sigma})^T}\right)\right|.$$

The determinant on the right-hand side is

$$\det\begin{pmatrix} 1/c & 0 \\ 0 & 1/c \end{pmatrix} = c^{-2}.$$

Hence we have the equation

$$\tilde{\pi}(\tilde{\mu},\tilde{\sigma}) = c^{-2}\pi\left(\frac{\tilde{\mu} - b}{c}, \frac{\tilde{\sigma}}{c}\right).$$

The relation $\Pi \circ L_t^{-1} = \Pi$ implies $\tilde{\pi}$ and π are the same function for all b, c; that is,

$$c^{-2}\pi\left(\frac{\mu - b}{c}, \frac{\sigma}{c}\right) = \pi(\mu,\sigma)$$

for all $\mu, b \in \mathbb{R}$, $c, \sigma \in (0,\infty)$. Take $\mu = 0, c = 1$, and we have

$$c^{-2}\pi\left(-\frac{b}{c}, \frac{1}{c}\right) = \pi(0,1).$$

Let $\mu = -b/c$, $\sigma = 1/c$. Then $c = 1/\sigma$, $b = -\mu/\sigma$, and

$$\sigma^2\pi(\mu,\sigma) = \pi(0,1) \Rightarrow \pi(\mu,\sigma) = \pi(0,1)/\sigma^2.$$

So, the left Haar measure has density proportional to $1/\sigma^2$.

To determine the right Haar measure, let

$$(\tilde{\mu},\tilde{\sigma}) = R_{b,c}(\mu,\sigma) = g_{\mu,\sigma}(b,c) = (\sigma b + \mu, 1/\sigma).$$

Then

$$(\mu,\sigma) = R_{b,c}^{-1}(\tilde{\mu},\tilde{\sigma}) = \left(\tilde{\mu} - \frac{b\tilde{\sigma}}{c}, \frac{\tilde{\sigma}}{c}\right).$$

The density of $(\tilde{\mu},\tilde{\sigma})$ is

$$\tilde{\pi}(\tilde{\mu},\tilde{\sigma}) = \pi\left(\tilde{\mu} - \frac{b\tilde{\sigma}}{c}, \frac{\tilde{\sigma}}{c}\right)\left|\det\left(\frac{\partial(\tilde{\mu} - b\tilde{\sigma}/c, \tilde{\sigma}/c)}{\partial(\tilde{\mu},\tilde{\sigma})^T}\right)\right|,$$

where the determinant on the right-hand side is

$$\det\begin{pmatrix}1 & -b/c\\ 0 & 1/c\end{pmatrix} = c^{-1}.$$

Hence the density of $(\tilde{\mu},\tilde{\sigma})$ reduces to

$$\tilde{\pi}(\tilde{\mu},\tilde{\sigma}) = c^{-1}\pi\left(\tilde{\mu} - \frac{b\tilde{\sigma}}{c}, \frac{\tilde{\sigma}}{c}\right).$$

The relation $\Pi \circ R_t^{-1} = \Pi$ implies $\tilde{\pi}$ and π are the same function for all b, c; that is,

$$c^{-1}\pi\left(\mu - \frac{b\sigma}{c}, \frac{\sigma}{c}\right) = \pi(\mu,\sigma)$$

for all $\mu, b \in \mathbb{R}$, $c, \sigma \in (0,\infty)$. Taking $\mu = 0, c = 1$, we have

$$c^{-1}\pi\left(-\frac{b}{c}, \frac{1}{c}\right) = \pi(0,1).$$

Let $\mu = -b/c$, $\sigma = 1/c$. Then

$$\pi(\mu,\sigma) = \pi(0,1)/\sigma.$$

So, the right Haar measure has density proportional to $1/\sigma$. □

5.6.3 Jeffreys prior

Another important class of noninformative prior is the Jeffreys priors, which is defined as

$$\pi_\Theta(\theta) \propto [\det I(\theta)]^{1/2},$$

where $I(\theta)$ is the Fisher information

$$I(\theta) = -E_\theta\left[\frac{\partial^2 \log f_{X|\Theta}(x|\theta)}{\partial\theta\partial\theta^T}\right].$$

A useful feature of Jeffreys prior is that it satisfies the usual transformation law of measure. Suppose that $\phi = h(\theta)$ is a one-to-one and differentiable transformation of θ. If P_Θ is a noninformative prior distribution we assign to Θ and P_Φ is the noninformative prior distribution we assign to $\Phi = h(\Theta)$. Then it is desirable to have

$$P_\Phi = P_\Theta \circ h^{-1}.$$

Contrary to intuition, this is not automatically satisfied. This is because P_Φ is not defined inherently by $P_\Theta \circ h^{-1}$, but rather it is subjectively assigned, possibly without regard to this transformation law. Let π_Θ and π_Φ be the densities of P_Θ and P_Φ with respect to the Lebesgue measure. Then, in terms of these densities, the above transformation rule is

$$\pi_\Phi(\phi) = \pi_\Theta(h^{-1}(\phi))|\det(\partial h^{-1}(\phi)/\partial \phi^T)|. \tag{5.26}$$

The next theorem shows that Jeffreys prior satisfies the above transformation rule.

Theorem 5.8 *Let π_Θ be Jeffreys prior density for Θ. Let $\Phi = h(\Theta)$, where h is a one-to-one differentiable function. Let π_Φ be the Jeffreys prior density for Φ. Then π_Θ and π_Φ satisfy the tranformation rule (5.26).*

Proof. Let $f_{X|\Phi}(x|\phi)$ be the conditional density of X expressed in ϕ. That is

$$f_{X|\Phi}(x|\phi) = f_{X|\Theta}(x|h^{-1}(\phi)).$$

Take the second partial derivatives with respect to ϕ_i, ϕ_j, $i, j = 1, \ldots, p$, to obtain

$$\begin{aligned}\frac{\partial^2 \log f_{X|\Phi}(x|\phi)}{\partial \phi_i \partial \phi_j} \\ = \sum_{k=1}^{p}\sum_{\ell=1}^{p} \frac{\partial^2 \log f_{X|\Theta}(x|\theta)}{\partial \theta_k \partial \theta_\ell}\frac{\partial \theta_k}{\partial \phi_i}\frac{\partial \theta_\ell}{\partial \phi_j} + \sum_{k=1}^{p} \frac{\partial \log f_{X|\Theta}(x|\theta)}{\partial \theta_k}\frac{\partial^2 \theta_k}{\partial \phi_i \partial \phi_j}.\end{aligned}$$

Thus, taking conditional expectation $E(\cdot|\theta)$ (or equivalently $E(\cdot|\phi)$), and noticing that the second term on the right hand side vanishes after taking this conditional expectation, we have

$$E\left[\frac{\partial^2 \log f_{X|\Phi}(X|\phi)}{\partial \phi_i \partial \phi_j}\middle|\phi\right] = \sum_{k=1}^{p}\sum_{\ell=1}^{p} E\left[\frac{\partial^2 \log f_{X|\Theta}(X|\theta)}{\partial \theta_k \partial \theta_\ell}\middle|\theta\right]\frac{\partial \theta_k}{\partial \phi_i}\frac{\partial \theta_\ell}{\partial \phi_j}.$$

If we let $\partial \theta^T/\partial \phi$ denote the matrix $A_{ik} = \partial \theta_k/\partial \phi_i$ and let $\partial \theta/\partial \phi^T$ denote the transpose of $\partial \theta^T/\partial \phi$, then the above equation becomes

$$I(\phi) = (\partial \theta^T/\partial \phi)I(\theta)(\partial \theta/\partial \phi^T).$$

It follows that

$$\det(I(\phi)) = \det(\partial\theta^T/\partial\phi)\det(I(\theta))\det(\partial\theta/\partial\phi^T) = \det(I(\theta))\left[\det(\partial\theta/\partial\phi^T)\right]^2.$$

Now take square root on both sides the equation to verify (5.26). □

Example 5.10 Suppose that $X_1, \ldots, X_n$ are i.i.d. $N(\theta, \sigma^2)$ variables where σ^2 is known. Then the Fisher information is $I(\theta) = n/\sigma^2$. Hence Jeffreys prior is $\sqrt{n}/\sigma$, which is proportional to the Lebesgue measure. If both μ and σ^2 are unknown, then the Fisher information is the matrix

$$I(\theta) = \begin{pmatrix} n/\sigma^2 & 0 \\ 0 & n/(2\sigma^4) \end{pmatrix}$$

So $\det(I(\theta)) = n^2/(2\sigma^6)$, and the Jeffreys prior is proportional to $1/\sigma^3$. □

5.7 Statistical decision theory

Many topics in Bayesian statistical inference, such estimation, testing, and classification, can be efficiently described within the framework of statistical decision theory Berger (1985).

The statistical decision theory consists of several elements. Let $(\Omega_A, \mathcal{F}_A)$ be a measurable space, where Ω_A is called the action space. For example, if we estimate a parameter Θ, then Ω_A is typically the same space as Ω_Θ. If our goal is to test a hypothesis, then Ω_A is $\{0, 1\}$, where 1 represents rejection. If our goal is classification, then Ω_A is a list of categories.

Any mapping $d : \Omega_X \to \Omega_A$ that is measurable with respect to $\mathcal{F}_X/\mathcal{F}_A$, and such that $L(\theta, d(X))$ is integrable with respect to $P_{X|\Theta}(\cdot|\theta)$ for any $\theta \in \Omega_\Theta$ is called a decision rule. A decision rule is a statistic — typically an estimate, a test, or a classifier. The class of all decision rules is written as $\mathcal{D}$.

A mapping $L : \Omega_\Theta \times \Omega_A \to \mathbb{R}$ is called a loss function. It represents the error one makes by taking a certain action in Ω_A. For example, if our action is to reject or accept a hypothesis, then it is either right or wrong; In this case L takes values in $\{0, 1\}$, representing right and wrong. If our action is to estimate Θ, then the loss may be Euclidean distance $\|\Theta - a\|$.

A loss is something that has already happened: if we took an action $a \in \Omega_A$, and it turned out the value θ of Θ is quite different from our action a, and then we lose, or lose a certain amount. Of course a statistician's job is to prevent or to minimize the loss *before* it happens. That means we need to be able to access a loss before it happens. This leads us to the notion of the risk, which is the expectation of loss. There are three ways of taking this expectation. The first is to condition on Θ:

$$R(\theta, d) = E[L(\Theta, d(X)|\Theta)]_\theta.$$

This is called the frequentist risk. It is a mapping from $\Omega_\Theta \times \mathcal{D} \to \mathbb{R}$. The second is to take expectation conditioning on X:

$$\rho(x, a) = E[L(\Theta, a)|X]_x.$$

This is called the posterior expected loss. It is a mapping from $\Omega_X \times \Omega_A \to \mathbb{R}$. Finally we can take the unconditional expectation

$$r(d) = E[L(\Theta, d(X))] = E[ER(\Theta, d(X)|\Theta)].$$

This is called the Bayes risk. It is a mapping from $\mathcal{D} \to \mathbb{R}$.

One of the most important principles for choosing a decision is the Bayes rule, defined as follows.

Definition 5.9 *The Bayes rule d_B is defined as*

$$d_B = \text{argmin}\{r(d) : d \in \mathcal{D}\}.$$

If P_Θ is improper, then this rule is called the generalized Bayes rule.

In appearance the Bayes rule is the minimizer of $r(d)$ over a class of functions $\mathcal{D}$, which is in general a difficult problem. However, under mild conditions this can be converted to a finite dimensional minimization problem using Fubini's theorem.

Theorem 5.9 *If $L(\theta, a) \geq C$ for some $C > -\infty$ for all $\theta \in \Omega_\Theta$ and $a \in \Omega_A$, then the decision rule*

$$\Omega_X \to \Omega_A, \quad x \mapsto \text{argmin}\{\rho(x, a) : a \in \Omega_A\} \tag{5.27}$$

is a Bayes rule.

Proof. By definition,

$$r(d) = \int_{\Omega_\Theta} \int_{\Omega_X} L(\theta, d(x)) f_{X|\Theta}(x|\theta) d\mu_X(x) \pi(\theta) d\mu_\Theta(\theta).$$

Since the loss function is bounded from below, we interchange the order of the integrals by Tonelli's theorem. That is

$$\begin{aligned} r(d) &= \int_{\Omega_X} \int_{\Omega_\Theta} L(\theta, d(x)) \pi_{\Theta|X}(\theta|x) d\mu_\Theta(\theta) f_X(x) d\mu_X(x) \\ &= \int_{\Omega_X} \rho(x, d(x)) f_X(x) d\mu_X(x). \end{aligned}$$

Now let $d_0 : \Omega_X \to \Omega_A$ be the decision rule

$$d_0(x) = \text{argmin}\{\rho(x, a) : a \in \Omega_A\}.$$

Then for any $d \in \mathcal{D}$,

$$\begin{aligned} r(d_0) &= \int_{\Omega_X} \rho(x, d_0(x)) f_X(x) d\mu_X(x) \\ &\le \int_{\Omega_X} \rho(x, d(x)) f_X(x) d\mu_X(x) = r(d). \end{aligned}$$

In other words, d_0 is a Bayes rule. □

Note that, to compute the posterior expected loss, we do not need the marginal density $f_X(x)$, because

$$E(L(\Theta, a)|X)_x \propto \int_{\Omega_X} L(\theta, a) f_{X|\Theta}(x|\theta) \pi_\Theta(\theta) d\mu_\Theta(\theta)$$

with a proportional constant $(1/f_X(x))$ that does not depend on a. So it is equivalent to define the d_0 in Theorem 5.9 as

$$\text{argmin}\{\int_{\Omega_X} L(\theta, a) f_{X|\Theta}(x|\theta) \pi_\Theta(\theta) d\mu_\Theta(\theta) : \ a \in \Omega_A\}. \tag{5.28}$$

The generalized Bayes rule can also be computed using (5.28).

Another optimal criterion in decision theory is admissibility. In the following we assume that every decision rule in $\mathcal{D}$ has an integrable risk $R(\theta, d)$ with respect to π_Θ, where π_Θ can be a proper or improper.

Definition 5.10 *A decision rule is $d \in \mathcal{D}$ is inadmissible if there is a decision rule d_1 such that*

$$\begin{aligned} &R(\theta, d_1) \le R(\theta, d) \text{ for all } \theta \in \Omega_\Theta, \\ &R(\theta, d_1) < R(\theta, d) \text{ for some } \theta \in \Omega_\Theta. \end{aligned}$$

A decision rule is admissible if it is not inadmissible.

Admissibility is a uniform (in θ) property; whereas Bayes is an average property. Admissibility is a weak optimality property, because it only presents itself from being uniformly worse than any other decision rules. Bayes rule, on the other hand, does require itself to be better than all other rules, albeit according to a criterion weaker than the uniform criterion used in admissibility. It is then not surprising that under some mild conditions, a Bayes rule is admissible.

Theorem 5.10 *Suppose*

1. *for each $d \in \mathcal{D}$, $R(\theta, d)$ is integrable with respect to P_Θ;*
2. *for any $d_1, d_2 \in \mathcal{D}$,*

$$\begin{aligned} &R(\theta, d_2) - R(\theta, d_1) < 0 \text{ for some } \theta \in \Omega_\Theta \\ &\qquad \Rightarrow P_\Theta(R(\theta, d_2) - R(\theta, d_1) < 0) > 0. \end{aligned}$$

Then any Bayes or generalized Bayes rule is admissible.

Proof. Suppose d_1 is an inadmissible Bayes rule in $\mathcal{D}$. Then there is $d_2 \in \mathcal{D}$ such that $R(\theta, d_2)$ is no more than $R(\theta, d_1)$ for all $\theta \in \Omega_\Theta$ and is strictly less than $R(\theta, d_1)$ for some θ. Hence

$$\int_{\Omega_\Theta} [R(\theta, d_2) - R(\theta, d_1)] dP_\Theta = \int_{R(\theta,d_2)-R(\theta,d_1)<0} [R(\theta, d_2) - R(\theta, d_1)] dP_\Theta$$

Since $P_\Theta(R(\theta, d_2) - R(\theta, d_1) < 0) > 0$, the right hand side is negative, which contradicts the assumption that d_1 is Bayes. □

Here, we have implicitly used assumption 1, because without it the difference

$$\int_{\Omega_\Theta} [R(\theta, d_2) - R(\theta, d_1)] dP_\Theta$$

may not be defined. Assumption 2 of the theorem covers two interesting cases. First, suppose $\Omega_\Theta = \{\theta_1, \theta_2, \ldots\}$ is countable and π_Θ is positive for each θ_i, then this assumption is obviously satisfied. Second, if $R(\theta, d)$ is continuous in θ for each d, and $P_\Theta(A) > 0$ for any nonempty open set, then it is also satisfied. See Problem 5.35.

Problems

5.1. Let $(\Omega_X, \mathcal{F}_X)$, $(\Omega_\Theta, \mathcal{F}_\Theta)$ be measurable spaces. Let

$$\Omega = \Omega_X \times \Omega_\Theta, \quad \mathcal{F} = \mathcal{F}_X \times \mathcal{F}_\Theta.$$

So $(\Omega, \mathcal{F})$ is a measurable space. Let X be the random element

$$X : \Omega \to \Omega, \quad (x, \theta) \mapsto x.$$

Show that $\sigma(X) = \{A \times \Omega_\Theta : A \in \mathcal{F}_X\}$.

5.2. Let Ω_1 and Ω_2 be two sets, and $T : \Omega_1 \to \Omega_2$ be any function. Show that

1. $T^{-1}(\Omega_2) = \Omega_1$;
2. for any $A \subseteq \Omega_2$, $T^{-1}(A^c) = (T^{-1}(A))^c$;
3. if $A_1, A_2, \ldots$ is a sequence of subsets of Ω_2, then

$$\cup_{n=1}^\infty T^{-1}(A_n) = T^{-1}(\cup_{n=1}^\infty A_n).$$

5.3. Let Ω_1, Ω_2 be two sets, and $\mathcal{G}_1$ be a σ-field in Ω_1. Let $T : \Omega_1 \to \Omega_2$ be any function. Let $\mathcal{B} = \{B : T^{-1}(B) \subseteq \mathcal{G}_1\}$. Show that $\mathcal{B}$ is a σ-field in Ω_2.

5.4. Let $(\Omega_1, \mathcal{F}_1)$ and $(\Omega_2, \mathcal{F}_2)$ be measurable space, and $T : \Omega_1 \to \Omega_2$ be a surjection that is measurable $\mathcal{F}_1/\mathcal{F}_2$. Let $\mathcal{A}$ be a subclass of $\mathcal{F}_2$ such that $\sigma(\mathcal{A}) = \mathcal{F}_2$. Show that $\sigma(T^{-1}(\mathcal{A})) = T^{-1}(\sigma(\mathcal{A}))$.

5.5. Let $(\Omega_X, \mathcal{F}_X)$, $(\Omega_\Theta, \mathcal{F}_\Theta)$ be measurable spaces. Let

$$\Omega = \Omega_X \times \Omega_\Theta, \quad \mathcal{F} = \mathcal{F}_X \times \mathcal{F}_\Theta.$$

So $(\Omega, \mathcal{F})$ is a measurable space. Let X be the random element

$$X : \Omega \to \Omega_X, \quad (x, \theta) \mapsto x.$$

Show that $\sigma(X) = \{A \times \Omega_\Theta : A \in \mathcal{F}_X\}$.

5.6. Let $(\Omega_X, \mathcal{F}_X)$, $(\Omega_\Theta, \mathcal{F}_\Theta)$, $(\Omega, \mathcal{F})$, and X be as defined in the previous problem. Let $(\Omega_T, \mathcal{F}_T)$ be another measurable space. Let $T : \Omega_X \to \Omega_T$ be a function measurable $\mathcal{F}_X / \mathcal{F}_T$.

1. Show that

$$\sigma(T \circ X) = \{T^{-1}(A) \times \Omega_\Theta : A \in \mathcal{F}_X\}.$$

2. Suppose, in addition, T is surjective. Let Θ be the random element

$$\Theta : \Omega \to \Omega_\Theta, \quad (x, \theta) \mapsto \theta.$$

Let $(T \circ X, \Theta)$ be the random element

$$(T \circ X, \Theta) : \Omega \to \Omega_T \times \Omega_\Theta, \quad (x, \theta) \mapsto (T(x), \theta).$$

Show that

$$\sigma(T \circ X, \Theta) = \sigma\{T^{-1}(C) \times D : C \in \mathcal{F}_T, D \in \mathcal{F}_\Theta\}.$$

3. Let

$$\mathcal{P} = \{T^{-1}(C) \times D : C \in \mathcal{F}_T, D \in \mathcal{F}_\Theta\}.$$

Sow that $\mathcal{P}$ is a π-system.

5.7. Let $(\Omega_X, \mathcal{F}_X)$, $(\Omega_\Theta, \mathcal{F}_\Theta)$, $(\Omega_T, \mathcal{F}_T)$, $(\Omega, \mathcal{F})$, X, Θ, T, and $\mathcal{P}$ be as defined in the previous problem. Let $A \in \mathcal{F}_X$. Define

$$Q_1(B) = E[I_B E(X^{-1}(A) | T \circ X, \Theta)], \quad Q_2(B) = E[I_B E(X^{-1}(A) | T \circ X)].$$

Show that $\mathcal{L} = \{B \in \sigma(\mathcal{P}) : Q_1(B) = Q_2(B)\}$ is a λ-system.

5.8. Let $(\Omega_X, \mathcal{F}_X, \mu_X)$, $(\Omega_\Theta, \mathcal{F}_\Theta, \mu_\Theta)$ be σ-finite measure spaces. Let P be a probability measure defined on $(\Omega_X \times \Omega_\Theta, \mathcal{F}_X \times \mathcal{F}_\Theta)$ with $P \ll \mu_X \times \mu_\Theta$.

1. Show that $P \circ X^{-1} \ll \mu_X$ and $P \circ \Theta^{-1} \ll \mu_\Theta$.

2. Let

$$f_{X|\Theta}(x|\theta) = \begin{cases} [dP/d(\mu_X \times \mu_\Theta)]/[dP\circ X^{-1}/d\mu_X] & \text{if } dP\circ X^{-1}/d\mu_X = 0 \\ 0 & \text{if } dP\circ X^{-1}/d\mu_X = 0 \end{cases}$$

Show that, for each $A \in \mathcal{F}_X$, the function

$$\theta \mapsto \int_A f_{X|\Theta}(x|\theta) d\mu_X(x)$$

is (a version of) $P(A|\Theta)$.

5.9. Suppose that $\Theta = (\Psi, \Lambda)$. Suppose $T(X)$ is a statistic that is sufficient for λ for each fixed ψ. Show that, for any $G \in \mathcal{F}_\Theta$,

$$P(\Theta^{-1}(G)|X, \Psi) = \pi(\Theta^{-1}(G)|T(X), \Psi).$$

5.10. Let $P_\theta = P_{X|\Theta}(\cdot|\theta)$. Show that the mapping

$$(x, \theta) \mapsto P_\theta(A|T\circ X)_x$$

is a version of the conditional probability

$$(x, \theta) \mapsto P(A|T\circ X, \Theta)_{(x,\theta)}.$$

5.11. Suppose that $X|\theta$ is a p-dimensional random vector with distribution $N(\theta, \Sigma)$, where Σ is a p by p positive definite matrix. This matrix is treated as the non-random parameter. The random parameter Θ is also distributed as p-dimensional multivariate normal $N(\mu, \Omega)$, where Ω is a p by p positive definite matrix. Here μ and Ω are treated as non-random parameters. Find the distribution of $\Theta|X$ as well as the marginal distribution of X.

5.12. Suppose that $X_1, \ldots, X_n|\theta$ are independent p-dimensional random vectors distributed as $N(\theta, \Sigma)$, where $\Sigma > 0$ (this means Σ is positive definite). Suppose that θ is distributed as $N(\mu, \Omega)$ where $\Omega > 0$. Find the posterior distribution $\theta|X_1, \ldots, X_n$. Write your result in its most interpretable form.

5.13. Let X and Y be two random elements. Suppose that the conditional density of $Y|X$ does not depend on X; that is $f(y|x) = h(y)$ for some function h. Then X and Y are independent.

5.14. Suppose that $\Phi \sim \chi_\nu^{-2}$ and $\Lambda|\phi \sim N(0, \phi)$.

1. Show that $\sqrt{\nu}\Lambda \sim t_{(\nu)}$.
2. Show that $\Phi|\lambda \sim (1 + \lambda^2)\chi_{\nu+1}^{-2}$ and that $\Phi/(1 + \Lambda^2) \perp\!\!\!\perp \Lambda$.

5.15. Show that, if

$$T|\phi \sim \phi\chi^2_{(m)}, \quad \Phi \sim \tau\chi^{-2}_{(k)},$$

then $\Phi|t \sim (t+\tau)\chi^{-2}_{(m+k)}$. From this deduce that if $X_1, \ldots, X_n$ are i.i.d. $N(\lambda, \phi)$, where λ is treated as a constant, and if $\phi \sim \tau\chi_k^{-2}$ then

$$\Phi|(x_1, \ldots, x_n) \sim \left(\sum_{i=1}^n (x_i - \lambda)^2 + \tau \right) \chi^{-2}_{(n+k)}.$$

5.16. Prove Proposition 5.1.

5.17. Prove Proposition 5.5.

5.18. Prove Proposition 5.6.

5.19. Prove Proposition 5.7.

5.20. Show that, if

$$T|\phi \sim \phi\chi^2_{(m)}, \quad \pi_\Phi(\phi) = 1/\phi, \ \phi > 0, \ k = 1, 2, \ldots$$

then $\Phi|t \sim t\chi^{-2}_{(m)}$.

5.21. Show that, in general, if

$$T|\phi \sim \phi\chi^2_{(m)}, \quad \pi_\Phi(\phi) = 1/\phi^{k/2}, \ \phi > 0, \ k = 1, 2, \ldots$$

then $\Phi|t \sim t\chi^{-2}_{(m+k-2)}$.

5.22. Suppose $X_1, \ldots, X_n|\lambda, \phi$ are i.i.d. $N(\lambda, \phi)$ random variables, where both λ and ϕ are unknown. Let (S, T) be as defined in (5.16). Let π_Θ be the invariant noninformative prior

$$\pi_\Theta(\lambda, \phi) = 1/\phi^{3/2}.$$

1. Find the marginal posterior distribution for $\Phi|X$.
2. Find the conditional posterior distribution for $\Lambda|\Phi, X$.
3. Find the marginal posterior distribution for $\Lambda|X$.

5.23. Suppose we have the following linear regression model

$$Y_i = x_i B + \varepsilon_i, \quad i = 1, \ldots, n,$$

where $\varepsilon_1, \ldots, \varepsilon_n|\phi$ are i.i.d. $N(0, \phi)$ variables, and $x_1, \ldots, x_n$ are nonrandom constants. Let

$$T = \sum_{i=1}^n x_i Y_i / \sum_{i=1}^n x^2, \quad S = \sum_{i=1}^n (Y_i - Tx_i)^2.$$

1. Show that $S \perp\!\!\!\perp T|(B, \Phi)$ and $S \perp\!\!\!\perp B|\Phi$.
2. Find a conjugate prior for $f_{S|\Phi}$.

3. For each fixed $\phi \in \Omega_\Phi$, find a conjugate prior for the conditional density $(t, \lambda) \mapsto f_{T|\Phi\Lambda}(t|\phi, \lambda)$.
4. Find the joint posterior distribution of (Λ, Φ).
5. Find the marginal posterior distribution of Λ.

5.24. Suppose

$$\Phi|(S = s) \sim s\chi^{-2}_{(m)}, \quad T|(\Phi = \phi) \sim N(a, c\phi),$$

where $a \in \mathbb{R}$, $c > 0$. Then

$$\pi_{\Phi|S}(\phi|s) f_{T|\Phi}(t|\phi) \propto [s + (t-a)^2/c] g(\phi),$$

where $g(\phi)$ is the density of $\chi^{-2}_{(m+1)}$, and the proportionality constant may depend on s, t but does not depend on ϕ.

5.25. Suppose X is a random variable with density of the form

$$\frac{1}{\sqrt{\phi}} f\left(\frac{x^2}{\phi}\right),$$

where $x \in \mathbb{R} = \Omega_X$ and $\phi \in (0, \infty) = \Omega_\Phi$. Consider the set of transformations

$$h_c(x) = cx, \quad c \in (0, \infty).$$

1. Show that $\mathcal{H} = \{h_c : c \in (0, \infty)\}$ is a group of transformations from Ω_X to Ω_X.
2. Determine the transformation g_c from Ω_Φ to Ω_Φ induced by the transformation h_c.
3. Show that the set of transformations $\mathcal{G} = \{g_c : c \in (0, \infty)\}$ is a group.
4. Find the left and right Haar measures for the group $\mathcal{G}$.

5.26. Suppose X is a random variable with density of the form

$$\frac{1}{\sqrt{\phi}} f\left(\frac{(x-\mu)^2}{\phi}\right).$$

Let $\theta = (\mu, \phi)$ and $\Omega_\Theta = \mathbb{R} \times \mathbb{R}^+$, where $\mathbb{R}^+ = (0, \infty)$, be the parameter space. Consider the set of transformations

$$\mathcal{H} = \{h_{b,c}(x) = cx + b : \quad b \in \mathbb{R}, \ c \in \mathbb{R}^+\}.$$

1. Show that $\mathcal{H}$ is a group of transformations from Ω_X to Ω_X.
2. Determine the transformation $g_{b,c}$ from Ω_Θ to Ω_Θ induced by the transformation $h_{b,c} \in \mathcal{H}$.
3. Show that $\mathcal{G} = \{g_{b,c} : b \in \mathbb{R}, c \in \mathbb{R}^+\}$ is a group of transformations from Ω_Θ to Ω_Θ.
4. Find the left and right Haar measures for the group $\mathcal{G}$.

5.27. Suppose X is a random vector of dimension p with density of the form

$$\frac{1}{\det(\Sigma)^{1/2}} f\left(x^T \Sigma^{-1} x\right),$$

where Σ is a member of $\mathbb{R}_+^{p \times p}$, the set of all positive definite matrices. Consider the set of transformations of the form

$$\mathcal{H} = \{h_C(x) = Cx : C \in \mathbb{R}_+^{p \times p}\}.$$

1. Show that $\mathcal{H}$ is a group of transformations from Ω_X to Ω_X.
2. Determine the transformation g_C induced by a transformation $h_C \in \mathcal{H}$.
3. Show that the set of transformations $\mathcal{G} = \{g_C : C \in \mathbb{R}_+^{p \times}\}$ is a group.
4. Find the left and right Haar measures for this group.

(Hint: To solve this problem, we need to take derivative of vec(ABC) with respect to vec(B), where A, B, C are matrices, and vec is the vectorization operator that stacks the columns of the argument matrix. Since vec$(ABC) = (C \otimes A^T)$vec(B), where $\otimes$ is the Kronecker product between matrices, we have $\partial \text{vec}(ABC)/\partial \text{vec}(B)^T = (C \otimes A^T)$. Also, the following identity will be useful: if A and B are square matrices with dimensions n_1 and n_2, respectively, then $\det(A \otimes B) = \det(A)^{n_1} \times \det(B)^{n_2}$.)

5.28. Suppose X is a random vector of dimension p with density of the form

$$\frac{1}{\det(\Sigma)^{1/2}} f\left((x-\mu)^T \Sigma^{-1}(x-\mu)\right),$$

where $\mu \in \mathbb{R}$, $\Sigma \in \mathbb{R}_+^{p \times p}$. Consider the set of transformations

$$\mathcal{H} = \{h_{b,C}(x) = Cx + b : b \in \mathbb{R}^p,\ C \in \mathbb{R}_+^{p \times p}\}.$$

Let $\theta = (\mu, \Sigma)$ be the whole parameter and let $\Omega_\Theta = \mathbb{R}^p \times \mathbb{R}_+^{p \times p}$ be the parameter space.

1. Show that $\mathcal{H}$ is a group of transformations from Ω_X to Ω_X.
2. Determine the transformation $g_{b,C}$ from Ω_Θ to Ω_Θ induced by a transformation $h_{b,C} \in \mathcal{H}$.
3. Show that the set of transformations $\mathcal{G} = \{g_{b,C} : b \in \mathbb{R}, C \in \mathbb{R}_+^{p \times p}\}$ is a group.
4. Calculate the left Haar measure for this group.

5.29. Let $\Omega_\Gamma = \mathbb{R}_+^{p \times p}$, the set of all positive definite matrices in $\mathbb{R}^{p \times p}$. Consider the set of transformations on Ω_Γ defined as

$$\mathcal{G} = \{(\Gamma \mapsto C^{1/2} \Gamma C^{1/2}) : C \in \mathbb{R}_+^{p \times p}\}.$$

1. Show that $\mathcal{G}$ is a group of transformations on Ω_Γ;
2. Find the left Haar measure on Ω_Γ with respect to $\mathcal{G}$;

3. Find the right Haar measure on Ω_Γ with respect to $\mathcal{G}$.

5.30. Let $\Omega_\Gamma = \{(\mu, \Gamma) : \mu \in \mathbb{R}^p, \ \Gamma \in \mathbb{R}_+^{p\times p}\}$. Consider the set of transformations on Ω_Γ defined as

$$\mathcal{G} = \{(\mu, \Gamma) \mapsto (C\mu + d), C^{1/2}\Gamma C^{1/2}) : \ d \in \mathbb{R}, C \in \mathbb{R}_+^{p\times p}\}.$$

1. Show that $\mathcal{G}$ is a group of transformations on Ω_Γ;
2. Find the left Haar measure on Ω_Γ with respect to $\mathcal{G}$;
3. Find the right Haar measure on Ω_Γ with respect to $\mathcal{G}$.

5.31. Let X be a random vector defined on $(\mathbb{R}^p, \mathcal{R}^p)$. Suppose the family of distributions of X has densities of the form

$$[1/\det(\Lambda)] f(\Gamma\Lambda\Gamma^T(x-\mu)), \quad \mu \in \mathbb{R}^p, \ \Lambda \in \mathbb{D}^{p\times p}, \ \Sigma > 0,$$

where $\mathbb{D}^{p\times p}$ is the class of all $p \times p$ diagonal matrices with positive diagonal entries, and Γ is a known orthogonal matrix. The parameter of this model is $\Theta = (\mu, \mathrm{diag}(\Lambda))$, where $\mathrm{diag}(\Lambda)$ is the vector of diagonal entries of Λ. Let $\mathcal{A}$ be a class of matrices of the form

$$\{\Gamma\Delta\Gamma^T : \ \Delta \in \mathbb{D}\}.$$

Let $\mathcal{G}$ be the class of all transformations from $\mathbb{R}^p$ to $\mathbb{R}^p$ defined

$$x \mapsto Ax + b, \quad A \in \mathcal{A}, \ b \in \mathbb{R}^p.$$

1. Show that $\mathcal{G}$ is a group.
2. Show that the transformation $\tilde{g} : \Omega_\Theta \to \Omega_\Theta$ induced by the transformation $g : \Omega_X \to \Omega_X$, $g(x) = cx + b$, $c > 0$, is of the form

$$\tilde{g}(\mu, \mathrm{diag}(\Lambda)) = (A\mu + b, \mathrm{diag}(D) \odot \mathrm{diag}(\Lambda)),$$

   where $\odot$ is the Hadamard, or entry-wise, product.
3. Derive the left and right Haar measures for the group of transformations in part 2.

5.32. Suppose $X_1, \ldots, X_n$ are i.i.d. $N(\mu, \sigma^2)$ where μ is known. Derive Jeffreys prior for σ^2.

5.33. Suppose $X_1, \ldots, X_n$ are i.i.d. p-dimensional multivariate Normal $N(\mu, \Sigma)$ where $\mu \in \mathbb{R}^p$ and $\Sigma \in \mathbb{R}_+^{p\times p}$. Derive Jeffreys prior for (μ, Σ).

5.34. 1. If X has a Gamma distribution with parameterization

$$\frac{1}{\Gamma(k)\theta]^k} x^{k-1} e^{-x/\theta}, \quad x > 0, \ \theta > 0, \ k > 0,$$

find the Jeffreys prior for (θ, k).

2. If X has a Gamma distribution with parameterization

$$\frac{\beta^\alpha}{\Gamma(\alpha)} x^{\alpha-1} e^{-\beta x}, \quad x > 0,\ \alpha > 0,\ \beta > 0,$$

find the Jeffreys prior for (α, β).

5.35. Show that, if $R(\theta, d)$ is continuous in θ for each $d \in \mathcal{D}$, and $P_\Theta(A) > 0$ for any nonempty open set $A \in \mathcal{F}_\Theta$, then, for any $d_1, d_2 \in \mathcal{D}$,

$$\begin{aligned} &R(\theta, d_2) - R(\theta, d_1) < 0 \text{ for some } \theta \in \Omega_\Theta \\ &\qquad \Rightarrow P_\Theta(R(\theta, d_2) - R(\theta, d_1) < 0) > 0. \end{aligned}$$

References

Berger, J. O. (1985). *Statistical decision theory and Bayesian analysis.* Second edition. Springer, New York.

Lee, P. M. (2012). *Bayesian Statistics: An Introduction, Fourth Edition.* Wiley.

Mardia, K. V., Kent, J. T. and Bibby, J. M. (1979). *Multivariate Analysis.* Academic.

O'Hagan, A. (1994). *Kendall's Advanced Theory of Statistics: Bayesian Inference, Volume 2B.* Edward Arnold.

6

Bayesian Inference

Based on the concepts and preliminary results introduced in the last chapter, we develop the Bayesian methods for statistical inference, including estimation, testing, and classification, in this chapter. We will focus on Bayesian rules under different settings. All three problems can be formulated as decision theoretic problems described in Section 5.7, with different parameter spaces, action space, and loss functions. For estimation, we take $\Omega_A = \Omega_\Theta = \mathbb{R}^p$; for hypothesis testing, we take Ω_A as a set of two elements — accept or reject; for classification, we take $\Omega_\Theta = \Omega_A$ as a finite set representing a list of categories. We will also explore some important special topics in Bayesian analysis, such as empirical Bayes and Stein's estimator.

6.1 Estimation

In a Bayesian estimation problem, the parameter Θ is typically a random vector whose distribution P_Θ is dominated by the Lebesgue measure. Also, it is natural to assume $\Omega_\Theta = \Omega_A$. The most commonly used loss function is the L_2-loss function, as defined below.

Definition 6.1 *The L_2-loss function is defined as*

$$L(\Theta, a) = (\Theta - a)^T W(\Theta)(\Theta - a), \tag{6.1}$$

where $W(\theta) \in \mathbb{R}^{p\times p}$ is a positive definite matrix for all $\theta \in \Omega_\Theta$.

Recall that a Bayes rule can be calculated by minimizing the posterior expected loss. For L_2 loss this has an explicit solution, described in the next Theorem.

Theorem 6.1 *Suppose*

1. $E[\Theta^T W(\Theta)\Theta|X] < \infty \quad [P]$;

B. Li and G. J. Babu, *A Graduate Course on Statistical Inference*, Springer Texts in Statistics, https://doi.org/10.1007/978-1-4939-9761-9_6

2. $E[W(\Theta)|X] > 0 \ [P]$.

Then the Bayes rule with respect to loss function (6.1) is

$$d_B(X) = [E(W(\Theta)|X)]^{-1}E[W(\Theta)\Theta|X]. \tag{6.2}$$

Proof. By definition,

$$E[W(\Theta)(\Theta - d_B(X))|X] = E[W(\Theta)\Theta|X] - E[W(\Theta)|X]d_B(X) = 0.$$

Hence, for any $a \in \Omega_A$,

$$\begin{aligned} &E[(d_B(X) - a)^T W(\Theta)(\Theta - d(X))|X] \\ &\qquad = (d_B(X) - a)^T E[W(\Theta)(\Theta - d_B(X))|X] = 0. \end{aligned}$$

Consequently,

$$\begin{aligned} &E[L(\Theta, a)|X] \\ &\quad = E[(\Theta - d_B(X))^T W(\Theta)(\Theta - d_B(X))|X] \\ &\qquad + E[(d_B(X) - a)^T W(\Theta)(d_B(X) - a)|X] \geq E[L(\Theta, d_B(X))|X]. \end{aligned}$$

That is, $d_B(X)$ is a Bayes rule. □

It can be shown that the Bayes rule for the L_2-loss (6.1) is unique modulo P. See Problem 6.3. Now let us turn to the L_1-loss function. We first consider the one-dimensional case.

Definition 6.2 *If* $p = 1$, *then the* L_1*-loss is the function*

$$L(\theta, a) = |\theta - a|. \tag{6.3}$$

We would like to minimize the posterior expected loss

$$E(|\Theta - a|\,|X).$$

We will show that the minimizer is the posterior median. We first give a general definition of median.

Definition 6.3 *Let* U *be a random variable with a cumulative distribution function* F. *Then any number* m *satisfying*

$$F(m) \geq 1/2, \quad F(m-) \leq 1/2$$

is called a median of U.

In the next theorem U is a generic random variable that takes values in Ω_U.

Theorem 6.2 *If U is integrable and m is a median of U, then*

$$\int |U - m| dP \leq \int |U - a| dP$$

for all $a \in \Omega_U$.

Proof. Suppose $a > m$, $a \in \Omega_U$. Then

$$\begin{aligned}
&\int_{\Omega_U} (|U - m| - |U - a|)\, dP \\
&\quad = \int_{U \leq m} ((m - U) - (a - U))\, dP + \int_{m < U \leq a} ((U - m) - (a - U))\, dP \\
&\qquad + \int_{U > a} ((U - m) - (U - a))\, dP \\
&\quad = (m - a)P(U \leq m) + (a - m)P(U > a) + \int_{m < U \leq a} (2U - m - a)\, dP.
\end{aligned}$$

Because the last term on the right side is no more than $(a - m)P(m < U \leq a)$, $E|U - m| - E|U - a|$ is no more than

$$\begin{aligned}
&(m - a)\left[P(U \leq m) - P(U > a) - P(m < U \leq a)\right] \\
&\qquad = (m - a)\left[P(U \leq m) - P(U > m)\right] = (m - a)\left[2F(m) - 1\right].
\end{aligned}$$

Similarly, for $a < m$, $a \in \Omega_U$, we have

$$\begin{aligned}
&\int_{\Omega_U} (|U - m| - |U - a|)\, dP \\
&\quad = \left(\int_{U < a} + \int_{a \leq U < m} + \int_{U \geq m}\right) (|U - m| - |U - a|)\, dP \\
&\quad = (m - a)P(U < a) + (a - m)P(U \geq m) + \int_{a \leq U < m} (m + a - 2U)\, dP \\
&\quad \leq (a - m)\left[1 - 2F(m-)\right].
\end{aligned}$$

To summarize, we have

$$E|U - m| - E|U - a| \leq \begin{cases} (m - a)\left[2F(m) - 1\right] & \text{if } m < a \\ (a - m)\left[1 - 2F(m-)\right] & \text{if } a < m \end{cases}$$

Because m is a median of U,

$$2F(m) - 1 \geq 0, \quad 1 - 2F(m-) \geq 0.$$

Hence $E|U - m| - E|U - a| \leq 0$ for all $a \in \Omega_U$. □

Since the L_1-loss is bounded from below, by Theorem 5.8 the posterior median $m(\Theta|X)$ is a Bayes rule.

Corollary 6.1 *The posterior median $m(\theta|X)$ is a Bayes rule with respect to the loss $L(\theta, a) = |\theta - a|$.*

When Θ is a vector, there are more than one way to define a L_1-loss. The one possibility is to use the Euclidean norm

$$L(\theta, a) = \|\theta - a\|.$$

The minimizer of the expectation of this loss is called the *geometric median.* Thus the Bayes rule based on this loss is the posterior geometric median. Another possibility is

$$L(\theta, a) = |\theta_1 - a_1| + \cdots + |\theta_p - a_p|.$$

Since the function is additive, with each term involving on θ_i, the minimization of the posterior expectation is equivalent to the minimization of each $E(|\theta_i - a_i|\,|X)$, so that the Bayes rule is simply the stacked marginal posterior medians. For other choices of multivariate L_1-loss, see Oja (1983); Hettmansperger and Randles (2002).

In the univariate case, one can generalize L_1-loss to the "check function", defined as

$$L(\theta, a) = \begin{cases} \alpha_1(\theta - a) & \text{if } \theta - a \geq 0 \\ \alpha_2(a - \theta) & \text{if } \theta - a < 0 \end{cases} \tag{6.4}$$

where $\alpha_1 > 0$ and $\alpha_2 > 0$. Note that the L_1-loss function is the special case where $\alpha_1 = \alpha_2 = 1$. It can be shown, using the similar argument for Theorem 6.2, that the $\alpha_1/(\alpha_1 + \alpha_2)$th percentile is the Bayes rule with respect to this loss function. See Problem 6.4.

Another commonly used Bayesian estimator is the mode of the posterior distribution, which is also called the generalized maximum likelihood estimator (Berger, 1985, Chapter 4).

Definition 6.4 *The generalized maximum likelihood estimator is*

$$\hat{\theta} = \text{argmax}\{\pi_{\Theta|X}(\theta|x) : \theta \in \Omega_\Theta\}.$$

The generalized maximum likelihood estimator is the point in the parameter space that is most likely to happen according to the posterior distribution.

We now illustrate these estimators by the constrained linear regression. In some cases we know a priori that there should be some restrictions on the parameter Θ. For example, if Θ represents height then we know it is nonnegative. In Bayesian analysis this is handled by restricting the support of the prior distribution according the desired constraint.

Example 6.1 Consider the linear regression model

$$Y_i = \Theta x_i + \varepsilon_i, \; i = 1, \ldots, n,$$

where, $\varepsilon_1, \ldots, \varepsilon_n|\theta$ are i.i.d. $N(0, \sigma^2)$ variables with a known σ^2, and $x_1, \ldots, x_n$ are nonrandom constants. Suppose we know the slope Θ is nonnegative and would like to take this prior knowledge into account in analyzing the data. Then it is natural to assign the improper prior

$$\pi_\Theta(\theta) = I_{[0,\infty)}(\theta).$$

Let $x = (x_1, \ldots, x_n)$ and $y = (y_1, \ldots, y_n)$. By simple algebra, we can show that the likelihood function in this case is proportional to

$$\exp\left[-\frac{1}{2}\left(\frac{x^T x}{\sigma^2}\theta^2 - 2\frac{x^T y}{\sigma^2}\theta\right)\right].$$

By completing the square, we can rewrite this as

$$e^{-(\theta-\nu(x,y))^2/(2\tau^2(x))}$$

where $\nu(x, y) = x^T y / x^T x$, and $\tau^2(x) = \sigma^2 / x^T x$. Thus the posterior density of Θ is

$$\frac{N(\nu(x,y), \tau^2(x))}{\int_0^\infty N(\nu(x,y), \tau^2(x))d\theta} = \frac{N(\nu(x,y), \tau^2(x))}{\int_{-\nu(x,y)/\tau(x)}^\infty N(0,1)d\gamma} = \frac{N(\nu(x,y), \tau^2(x))}{\Phi(\nu(x,y)/\tau(x))}$$

where $N(a, b)$ denotes the density of the Normal distribution $N(a, b)$, and Φ denotes the c.d.f. of $N(0, 1)$. The posterior density of $\Gamma = (\Theta - \nu(x, y))/\tau(x))$ is

$$\frac{N(0,1)}{\Phi(\nu(x,y)/\tau(x))} I_{[0,\infty)}(\gamma).$$

The posterior mean of Γ is then

$$E(\Gamma|Y)_y = \frac{\int_{-\nu(x,y)/\tau(x)}^\infty \gamma N(0,1)d\gamma}{\Phi(\nu(x,y)/\tau(x))} = \frac{e^{-\nu^2(x,y)/(2\tau^2(x))}}{\sqrt{2\pi}\Phi(\nu(x,y)/\tau(x))}.$$

Hence the posterior mean of Θ is

$$E(\Theta|Y)_y = \tau(x)\frac{e^{-\nu^2(x,y)/(2\tau^2(x))}}{\sqrt{2\pi}\Phi(\nu(x,y)/\tau(x))} + \nu(x,y).$$

The posterior median is determined by the equation

$$\int_0^m N(\nu(x,y), \tau^2(x))d\theta = (1/2)\int_0^\infty N(\gamma(x,y), \tau^2(x))d\theta,$$

which is equivalent to

$$\int_{-\nu(x,y)/\tau(x)}^{(m-\nu(x,y))/\tau(x)} N(0,1)d\gamma = (1/2)\int_{-\nu(x,y)/\tau(x)}^\infty N(0,1)d\gamma.$$

Solve this equation to obtain

$$m = \tau(x)\Phi^{-1}(1 - \Phi(\nu(x,y)/\tau(x))/2) + \nu(x,y).$$

The generalized maximum likelihood estimate of Θ is the maximizer of

$$N(\nu(x,y), \tau^2(x))$$

subject to $\theta \geq 0$. So it is $\max(\gamma(x,y), 0)$. □

6.2 Bayes rule and unbiasedness

Recall from Chapter 3 that an estimator $d(X)$ of θ (lower case, as in the frequentist setting) is unbiased if $E_\theta d(X) = \theta$ for all θ. In the Bayesian setting, this means $E(d(X)|\Theta)_\theta = \theta$ for all $\theta \in \Omega_\Theta$. In fact, since conditional expectation is unique modulo P, it is more accurate to write unbiasedness in the Bayesian setting as

$$E(d(X)|\Theta) = \Theta \;\; [P].$$

It is then natural to ask: is the Bayes rule unbiased? The answer is, somewhat surprisingly, no.

Theorem 6.3 *Suppose that $d_B(X)$ is a Bayes rule with respect to the L_2-loss (6.1) and suppose $d_B(X)$ is unbiased. Then*

$$d_B(X) = \Theta \;\; [P].$$

This theorem says that, unless $d_B(X) = \Theta$ modulo P, a Bayes rule with respect to the L_2-loss cannot be unbiased. But $d_B(X) = \Theta$ $[P]$ means we can estimate Θ perfect modulo P — a clearly unrealistic premise. Thus, the theorem implies that, for all practical purposes, the Bayes rule is always biased.

Proof. By Theorem 6.1 and Problem 6.3, d_B takes the form

$$d_B(X) = [E(W(\Theta)|X)]^{-1} E[W(\Theta)\Theta|X] \;\; [P]. \tag{6.5}$$

It suffices to show that $r(d_B) = 0$. In the following we will abbreviate $d(X)$ by d and $W(\Theta)$ by W. Using the iterative law of conditional expectations, we have

$$\begin{aligned} r(d_B) &= E\{E[(\Theta - d_B)^T W(\Theta - d_B)|X]\} \\ &= E[E(\Theta^T W\Theta|X) - 2d_B^T E(W\Theta|X) + d_B^T E(W|X) d_B] \end{aligned}$$

Substitute (6.5) for the second d_B in the last term on the right hand side, to obtain

$$r(d_B) = E[E(\Theta^T W\Theta|X) - d_B^T E(W\Theta|X)] = E(\Theta^T W\Theta - d_B^T W\Theta).$$

Now apply the iterative law of conditional expectations in the reversed order, to obtain

$$r(d_B) = E[\Theta^T W\Theta - E(d_B^T|\Theta)W\Theta] = E(\Theta^T W\Theta - \Theta^T W\Theta) = 0.$$

where the second equality holds because, by unbiasedness of d_B, $E(d_B|\Theta) = \Theta$. □

This theorem is by no means suggests that the Bayes rule with respect to the L_2-loss is undesirable. In fact, many useful estimators are not unbiased, and the unbiasedness seems to be a too strict requirement for many practical purposes.

6.3 Error assessment of estimators

Conceptually, the error of an estimate has very different meanings in the Bayesian and the frequentist setting. In the frequentist setting, θ is fixed and the error of $d(X)$ is measured by how closely the distribution of $d(X)$ clusters around θ. In Bayesian analysis, $d(x)$ is fixed, and its error is measured by how closely the posterior distribution of $\Theta|X$ clusters around $d(x)$. One such measure is the posterior variance $\text{var}(\Theta|X)$. As usual, when Θ is a vector, $\text{var}(\Theta|X)$ is a matrix. Another measurement of error is the posterior mean squared error, which is the matrix

$$\text{mse}_d(\Theta|X) = E[(d(X) - \Theta)(d(X) - \Theta)^T|X].$$

Note that the posterior variance is not estimator specific — it only applies to the Bayes rule d_B for the L_2-loss. In fact, we have

$$\text{var}(\Theta|X) = \text{mse}(\Theta, d_B|X).$$

The posterior mean squared error is specific to the estimator d, and is used as a measurement of error of $d(X)$. It can be shown that (Problem 6.6) the posterior mean squared error and the posterior variance are related by

$$\text{mse}_d(\Theta|X) = \text{var}(\Theta|X) + (d(X) - d_B(X))(d(X) - d_B(X))^T. \tag{6.6}$$

The next example illustrates the calculation of $\text{var}(\Theta|X)$ and $\text{mse}_d(\Theta|X)$.

Example 6.2 Suppose $X_1, \ldots, X_n|\phi$ are i.i.d. $N(0, \phi)$ variables and $\Phi \sim \tau\chi^{-2}_{(\nu)}$. Then, by Problem 5.14,

$$\Phi|x \sim (S + \tau)\chi^{2}_{(\nu+n)}$$

where $S = \sum_{i=1}^n x_i^2$. By Problem 6.5,

$$E(\Phi|X) = \frac{S+\tau}{\nu+n-2}, \quad \mathrm{var}(\Phi|X) = \frac{2(S+\tau)^2}{(\nu+n-2)^2(\nu+n-4)}.$$

Let $d_1(X)$ be the usual unbiased estimator of Φ; that is, $d_1(X) = S/(n-1)$. Then

$$\mathrm{mse}_{d_1}(\Phi|X) = \frac{2(S+\tau)^2}{(\nu+n-2)^2(\nu+n-4)} + \left(\frac{S}{n-1} - \frac{S+\tau}{\nu+n-2}\right)^2.$$

Let $d_2(X)$ be the generalized maximum likelihood estimate, which, by Problem 6.5 is of the form $(S+\tau)/(n+\nu+2)$. Then

$$\mathrm{mse}_{d_2}(\Phi|X) = \frac{2(S+\tau)^2}{(\nu+n-2)^2(\nu+n-4)} + \left(\frac{S+\tau}{n+\nu+2} - \frac{S+\tau}{\nu+n-2}\right)^2.$$

It is often easier to compute mse through relation (6.6) than to compute it directly. □

6.4 Credible sets

The credible set is the Bayesian counterpart of the frequentist confidence set, and has a more direct interpretation. Recall that, in the frequentist setting, a confidence set is a random set that covers a nonrandom parameter with certain probability. In Bayesian analysis, since Θ is random, the credible set is directly defined as a set in $\mathcal{F}_\Theta$ such that Θ falls into it with a certain probability.

Definition 6.5 *A* $100(1-\alpha)\%$ *credible set for* Θ *is any* $C \in \mathcal{F}_\Theta$ *such that* $P(\Theta^{-1}(C)|x) \geq 1-\alpha$.

According to this definition there can be infinitely many $100(1-\alpha)\%$ credible sets. But, just as in the frequentist setting (Section 4.8), we would like to minimize the size of the credible set. That gives rise to another criterion for constructing credible sets – the measure of the credible set. The smaller the measure of the credible set, the better. It turns out that the credible set that has the highest posterior density has the smallest measure. We first define the highest posterior density credible set and then prove its optimality.

Definition 6.6 *The* $100(1-\alpha)\%$ *highest posterior density (HPD) credible set for* Θ *is a* $C \in \mathcal{F}_\Theta$ *of the form*

$$C = \{\theta \in \Omega_\Theta : \pi_{\Theta|X}(\theta|x) \geq \kappa_\alpha\}$$

where κ_α *is the largest constant such that* $P(\Theta^{-1}(C)|x) \geq 1-\alpha$*; that is,*

$$\kappa_\alpha = \sup\{\kappa : P_{\Theta|X}(C(\kappa)|x) \geq 1-\alpha\}$$

where $C(\kappa)$ *is the set* $\{\theta : \pi_{\Theta|X}(\theta|x) \geq \kappa\}$.

Intuitively, when κ increases, the posterior probability of C decreases. The critical point κ_α is the largest value of κ before $P(C|x)$ drops below $1-\alpha$. The following theorem shows that the measure of the HPD credible set is minimal. The argument is somewhat similar to the Neyman Pearson Lemma.

Theorem 6.4 *Suppose that* $\pi_\Theta(\theta) > 0$ *on* Ω_Θ. *Let* C^*_α *be a* $100(1-\alpha)\%$ *HPD credible set and* C_α *be any* $100(1-\alpha)\%$ *credible set. Furthermore, assume that* $P_{\Theta|X}(C^*_\alpha|x) = 1-\alpha$. *Then* $\mu_\Theta(C^*_\alpha) \le \mu_\Theta(C_\alpha)$.

Proof. It suffices to show that any $C \in \mathcal{F}_\Theta$ with $\mu_\Theta(C) < \mu_\Theta(C^*_\alpha)$ cannot be a $100(1-\alpha)\%$ credible set. That is,

$$\mu_\Theta(C) < \mu_\Theta(C^*_\alpha) \Rightarrow P_{\Theta|X}(C|x) < 1-\alpha.$$

Let C be a set in $\mathcal{F}_\Theta$ such that $\mu_\Theta(C) < \mu_\Theta(C^*_\alpha)$. Then

$$\begin{aligned}\mu_\Theta(C^*_\alpha) &= \mu_\Theta(C^*_\alpha \cap C) + \mu_\Theta(C^*_\alpha \setminus C),\\ \mu_\Theta(C) &= \mu_\Theta(C^*_\alpha \cap C) + \mu_\Theta(C \setminus C^*_\alpha).\end{aligned}$$

Because $\mu_\Theta(C) < \mu_\Theta(C^*_\alpha)$ we see that $\mu_\Theta(C \setminus C^*_\alpha) < \mu_\Theta(C^*_\alpha \setminus C)$. Moreover, by construction, $\pi_{\Theta|X}(\theta|x) \ge \kappa_\alpha$ on $C^*_\alpha \setminus C$ and $\pi_{\Theta|X}(\theta|x) \le \kappa_\alpha$ on $C \setminus C^*_\alpha$. Hence

$$\begin{aligned}P_{\Theta|X}(C^*_\alpha \setminus C|x) =& \int_{C^*_\alpha \setminus C} \pi_{\Theta|X}(\theta|x) d\mu_\Theta(\theta)\\ \ge& \kappa_\alpha \mu_\Theta(C^*_\alpha \setminus C)\\ >& \kappa_\alpha \mu_\Theta(C \setminus C^*_\alpha)\\ \ge& \int_{C \setminus C^*_\alpha} \pi_{\Theta|X}(\theta|x) d\mu_\Theta(\theta)\\ =& P_{\Theta|X}(C \setminus C^*_\alpha|x).\end{aligned} \tag{6.7}$$

Because

$$\begin{aligned}P_{\Theta|X}(C^*_\alpha|x) &= P_{\Theta|X}(C^*_\alpha \cap C|x) + P_{\Theta|X}(C^*_\alpha \setminus C|x),\\ P_{\Theta|X}(C|x) &= P_{\Theta|X}(C^*_\alpha \cap C|x) + P_{\Theta|X}(C \setminus C^*_\alpha|x),\end{aligned}$$

the inequality (6.7) implies $P_{\Theta|X}(C|x) < P_{\Theta|X}(C^*_\alpha|x) = 1-\alpha$. So C is not a $(1-\alpha) \times 100\%$ credible set. □

Example 6.3 Consider the model in Example 5.5, where $X_1, \ldots, X_n|\lambda, \phi$ are i.i.d. $N(\lambda, \phi)$ random variables, and the prior distribution of (Λ, Φ) is determined by

$$\Lambda|\phi \sim N(a, \phi/m), \quad \Phi \sim \tau\chi^{-2}_{(\kappa)}.$$

In Example 5.5 we showed that

$$\frac{\sqrt{n+m}(\Lambda - m(X))}{\sqrt{\tau(X)/(n+\kappa)}}|x \sim t_{(n+k)},$$

where

$$\begin{aligned}\tau(x) &= \sum_{i=1}^{n}(x_i - \bar{x})^2 + \tau + (\bar{x} - a)^2(m^{-1} + n^{-1})\\ m(x) &= (n\bar{x} + ma)/(n+m).\end{aligned}$$

So the $100(1-\alpha)\%$ HPD credible set is

$$\left\{\lambda : -t_{(n+\kappa)}(\alpha/2) < \frac{\sqrt{n+m}(\Lambda - m(X))}{\sqrt{\delta(X)/(n+\kappa)}} < t_{(n+\kappa)}(\alpha/2)\right\}.$$

It can be written as the interval

$$\Big(m(x) - t_{(n+\kappa)}(\alpha/2)\sqrt{\delta(x)/[(n+\kappa)(n+m)]},\\ m(x) + t_{(n+\kappa)}(\alpha/2)\sqrt{\delta(x)/[(n+\kappa)(n+m)]}\Big).$$

Also, in Example 5.5 we derived that

$$\frac{\Phi}{\tau(X)}|x \sim \chi^{-2}_{(n+\kappa)}$$

So, if we let $h(\phi)$ denote the density of $\chi^{-2}_{(n+\kappa)}$, then the $100(1-\alpha)$ percent credible set has the form

$$(c_1\delta(x), c_2\tau(x))$$

where c_1, c_2 are the solutions to the equations

$$h(c_1) = h(c_2), \quad \int_{c_1}^{c_2} h(\phi)d\phi = 1 - \alpha,$$

which can be solved numerically. □

6.5 Hypothesis test

For hypothesis test the action space Ω_A consists of only two actions, to accept or to reject, which we denote by $\{a_0, a_1\}$. Similar to the frequentist setting, the hypotheses are

$$H_0 : \Theta \in \Omega_\Theta^{(0)} \quad \text{vs} \quad H_1 : \Theta \in \Omega_\Theta^{(1)},$$

where $\Omega_\Theta^{(0)} \in \mathcal{F}_\Theta$ and $\Omega_\Theta^{(0)} \cap \Omega_\Theta^{(1)} = \Omega_\Theta$. A commonly used loss function is the 0-1 loss. That is, the loss is 1 if we make a wrong decision, and 0 if we make a right decision. In symbols,

$$L(\theta, a) = \begin{cases} 0 \text{ if } (\theta, a) \in (\Omega_\Theta^{(0)} \times \{a_0\}) \cup (\Omega_\Theta^{(1)} \times \{a_1\}) \\ 1 \text{ if } (\theta, a) \in (\Omega_\Theta^{(0)} \times \{a_1\}) \cup (\Omega_\Theta^{(1)} \times \{a_0\}) \end{cases} \tag{6.8}$$

A more nuanced loss function assigns different costs for two types of errors: rejecting H_0 when it is right (false positive), or accepting H_0 when it is wrong (false negative). In symbols,

$$L(\theta, a) = \begin{cases} 0 & \text{if } (\theta, a) \in (\Omega_\Theta^{(0)} \times \{a_0\}) \cup (\Omega_\Theta^{(1)} \times \{a_1\}) \\ c_0 & \text{if } (\theta, a) \in (\Omega_\Theta^{(0)} \times \{a_1\}) \\ c_1 & \text{if } (\theta, a) \in (\Omega_\Theta^{(1)} \times \{a_0\}) \end{cases} \tag{6.9}$$

where $c_0 > 0$ and $c_1 > 0$ represent, respectively, the false positive cost and false negative cost. Note that (6.8) is a special case of (6.9) with $c_0 = c_1 = 1$. The next theorem gives the Bayes rule for the loss function (6.9).

Theorem 6.5 *Suppose* $0 < P_{\Theta|X}(\Omega_\Theta^{(0)}|x) < 1$ *for all* $x \in \Omega_X$. *Then the Bayes rule for loss function (6.9) is*

$$d_B(x) = \begin{cases} a_0 & \textit{if } c_1 P_{\Theta|X}(\Omega_\Theta^{(1)}|x) \leq c_0 P_{\Theta|X}(\Omega_\Theta^{(0)}|x) \\ a_1 & \textit{if } c_1 P_{\Theta|X}(\Omega_\Theta^{(1)}|x) > c_0 P_{\Theta|X}(\Omega_\Theta^{(0)}|x) \end{cases} \tag{6.10}$$

Proof. Since the loss is bounded, by Theorem 5.8 we can calculate the Bayes rule by minimizing the posterior expected loss (Section 5.6). For $a = a_0$,

$$\begin{aligned} \rho(x, a_0) &= E[L(\Theta, a_0)|X]_x \\ &= \int_{\Omega_\Theta^{(0)}} 0 dP_{\Theta|X}(\theta|x) + \int_{\Omega_\Theta^{(1)}} c_1 dP_{\Theta|X}(\theta|x) \\ &= c_1 P_{\Theta|X}(\Omega_\Theta^{(1)}|x). \end{aligned}$$

Similarly, $\rho(x, a_1) = c_0 P_{\Theta|X}(\Omega_\Theta^{(0)}|x)$. So the Bayes rule for this problem is

$$d_B(x) = \text{argmin}\{\rho(x, a) : a \in \{a_0, a_1\}\},$$

which is the same as (6.10). □

Since $P_{\Theta|X}(\Theta_0|x) = 1 - P_{\Theta|X}(\Theta_1|x)$, we can rewrite the Bayes rule (6.10) as

$$d_B(x) = \begin{cases} a_0 & \text{if } P_{\Theta|X}(\Omega_\Theta^{(1)}|x) \leq c_0/(c_0 + c_1) \\ a_1 & \text{if } P_{\Theta|X}(\Omega_\Theta^{(1)}|x) > c_0/(c_0 + c_1) \end{cases} \tag{6.11}$$

In some cases, $\Omega_\Theta^{(0)}$ is a singleton or a finite set. For example for the two sided hypothesis

$$H_0 : \Theta = \theta_0 \quad \text{vs} \quad H_1 : \Theta \neq \theta_0, \tag{6.12}$$

$\Omega_\Theta^{(0)}$ is the singleton $\{\theta_0\}$. Any measure μ_Θ dominated by the Lebesgue measure will entail $P_{\Theta|X}(\Omega_\Theta^{(0)}|x) = 0$, which violates the assumptions in Theorem 6.5. Thus to avoid this difficulty we must use a prior distribution not dominated by the Lebesgue measure.

Definition 6.7 *Let $(\Omega, \mathcal{F})$ be a measurable space and $a \in \Omega$. The Dirac measure for a is the set function*

$$\delta_a(B) = \begin{cases} 0 & \text{if } a \notin B \\ 1 & \text{if } a \in B \end{cases}$$

It is left as an exercise to show that δ_a is indeed a measure on $(\Omega, \mathcal{F})$, and for any $\mathcal{F}$-measurable function we have

$$\int_\Omega f(\omega) d\delta_a(\omega) = f(a).$$

Using the Dirac measure, we can construct a prior distribution P_Θ with nonzero mass at θ_0, which gives rise to a posterior distribution that gives nonzero mass at θ_0. Let ν_Θ be a σ-finite measure on $(\Omega_\Theta, \mathcal{F}_\Theta)$: the measure we have in mind is one that is dominated by the Lebesgue measure. Suppose Q_Θ is a probability measure on $(\Omega_\Theta, \mathcal{F}_\Theta)$ dominated by ν_Θ. Let $\pi_\Theta = dQ_\Theta / d\nu_\Theta$. Let P_Θ be the measure on $(\Omega_\Theta, \mathcal{F}_\Theta)$ defined by

$$dP_\Theta = (1 - \epsilon) dQ_\Theta + \epsilon d\delta_{\theta_0}. \tag{6.13}$$

As we will show in the next theorem, this prior distribution gives rise to a posterior distribution that assign nonzero mass at θ_0.

Theorem 6.6 *Suppose the prior distribution P_Θ is defined by (6.13), where ν_Θ is dominated by the Lebesgue measure. Then the posterior probability of the null set in the hypothesis (6.12) is*

$$P_{\Theta|X}(\{\theta_0\}|x) = \frac{\epsilon f_{X|\Theta}(x|\theta_0)}{(1-\epsilon)\int_{\Omega_\Theta} f_{X|\Theta}(x|\theta)\pi_\Theta(\theta) d\nu_\Theta(\theta) + \epsilon f_{X|\Theta}(x|\theta_0)}$$

Proof. Let

$$\tau_\Theta(\theta) = \begin{cases} \pi_\Theta(\theta) & \text{if } \theta \neq \theta_0 \\ 1 & \text{if } \theta = \theta_0 \end{cases}, \quad \mu_\Theta = (1-\epsilon)\nu_\Theta + \epsilon\delta_{\theta_0}.$$

Then, for any $B \in \mathcal{F}_\Theta$,

$$\int_B \tau_\Theta(\theta)d\mu_\Theta(\theta) = (1-\epsilon)\int_B \tau_\Theta(\theta)d\nu_\Theta(\theta) + \epsilon\delta_{\theta_0}(B).$$

Since ν_Θ is dominated by the Lebesgue measure, $\int_B \tau_\Theta d\nu_\Theta = \int_B \pi_\Theta d\nu_\Theta$, and the right hand side above becomes

$$(1-\epsilon)\int_B \pi_\Theta(\theta)d\nu_\Theta(\theta) + \epsilon\delta_{\theta_0}(B) = P_\Theta(B).$$

This means $P_\Theta \ll \mu_\Theta$ and $dP_\Theta/d\mu_\Theta = \tau_\Theta$. That is, τ_Θ is the prior density of P_Θ with respect to μ_Θ. The posterior density derived from τ_Θ and the likelihood $f_{X|\Theta}(\theta|x)$ is then

$$\tau_{\Theta|X}(\theta|x) = \frac{f_{X|\Theta}(\theta|x)\tau_\Theta(\theta)}{\int_{\Omega_\Theta} f_{X|\Theta}(\theta|x)\tau_\Theta(\theta)d\mu_\Theta(\theta)}, \tag{6.14}$$

from which it follows that

$$P_{\Theta|X}(\{\theta_0\}|x) = \frac{\int_{\{\theta_0\}} f_{X|\Theta}(\theta|x)\tau_\Theta(\theta)d\mu_\Theta(\theta)}{\int_{\Omega_\Theta} f_{X|\Theta}(\theta|x)\tau_\Theta(\theta)d\mu_\Theta(\theta)}. \tag{6.15}$$

Because τ_Θ is dominated by the Lebesgue measure we have the following equalities:

$$\begin{aligned}&\int_{\Omega_\Theta} f_{X|\Theta}(\theta|x)\tau_\Theta(\theta)d\mu_\Theta(\theta)\\ &\qquad = (1-\epsilon)\int_{\Omega_\Theta} f_{\Theta|X}(\theta|x)\pi_\Theta(\theta)d\nu_\Theta(\theta) + \epsilon f_{X|\Theta}(x|\theta_0)\end{aligned} \tag{6.16}$$

and

$$\int_{\{\theta_0\}} f_{X|\Theta}(\theta|x)\tau_\Theta(\theta)d\mu_\Theta(\theta) = \epsilon f_{\Theta|X}(x|\theta_0). \tag{6.17}$$

Substitute (6.16) and (6.17) into (6.15) to complete the proof. □

We can now construct the Bayes rule for the prior distribution defined by (6.13) and the two-sided hypothesis (6.12).

Corollary 6.2 *If the loss function is defined by (6.9) and the prior distribution P_Θ is defined by (6.13) where ν_Θ is dominated by the Lebesgue measure, then the Bayes for testing hypothesis (6.12) is defined by the rejection region*

$$\frac{\epsilon f_{X|\Theta}(x|\theta_0)}{(1-\epsilon)\int_{\Omega_\Theta} f_{X|\Theta}(\theta|x)\pi_\Theta(\theta)d\nu_\Theta(\theta) + \epsilon f_{X|\Theta}(x|\theta_0)} < \frac{c_1}{c_0+c_1} \tag{6.18}$$

We can also express the rejection rule (6.18) in terms of the posterior density. Let $\pi_{\Theta|X}$ denote the posterior density derived from π_Θ and $f_{X|\Theta}$.

Note that this is not the true posterior density, but rather the posterior density derived from the continuous component of τ_Θ. In practice, this is usually the continuous prior we assign when probability at θ_0 being 0 is not of a concern. We can now express the rejection rule (6.18) in terms on the $\pi_{\Theta|X}$.

Corollary 6.3 *If $\pi(\theta_0) \neq 0$, then the rejection region (6.18) can be expressed as*

$$\frac{\epsilon \pi_{\Theta|X}(\theta_0|x)}{(1-\epsilon)\pi_\Theta(\theta_0) + \epsilon \pi_{\Theta|X}(\theta_0|x)} < \frac{c_0}{c_0 + c_1}. \tag{6.19}$$

Proof. Rewrite the density (6.14) as

$$\tau_{\Theta|X}(\theta|x) = \frac{f_{X|\Theta}(\theta|x)\pi_\Theta(\theta)(1 - I_{\{\theta_0\}})(\theta) + f_{X|\Theta}(x|\theta_0) I_{\{\theta_0\}}}{(1-\epsilon)\int_{\Omega_\Theta} f_{X|\Theta}(\theta|x)\pi_\Theta(\theta) d\nu_\Theta(\theta) + \epsilon f_{X|\Theta}(x|\theta_0)}. \tag{6.20}$$

Let f_X be the marginal density of X derived from π_Θ and $f_{X|\Theta}$; that is,

$$f_X(x) = \int_{\Omega_\Theta} f_{X|\Theta}(\theta|x)\tau_\Theta(\theta) d\mu_\Theta(\theta).$$

Again, this is not the true marginal density, which is based on τ_Θ and $f_{X|\Theta}$. Divide the numerator and denominator of (6.20) by $f_X(x)$, to obtain

$$\tau_{\Theta|X}(\theta|x) = \frac{\pi_{\Theta|X}(\theta|x)(1 - I_{\{\theta_0\}}(\theta)) + [\pi_{\Theta|X}(\theta_0|x)/\pi_\Theta(\theta_0)] I_{\{\theta_0\}}(\theta)}{(1-\epsilon) + \epsilon \pi_{\Theta|X}(\theta_0|x)/\pi_\Theta(\theta_0)}.$$

Now take the integral $\int_{\{\theta_0\}}(\cdots) d\tau_\Theta$ on both sides of the equation to complete the proof. □

The next example illustrates the construction of one-sided and two-sided tests.

Example 6.4 Suppose that $X_1, \ldots, X_n|\theta$ are i.i.d. Exp(θ) variables. Then the likelihood function is

$$f_{X|\Theta}(x|\theta) = \theta^{-n} e^{-t(x)/\theta}, \quad \text{where } \theta > 0 \text{ and } t(x) = \textstyle\sum_{i=1}^n x_i.$$

Suppose $\Theta \sim \tau \chi^{-2}_{(m)}$. Then it can be shown that (Problem 6.7)

$$\Theta|x \sim (2t(x) + \tau)\chi^{-2}_{(2n+m)}.$$

Suppose we want to test the one-sided hypothesis

$$H_0 : \Theta \leq a \quad \text{vs} \quad H_1 : \Theta > a.$$

By (6.11), the Bayes rule has rejection region

$$1 - F\left(\frac{a}{2t(x) + \tau}\right) > \frac{c_1}{c_0 + c_1},$$

where F is the c.d.f. of $\chi^{-2}_{(2n+m)}$. An alternative way to write this rule is

$$a < (\tau + 2t(x))\chi^{-2}_{(2n+m)}(c_1/(c_0 + c_1)).$$

To test the two sided hypothesis

$$H_0 : \Theta = a \quad \text{vs} \quad H_1 : \Theta \neq a,$$

it is more convenient to use (6.19) than (6.18), because the form of $\pi_{\theta|X}$ is known. To use this rule we simply substitute $\pi_\Theta(\theta_0)$ and $\pi_{\Theta|X}$ by the densities of $\tau\chi^{-2}_{(m)}$ and $(2t(x) + \tau)\chi^{-2}_{(2n+m)}$. □

6.6 Classification

In a classification problem both the parameter space and the action space are finite:

$$\Omega_\Theta = \{1, \ldots, k\}, \quad \Omega_A = \{a_1, \ldots, a_k\}.$$

As before, X is a p-dimensional random vector. Conditioning on $\Theta = \theta$, X has distribution $P_{X|\Theta}(\cdot|\theta)$ for $\theta = 1, \ldots, k$. The likelihoods $P_{X|\Theta}(\cdot|\theta)$, $\theta = 1, \ldots, k$, represent "classes" or "sub-populations". These distributions describe k clusters or data clouds in a p-dimensional space. Estimating Θ amounts to determining to which cluster a newly observed X belongs. An action $a_\theta \in \Omega_A$ is the action of choosing class θ. For simplicity, we write

$$\Omega_A = \{1, \ldots, k\}$$

where each $a \in \Omega_A$ is to be interpreted as "choosing class a".

A distinct feature of the classification problem is that both the prior distribution and the likelihood function are to be estimated from a training sample or training data set, whereas the decision rule is to be applied to a separate testing sample or testing data set. Because of this one might argue that a classification procedure described here is not a truly Bayesian procedure. Nevertheless, if we regard the training data as prior knowledge, then it is Bayesian relative to this prior knowledge. Also note that this feature (of having to estimate prior and likelihood function) should not be confused with the empirical Bayes procedure to be discussed in the next section: in the latter case, the prior is estimated by the same sample that is used for parameter estimation.

The loss function in this setting can be represented by a matrix:

$$L(i,j) = c_{ij}, \quad \text{where} \quad c_{ij} \begin{cases} = 0 & \text{if } i = j \\ > 0 & \text{if } i \neq j \end{cases} \tag{6.21}$$

The next theorem gives the Bayes rule for this problem.

Theorem 6.7 *The Bayes rule with respect to the loss function (6.21) is*

$$d_B(x) = \text{argmin}\{a : \sum_{\theta=1}^{k} c_{\theta a} f_{X|\Theta}(x|\theta)\pi_\Theta(\theta)\}. \tag{6.22}$$

Proof. Because the loss function is bounded, by Theorem 5.8 $d_B(x)$ is the minimizer of the posterior expected loss $\rho(x,a)$, which in this context takes the form

$$\rho(x,a) = E(L(\Theta,a)|X)_\theta = \sum_{\theta=1}^{k} c_{\theta a}\pi_{\Theta|X}(\theta|x).$$

Thus

$$d_B(x) = \text{argmin}\{a : \rho(x,a)\} = \text{argmin}\{a : \sum_{\theta=1}^{k} c_{\theta a}\pi_{\Theta|X}(\theta|x)\}.$$

Because $\pi_{\Theta|X}(\theta|x)$ is proportional to $f_{X|\Theta}(x|\theta)\pi_\Theta(\theta)$, where the proportionality constant is independent of a, the above minimizer can be rewritten as (6.22). □

To use the Bayes rule (6.22) we need to know the likelihood function $f_{X|\Theta}$ and the prior density π_Θ. This is where the classification problem differs from a typical Bayesian procedure. In actual classification problems the class label does not fully specify the distribution of that class. We can think of k data clouds situated in a p-dimensional space, with each cloud having a different shape. The labels can only tell us cloud A, cloud B,..., but does not carry the information about the shape of each cloud. Thus the distribution $P_{X|\Theta}(\cdot|\theta)$ requires an additional parameter (real- or vector-valued), say Ψ, to specify itself. Let us then denote the likelihood as $P_{X|\Theta\Psi}$, with Θ representing class label and Ψ representing the additional parameter.

A training data set is one for which the class label for each X is known. Thus, we have

$$X_{\theta 1}, \ldots, X_{\theta n_\theta} \text{ is an i.i.d. sample from } P_{X|\Theta\Psi}(\cdot|\theta,\psi), \; \theta = 1,\ldots,k.$$

We can then use the training data set

$$\{X_{\theta 1}, \ldots, X_{\theta n_\theta}\}, \quad \theta = 1,\ldots,k,$$

to estimate Ψ, by either a Bayesian or a frequentist method. Denote this estimate by $\hat{\psi}$. The training sample also allows us to estimate the prior probability of each class. For example, a natural estimate is the proportion of the sample size of a class relative to the total sample size:

$$\hat{\pi}_\Theta(\theta) = n_\theta/(n_1 + \cdots + n_k).$$

We can then substitute $P_{X|\Theta\Psi}(\cdot|\theta, \hat{\psi})$ and $\hat{\pi}_\Theta(\theta)$ into the Bayes rule (6.22) to perform classification. That is,

$$\hat{d}_B(x) = \text{argmin}\{a : \sum_{\theta=1}^{k} c_{\theta a} f_{X|\Theta\Psi}(x|\theta, \hat{\psi})\hat{\pi}_\Theta(\theta)\}, \tag{6.23}$$

where $f_{X|\Theta\Psi}$ is the density of $P_{X|\Theta\Psi}$.

Commonly used models for $f_{X|\Theta\Psi}$ are multivariate Normal distributions with equal or unequal covariance matrices. That is,

$$\begin{cases} N(\mu_\theta, \Sigma) \\ N(\mu_\theta, \Sigma_\theta) \end{cases}, \quad \theta = 1, \ldots, k, \tag{6.24}$$

where μ_θ are p-dimensional vectors and Σ_θ, Σ are $p \times p$ positive definite matrices. The Bayes rule based on the first model (with common variance matrix) is called *linear discriminant analysis* (LDA); that based on the second model (with different variance matrices) is called *quadratic discriminant analysis* (QDA). For example, if we use the UMVU estimates, then the estimates of μ_θ and Σ for LDA are:

$$\begin{aligned} \hat{\mu}_\theta &= n_\theta^{-1} \sum_{i=1}^{n_\theta} X_{\theta i}, \\ \hat{\Sigma} &= \left(\sum_{\theta=1}^{k} n_\theta - k \right)^{-1} \sum_{\theta=1}^{k} \sum_{i=1}^{n_\theta} (X_{\theta i} - \hat{\mu}_\theta)(X_{\theta i} - \hat{\mu}_\theta)^T, \end{aligned} \tag{6.25}$$

and the estimates of μ_θ and Σ_θ for QDA are:

$$\begin{aligned} \hat{\mu}_\theta &= n_\theta^{-1} \sum_{i=1}^{n_\theta} X_{\theta i}, \\ \hat{\Sigma}_\theta &= (n_\theta - 1)^{-1} \sum_{i=1}^{n_\theta} (X_{\theta i} - \hat{\mu}_\theta)(X_{\theta i} - \hat{\mu}_\theta)^T. \end{aligned} \tag{6.26}$$

It is left as an exercise (6.10) to show that these two sets of estimators are indeed the UMVU estimators for the two models in (6.24). The reason for the qualifiers *linear* and *quadratic*, is that, for LDA, $\hat{d}_B(x)$ can be expressed in terms of a linear function of x when $k = 2$; whereas for QDA, $\hat{d}_B(x)$ can be expressed in terms of a quadratic function of x when $k = 2$ (see Problems 6.8 and 6.9).

LDA and QDA each have their own advantages for different distribution shapes. If we imagine the k data clouds are all watermelon (or football) shaped, then LDA works the best when these watermelons all have the same orientation and shape (but they can have different sizes), and QDA works the best when these watermelons have different orientations, and have different shapes (some longer, some rounder); they can also have different sizes.

6.7 Stein's phenomenon

A phenomenon discovered by Stein in a short paper (Stein, 1956) turned out to be highly influential in a wide range of subsequent developments, such as shrinkage estimation and empirical Bayes. A lemma implicitly used in that paper (see also Stein, 1981), the famous Stein's lemma, is frequently used in the area of sufficient dimension reduction (Li, 1992) and several other areas. We devote this section to that result.

The phenomenon concerns the inadmissibility of a maximum likelihood estimate based on the multivariate Normal distribution. We first state Stein's Lemma Stein (1981), which is a convenient method to demonstrate Stein's phenomenon. Although the lemma can be proved by a single line of integration by parts, such a proof requires conditions quite strong. We will instead use the original argument of Stein (1981) that relies on Fubini's theorem, which requires weaker assumptions than those used in the proof via integration by parts.

Since the discussion of this section is largely in the realm of frequentist decision theory, we tentatively revert to the frequentist notation. For example, we use P_θ instead of $P_{X|\Theta}(\cdot|\theta)$ to denote the conditional distribution of X given θ.

Lemma 6.1 *Suppose X is a Normally distributed random variable. Let g : real $\to \mathbb{R}$ be a function differentiable $[\lambda]$, λ being the Lebesgue measure, and its derivative $\dot{g}$ satisfies $E|\dot{g}(X)| < \infty$. Then*

$$\mathrm{cov}(X, g(X)) = \mathrm{var}(X) E\dot{g}(X). \tag{6.27}$$

Proof. First, assume $X \sim N(0, 1)$. Let ϕ denote the standard Normal density. Let a be a point in $\mathbb{R}$ at which g is differentiable. Recall that $\dot{\phi}(y) = -y\phi(y)$. We have

$$\begin{aligned}\int_{-\infty}^{\infty} \dot{g}\phi dx &= \int_{-\infty}^{a} \dot{g}\phi dx + \int_{a}^{\infty} \dot{g}\phi dx \\ &= \int_{-\infty}^{a} \dot{g}(x) \int_{-\infty}^{x} \dot{\phi}(z) dz dx - \int_{a}^{\infty} \dot{g}(x) \int_{x}^{\infty} \dot{\phi}(z) dz dx.\end{aligned}$$

By Fubini's theorem, the right hand side is

$$\int_{-\infty}^{a} \dot{\phi}(z) \int_{z}^{a} \dot{g}(x)dxdz - \int_{a}^{\infty} \dot{\phi}(z) \int_{a}^{z} \dot{g}(x)dxdz$$
$$= \int_{-\infty}^{a} \dot{\phi}(z)(g(a) - g(z))dz - \int_{a}^{\infty} \dot{\phi}(z)(g(z) - g(a))dz$$
$$= -\int_{-\infty}^{\infty} \dot{\phi}(z)g(z)dz$$
$$= \int_{-\infty}^{\infty} z\phi(z)g(z)dz.$$

This proves the theorem for the standard Normal case. Now suppose $X \sim N(\mu, \sigma^2)$. Then $Z = (X - \mu)/\sigma$ has a standard Normal distribution. Hence, if we let $g_1(z) = g(\sigma z + \mu)$, then

$$\begin{aligned}\text{cov}(X, g(X)) &= \text{cov}(\sigma Z + \mu, g_1(Z)) \\ &= \sigma \text{cov}(Z, g_1(Z)) = \sigma E\dot{g}_1(Z) = \sigma^2 E\dot{g}(X) = \text{var}(X)E\dot{g}(X),\end{aligned}$$

as to be demonstrated. □

We now generalize this lemma to the multivariate case.

Lemma 6.2 *Suppose X is a p-dimensional random vector distributed as $N(\theta, I_p)$. Suppose $g : \Omega_X \to \mathbb{R}^q$ is a differentiable function such that the components of $\partial g/\partial x^T$ are integrable. Then*

$$E[(X - EX)g^T(X)] = E[\partial g^T(X)/\partial x].$$

Proof. The (i, j)th entry of the matrix on the left is

$$E[(X_i - EX_i)^T g_j(X)] = E\{E[(X_i - EX_i)^T g_j(X)|X_\ell, \ell \neq i]\}.$$

Consider the conditional expectation

$$E[(X_i - EX_i)g_j(X)|X_\ell = x_\ell, \ell \neq i].$$

Because X_i is independent of $\{X_\ell : \ell \neq i\}$, the above conditional expectation is the unconditional expectation

$$E[(X_i - EX_i)g_j(x_1, \ldots, x_{i-1}, X_i, x_{i+1}, \ldots, x_p)],$$

which only involves one random variable X_i. By Lemma 6.1, the above expectation is the same as

$$E[\partial g_j(x_1, \ldots, x_{i-1}, X_i, x_{i+1}, \ldots, x_p)/\partial x_i] = E[\partial g_j(X)/\partial x_i|X_\ell = x_\ell, \ell \neq i].$$

In other words,

$$E[(X_i - EX_i)g_j(X)|X_\ell, \ell \neq i] = E[(\partial g_j(X)/\partial x_i|X_\ell, \ell \neq i].$$

Now take expectation on both sides to complete the proof. □

We are now ready to state Stein's paradox. Recall that, if $X \sim N(\theta, \Sigma)$, then X is the maximum likelihood estimate of θ. In fact, X is also the UMVU estimator of θ. Stein (1956) showed that there is an estimator that is uniformly better than X in terms of the frequentist risk $R(\theta, d)$ for the L_2-loss.

Theorem 6.8 *Suppose $X \sim N(\theta, I_p)$, $p \geq 3$, $L(\theta, a) = \|\theta - a\|^2$, and $d(x) = x$. Then there is a decision rule $d_1 \in \mathcal{D}$ such that*

$$R(\theta, d_1) < R(\theta, d), \quad \textit{for all } \theta \in \mathbb{R}.$$

Proof. Consider the alternative estimate

$$d_1(x) = [1 - h(x)]x,$$

where $h : \mathbb{R}^p \to \mathbb{R}$ is a function to be specified later. For now we assume nothing about h except that it must make d_1 square integrable P_θ, and that $h(x)x$ must satisfy the conditions for g in Lemma 6.2. We have

$$\begin{aligned} R(\theta, d_1) &= E\|X - \theta - h(X)X\|^2 \\ &= R(\theta, d) - 2E[(X - \theta)^T h(X)X] + E[h^2(X)\|X\|^2]. \end{aligned} \tag{6.28}$$

By Lemma 6.2,

$$\begin{aligned} E[(X - \theta)^T h(X)X] &= \mathrm{tr}[E(X - \theta)h(X)X^T] \\ &= \mathrm{tr}\{XE[\partial h(X)/\partial x^T + h(X)I_p]\} \\ &= E[(\partial h(X)/\partial x^T)X] + pEh(X). \end{aligned}$$

Hence

$$R(\theta, d_1) = R(\theta, d) - 2E[(\partial h/\partial x^T)X] - 2pEh(X) + E[h^2(X)\|X\|^2].$$

Let $h(x) = \alpha/\|x\|^2$. Then h is differentiable and the derivative is integrable, so that the above equality holds for this particular choice. Using the form of h we find, after some elementary computation,

$$-2(\partial h/\partial x^T)x - 2ph(x) + h^2(x)\|x\|^2 = \frac{\alpha(4 - 2p + \alpha)}{\|x\|^2}.$$

Substitute this into the (6.28) to obtain

$$R(\theta, d_1) - R(\theta, d) = \alpha(4 - 2p + \alpha)E(\|X\|^{-2}).$$

The right hand side is negative if $0 < \alpha < 2p - 4$, which is possible because $p \geq 3$. For example, if we take $\alpha = p - 2$, then $R(\theta, d_1) < R(\theta, d)$ for all $\theta \in \mathbb{R}^p$. □

The choice $\alpha = p - 2$ in the proof leads to the estimator

$$d_1(x) = \left(1 - \frac{p - 2}{\|x\|^2}\right) x, \tag{6.29}$$

which is called the James-Stein estimator (James and Stein, 1961).

6.8 Empirical Bayes

In the Bayes procedures described in the previous sections the prior distribution P_Θ is assumed known — except for the classification problem in Section 6.6 where it is estimated from the training set. Even if we think of the training set as prior knowledge, P_Θ has a known form. In the empirical Bayes method the prior distribution is estimated from the same sample that will be used for estimation of the parameter Θ. Specifically, suppose

$$\{P_{\Theta,b}: \ b \in \Omega_B\}$$

is a parametric family of prior distributions defined on the sample space Ω_B. Then, for a $b \in \Omega_B$, the density of X is

$$f_{X,b}(x) = \int_{\Omega_\Theta} f_{X|\Theta}(x|\theta)\pi_{\Theta,b}(\theta) d\mu_\Theta(\theta).$$

Thus we have a parametric family of probability distributions indexed by b:

$$\{f_{X,b}(x): \ b \in \Omega_B\}.$$

We can then use a frequentist method to estimate b from this family. For example, we can estimate b by the maximum likelihood estimate, the moment estimate, or the UMVU estimate described in Chapter 2. Once an estimate $\hat{b}$ is obtained in this way, we then draw statistical inference about Θ using the posterior density

$$\pi_{\Theta|X,b} = \frac{f_{X|\Theta}(\theta|x)\pi_{\Theta,b}(\theta)}{f_{X,b}(x)},$$

with b replaced by $\hat{b}$. This procedure is called the Empirical Bayes procedure. See Robbins (1955); Efron and Morris (1973), and Berger (1985, Section 4.5).

The empirical Bayes method is especially useful when we have a large number of parameters — as many parameters as the sample size n or even more. These parameters cannot be estimated accurately unless we introduce some structures to them. Assigning these parameters a prior distribution with a common parameter b is an effective way of building such a structure.

The next example shows that the James-Stein estimator described in the last section can be derived as an empirical Bayes estimator with b estimated by the UMVUE (Efron and Morris, 1973).

Example 6.5 Suppose that $X_1, \ldots, X_p|\Theta$ are independent random variables, where $\Theta = (\Theta_1, \ldots \Theta_p)^T$. Also assume:

1. $\Theta_1 \perp\!\!\!\perp \cdots \perp\!\!\!\perp \Theta_p$;
2. $X_i|\Theta = \theta \sim N(\theta_i, 1)$;
3. $\Theta_i \sim N(0, b)$.

Note that assumption 2 also implies $X_i \perp\!\!\!\perp \Theta_{(-i)}|\Theta_i$, where $\Theta_{(-i)}$ denote the vector Θ with its i-th component removed. Under these assumptions the likelihood function and the prior density can be simplified as

$$f_{X|\Theta}(x|\theta) = \prod_{i=1}^{p} f_{X_i|\Theta}(x_i|\theta) = \prod_{i=1}^{p} f_{X_i|\Theta_i}(x_i|\theta_i),$$

$$\pi_{\Theta,b} = \prod_{i=1}^{p} \pi_{\Theta_i,b}(\theta_i).$$

Hence the marginal density $f_{X,b}(x)$ is

$$f_{X,b}(x) = \prod_{i=1}^{p} \int_{\mathbb{R}} f_{X_i|\Theta_i}(x_i|\theta_i)\pi_{\Theta_i,b}(\theta_i)dP_{\Theta_i,b} = \prod_{i=1}^{p} f_{X_i,b}(x_i).$$

The posterior density is

$$\pi_{\Theta|X,b}(\theta|X) = \prod_{i=1}^{p} f_{X_i|\Theta_i}(x_i|\theta_i)\pi_{\Theta_i,b}(\theta_i)/f_{X_i,b}(x_i) = \prod_{i=1}^{p} \pi_{\Theta|X,b}(\theta_i|x_i, b).$$

The Bayes rule with respect to the L_2-loss $L(\theta, a) = \|\theta - a\|^2$ loss is

$$d_B(x) = E(\Theta|X)_x = (E(\Theta_1|X)_x, \ldots, E(\Theta_p|X)_x),$$

where each component is

$$\begin{aligned} E(\Theta_j|X)_x &= \int_{\mathbb{R}^p} \theta_j \prod_{i=1}^{p} \pi_{\Theta|X,b}(\theta_i|x_i)d\theta_1 \cdots d\theta_p \\ &= \int_{\mathbb{R}} \theta_j \pi_{\Theta_j|X_j,b}(\theta_j|x_j, b)d\theta_j \\ &= E(\Theta_j|X_j)_{x_j}. \end{aligned}$$

By Example 5.1,

$$\Theta_j|x_j \sim N\left(\frac{x_j}{b^{-1}+1}, \frac{1}{b^{-1}+1}\right).$$

So the Bayes rule is

$$d_B(x) = \left(\frac{1}{b^{-1}+1}\right)x = \left(1 - \frac{1}{1+b}\right)x. \tag{6.30}$$

Now let us derive the UMVU estimate for b. By Example 5.1 again, the marginal distribution of X_i is $N(0, b+1)$. Hence $S = \sum_{i=1}^{p} X_i^2$ is complete and sufficient for b. Moreover, $S/(b+1) \sim \chi^2_{(p)}$, which implies $(b+1)/S \sim \chi^{-2}_{(p)}$. Hence, by Problem 6.5,

$$E((b+1)/S) = 1/(p-2) \Rightarrow E((p-2)/S) = 1/(p-2).$$

Because S is complete and sufficient, by Corollary 2.2, $(p-2)/S$ is the UMVUE for $1/(b+1)$. Substituting this estimate for $1/(b+1)$ into the Bayes rule (6.30) yields the James-Stein estimate (6.29). □

This method can be generalized to $\theta_i \sim N(b_1, b_2)$, where $b_1 \in \mathbb{R}$, $b_2 > 0$ are unknown, as well as to the regression setting. Interested readers can consult Morris (1983) and Berger (1985, Section 4.5). Some of these generalizations are given as exercises.

Problems

6.1. Suppose $X_1, \ldots, X_n$ are an independent sample from $U(0, \theta)$. Let $\pi(\theta)$ be the Pareto(θ_0, α) distribution, defined by

$$\pi(\theta) = \frac{\alpha}{\theta_0}\left(\frac{\theta_0}{\theta}\right)^{\alpha+1} I_{(\theta_0,\infty)}(\theta), \quad \alpha > 0, \quad \theta_0 > 0.$$

1. Find the posterior density $\pi(\theta|X_1, \ldots, X_n)$.
2. Assuming $\alpha > 2$, find the Bayes estimate of θ with respect to the L_2 loss $L(\theta, a) = (\theta - a)^2$.
3. Assuming $\alpha > 1$, find the Bayes estimate of θ with respect to the L_1-loss $L(\theta, a) = |\theta - a|$.
4. Find the Generalized MLE for θ.
5. For what values of α are the Generalized MLE, the Bayes estimates based on L_1 and L_2 losses approximately the same?
6. Let $0 < \gamma < 1$. Find the $(1-\gamma)$-level HPD credible set for θ.
7. Derive the Bayes rule (in the form of rejection region) for testing the hypothesis

$$H_0 : \theta \leq 2\theta_0 \quad \text{vs} \quad H_1 : \theta > 2\theta_0.$$

   Use the loss function (6.9) with $c_0 = 0$, $c_1 = 1$, $c_2 = 2$. Make the answer as explicit as possible.
8. Derive the Bayes rule for testing the hypothesis

$$H_0 : \theta = 2\theta_0 \quad \text{vs} \quad H_1 : \theta \neq 2\theta_0$$

   use the same loss function and the prior distribution (6.13).

6.2. Let $X_1, \ldots, X_n$ be an i.i.d. sample from $N(\mu, \phi)$, and let $Y_1, \ldots, Y_m$ be an i.i.d. sample from $N(\nu, \psi)$. Here, μ and ν are treated as parameters, and ϕ and ψ as known constants. Suppose we know $\mu \leq \nu$, and would like to incorporate this prior knowledge into the estimation process. For example, if we are to estimate the population means of men's heights and women's heights, as represented by μ and ν respectively, then it is quite reasonable to assume $\mu \geq \nu$. Assign (μ, ν) the improper prior

$$\pi_\Theta(\mu, \nu) = I(\mu \leq \nu).$$

Derive the Bayes rule for estimating (μ, ν).

Hint: Let $\delta = \nu - \mu$. Then the problem becomes that of deriving $E(\mu|X)$ and $E(\mu + \delta|X)$. The corresponding improper prior is

$$\pi_\Theta(\mu, \delta) = I(\delta \geq 0).$$

6.3. Prove that, under the assumptions of Theorem 6.1, the Bayes estimator (6.2) is unique modulo P.

6.4. Suppose U is a random variable defined on $(\Omega, \mathcal{F}, P)$ taking values in Ω_U, and U is integrable with respect to P. Let $L(\theta, a)$ be the loss function (6.4). Show that, if q satisfies

$$F(q) \geq \alpha_1/(\alpha_1 + \alpha_2), \quad F(q-) \leq \alpha_1/(\alpha_1 + \alpha_2),$$

then

$$\int_{\Omega_U} |q - U| dP \leq \int_{\Omega_U} |a - U| dP$$

for all $a \in \Omega_U$.

6.5. Suppose $X \sim \chi^{-2}_{(m)}$.

1. Show that if $m > 2$ then X is integrable and $E(X) = (m-2)^{-1}$.
2. Show that if $m > 4$ then X has finite variance and

$$\mathrm{var}(X) = 2(m-2)^{-2}(m-4)^{-1}.$$

3. Show that the mode of the density of X is $(m+2)^{-1}$.

6.6. Suppose Θ is square integrable with respect to P and $d \in \mathcal{D}$. Show that (6.6) holds.

6.7. Suppose $X_1, \ldots, X_n|\theta$ are i.i.d. $\mathrm{Exp}(\theta)$ random variables and $\Theta \sim \tau\chi^{-2}_{(m)}$. Show that

$$\Theta|x \sim (2t(x) + \tau)\chi^{-2}_{(2n+m)},$$

where $t(x) = \sum_{i=1}^n x_i$.

6.8. Show that, under the first model in (6.24) and $k = 2$, the Bayes rule $\hat{d}_B$ based on (6.25) can be rewritten as

$$\hat{d}_B(x) = \begin{cases} 1 & \text{if } (x - (\hat{\mu}_1 + \hat{\mu}_2)/2)^T \hat{\Sigma}^{-1}(\hat{\mu}_2 - \hat{\mu}_1) \leq \log[c_{21}n_1/(c_{12}n_2)] \\ 2 & \text{if } (x - (\hat{\mu}_1 + \hat{\mu}_2)/2)^T \hat{\Sigma}^{-1}(\hat{\mu}_2 - \hat{\mu}_1) > \log[c_{21}n_1/(c_{12}n_2)]. \end{cases}$$

Note that, when

$$c_{21}n_1 = c_{12}n_2, \tag{6.31}$$

the Bayes rule reduces to

$$\hat{d}_B(x) = 1 \text{ iff } g(x) \leq g((\hat{\mu}_1 + \hat{\mu}_2)/2),$$

where

$$g(x) = x^T \hat{\Sigma}^{-1}(\hat{\mu}_2 - \hat{\mu}_1).$$

This function is linear in x, an is called Fisher's linear discriminant function (Fisher, 1935).

6.9. Show that, under the second model in (6.24), $k = 2$, and (6.31), the Bayes rule $\hat{d}_B(x)$ based on (6.26) takes action 1 if and only

$$(x - \hat{\mu}_1)^T \hat{\Sigma}_1^{-1}(x - \hat{\mu}_1) - (x - \hat{\mu}_2)^T \hat{\Sigma}_1^{-1}(x - \hat{\mu}_2) \leq \det(\hat{\Sigma}_2) - \det(\hat{\Sigma}_2).$$

The left hand side is a quadratic function of x, and is called the quadratic discriminant function.

6.10. Let

$$X_{i1}, \ldots, X_{in_i}, \quad i = 1, \ldots, k$$

be p dimensional random vectors. Assume:

a. for each i, $X_{i1}, \ldots, X_{in_i}$ are an i.i.d. $N(\mu_i, \Sigma_i)$;
b. for $i \neq j$,

$$\{X_{i1}, \ldots, X_{in_i}\} \perp\!\!\!\perp \{X_{j1}, \ldots, X_{jn_j}\}.$$

For each $i = 1, \ldots, k$, let

$$\hat{\mu}_i = n_i^{-1} \sum_{\ell=1}^{n_i} X_{i\ell}, \quad \hat{\Sigma}_i = (n_i - 1)^{-1} \sum_{\ell=1}^{n_i} (X_{i\ell} - \hat{\mu}_i)(X_{i\ell} - \hat{\mu}_i)^T.$$

Also, let $n = n_1 + \cdots + n_k$. Prove the following statements:

1. The statistic

$$\{(\hat{\mu}_i, \hat{\Sigma}_i) : i = 1, \ldots k\}$$

is the UMVU estimator of $\{(\mu_i, \Sigma_i) : i = 1, \ldots, k\}$.

2. If $\Sigma_1 = \cdots = \Sigma_k$, then the statistic

$$\left(\hat{\mu}_1, \ldots, \hat{\mu}_k, \quad \frac{\sum_{\theta=1}^{k}(n_\theta - 1)\hat{\Sigma}_\theta}{\sum_{\theta=1}^{k} n_\theta - 1}\right)$$

is the UMVU estimator for $(\mu_1, \ldots, \mu_k, \Sigma)$.

6.11. Prove the following variations of Stein's lemma. Suppose that X is a random variable distributed as $N(\theta, 1)$. Then the following equalities hold.

1. If $g : \mathbb{R} \to \mathbb{R}$ is twice differentiable $[\lambda]$, λ being the Lebesgue measure, and $|g(X)|$ and $|g^{(2)}|$ have finite expectations, then

$$E[(X - EX)^2 g(X)] = E[g(X) + g^{(2)}(X)].$$

2. If, in addition, g is four times differentiable $[\lambda]$ and $|g^{(4)}|$ has a finite expectation, then

$$E[(X - \theta)^4 g(X)] = E[3g(X) + 6g^{(2)} + g^{(4)}(X)].$$

6.12. Suppose X is a p-dimensional multivariate Normal random vector and $g : \mathbb{R}^p \to \mathbb{R}^p$ is differentiable modulo λ, the Lebesgue measure on $(\mathbb{R}^p, \mathcal{R}^p)$. Suppose the entries of $\partial g^T / \partial x$ are integrable. Then

$$\text{cov}[X, g(X)] = \text{var}(X) E[\partial g^T(X)/\partial x].$$

6.13. The Beta prime distribution is defined by the density function

$$f(x) = x^{\alpha - 1}(x + 1)^{-\alpha - \beta}, \quad x > 0, \; \alpha > 0, \; \beta > 0,$$

which we write as $\text{Beta}'(\alpha, \beta)$.

1. Show that if $X \sim \text{Beta}'(\alpha, \beta)$ and $\beta > 1$, then X is integrable and

$$E(X) = \alpha/(\beta - 1).$$

2. Show that if $X \sim \text{Beta}'(\alpha, \beta)$ and $\beta > 2$ then X has finite variance and

$$\text{var}(X) = \frac{\alpha(\alpha + \beta - 1)}{(\beta - 2)(\beta - 1)^2}.$$

3. Show that, if $S|\phi \sim \phi\chi^2_{(n)}$ and $\Phi \sim \tau\chi^{-2}_{(m)}$, then

$$\frac{S}{\tau} \sim \text{Beta}'(n/2, m/2).$$

6.14. Suppose $\phi = (\phi_1, \ldots, \phi_n)$, where the n components are i.i.d. $\tau\chi^{-2}_{(m)}$ random variables. Suppose $S_1, \ldots, S_n|\phi$ are independent random variables with $S_i|\phi \sim \phi_i\chi^{-2}_{(n_i - 1)}$. Here S_i may be the sum of the squared errors of an i.i.d. sample $X_{i1}, \ldots, X_{in_i} \sim N(\mu_i, \phi_i)$ from the ith group.

1. Assuming τ is known, derive a Bayes rule for estimating ϕ.
2. Use the result of Problem 6.13 to find an unbiased estimate of τ of the form
$$\hat{\tau} = \sum_{i=1}^{n} w_i S_i.$$
3. Construct an empirical Bayes estimate of ϕ based on parts 1 and 2.

6.15. Suppose $\theta_1, \ldots, \theta_n$ are i.i.d. $N(b_1, b_2)$ random variables, where $b_1 \in \mathbb{R}$ and $b_2 > 0$. Let $\theta = (\theta_1, \ldots, \theta_n)$. Suppose $X_1, \ldots, X_n|\theta$ are independent and $X_i|\theta \sim N(\theta_i, 1)$.

1. Assuming b_1 and b_2 are known, derive a Bayes rule for estimating θ.
2. Find the marginal distribution of $(X_1, \ldots, X_n)$ for a fixed (b_1, b_2). Based on this distribution find the UMVU estimates of (b_1, b_2).
3. Construct an empirical Bayes rule for estimating θ based on the above two parts.

6.16. Suppose $\theta_1, \ldots, \theta_n$ are independent random variables with $\theta_i \sim N(\beta x_i, \tau)$, where β is a parameter, $\tau > 0$ is a known constants, and $x_1, \ldots, x_n$ are constants. Suppose $Y_1, \ldots, Y_n|\theta$ are independent with $Y_i|\theta \sim N(\theta_i, \phi)$, where $\phi > 0$ is known.

1. Pretending β is known, derive a Bayes rule for estimating θ.
2. For each fixed β, derive the marginal distribution of $(Y_1, \ldots, Y_n)$. Based on this distribution derive the UMVU estimate for β.
3. Construct an empirical Bayes estimate for θ based on Parts 1 and 2.
4. Construct a test for the hypothesis $H_0 : \theta_1 \leq 0$ vs $H_1 : \theta_1 > 0$.
5. Construct a test for the hypothesis $H_0 : \theta_1 = 0$ vs $H_1 : \theta_1 \neq 0$.

6.17. Suppose

1. $\lambda|\phi \sim N(b_1, \phi/m), \quad b_1 \in \mathbb{R}$.
2. $\phi \sim b_2 \chi^{-2}_{(k)}, \quad b_2 > 0$.
3. $T|\phi, \lambda \sim N(\lambda, \phi/n)$.
4. $S|\phi, \lambda \sim \phi \chi^2_{(n-1)}$.
5. $S \perp\!\!\!\perp T|\lambda, \phi$.

Prove the following statements:

1. $T \perp\!\!\!\perp S|\phi$.
2. $T|s \sim t_{(n-1+k)}(b_1, (n^{-1} + m^{-1})s/(n-1))$.
3. $S \sim b_2 \,\mathrm{Beta}'((n-1)/2, k/2)$.

6.18. Suppose $\theta_1, \ldots, \theta_n$ are independent 2-dimensional random vectors. Let $\theta_i = (\lambda_i, \phi_i)$. Let
$$\theta = (\theta_1, \ldots, \theta_n), \quad \phi = (\phi_1, \ldots, \phi_n), \quad \lambda = (\lambda_1, \ldots, \lambda_n).$$

Suppose

$$\Lambda_i|\phi_i \sim N(b_1, \phi_i/m), \quad \Phi_i \sim b_2\chi^{-2}_{(k)}, \quad b_1 \in \mathbb{R},\ b_2 > 0.$$

Suppose $(S_1, T_1), \ldots, (S_n, T_n)$ are 2-dimensional random vectors satisfying

1. $(S_1, T_1) \perp\!\!\!\perp \cdots \perp\!\!\!\perp (S_n, T_n)|\theta$.
2. $(S_i, T_i) \perp\!\!\!\perp \theta_{(-i)}|\theta_i$.
3. $S_i|\theta_i \sim \phi_i\chi^2_{(n_i-1)}$.
4. $T_i|\theta_i \sim N(\lambda_i, \phi_i/n_i)$.

Solve the following problems.

1. Assuming b_1, b_2 are known, derive a Bayes rule for estimating θ.
2. Use the result of Problem 6.17 to derive the joint distribution of

$$(S_1, T_1), \ldots, (S_1, T_1)$$

   (this is a distribution not conditioned on θ).
3. Use part 2 to derive an unbiased estimate of b_1, b_2.
4. Derive the empirical Bayes estimate of θ.

References

Berger, J. O. (1985). *Statistical decision theory and Bayesian analysis.* Second edition. Springer, New York.

Efron, B. and Morris, C. (1973). Stein's estimation rule and its competitors – an empirical Bayes approach. *Journal of the American Statistical Association*, **68**, 117–130.

Fisher, R. A. (1935). Theory of Statistical Estimation. *Mathematical Proceedings of the Cambridge Philosophical Society*, **22**, 700–725.

Hettmansperger, T. P. and Randles, R. H. (2002). A practical affine equivariant multivariate median. *Biometrika*, **89**, 851–860.

James, W. and Stein, C. (1961). Estimation with quadratic loss. *Proceeding of the Fourth Berkeley Symposium on Mathematical Statistics and Probability*, 361–379.

Li, K.-C. (1992). On Principal Hessian Directions for Data Visualization and Dimension Reduction: AnotherApplication of Stein's Lemma. *Journal of the American Statistical Association*, **87**, 1025–1039.

Morris, C. N. (1983). Parametric empirical Bayes inference: theory and applications. *Joural of the American Statistical Association*, **78**, 47–55.

Oja, H. (1983). Descriptive statistics for multivariate distributions. *Statistics and Probability Letters*, **1**, 327–332.

Robbins, H. (1955). An empirical Bayes approach to Statistics. *Proceedings of the Third Berkeley Symposium on Mathematical Statistics and Probability, Volumn 1: Contributions to the Theory of Statistics*, 157–163.

Stein, C. (1956). Inadmissibility of the usual estimator for the mean of a multivariate Normal distribution. *Berkeley Symposium on Mathematical Statistics and Probability,* 197–206.

Stein, C. (1981). Estimation of the mean of a multivariate Normal distribution. *The Annals of Statistics*, **9**, 1135–1151.

7

Asymptotic tools and projections

In this chapter we review some crucial results about limit theorems in probability theory, which will be used repeatedly in the rest of the book. The limit theorems include those about convergence in probability, almost everywhere convergence, and convergence in distribution. For further information about the content of this chapter, see Serfling (1980) and Billingsley (1995).

We shall also review some basic results on projections in Hilbert spaces.

7.1 Laws of Large Numbers

Let $\|\cdot\|$ denote the Euclidean norm.

Definition 7.1 *The sequence X_n of random vectors is said to converge in probability to a random vector X if for any $\epsilon > 0$*

$$\lim_{n\to\infty} P(\|X_n - X\| > \epsilon) = 0.$$

This convergence is expressed as $X_n \xrightarrow{P} X$.

Thus convergence in probability means that, as $n \to \infty$, the distributions of the random vectors $X_n - X$ are increasingly concentrated at the origin: the probability of $\|X_n - X\|$ escaping outside of any interval $(-\epsilon, \epsilon)$ goes to 0.

In most cases, the limit random vector X is a constant vector. Convergence in probability in these cases is often proved by the Weak Law of Large Numbers (WLLN). A proof of this depends on few useful probability inequalities, which are presented below.

Lemma 7.1 (Markov's inequality) *If $U \geq 0$ is a random variable with $E(U) < \infty$, then for any $\epsilon > 0$,*

$$P(U > \epsilon) \leq \epsilon^{-1} E(U). \tag{7.1}$$

B. Li and G. J. Babu, *A Graduate Course on Statistical Inference*,
Springer Texts in Statistics, https://doi.org/10.1007/978-1-4939-9761-9_7

Proof. For any $\epsilon > 0$,

$$P(U > \epsilon) = \int_{U>\epsilon} dP \leq \int_{U>\epsilon} (U/\epsilon)\, dP \leq \epsilon^{-1} E(U),$$

as desired. □

Chebyshev's inequality given in the next lemma is an easy consequence of lemma 7.1.

Lemma 7.2 (Chebyshev's inequality) *Suppose X is a random variable with finite variance. Then* $P(|X - E(X)| > \epsilon) \leq \epsilon^{-2}\,\mathrm{var}(X)$. □

From this inequality we can easily derive the Chebyshev's Weak Law of Large Numbers for random variables. A similar result for random vectors follow easily by applying the result on random variables coordinate-wise.

Theorem 7.1 (Weak Law of Large Numbers) *Let $\{X_1, X_2, \ldots\}$ be a sequence of uncorrelated random variables with finite second moments; that is,*

$$E(X_i^2) < \infty, \quad \mathrm{cov}(X_i, X_j) = 0, \quad i \neq j.$$

If $\lim_{n\to\infty} n^{-2} \sum_{i=1}^n \mathrm{var}(X_i) = 0$, *then*

$$\frac{1}{n}\sum_{i=1}^n (X_i - E(X_i)) \xrightarrow{P} 0.$$

Proof. Let $Y_i = X_i - E(X_i)$. Then

$$\left(\frac{1}{n}\sum_{i=1}^n Y_i\right)^2 = \frac{1}{n^2}\sum_{i=1}^n Y_i^2 + \frac{2}{n^2}\sum_{i<j}^n Y_i Y_j.$$

Since X_i and X_j are uncorrelated, Y_i and Y_j are also uncorrelated. Thus

$$\mathrm{var}\left(\frac{1}{n}\sum_{i=1}^n X_i\right) = \frac{1}{n^2}\sum_{i=1}^n \mathrm{var}(X_i) \to 0.$$

The theorem now follows from Lemma 7.2. □

A stronger mode of convergence of sequence of random variables is the almost everywhere convergence.

Definition 7.2 *A sequence of random variables X_n converges almost everywhere to a random variable X if* $P(\lim_{n\to\infty} |X_n - X| = 0) = 1$.

If this definition is satisfied then we write $X_n \to X$ $[P]$.

Before focusing on almost everywhere convergence results for sample means, let us first review limits of sets and the Borel-Cantelli Lemma.

Definition 7.3 *For a sequence A_n of measurable sets,*

$$\limsup_{n\to\infty} A_n = \{\omega : \omega \in A_k, \text{ for infinitely many } k\},$$
$$\liminf_{n\to\infty} A_n = \{\omega : \omega \in A_k, \text{ for all but finitely many } k\}.$$

It is easy to establish that

$$\liminf_{n\to\infty} A_n = \bigcup_{n=1}^{\infty} \bigcap_{k=n}^{\infty} A_k \subset \limsup_{n\to\infty} A_n = \bigcap_{n=1}^{\infty} \bigcup_{k=n}^{\infty} A_k.$$

Lemma 7.3 (Borel-Cantelli Lemma) *If $\sum_{n=1}^{\infty} P(A_n) < \infty$, then*

$$P\left(\limsup_{n\to\infty} A_n\right) = 0. \tag{7.2}$$

Proof. Note that the probability on the left hand side of (7.2) is no more than $P(\cup_{k=n}^{\infty} A_k)$ for any n, which is bounded from above by the sum $\sum_{k=n}^{\infty} P(A_k)$. This sum converges to 0 provided $\sum_{n=1}^{\infty} P(A_n) < \infty$. □

Results concerning almost everywhere convergence of sums of random elements are called strong laws of large numbers (SLLN). A version of SLLN can be proved along the lines of the proof for the weak law, but it is not the most general one. Because the method is simple and exemplifies the use of Markov inequality combined with Borel-Cantelli Lemma, we choose to prove this version. We will state a more general version without proof later.

Theorem 7.2 (Strong Law of Large Numbers) *If $X_1, X_2, \ldots$ is a sequence of independent and identically distributed (i.i.d.) random variables and X_i has finite fourth moment, then $n^{-1}\sum_{i=1}^{n} X_i \to E(X_1)$ almost everywhere.*

Henceforth, we will use $E_n X$ to denote the sample average $n^{-1}\sum_{i=1}^{n} X_i$. This notation is motivated by the fact that the sample average is in fact the expectation of X with respect to the empirical distribution that assign probability mass $1/n$ at each observation X_i.

Proof. If necessary by subtracting its expected value from X_i, we assume, without loss of generality that $E(X_i) = 0$. Note that a sequence of numbers, say x_n, converges to 0 if there is a nonnegative sequence of numbers, say α_n such that $\alpha_n \to 0$ and, for all large n, $|x_n| \le \alpha_n$. So it suffices to show that there is a nonnegative sequence α_n, which converges to 0, such that

$$P(|E_n(X)| > \alpha_n \text{ for infinitely many } n) = 0. \tag{7.3}$$

By Markov's inequality, we have

$$P(|E_n(X)| > \alpha_n) \le P(|(E_n(X))^4 > \alpha_n^4) \le \alpha_n^{-4} E(E_n(X))^4. \tag{7.4}$$

To get an upper bound for $E(E_n(X))^4$, let $S_n = \sum_{i=1}^n X_i$. Since X_i are i.i.d. with mean 0, we have $E(X_n S_{n-1}^3) = 0 = E(X_n^3 S_{n-1})$,

$$E(X_n^2 S_{n-1}^2) = E(X_n^2)E(S_{n-1}^2), \text{ and for any } m, E(S_m^2) = mE(X_1^2).$$

Hence for some constant $K > 0$,

$$\begin{aligned} E(S_n^4) &= E(S_n S_n^3) = \textstyle\sum_{i=1}^n E(X_i S_n^3) = nE(X_n S_n^3) \\ &= nE(X_n(X_n + S_{n-1})^3) \\ &= nE(X_n(X_n^3 + 3X_n S_{n-1}^2 + 3X_n^2 S_{n-1} + S_{n-1}^3) \\ &= nE(X_n^4) + 3nE(X_n^2)E(S_{n-1}^2) \\ &= nE(X_1^4) + 3nE(X_1^2)(n-1)E(X_1^2) \\ &= nE(X_1^4) + 3n(n-1)(E(X_1^2))^2 < Kn^2. \end{aligned} \tag{7.5}$$

By taking $\alpha_n = n^{-1/8}$, it follows from equations (7.4) and (7.5) that

$$\sum_{n=1}^{\infty} P(|E_n(X)| > \alpha_n) \le \sum_{n=1}^{\infty} K\alpha_n^{-4} n^{-2} \le K \sum_{n=1}^{\infty} n^{-3/2} < \infty.$$

The theorem now follows from Borel-Cantelli lemma and (7.3). □

Theorem 7.2 can be established under a weaker assumption of finite first moment. A proof of the next theorem, called Kolmogorov's SLLN, can be found in Billingsley (1995, page 282).

Theorem 7.3 *If $X_1, X_2, \ldots$ are i.i.d. random variables with $E|X_i| < \infty$, then $E_n(X) \to E(X)$ almost everywhere.*

By considering coordinate-wise convergence, the above theorem can be extended to random vectors

Theorem 7.4 *If $X_1, X_2, \ldots$ are i.i.d. random vectors with $E\|X_i\| < \infty$, then $E_n(X) \to E(X)$ almost everywhere.*

It is easy to see that convergence almost everywhere implies convergence in probability (see Problem 7.3). We record this fact below for future reference.

Corollary 7.1 *Let $X_1, X_2, \ldots$ be an i.i.d. sequence or random vectors, where $E\|X_1\| < \infty$. Then $E_n X \xrightarrow{P} E(X_1)$.*

7.2 Convergence in distribution

Convergence of a sequence random vectors means, roughly, that the distribution of X_n converges to that of X. However, this intuitive statement has severe limitations. Consider the sequence of random variables $\{X_n\}$ that is distributed as $N(0, 1/n)$. Here, it is intuitively clear that these distributions converge to a probability measure assigning mass 1 at 0. Hence the limiting distribution should be $F(x) = I_{[0,\infty)}(x)$. However, if we denote by F_n the cumulative distribution function of $N(0, 1/n)$, then $F_n(0) = 1/2$ for all n, and therefore does not tend to $F(0)$, which equals 1. So a precise definition of convergence in distribution or weak convergence should take care of this caveat.

We first define weak convergence of a sequence of probability measures. As before, let $\mathbb{N}$ denote the set of positive integers $\{1, 2, \ldots\}$.

Definition 7.4 *Let $\{\mu_n : n \in \mathbb{N}\}$ and μ be probability measures on $(\mathbb{R}^p, \mathcal{R}^p)$. We say that μ_n converges weakly to μ and write $\mu_n \Rightarrow \mu$ if, for any bounded and continuous function f on $\mathbb{R}^p$, $\int f d\mu_n \to \int f d\mu$.*

For a set A, let $\partial A = \bar{A} \setminus A^\circ$ denote the boundary of A, where $\bar{A}$ is the closure of A, and A° is the interior of A. As will be seen in the Portmanteau Theorem stated later, μ_n converges weakly to μ if and only if, for any measurable set A with $\mu(\partial A) = 0$, $\mu_n(A) \to \mu(A)$. This successfully avoids the caveat we mentioned at the beginning of this section: if $A = \{0\}$, then $\partial A = \{0\}$, and $\mu(\partial A) = 1$. It is easy to check in that example that for any set $A \in \mathcal{R}$ except $A = \{0\}$ we have $\mu_n(A) \to \mu(A)$. Thus $N(0, 1/n)$ indeed converges weakly to the point mass at 0 even though the c.d.f. of $N(0, 1/n)$ at 0 does not converge to 1.

A sequence of p-dimensional random vectors $\{X_n\}$ defined on probability spaces $(\Omega_n, \mathcal{F}_n, P_n)$ converges in distribution to a random vector X defined on a probability space $(\Omega, \mathcal{F}, P)$ if and only if the probability distributions $P_n \circ X_n^{-1}$ on $(\mathbb{R}^p, \mathcal{R}^p)$ converges weakly to $P \circ X^{-1}$. In symbols, $X_n \xrightarrow{\mathcal{D}} X$ if and only if $P_n \circ X_n^{-1} \Rightarrow P \circ X^{-1}$. If Q is a probability measure on $(\mathbb{R}^p, \mathcal{R}^p)$, and if $P_n \circ X_n^{-1} \Rightarrow Q$, then we also use the notation $X_n \xrightarrow{\mathcal{D}} Q$.

A special case of convergence in distribution is $X_n \xrightarrow{\mathcal{D}} X$, where X is a degenerate random variable; that is, $X = a$ almost everywhere for some constant a. Then it can be shown that $P(X = a) = 1$ if and only if the distribution $P \circ X^{-1}$ is δ_a, the Dirac measure at a. Thus, the meaning of $X_n \xrightarrow{\mathcal{D}} X$ in this case is simply $X_n \xrightarrow{\mathcal{D}} \delta_a$. Furthermore, it can be shown that $X_n \xrightarrow{\mathcal{D}} \delta_a$ if and only if $X_n \xrightarrow{P} a$. See Problem 7.4.

The spaces $(\Omega_n, \mathcal{F}_n, P_n)$ and $(\Omega, \mathcal{F}, P)$ do not appear directly, but only by way of the distributions they induce on the range space. In view of this, there is no source of confusion if we drop the subscripts from P_n and E_{P_n}. So from now on, for example, we simply write $E(X_n)$ for expectation instead of $E_{P_n}(X_n)$. We return to explicit mention of P_n and $E_{P_n}(X_n)$ in Chapter 10.

In general, convergence in distribution is weaker than convergence in probability. That is, if $X_n \xrightarrow{P} X$ for some random vector X, then $X_n \xrightarrow{\mathcal{D}} X$. See Problem 7.5.

Often, results on weak convergence of random variables can be simplified by using the next theorem, known as Skorohod's Theorem. See Billingsley (1995, Theorem 25.6).

Theorem 7.5 (Skorohod Theorem) *Suppose a sequence of probability measures $\{\nu_n\}$ on the real-line converges weakly to a probability measure ν. Then there exist random variables Y_n and Y defined on a common probability space $(\Omega, \mathcal{F}, P)$ such that Y_n has distribution ν_n, Y has distribution ν, and $Y_n(\omega) \to Y(\omega)$ for all $\omega \in \Omega$.*

In particular, if a sequence of random variables $X_n \xrightarrow{\mathcal{D}} X$, where X_n has distribution F_n and X has distribution F, then there exist random variables Y_n and Y defined on a common space Ω such that Y_n has distribution F_n, Y has distribution F, and $Y_n(\omega) \to Y(\omega)$ for all $\omega \in \Omega$. Combining this with Fatou's Lemma 1.6 and Bounded Convergence Theorem (Theorem 1.8), we get the following result. Proof is left as an exercise.

Lemma 7.4 *Let $\{X_n\}$ be a sequence of random variables such that $X_n \xrightarrow{\mathcal{D}} X$. Then*

$$E(|X|) \leq \liminf_{n\to\infty} E(|X_n|).$$

If, in addition, $|X_n|$ are uniformly bounded (that is, for some $M > 0$, then by $|X_n| \leq M$ for all n), then

$$E(X_n) \to E(X).$$

Skorohod's Theorem also holds for random vectors.

There are several equivalent conditions to the definition of weak convergence, and depending on context, each may be more useful than the others, either as a tool to work with, or as a goal to work towards. These statements are collectively called the Portmanteau theorem (See Billingsley, 1999, pages 16 and 26). Let h be a real-valued function. Let D_h be the collection of points at which h is discontinuous, that is, $D_h = \{x : h \text{ is discontinuous at } x\}$. We say that a set $A \in \mathcal{R}^p$ is a $P \circ X^{-1}$-continuity set of $P \circ X^{-1}(\partial A) = 0$. We first define the upper and lower semi-continuity.

Definition 7.5 *A function f is said to be upper semi-continuous at x if, for any sequence $x_n \to x$, $\limsup_{n\to\infty} f(x_n) \leq f(x)$.*
A function f is called lower semi-continuous at x if, for any sequence $x_n \to x$, $\liminf_{n\to\infty} f(x_n) \geq f(x)$.

Theorem 7.6 (Portmanteau Theorem) *Let $\{X_n : n \in \mathbb{N}\}$ be a sequence of random vectors, and X a random vector. The following statements are equivalent:*

1. $X_n \xrightarrow{\mathcal{D}} X$;
2. $Eh(X_n) \to Eh(X)$ *for any bounded and uniformly continuous function h;*

3. *For any* $P \circ X^{-1}$*-continuity set* A, $\lim_{n\to\infty} P(X_n \in A) = P(X \in A)$;
4. *For any closed set* F, $\limsup_{n\to\infty} P(X_n \in F) \leq P(X \in F)$;
5. *For any open set* G, $\liminf_{n\to\infty} P(X_n \in G) \geq P(X \in G)$;
6. *for any upper semi-continuous function* h *that is bounded from above, we have* $\limsup_{n\to\infty} Eh(X_n) \leq Eh(X)$;
7. *For any lower semi-continuous function* h *that is bounded from below, we have* $\liminf_{n\to\infty} Eh(X_n) \geq Eh(X)$.

Statements 5 and 7 are easy consequences of Lemma 7.4 under condition 1. The statements in the above theorem are in terms of convergence in distribution of X_n to X, but they can be equivalently stated in terms of weak convergence of probability measures P_n to P defined on $(\mathbb{R}^p, \mathcal{R}^p)$. For example, the first five statements can be reformulated as

1. $P_n \Rightarrow P$;
2. $\int f dP_n \to \int f dP$ for every bounded and uniformly continuous function f;
3. $\lim_{n\to\infty} P_n(A) = P(A)$ for any $A \in \mathcal{R}^p$ with $P(\partial A) = 0$;
4. $\limsup_{n\to\infty} P_n(F) \leq P(F)$ for any closed set F in $\mathbb{R}^p$;
5. $\liminf_{n\to\infty} P_n(G) \geq P(G)$ for any open set G in $\mathbb{R}^p$.

The rest of the theorem can be similarly translated in terms of probability measures.

The Portmanteau theorem is a fundamental result and is extremely useful. Many important results in asymptotic analysis can be derived from them. Below we derive several of these results, both because of their importance in future discussion and as exercises to practice the use of the Portmanteau theorem.

The first result is the Continuous Mapping Theorem, which says that if $X_n \xrightarrow{\mathcal{D}} X$ and h is a continuous function then $h(X_n) \xrightarrow{\mathcal{D}} h(X)$. This is easily seen using the Portmanteau theorem. Since h is continuous, $h^{-1}(G)$ is open whenever G is open. Thus

$$\begin{aligned}\liminf_{n\to\infty} P(h(X_n) \in G) &= \liminf_{n\to\infty} P(X_n \in h^{-1}(G)) \\ &\geq P(X \in h^{-1}(G)) = P(h(X) \in G).\end{aligned}$$

Hence $h(X_n) \xrightarrow{\mathcal{D}} h(X)$ by the Portmanteau theorem.

We record below without proof a more general version of the Continuous Mapping Theorem. The proof is in the same spirit as the last paragraph.

Proposition 7.1 (Continuous Mapping Theorem) *Suppose* h *is a vector valued function such that* $P(X \in D_h) = 0$.

1. *If* $X_n \to X$ *almost everywhere, then* $h(X_n) \to h(X)$ *almost everywhere.*
2. *If* $X_n \xrightarrow{\mathcal{D}} X$, *then* $h(X_n) \xrightarrow{\mathcal{D}} h(X)$.
3. *If* $X_n \xrightarrow{P} X$, *then* $h(X_n) \xrightarrow{P} h(X)$.

The proof of part 1 of this proposition is relatively straightforward, and is left as an exercise. Part 2 follows from part 1 and k-dimensional version of Skorohod Theorem (see Theorem 7.5). A complete proof of the above theorem can be found in Serfling (1980).

Sometimes we want to know whether

$$X_n \stackrel{\mathcal{D}}{\to} X, \quad Y_n \stackrel{\mathcal{D}}{\to} Y \quad \Rightarrow \quad (X_n, Y_n) \stackrel{\mathcal{D}}{\to} (X, Y).$$

Obviously this cannot be true generally. In fact, we don't even know what (X, Y) means — even if X and Y are well defined individually, it does not specify anything about what happens between them. However, in some special cases X and Y does specify (X, Y). For example, if Y is a constant vector then (X, Y) is well defined. The question now is whether $(X_n, Y_n) \stackrel{\mathcal{D}}{\to} (X, Y)$ in this case. We are interested in this because many statistics we use, such as the studentized statistics, involve two sequences with one converging to a random vector and another converging to a constant vector. Using the Portmanteau Theorem we can answer this question reasonably easily.

Theorem 7.7 *Let $\{X_n\}$ be a sequence of random vectors in $\mathbb{R}^p$ and let $\{Y_n\}$ be a sequence of random vectors in $\mathbb{R}^q$.*

1. *If $p = q$, $X_n \stackrel{\mathcal{D}}{\to} X$ and $\|X_n - Y_n\| \stackrel{P}{\to} 0$, then $Y_n \stackrel{\mathcal{D}}{\to} X$.*
2. *If $X_n \stackrel{\mathcal{D}}{\to} X$ and $Y_n \stackrel{P}{\to} c$ for some constant $c \in \mathbb{R}^q$, then $(X_n, Y_n) \stackrel{\mathcal{D}}{\to} (X, c)$.*

Proof. 1. We will use the third assertion of the Portmanteau theorem. Let F be a closed set, we will show that $\limsup_{n\to\infty} P(Y_n \in F) \le P(X \in F)$. Let $\epsilon > 0$. Then

$$P(Y_n \in F) = P(Y_n \in F, \|X_n - Y_n\| < \epsilon) + P(Y_n \in F, \|X_n - Y_n\| \ge \epsilon).$$

The second probability on the right hand side goes to 0 as $n \to \infty$. The event inside the first probability on the right hand side implies $X_n \in A(\epsilon)$, where $A(\epsilon) = \{x : \sup_{y\in F} \|x - y\| < \epsilon\}$. Hence the first term on the right is no more than $P(X_n \in A(\epsilon))$. Because F is closed, $\cap_{k=1}^{\infty} A(1/k) = F$. By continuity of probability, for any $\delta > 0$ we can select a k so large that $P(X_n \in A(1/k))$ is no more than $P(X_n \in F) + \delta$. Take $\epsilon < 1/k$. Then

$$P(Y_n \in F) \le P(X_n \in F) + \delta + P((\|X_n - Y_n\| \ge \epsilon).$$

Consequently $\limsup_{n\to\infty} P(Y_n \in F) \le P(X \in F) + \delta$ for any δ, which implies $\limsup_{n\to\infty} P(Y_n \in F) \le P(X \in F)$.

2. We will use the first assertion of the Portmanteau theorem. Note that $\|(X_n, Y_n) - (X_n, c)\| = \|Y_n - c\| \stackrel{P}{\to} 0$. By the first part of this theorem it suffices to show that $(X_n, c) \stackrel{\mathcal{D}}{\to} (X, c)$. Let h be a bounded and real-valued continuous function. Then g defined by $g(x) = h(x, c)$ is a bounded continuous function. Because $X_n \stackrel{\mathcal{D}}{\to} X$, we have $Eg(X_n) \to Eg(X)$, and hence

$Eh(X_n, c) \to Eh(X, c)$. □

The next result is the well known Slutsky's theorem, which combines the second assertion of the above theorem with the continuous mapping theorem.

Corollary 7.2 (Slutsky's Theorem) *Suppose that* $X_n \xrightarrow{\mathcal{D}} X$, $Y_n \xrightarrow{\mathcal{D}} c$, *and* h *is a vector-valued function with* $P((X, c) \in D_h) = 0$. *Then* $h(X_n, Y_n) \xrightarrow{\mathcal{D}} h(X, c)$.

In some text books Slutsky's theorem refers to the special case of the above theorem when h is a rational function and X_n and Y_n are random variables.

Corollary 7.3 *Suppose that* $\{(X_n, Y_n)\}$ *is a sequence of bivariate random vectors such that* $X_n \xrightarrow{\mathcal{D}} X$, *and* $Y_n \xrightarrow{P} c$. *Then*

1. $X_n + Y_n \xrightarrow{\mathcal{D}} X + c$.
2. $X_n Y_n \xrightarrow{\mathcal{D}} cX$.
3. *If* $c \neq 0$, *then* $X_n / Y_n \xrightarrow{\mathcal{D}} X/c$.

Besides the equivalent statements of convergence in distribution given in the Portmanteau theorem, there is another such statement using characteristic functions. A characteristic function of a random vector X is its Fourior transform with respect to the measure P; that is $\phi_X(t) = E(e^{it^T X})$, where $i = \sqrt{-1}$ and t is any vector in $\mathbb{R}^p$.

Proposition 7.2 *A sequence of random vectors* X_n *converges in distribution to* X *if and only if* $\phi_{X_n}(t) \to \phi_X(t)$ *for all* $t \in \mathbb{R}^p$.

Characteristic functions can be used to prove the Central Limit Theorems. They can also be used to extend convergence in distribution for a sequence of random variables to a sequence of random vectors. This is called the Cramér-Wold device, introduced by Cramér and Wold (1936), see also Billingsley (1995, page 383). For example, in the next section, we will use this device to extend the Lindeberg Theorem to sequences of random vectors.

Theorem 7.8 (Cramér-Wold Device) *Let* X, $\{X_n : n = 1, 2, \ldots\}$ *be random random vectors of dimension* p. *Suppose that, for any* $a \in \mathbb{R}^p$, $a^T X_n \xrightarrow{\mathcal{D}} a^T X$. *Then* $X_n \xrightarrow{\mathcal{D}} X$.

Proof. Because $a^T X_n \xrightarrow{\mathcal{D}} a^T X$, the characteristic function of $a^T X_n$ converges to that of $a^T X$. That is, for any $t \in \mathbb{R}$,

$$\phi_{a^T X_n}(t) = Ee^{it(a^T X_n)} \to Ee^{it(a^T X)} = \phi_{a^T X}(t).$$

Take $t = 1$. Then $Ee^{ia^T X_n} \to Ee^{ia^T X}$. In other words, $\phi_{X_n}(a) \to \phi_X(a)$ for all $a \in \mathbb{R}^p$. Hence, by Proposition 7.2, $X_n \xrightarrow{\mathcal{D}} X$. □

7.3 Argument via subsequences

In this section we introduce two theorems that are useful for proving weak convergence. Recall that, in calculus, a sequence of numbers $\{a_n\}$ converges to a number a if and only if every subsequence of $\{a_n\}$ contains a further subsequence that converges to a. This equivalence also applies to convergence in distribution.

Theorem 7.9 *A sequence of random vectors X_n converges in distribution to X if and only if every subsequence $X_{n'}$ contains a further subsequence $X_{n''}$ that converges in distribution to X.*

A weaker condition than "every subsequence of $\{X_n\}$ contains a further subsequence that converges to X" is "every subsequence of $\{X_n\}$ contains a further subsequence that converges to *some* random vector". In the latter statement, the random vector that the further subsequence converges to may depends on the subsequence and further subsequence involved. This property is known as relative compactness. Prohorov's theorem below gives a sufficient condition for relative compactness: tightness.

Tightness is an extension of the boundedness in probability to metric spaces. Since in this book we focus on random vectors in Euclidean spaces, we do not need the more general meaning of tightness. Nevertheless, following the standard usage in this area we will use the term tightness rather than boundedness in probability. We first give the formal definition of tightness.

Definition 7.6 *A family Π of probability measures on $(\mathbb{R}^p, \mathcal{R}^p)$ is tight, if for every $\epsilon > 0$, there exists a compact set $K_\epsilon \subseteq \mathbb{R}^p$ such that $P(K_\epsilon) > 1 - \epsilon$ for all $P \in \Pi$.*

Since a compact set is a bounded and closed set in a Euclidean space, tightness is equivalent to boundedness in probability in the case of Euclidean spaces. For a sequence $\{U_n\}$ of random vectors, tightness translates to the property that for every $\epsilon > 0$, there exists a real number M_ϵ, sucht that $P(\|U_n\| > M_\epsilon) < \epsilon$ for all n. The next lemma asserts that marginal tightness implies joint tightness, and marginal stochastic smallness implies joint stochastic smallness.

Lemma 7.5 *If $\{U_n\}$ and $\{V_n\}$ are tight, then $\{(U_n^T, V_n^T)^T\}$ is tight. Moreover, if $U_n \xrightarrow{P} 0$ and $V_n \xrightarrow{P} 0$, then $(U_n^T, V_n^T)^T \xrightarrow{P} 0$.*

Proof. We note that

$$\|(U_n^T, V_n^T)^T\|^2 = \|U_n\|^2 + \|V_n\|^2.$$

Thus, if $\|U_n\| \leq K$ and $\|V_n\| \leq K$ then $\|(U_n^T, V_n^T)^T\| \leq \sqrt{2}K$. Let $\epsilon > 0$ be any fixed constant. Let K be sufficiently large so that $P(\|U_n\| > K) < \epsilon/2$ and $P(\|V_n\| > K) < \epsilon/2$. Then

$$P(\|(U_n^T, V_n^T)^T\| > \sqrt{2}K) \le P(\|U_n\| > K) + P(\|V_n\| > K) < \epsilon.$$

Thus (U_n^T, V_n^T) is tight. The second statement can be proved similarly. □

Argument via subsequenes is implied by tightness. This is the Prohorov's Theorem. See, for example, Billingsley (1999, page 57).

Theorem 7.10 (Prohorov's theorem) *If a sequence of random vectors $\{U_n\}$ is tight, then every subsequence $U_{n'}$, contains a further subsequence $\{U_{n''}\}$ such that $U_{n''}$ converges in distribution to a random vector.*

Theorems 7.9 and 7.10 are often used together to show that a tight sequence of random vectors converges in distribution. If we are given a tight sequence, say $\{U_n\}$, then, by Theorem 7.10, every subsequence $\{U_{n'}\}$ contains a further subsequence $\{U_{n''}\}$ that converges in distribution to some random vector U. If we can further show that this U does not depend on the subsequence $\{n'\}$ and the further subsequence $\{n''\}$, then, by Theorem 7.9, the entire sequence $\{U_n\}$ converges in distribution to U. For easy reference, we refer to this method as the *argument via subsequences.*

Sometimes we have two sequences of random vectors that are tight under different distributions. Specifically, suppose, for each n, U_n is a random vector distributed as P_n, and V_n is a random vector distributed as Q_n. If $\{U_n\}$ is tight with respect to $\{P_n\}$ and $\{V_n\}$ is tight with respect to $\{Q_n\}$ then by Theorem 7.10, for any subsequence n', there is a subsequence n'' such that $U_{n''}$ converges in distribution under $P_{n''}$, and there is another subsequence n''' such that $V_{n'''}$ converge in distribution under $Q_{n'''}$. The question is, can n'' and n''' be taken as the same subsequence? The next lemma answers this question.

Lemma 7.6 *If $\{U_n\}$ is tight under $\{P_n\}$ and $\{V_n\}$ is tight under $\{Q_n\}$, then, for any subsequence $\{n'\}$, there exist a further subsequence $\{n''\}$, and random vectors U and V, such that*

$$U_{n''} \xrightarrow{\mathcal{D}} U, \quad V_{n''} \xrightarrow{\mathcal{D}} V. \tag{7.6}$$

Proof. Let n' be any subsequence. Because $\{U_n\}$ is tight under $\{P_n\}$, there is a subsequence m' of n' and a random vector U such that

$$U_{m'} \xrightarrow{\mathcal{D}} U.$$

Because V_n is tight under Q_n and $\{m'\}$ is a subsequence of $\{n\}$, there is a further subsequence $\{n''\}$ of $\{m'\}$ and a random vector V such that

$$V_{n''} \xrightarrow{\mathcal{D}} V.$$

Because $U_{m'}$ converges in distribution to U under $P_{m'}$, it also converges in distribution to U along the subsequence $\{n''\}$. Thus we have (7.6). □

7.4 Argument via simple functions

Let μ_1 and μ_2 be two measures defined on a measurable space $(\Omega, \mathcal{F})$. Let $g_1 \geq 0$ and $g_2 \geq 0$ be measurable functions on the same space. In this section we develop a general method to show that

$$\int f g_1 \, d\mu_1 = \int f g_2 \, d\mu_2 \tag{7.7}$$

holds for an arbitrary f.

For this we need the following fact: for any nonnegative and $\mathcal{F}$-measurable function f, there is a sequence of nonnegative simple functions $\{f_n\}$ such that $0 \leq f_n \uparrow f$. Such a sequence can be constructed as follows. For each $n = 1, 2, \ldots$, we divide $[0, \infty)$ into $n2^n + 1$ left closed and right open intervals. The first $n2^n$ intervals are of length 2^{-n}, equally spaced between 0 and n; the last interval is $[n, \infty)$. If $f(\omega)$ falls in one of these intervals, say $[a, b)$, then the value $f_n(\omega)$ is defined as b. Specifically,

$$f_n(\omega) = \begin{cases} 0 & \text{if } 0 \leq f(\omega) < 2^{-n} \\ 2^{-n} & \text{if } 2^{-n} \leq f(\omega) < 2 \cdot 2^{-n} \\ \vdots & \\ (k-1)2^{-n} & \text{if } (k-1)2^{-n} \leq f(\omega) < k2^{-n} \\ \vdots & \\ n - 2^{-n} & \text{if } n - 2^{-n} \leq f(\omega) < n \\ n & \text{if } n \leq f(\omega) < \infty \end{cases}$$

For convenience, the collection $[0, 2^{-n}), \ldots, [n-2^{-n}, n), [n, \infty)$ is referred to as the nth generation intervals. For each n, the collection of $(n+1)$th generation intervals is a refinement of the collection of nth generation intervals; that is, each $(n+1)$th generation interval is contained in an nth generation interval. Consequently, if $f(\omega)$ is contained in an $(n+1)$th generation interval $[a, b)$, then $[a, b) \subseteq [c, d)$ for some nth generation interval $[c, d)$. Thus $f_{n+1}(\omega) = a \geq c = f_n(\omega)$. Also, by construction, for $f(\omega) < n$,

$$|f_n(\omega) - f(\omega)| \leq 2^{-n},$$

and for any $\omega \in \Omega$, $f(\omega) < n$ for large enough n. From this we see that $\lim_{n \to \infty} f_n(\omega) = f(\omega)$ for all $\omega \in \Omega$. Thus $\{f_n\}$ is a sequence of simple functions satisfying $0 \leq f_n \uparrow f$. These preliminaries help in establishing the next theorem.

Theorem 7.11 *If (7.7) holds for all $\mathcal{F}$-measurable indicator functions, then*

1. *it holds for all nonnegative measurable functions f;*
2. *it holds for all measurable f such that the integrals on both sides of (7.7) are finite.*

Proof. 1. Because equation (7.7) holds for all measurable indicator functions, it holds for all simple functions, and in particular all nonnegative simple functions. Now let f be a nonnegative measurable function and f_n a sequence of nonnegative simple functions such that $f_n \uparrow f$. Then $0 \le f_n g_1 \uparrow f g_1$ and $0 \le f_n g_2 \uparrow f g_2$. By the Monotone Convergence Theorem (see Theorem 1.5), as $n \to \infty$,

$$\int f_n g_1 \, d\mu_1 \to \int f g_1 \, d\mu_1, \quad \int f_n g_2 \, d\mu_2 \to \int f g_2 \, d\mu_1.$$

Because $\int f_n g_1 \, d\mu_1 = \int f_n g_2 \, d\mu_2$ holds for all n, equality (7.7) holds for f.

2. Let f be a measurable function such that $\int f^{\pm} g_1 \, d\mu_1$ and $\int f^{\pm} g_2 \, d\mu_2$ are finite. Then

$$\int f g_1 \, d\mu_1 = \int f^+ g_1 \, d\mu_1 - \int f^- g_1 \, d\mu_1$$
$$\int f g_2 \, d\mu_2 = \int f^+ g_2 \, d\mu_2 - \int f^- g_2 \, d\mu_2.$$

By part 1,

$$\int f^+ g_1 \, d\mu_1 = \int f^+ g_2 \, d\mu_2, \quad \int f^- g_1 \, d\mu_1 = \int f^- g_2 \, d\mu_2.$$

Hence (7.7) holds for f. □

7.5 The Central Limit Theorems

Convergence in distribution is usually established using the Central Limit Theorems. In the independent case, the most general version is the Lindeberg Theorem. This is concerned with a triangular array of random vectors:

$$\{X_{nk} : \; k = 1, \ldots, k_n, n = 1, 2, \ldots\} \tag{7.8}$$

Typically, $k_n = n$, in which case this array does look like a triangle. We first consider the scalar case; that is, X_{nk} are random variables.

Theorem 7.12 (Lindeberg Theorem) *Suppose*

1. $X_{n1}, \ldots, X_{nk_n}$ *are independent for each* n*;* $S_n = X_{n1} + \cdots + X_{nk_n}$*;*
2. $E(X_{nk}^2) < \infty$ *for each* n, k*;* $\mathrm{var}(S_n) > 0$*;* $U_{nk} = [X_{nk} - E(X_{nk})]/\sqrt{\mathrm{var}(S_n)}$*;*
3. *and for any* $\epsilon > 0$,

$$L_n(\epsilon) = \sum_{k=1}^{k_n} \int_{|U_{nk}|>\epsilon} U_{nk}^2 dP \to 0. \tag{7.9}$$

Then $(S_n - ES_n)/\sqrt{\text{var}(S_n)} \xrightarrow{\mathcal{D}} N(0,1)$.

We will omit the proof of this theorem. Interested readers can find a proof in Billingsley (1995, page 359). The sequence $L_n(\epsilon)$ in (7.9) is called the Lindeberg sequence, and the condition $L_n(\epsilon) \to 0$ is called the Lindeberg condition. The meaning of this condition is best seen through its special cases. The simplest special case is concerned with a sequence of independent and identically distributed random variables.

Corollary 7.4 (The Lindeberg-Levy Theorem) *Suppose that* $X_1, X_2, \ldots$ *is an i.i.d. sequence of random variables with a finite nonzero variance* σ^2. *Then*

$$\sqrt{n}(E_n(X) - E(X_1))/\sigma \xrightarrow{\mathcal{D}} N(0,1).$$

Proof. Let $\mu = E(X_1)$. Consider the triangular array $\{X_{nk} : k = 1, \ldots, n,\ n = 1, 2, \ldots\}$ defined by $X_{nk} = (X_k - \mu)/\sigma$. For this triangular array,

$$s_n^2 = \text{var}(X_{n1}) + \cdots + \text{var}(X_{nn}) = 1 + \cdots + 1 = n.$$

Since X_{nk} are identically distributed as $Z = (X_1 - \mu)/\sigma$, the Lindeberg number $L_n(\epsilon)$ is

$$\frac{1}{n}\sum_{k=1}^{n}\int_{|X_{nk}|>\epsilon\sqrt{n}} X_{nk}^2\, dP = \frac{1}{n}\sum_{k=1}^{n}\int_{|Z|>\epsilon\sqrt{n}} Z^2\, dP = \int_{|Z|>\epsilon\sqrt{n}} Z^2\, dP.$$

Because $E(Z^2) < \infty$, the right hand side of the above expression tends to 0 as $n \to \infty$. □

The next special case applies to triangular arrays where each $|X_{nk}|$ has slightly higher than second moment.

Corollary 7.5 (The Lyapounov Theorem) *Let* X_{nk} *be a triangular array, where the random variables in each row are independent. Suppose that,* $E|X_{nk}|^{2+\delta} < \infty$ *for some* $\delta > 0$ *and suppose, without loss of generality,* $E(X_{nk}) = 0$. *If*

$$M_n(\epsilon) = \sum_{k=1}^{k_n} \frac{1}{s_n^{2+\delta}} E\left(|X_{nk}|^{2+\delta}\right) \to 0, \ \textit{then}\ S_n/s_n \xrightarrow{\mathcal{D}} N(0,1).$$

Proof. It suffices to show that $L_n(\epsilon) \le cM_n(\epsilon)$ for some $c > 0$ that does not depend on n. We have

$$
\begin{aligned}
L_n(\epsilon) &= a_n \sum_{k=1}^{k_n} \int_{|X_{nk}|>\epsilon s_n} |X_{nk}/s_n|^2 dP \\
&\le \sum_{k=1}^{k_n} \int_{|X_{nk}|>\epsilon s_n} (|X_{nk}/s_n|/\epsilon)^\delta |X_{nk}/s_n|^2 dP \\
&\le \sum_{k=1}^{k_n} \int (|X_{nk}/s_n|/\epsilon)^\delta |X_{nk}/s_n|^2 dP = M_n(\epsilon)/\epsilon^\delta,
\end{aligned}
$$

as desired. □

Now consider the triangular arrays in which X_{nk} are p-dimensional vectors. We will use the Cramér-Wold device described in Theorem 7.8, which allows us to pass from the a central limit theorem for scalar random variables to random vectors. Let $\text{var}(X_{nk}) = \Sigma_{nk}$ and without loss of generality assume $E(X_{nk}) = 0$. Let $S_n = X_{n1} + \cdots + X_{nk_n}$ and $\Sigma_n = \Sigma_{n1} + \cdots + \Sigma_{nk_n}$. Define

$$
L_n^{(p)}(\epsilon) = \sum_{k=1}^{k_n} \int_{(X_{nk}^T \Sigma_n^{-1} X_{nk})^{1/2} > \epsilon} \left(X_{nk}^T \Sigma_n^{-1} X_{nk}\right) dP.
$$

Here, the superscript (p) of $L_n^{(p)}(\epsilon)$ indicates the dimension of X_{nk}. Note that when $p = 1$, this reduces to the usual Lindeberg sequence.

Theorem 7.13 *Suppose that $\{X_{nk}\}$ is a triangular array of p-dimensional random vectors with $E(X_{nk}) = 0$ and positive definite variance matrices Σ_{nk}. Suppose that the random vectors in each row are independent. If $L_n^{(p)}(\epsilon) \to 0$, then $\Sigma_n^{-1/2} S_n \xrightarrow{\mathcal{D}} N(0, I_p)$, where I_p is the $p \times p$ identity matrix.*

Proof. Applying the Cramér-Wold device, it suffices to show that for any $t \in \mathbb{R}^p$, $t \ne 0$,

$$
t^T \Sigma_n^{-1/2} S_n \xrightarrow{\mathcal{D}} N(0, \|t\|^2).
$$

To do so we need to verify that the Lindeberg sequence $L_n(\epsilon)$ for the triangular array $\{t^T X_{nk}\}$ converges to 0, where

$$
L_n(\epsilon) = \sum_{k=1}^{k_n} \int_{|t^T \Sigma_n^{-1/2} X_{nk}| > \epsilon} \left(t^T \Sigma_n^{-1/2} X_{nk}\right)^2 dP. \tag{7.10}
$$

Applying the Cauchy-Schwarz inequality (see Lemma 2.3), we obtain

$$
\left(t^T \Sigma_n^{-1/2} X_{nk}\right)^2 \le \|t\|^2 \left(X_{nk}^T \Sigma_n^{-1} X_{nk}\right). \tag{7.11}
$$

Consequently, we can replace the inequality that specifies the integral in (7.10) by $(X_{nk}^T \Sigma_n^{-1} X_{nk})^{1/2} > \epsilon/\|t\|$ and replace the integrand of (7.10) by the quantity on the right hand side of (7.11) without making the integral smaller. In other words,

$$L_n(\epsilon) \leq \|t\|^2 \sum_{k=1}^{k_n} \int_{(X_{nk}^T \Sigma_n^{-1} X_{nk})^{1/2} > \epsilon/\|t\|} (X_{nk}^T \Sigma_n^{-1} X_{nk}) \, dP.$$

However, right hand side is just $\|t\|^2 L_n^{(p)}(\epsilon/\|t\|)$, which converges to 0 by assumption. □

From this theorem we can easily generalize the Lyapounov and Lindeberg-Levy theorems from random variables to random vectors. The proofs will be left as exercises.

Corollary 7.6 *Suppose that $X_1, \ldots, X_n$ are independent and identically distributed random vectors in $\mathbb{R}^p$ with finite mean μ and positive definite variance matrix Σ. Then, $\sqrt{n} E_n(X - \mu) \xrightarrow{\mathcal{D}} N(0, \Sigma)$.*

Corollary 7.7 *Let $\{X_{nk} : k = 1, \ldots, k_n, n = 1, 2, \ldots\}$ be a triangular array of p-dimensional random vectors in which the random vectors in each row are independent. Suppose, without loss of generality, that $E(X_{nk}) = 0$. Let*

$$M_n(\epsilon) = \sum_{k=1}^{k_n} E\left(X_{nk}^T \Sigma_n^{-1} X_{nk}\right)^{1+\delta}.$$

If $M_n(\epsilon) \to 0$ for some $\delta > 0$, then $\Sigma_n^{-1/2} S_n \xrightarrow{\mathcal{D}} N(0, 1)$.

7.6 The δ-method

The δ-method is a convenient device that allows us to find the asymptotic distribution of a function of a random vector that converges in distribution to some random vector. Specifically, suppose X_n and U are p-dimensional random vectors, and g is a differentiable function taking values in $\mathbb{R}^m$, where $m \leq p$. The question is: if, for some positive sequence a_n and some $\mu \in \mathbb{R}^p$, $a_n(X_n - \mu)$ converges in distribution to a random vector U, then what is the limit of $a_n[g(X_n) - g(\mu)]$? The most commonly used form of this result is the case where U is the multivariate Normal random vector, $a_n = \sqrt{n}$, and $\mu = E(X)$. However, it is actually easier to prove the theorem in the general case.

In the following, we denote the ith component of g by g_i, and jth component of X_n by X_n^j. Furthermore, we let

$$\frac{g(x) - g(\mu)}{(x - \mu)^T}$$

denote the $m \times p$ matrix whose (i, j)th entry is $[g_i(x) - g_i(\mu)]/(x^j - \mu^j)$. Note that, in this notation, we have

$$g(x) - g(\mu) = \frac{g(x) - g(\mu)}{(x-\mu)^T}(x - \mu).$$

Moreover, if g is differentiable at μ, then

$$\lim_{x \to \mu} \frac{g(x) - g(\mu)}{(x-\mu)^T} = \frac{\partial g(\mu)}{\partial \mu^T}. \tag{7.12}$$

Theorem 7.14 (The δ-method) *Suppose $\{X_n : n = 1, 2, \ldots\}$ is a sequence of p-dimensional random vectors such that, for a positive sequence $a_n \to \infty$ and a vector $\mu \in \mathbb{R}^p$,*

$$a_n(X_n - \mu) \xrightarrow{\mathcal{D}} U. \tag{7.13}$$

If g is a differentiable function taking values in $\mathbb{R}^m$, where $m \le p$, then

$$a_n[g(X_n) - g(\mu)] \xrightarrow{\mathcal{D}} [\partial g(\mu)/\partial \mu^T]U.$$

Proof. Note that

$$a_n[g(X_n) - g(\mu)] = \frac{g(X_n) - g(\mu)}{(X_n - \mu)^T}[a_n(X_n - \mu)]. \tag{7.14}$$

Define $h : \mathbb{R}^p \to \mathbb{R}^{m \times p}$ to be the following function

$$h(t) = \begin{cases} [g(t) - g(\mu)]/(t-\mu)^T & t \ne \mu \\ \partial g(\mu)/\partial \mu^T & t = \mu \end{cases}$$

Then we can rewrite the identity (7.14) as

$$a_n[g(X_n) - g(\mu)] = h(X_n)\,[a_n(X_n - \mu)],$$

which also holds obviously when $X_n = \mu$. By (7.12), h is continuous at μ. By (7.13), $X_n \xrightarrow{P} \mu$. Hence, by the Continuous Mapping Theorem,

$$h(X_n) \xrightarrow{P} h(\mu) = \partial g(\mu)/\partial \mu^T.$$

By Slutsky's Theorem (see Corollary 7.2),

$$a_n[g(X_n) - g(\mu)] = h(X_n)a_n(X_n - \mu) \xrightarrow{\mathcal{D}} [\partial g(\mu)/\partial \mu^T]\,U,$$

as desired. □

In the multivariate Normal case the above reduces to the following familiar form.

Corollary 7.8 *Suppose $\{X_n : n = 1, 2, \ldots\}$ is a sequence of p-dimensional random vectors such that*

$$\sqrt{n}(X_n - \mu) \xrightarrow{\mathcal{D}} N(0, \Sigma),$$

where $\mu = E(X)$. If g is a differentiable function taking values in $\mathbb{R}^m$, where $m \le p$, then

$$\sqrt{n}[g(X_n) - g(\mu)] \xrightarrow{\mathcal{D}} N\left(0, \frac{g(\mu)}{\partial \mu^T}\Sigma\frac{g^T(\mu)}{\partial \mu}\right).$$

7.7 Mann-Wald notation for order of magnitude

Recall that, in calculus, the magnitude of a sequence of numbers is denoted by the little o or the big O notation. If a sequence of numbers $\{x_n\}$ is bounded, then we write $x_n = O(1)$; if the sequence converges to 0, then we write $x_n = o(1)$. Furthermore, if $\{a_n\}$ is another sequence such that $x_n/a_n = O(1)$, then we write $x_n = O(a_n)$ and say that the magnitude of x_n is no greater than that of a_n. If $x_n/a_n = o(1)$, then we write $x_n = o(a_n)$ and say that x_n is ignorable compared with a_n. A similar notational system can be applied to a sequence of random variables or random vectors. This notational system was introduced by Mann and Wald (1943). In the following, X_n are p-dimensional random vectors.

Definition 7.7 *A sequence of random vectors $\{X_n\}$ is said to be bounded in probability if, for any $\epsilon > 0$, there is a $K > 0$, such that*

$$P(\|X_n\| > K) < \epsilon. \tag{7.15}$$

for all n.

There are two more equivalent conditions for this definition: the first requires (7.15) to hold for all sufficiently large n. That is, there exists an n_0 such that (7.15) holds for all $n > n_0$; the second is

$$\limsup_{n\to\infty} P(\|X_n\| > K) < \epsilon.$$

This is a generalization of the notion of bounded sequence of numbers to a sequence of random variables. Using this notion of boundedness, we can extend the big O notation to sequence of random vectors.

Definition 7.8 *If a sequence of random vectors $\{X_n : n = 1, 2, \ldots\}$ is bounded in probability, then we write $X_n = O_P(1)$. Furthermore, if $\{a_n\}$ is a sequence of non-random positive constants, and if $X_n/a_n = O_P(1)$, then we $X_n = O_P(a_n)$.*

The interpretation of $X_n = O_P(a_n)$ is that the order of magnitude of the random sequence $\{X_n\}$ is not greater than that of the nonrandom sequence $\{a_n\}$. Similarly, we replace the deterministic convergence $x_n \to 0$ with the stochastic convergence $X_n \xrightarrow{P} 0$ extend the little o notation to a sequence of random vectors.

Definition 7.9 *If a sequence of random vectors $\{X_n : n = 1, 2, \ldots\}$ converges in probability to zero, then we write $X_n = o_P(1)$. Furthermore, $X_n = o_P(a_n)$ if $\{a_n\}$ is a sequence of non-random positive constants, and $X_n/a_n = o_P(1)$.*

The interpretation of $X_n = o_P(a_n)$ is that the random sequence $\{X_n\}$ is ignorable compared with nonrandom sequence $\{a_n\}$.

If we think of O as a nonzero constant such as 1, and o as 0, then the product between different types of O's obeys the same rules of the product of 0 and 1. That is, similar to

$$1 \times 1 = 1, \quad 1 \times 0 = 0, \quad 0 \times 0 = 0.$$

Thus, for two positive nonrandom sequences $\{a_n\}$ and $\{b_n\}$:

$$\begin{aligned} O(a_n)O(b_n) &= O(a_n b_n), \\ O(a_n)o(b_n) &= o(a_n b_n), \\ o(a_n)o(b_n) &= o(a_n b_n). \end{aligned}$$

These equalities should be interpreted in the following way. For example, the first equality means that if $x_n = O(a_n)$, $y_n = O(b_n)$, then $x_n y_n = O(a_n b_n)$. The above rules can be easily proved by the definitions of $O(a_n)$ and $o(a_n)$. A similar set of rules apply to O_P and o_P, as summarized by the next theorem.

Theorem 7.15 (The rules of Os)

$$\begin{aligned} O_P(a_n)O_P(b_n) &= O_P(a_n b_n) \\ O_P(a_n)o_P(b_n) &= o_P(a_n b_n) \\ o_P(a_n)o_P(b_n) &= o_P(a_n b_n). \end{aligned}$$

Again, these equalities should be interpreted in terms of the underlying sequences of random random variables. For example, the first equality should be interpreted as: if $X_n = O_P(a_n)$, $Y_n = O_P(b_n)$, then $X_n Y_n = O_P(a_n b_n)$. Note that, here, we assume X_n and Y_n to be numbers rather than vectors.

Proof. 1. If $X_n = O_P(a_n)$ and $Y_n = O_P(b_n)$, then, for any $\epsilon > 0$, there exist $K_1 > 0$ and $K_2 > 0$ such that

$$\limsup_n P(|X_n| > K_1) < \epsilon/2, \quad \limsup_n P(|Y_n| > K_2) < \epsilon/2.$$

Because $|X_n Y_n|/(a_n b_n) > K_1 K_2$ implies that at least one of the following inequalities hold

$$\frac{|X_n|}{a_n} > K_1, \quad \frac{|Y_n|}{b_n} > K_2,$$

we have

$$P\left(\frac{|X_n Y_n|}{a_n b_n} > K_1 K_2\right) \le P\left(\frac{|X_n|}{a_n} > K_1\right) + P\left(\frac{|Y_n|}{b_n} > K_2\right).$$

Hence

$$\limsup_{n\to\infty} P\left(\frac{|X_nY_n|}{a_nb_n} > K_1K_2\right) \leq \limsup_{n\to\infty} P\left(\frac{|X_n|}{a_n} > K_1\right) + \limsup_{n\to\infty} P\left(\frac{|Y_n|}{b_n} > K_2\right)$$
$$< \frac{\epsilon}{2} + \frac{\epsilon}{2} = \epsilon,$$

which means $|X_nY_n|/(a_nb_n)$ is bounded in probability.

2. Suppose that $X_n = O_P(a_n)$ and $Y_n = o_P(b_n)$. Let $\epsilon > 0,\ \delta > 0$ be constants. Let $K > 0$ be such that

$$P\left(\frac{|X_n|}{a_n} \geq K\right) < \delta.$$

Then

$$\begin{aligned}
P\left(\frac{|X_nY_n|}{a_nb_n} > K\right) &= P\left(\frac{|X_nY_n|}{a_nb_n} > \epsilon,\ \frac{|X_n|}{a_n} > K\right) + P\left(\frac{|X_nY_n|}{a_nb_n} > \epsilon,\ \frac{|X_n|}{a_n} \leq K\right) \\
&\leq P\left(\frac{|X_n|}{a_n} > K\right) + P\left(\frac{|Y_n|}{b_n} > \frac{\epsilon}{K}\right) \\
&\leq P\left(\frac{|Y_n|}{b_n} > \frac{\epsilon}{K}\right) + \delta.
\end{aligned}$$

Therefore,

$$\limsup_{n\to\infty} P\left(\frac{X_nY_n}{a_nb_n} > \epsilon\right) \leq \limsup_{n\to\infty} P\left(\frac{|Y_n|}{b_n} > \frac{\epsilon}{K}\right) + \delta = \delta.$$

Since $\delta > 0$ is arbitrary, we have

$$\limsup_{n\to\infty} P\left(\frac{X_nY_n}{a_nb_n} > \epsilon\right) = 0,$$

as desired.

3. Let $X_n/a_n = o_P(1)$ and $Y_n/b_n = o_P(1)$. It suffices to show $X_n/a_n = O_P(1)$ because, by part 2,

$$\frac{X_nY_n}{a_nb_n} = O_P(1)o_P(1) = o_P(1).$$

If $\epsilon > 0$, then

$$P\left(\frac{|X_n|}{a_n} > 1\right) = 0 < \epsilon.$$

So $X_n/a_n = O_P(1)$. □

We can also use Theorem 7.15 to evaluate the order of products such as a_nX_n, where a_n is fixed and X_n is random. This is because o or O are special cases of o_P or O_P, as the next proposition shows.

Proposition 7.3 *If $X_n = O(a_n)$, then $X_n = O_P(a_n)$; if $X_n = o(a_n)$, then $X_n = o_P(a_n)$.*

Proof. If $X_n = O(a_n)$, then $|X_n/a_n| \le K$ for some constant K. Let $\epsilon > 0$, then $P(|X_n/a_n| > K + 1) = 0 < \epsilon$. If $X_n = o(a_n)$, then $X_n/a_n \to 0$. Let $\epsilon > 0$. Then, for sufficiently large n, $|X_n/a_n| \le \epsilon$. So, for sufficiently large n, $P(|X_n/a_n| > \epsilon) = 0$. □

For example, by Theorem 7.15 and Proposition 7.3 we have the following relations:

$$\begin{aligned} O_P(a_n)O(b_n) &= O_P(a_n b_n), \\ O_P(a_n)o(b_n) &= o_P(a_n b_n), \\ O(a_n)o_P(b_n) &= o_P(a_n b_n), \\ o_P(a_n)o(b_n) &= o_P(a_n b_n). \end{aligned}$$

7.8 Hilbert spaces

The notion of projections in Hilbert spaces is used frequently later in this book. In this section we outline basic properties of Hilbert spaces. A Hilbert space is an extension of the Euclidean space $\mathbb{R}^p$. We begin with the definition of a vector space defined on the field of real numbers.

Definition 7.10 (Vector space) *A vector space is a set $\mathcal{V}$, together with an operation $+$ between elements in $\mathcal{V}$, and an operation $\cdot$ between numbers in $\mathbb{R}$ and elements $\mathcal{V}$ satisfying the following conditions.*

1. *Operation $+$:*
   - *1a. (closure) If $v_1, v_2 \in \mathcal{V}$ then $v_1 + v_2 \in \mathcal{V}$;*
   - *1b. (commutative law) $v_1 + v_2 = v_2 + v_1$;*
   - *1c. (associative law) $v_1 + (v_2 + v_3) = (v_1 + v_2) + v_3$;*
   - *1d. (zero element) There is a unique element $0 \in \mathcal{V}$ such that, for all $v \in \mathcal{V}$, $v + 0 = v$;*
   - *1e. (negative element) For each $v \in \mathcal{V}$, there is a $(-v) \in \mathcal{V}$ such that $v + (-v) = 0$.*
2. *Operation $\cdot$ between the members of $\mathbb{R}$ and $\mathcal{V}$:*
   - *2a. (closure) If $\lambda \in \mathbb{R}$, $v \in \mathcal{V}$, then $\lambda \cdot v \in \mathcal{V}$;*
   - *2b. (distributive law 1) If $\lambda \in \mathbb{R}$, $u, v \in \mathcal{V}$, then $\lambda \cdot (u + v) = \lambda \cdot u + \lambda \cdot v$;*
   - *2c. (distributive law 2) If $\lambda, \mu \in \mathbb{R}$ and $v \in \mathcal{V}$, then $(\lambda + \mu) \cdot v = \lambda \cdot v + \mu \cdot v$;*
   - *2d. (associative law) If $\lambda, \mu \in \mathbb{R}$ and $v \in \mathcal{V}$ then $\lambda \cdot (\mu \cdot v) = (\lambda\mu) \cdot v$;*
   - *2e. (unit element) For any $v \in \mathcal{V}$, $1 \cdot v = v$.*

Note that $0 \cdot v = 0$ for any $v \in \mathcal{V}$, because

$$v + 0 \cdot v = 1 \cdot v + 0 \cdot v = (1 + 0) \cdot v = 1 \cdot v = v.$$

Thus, $0 \cdot v$ is the zero element in $\mathcal{V}$. Also note that the zero element is unique. In fact, let $0_1, 0_2$ be members of $\mathcal{H}$ that satisfies $0_1 + v = v$ and $0_2 + v = v$ for all $v \in \mathcal{V}$. By taking $v = 0_1$ and $v = 0_2$ separately in $1d$ above, we get $0_1 + 0_2 = 0_2$, and $0_1 + 0_2 = 0_2 + 0_1 = 0_1$. Therefore, $0_1 = 0_2$.

For the rest of the book we will omit the dot in $\lambda \cdot v$ and write it simply as λv. To sum up, a vector space consists of four ingredients: a set $\mathcal{V}$, an operation $+$ between members of $\mathcal{V}$, an operation $\cdot$ between members of $\mathbb{R}$ and members of $\mathcal{V}$, and finally a zero element in $\mathcal{V}$. Thus, a rigorous notation of a vector space is $\{\mathcal{V}, +, \cdot, 0\}$. However, in most cases we simply denote a vector space by the set $\mathcal{V}$ without causing ambiguity. We now give some examples of a vector space.

Example 7.1 The space $\mathbb{R}^n$. Let $a, b \in \mathbb{R}^n$ and $\lambda \in \mathbb{R}$. Define

$$\begin{pmatrix} a_1 \\ \vdots \\ a_n \end{pmatrix} + \begin{pmatrix} b_1 \\ \vdots \\ b_n \end{pmatrix} = \begin{pmatrix} a_1 + b_1 \\ \vdots \\ a_n + b_n \end{pmatrix}, \quad \text{and } \lambda \begin{pmatrix} a_1 \\ \vdots \\ a_n \end{pmatrix} = \begin{pmatrix} \lambda a_1 \\ \vdots \\ \lambda a_n \end{pmatrix}.$$

Furthermore, define the zero element in $\mathbb{R}^n$ to be the vector $(0, \ldots, 0)^T$. Then, it is easy to verify that the conditions in Definition 7.10 are satisfied, which means $\mathbb{R}^n$ is a vector space. □

Example 7.2 Space of square-integrable functions. Let $(\Omega, \mathcal{F}, P)$ be a probability space. Let $L_2(P)$ be the set of all real-valued functions f such that $\int f^2 dP < \infty$. For $f_1, f_2 \in L_2(P)$ and $\lambda \in \mathbb{R}$ define $f_1 + f_2$ and λf_1 to be the following members of $L_2(P)$:

$$f_1 + f_2 : x \mapsto f_1(x) + f_2(x), \quad \lambda f_1 : x \mapsto \lambda f_1(x).$$

Furthermore, define the zero element in $L_2(P)$ to be the function $f(x) = 0$ almost everywhere P. Then it is easy to verify that the conditions in Definition 7.10 are satisfied. This space is called the L_2 space with respect to $(\Omega, \mathcal{F}, P)$, and is written as $L_2(P)$. □

An inner product space $\mathcal{V}$, or pre-Hilbert space, is a vector space together with a mapping $u : \mathcal{V} \times \mathcal{V} \to \mathbb{R}$ that is symmetric, bilinear, and positive definite.

Definition 7.11 *Suppose that $\mathcal{V}$ is a vector space. An inner product is a function $u : \mathcal{V} \times \mathcal{V} \mapsto \mathbb{R}$ such that for any $x, y, z \in \mathcal{V}$ and $\alpha, \beta \in \mathbb{R}$* we have

i. (symmetric) $u(x, y) = u(y, x)$,
ii. (bilinear) $u(\alpha x + \beta y, z) = \alpha u(x, z) + \beta u(y, z)$,
iii. (positive) $u(x, x) \geq 0$ for all $x \in \mathcal{V}$,
iv. (definite) $u(x, x) = 0$ implies $x = 0$.

A vector space $\mathcal{V}$, together with an inner product $u : \mathcal{V} \times \mathcal{V} \to \mathbb{R}$, is called an inner product space.

Note that, if $x = 0$, then it can be written as $0 \cdot y$ for some $y \in \mathcal{V}$. Therefore, $u(x, x) = u(x, 0 \cdot y) = 0 \times u(x, y) = 0$. Thus an inner product also satisfies the following condition

v. $x = 0$ *implies* $u(x, x) = 0$.

Also, by properties *i* and *ii* we see that

$$u(x, \alpha y + \beta z) = u(\alpha y + \beta z, x) = \alpha u(y, x) + \beta u(z, x) = \alpha u(x, y) + \beta u(x, z).$$

That is, u is in fact a bilinear function. We record this as the sixth property of an inner product:

vi. $u(x, \alpha y + \beta z) = \alpha u(x, y) + \beta u(x, z)$.

For the rest of the book we will write $u(x, y)$ as $\langle x, y \rangle_{\mathcal{V}}$. If there is no source of confusion, the subscript $\mathcal{V}$ is dropped and the inner product in $\mathcal{V}$ is simply written as $\langle x, y \rangle$. Two examples of inner product spaces are given below.

Example 7.3 Let $\mathcal{V}$ be the vector space in Example 7.1, and let A be an n by n positive definite matrix. Let u be the mapping

$$u : \mathbb{R}^n \times \mathbb{R}^n \to \mathbb{R}, \quad (x, y) \mapsto x^T A y.$$

Then it can be easily verified that u defines an inner product. This inner product space is called the n-dimensional Euclidean space. □

Example 7.4 Let $\mathcal{V}$ be the vector space $L_2(P)$ in Example 7.2. Define the mapping $u : \mathcal{V} \times \mathcal{V} \to \mathbb{R}$ by

$$(f, g) \mapsto \int f g \, dP.$$

The right-hand side is a finite number because, by Hölder's inequality,

$$\int |fg| dP \leq \left(\int f^2 dP \right)^{1/2} \left(\int g^2 dP \right)^{1/2}.$$

See, for example, Billingsley (1995). It is easy to check that the function u thus defined satisfies properties *i, ii, iii* in Definition 7.11. However, condition *iv* is in general not satisfied, because $\int f^2 dP = 0$ only implies $f = 0$ almost everywhere P.

To make u an inner product, we introduce the following equivalence relation $\sim$ in $\mathcal{V}$:

$$f \sim g \text{ if and only if } f = g \text{ almost everywhere } P.$$

It can be easily verified that $\sim$ thus defined is indeed an equivalence relation. Let $\mathcal{V}/\sim$ be the quotient space with respect to $\sim$ (see Kelley, 1955). For two members F and G of $\mathcal{V}/\sim$, let $F+G$ be the equivalent class of $f+g$, where f is any member of F and g is any member of G. For $\lambda \in \mathbb{R}$ and $F \in \mathcal{V}/\sim$, let $\lambda \cdot F$ be the equivalent class of $\lambda \cdot f$, where f is any member of F. Furthermore, define the zero element of $\mathcal{V}/\sim$ as the equivalent class of any function that is almost everywhere 0. It can then be shown that, with these definitions of $+$, $\cdot$, 0, $\mathcal{V}/\sim$ is indeed a vector space relative to $\mathbb{R}$.

Furthermore, we introduce the mapping $\tilde{u} : (\mathcal{V}/\sim) \times (\mathcal{V}/\sim) \to \mathbb{R}$ as follows. If F, G are members of $\mathcal{V}/\sim$, then

$$\tilde{u}(F, G) = u(f, g),$$

where f is any member of F and g is any member of G. Note that $u(f, g)$ is not affected by the choices of f and g. It can be shown that $\tilde{u}$ does satisfy all four conditions in Definition 7.11. Thus, it is a well defined inner product. In other words,

$$\{\mathcal{V}/\sim, +, \cdot, 0, \tilde{u}\}$$

forms an inner product space.

For our purpose, however, it will not cause serious ambiguity if we simply treat almost everywhere equal functions as the same function, and treat (simply) $\mathcal{V}/\sim$ as $\mathcal{V}$. This we will do throughout the rest of the book. □

A *norm* in a vector space $\mathcal{V}$ is a mapping $\rho : \mathcal{V} \to \mathbb{R}$ such that

i. $\rho(f) \geq 0$ for all $f \in \mathcal{V}$,
ii. for any $a \in \mathbb{R}$ and $f \in \mathcal{V}$, $\rho(af) = |a|\rho(f)$,
iii. for any $f, g \in \mathcal{V}$, $\rho(f + g) \leq \rho(f) + \rho(g)$,
iv. $\rho(f) = 0$ implies $f = 0$.

A norm is a generalization of the absolute value. In particular, $\rho(f - g)$ is a measure of distance between two members of $\mathcal{V}$, just like $|a - b|$ is a measure of distance between two numbers. For the rest of the book, we write $\rho(f)$ as $\|f\|_{\mathcal{V}}$ or simply $\|f\|$. A vector space $\mathcal{V}$, together with a norm $\rho : \mathcal{V} \to \mathbb{R}$, is called a normed space.

It can be shown that, if $(\mathcal{V}, u)$ is an inner product space, then the function

$$\rho : \mathcal{V} \to \mathbb{R}, \ f \mapsto [u(f, f)]^{1/2}$$

is a norm. Thus, an inner product space is also a normed space.

Using this norm we can define the notions of limit and completeness in an inner product space. A sequence $\{f_n : n = 1, 2, \ldots\}$ of elements of $\mathcal{V}$ is a *Cauchy sequence* if, for any $\epsilon > 0$, there exists an n_0, such that for all $m, n > n_0$ we have

$$\|f_n - f_m\| < \epsilon.$$

We say that a sequence f_n converges to a member f of $\mathcal{V}$ if $\lim_{n\to\infty} \|f_n - f\| = 0$.

Definition 7.12 *An inner product space $\mathcal{V}$ is complete if every Cauchy sequence $\{f_n\}$ in $\mathcal{V}$ converges to a member of $\mathcal{V}$. That is, $\lim_{n\to\infty} \|f_n - f\| = 0$ for some $f \in \mathcal{V}$. A complete inner product space is called a Hilbert space.*

If an inner product $\mathcal{V}$ is finite dimensional, then it is always complete. Also, the L_2-space with respect to a measure is always complete. In other words, the inner product spaces in Example 7.3 and Example 7.4 are Hilbert spaces.

Recall that, one of the defining assumptions of a norm is the triangular inequality $\|f + g\| \le \|f\| + \|g\|$. There is an inequality for inner product of similar importance, but, unlike the triangular inequality, it is a consequence of the four defining assumptions of the inner product. This is the Cauchy-Schwarz inequality.

Theorem 7.16 (The Cauchy-Schwarz inequality) *If $\langle \cdot, \cdot \rangle$ is an inner product in an inner product space $\mathcal{V}$, then, for any $f, g \in \mathcal{V}$,*

$$\langle f, g\rangle^2 \le \langle f, f\rangle\langle g, g\rangle. \tag{7.16}$$

Moreover, the equality holds if and only if f and g are proportional to each other.

Proof. First note that, if $\langle g, g\rangle = 0$, then $g = 0$ and consequently the inequality (7.16) holds. Now assume $\langle g, g\rangle \neq 0$. Let $\alpha \in \mathbb{R}$, $f, g \in \mathcal{V}$. Then

$$0 \le \langle f - \alpha g, f - \alpha g\rangle = \langle f, f\rangle - 2\alpha\langle f, g\rangle + \alpha^2\langle g, g\rangle \equiv F(\alpha).$$

Since $F(\alpha)$ is a quadratic polynomial, it can be easily verified that it achieves its minimum at $\alpha^* = \langle f, g\rangle / \langle g, g\rangle$. So we have

$$\begin{aligned} 0 \le F(\alpha^*) &= \langle f, f\rangle - 2\alpha^*\langle f, g\rangle + \alpha^{*2}\langle g, g\rangle \\ &= \langle f, f\rangle - 2\frac{\langle f, g\rangle}{\langle g, g\rangle}\langle f, g\rangle + \frac{\langle f, g\rangle^2}{\langle g, g\rangle^2}\langle g, g\rangle \\ &= \langle f, f\rangle - \frac{\langle f, g\rangle^2}{\langle g, g\rangle}, \end{aligned}$$

which is the inequality (7.16).

Now suppose the equality in (7.16) holds, then $F(\alpha^*) = 0$. Hence $\langle f - \alpha^* g, f - \alpha^* g\rangle = 0$, which implies $f = \alpha^* g$. Thus f and g are proportional to each other. Conversely, if f and g are proportional then it is obvious that the equality in (7.16) holds. □

Using the Cauchy-Schwarz inequality we can easily show that the mapping $\rho(f) = \langle f, f\rangle^{1/2}$ indeed satisfies the triangular inequality.

Corollary 7.9 *If $\langle \cdot, \cdot \rangle$ is an inner product in an inner product space $\mathcal{V}$ and $\rho(f) = \langle f, f \rangle^{\frac{1}{2}}$, then, for any $f, g \in \mathcal{V}$,*

$$\rho(f + g) \leq \rho(f) + \rho(g).$$

Proof. Note that

$$\begin{aligned} \rho(f+g)^2 &= \rho(f)^2 + 2\langle f, g \rangle + \rho(g)^2 \\ &\leq \rho(f)^2 + 2|\langle f, g \rangle| + \rho(g)^2 \\ &\leq \rho(f)^2 + 2\rho(f)\,\rho(g) + \rho(g)^2 \\ &= (\rho(f) + \rho(g))^2, \end{aligned}$$

where the second inequality follows from the Cauchy-Schwarz inequality. Now take square root on both sides to complete the proof. □

7.9 Multivariate Cauchy-Schwarz inequality

Recall that, in Lemma 2.3, we stated a version of the multivariate Cauchy-Schwarz inequality to establish the Cramér-Rao lower bound. In this section we further extend this inequality in terms of inner product matrices, which is useful for developing optimal estimating equations among other things. Let $\mathcal{H}$ be a Hilbert space, and let $\mathcal{H}^p$ be the p-fold Cartesian product:

$$\mathcal{H}^p = \underbrace{\mathcal{H} \times \ldots \times \mathcal{H}}_{p}.$$

For any two members S, G of $\mathcal{H}^p$, define their inner product matrix as

$$[S, G] = \begin{pmatrix} \langle s_1, g_1 \rangle & \cdots & \langle s_1, g_p \rangle \\ \vdots & \ddots & \vdots \\ \langle s_p, g_1 \rangle & \cdots & \langle s_p, g_p \rangle \end{pmatrix}.$$

The inner product matrix shares similar properties with an inner product, as shown by the next Proposition. The proof is left as an exercise.

Proposition 7.4 *Let $\mathcal{H}$ be a Hilbert space and $\mathcal{H}^p$ be its p-fold Cartesian product. Let $[\cdot, \cdot] : \mathcal{H}^p \times \mathcal{H}^p \to \mathbb{R}^{p \times p}$ be the inner product matrix in $\mathcal{H}^p$. Then*

1. *(symmetry after transpose) $[G_1, G_2] = [G_2, G_1]^T$;*
2. *(bilinear) If $G_1, G_2, G_3 \in \mathcal{H}^p$ and $a_1, a_2, a_3 \in \mathbb{R}$, then*

$$\begin{aligned} [a_1 G_1 + a_2 G_2, G_3] &= a_1 [G_1, G_3] + a_2 [G_2, G_3], \\ [G_1, a_2 G_2 + a_3 G_3] &= a_2 [G_1, G_3] + a_3 [G_1, G_3]; \end{aligned}$$

3. (positivity) for any $G \in \mathcal{H}^p$, $[G, G]$ *is positive semidefinite;*
4. (definiteness) $[G, G] = 0$ *implies* $G = 0$.

Another useful property for the inner product matrix is that, if $A, B \in \mathbb{R}^{p \times p}$ and $F, G \in \mathcal{H}^p$, then

$$[AG, BF] = A[G, F]B^T. \tag{7.17}$$

To see this, let $(AG)_i$ and $(BF)_i$ be the ith component of AG and BF. Then $(AG)_i = \text{row}_i(A)G$ and $(BF)_i = \text{row}_i(B)F$, where, for example, $\text{row}_i(A)$ means the ith row of A. We have

$$\begin{aligned}
[AG, BF] &= \begin{pmatrix} \langle \text{row}_1(A)G, \text{row}_1(B)F\rangle & \cdots & \langle \text{row}_1(A)G, \text{row}_p(B)F\rangle \\ \cdots & \ddots & \vdots \\ \langle \text{row}_p(A)G, \text{row}_1(B)F\rangle & \cdots & \langle \text{row}_p(A)G, \text{row}_p(B)F\rangle \end{pmatrix} \\
&= \begin{pmatrix} \text{row}_1(A) \\ \vdots \\ \text{row}_p(A) \end{pmatrix} [G, F] \left(\text{row}_1(B)^T, \cdots, \text{row}_p(B)\right) = A[G, F]B^T.
\end{aligned}$$

Definition 7.13 *For any two symmetric square matrices* A *and* B*, write* $A \succeq B$ *if* $A - B$ *is positive semi-definite. This partial ordering is called the Loewner ordering.*

In the special case where $p = 1$, G and S are simply members of $\mathcal{H}$, and the matrix inequality in Theorem 7.17 below reduces to the classical Cauchy-Schwarz inequality:

$$\langle G, S\rangle^2 \leq \langle G, G\rangle\langle S, S\rangle.$$

Theorem 7.17 (Multivariate Cauchy-Schwarz inequality) *If* S *and* G *are members of* $\mathcal{H}^p$ *and the matrix* $[G, G]$ *is invertible, then*

$$[S, S] \succeq [S, G][G, G]^{-1}[G, S].$$

Proof. By Proposition 7.4, $[G, G] \succeq 0$. Hence the matrix

$$[S - G[G, G]^{-1}G, S - G[G, G]^{-1}G]$$

is positive semi-definite. This matrix can be decomposed as

$$\begin{aligned}
&[S, S] - [S, G][G, G]^{-1}[G, S] - [S, G][G, G]^{-1}[G, S] \\
&+ [S, G][G, G]^{-1}[G, G][G, G]^{-1}[G, S] \\
&= [S, S] - [S, G][G, G]^{-1}[G, S].
\end{aligned}$$

Hence the desired inequality. □

The condition that $[G, G]$ is invertible in the above Proposition can be relaxed using the Moore-Penrose inverse – see Problem 7.22.

7.10 Projections

One of the most useful tools related to a Hilbert space is the notion of projection, which is based on orthogonality, as defined below.

Definition 7.14 (Orthogonality) *If $\mathcal{H}$ is a Hilbert space and if $f, g \in \mathcal{H}$, then f and g are orthogonal if $\langle f, g\rangle = 0$. We write this as $g \perp f$.*

An immediate consequence of orthogonality is the Pythagaras theorem, as given below. The proof is left as an exercise.

Proposition 7.5 *If $f_1, \ldots, f_n$ are pairwise orthogonal vectors in $\mathcal{H}$ then*

$$\|f_1 + \cdots + f_n\|^2 = \|f_1\|^2 + \cdots + \|f_n\|^2.$$

A more general version of Pythagoras theorem the parallelogram law. Again, the proof is left as an exercise.

Proposition 7.6 *If $\mathcal{H}$ is a Hilbert space and $f, g \in \mathcal{H}$, then*

$$\|f + g\|^2 + \|f - g\|^2 = 2(\|f\|^2 + \|g\|^2).$$

We now define the linear subspace. Intuitively, it is any hyperplane that passes through the origin.

Definition 7.15 *A subset $\mathcal{G}$ of a Hilbert space is called a linear manifold if it is closed under linear operation. That is, for any $f_1, f_2 \in \mathcal{G}$ and $c_1, c_2 \in \mathbb{R}$, we have $c_1 f_1 + c_2 f_2 \in \mathcal{G}$. A closed linear manifold is called a linear subspace.*

Note that a subspace $\mathcal{G}$ must contain the zero element of $\mathcal{H}$. This is because, for any $g \in \mathcal{G}$, $0 \cdot g = 0$ must be a member of $\mathcal{G}$. We now define projection. For a member f of $\mathcal{H}$ and a subspace $\mathcal{G}$ of $\mathcal{H}$, the member of $\mathcal{G}$ that is nearest to f is the projection of f on to $\mathcal{G}$. Intuitively, if f^* is the projection of f on to $\mathcal{G}$, then the vector $f - f^*$ should be orthogonal to $\mathcal{G}$.

Theorem 7.18 (Projection theorem) *If $\mathcal{H}$ is a Hilbert space and $\mathcal{G}$ is a linear subspace then, for any $f \in \mathcal{H}$, there is a unique element $f_0 \in \mathcal{G}$ such that*

$$\|f - f_0\| \leq \|f - g\|$$

for all g in $\mathcal{G}$. furthermore, a vector f_0 satisfies the above relation if and only if it satisfies

$$\langle f - f_0, g\rangle = 0 \tag{7.18}$$

for all $g \in \mathcal{G}$.

The vector f_0 is called the orthogonal projection of f on to $\mathcal{G}$, and is written as $P_{\mathcal{G}}(f)$. Or, if there is no ambiguity we will simply write this as Pf. The operator $P : \mathcal{H} \to \mathcal{H}$ thus defined is called a projection operator. It can be shown that P is a idempotent and self-adjoint linear operator; that is, for any $f, g \in \mathcal{H}$, $\alpha, \beta \in \mathbb{R}$, we have

$$P(Pf) = Pf, \quad \langle f, Pg \rangle = \langle Pf, g \rangle, \quad P(\alpha f + \beta g) = \alpha Pf + \beta Pg.$$

Conversely, if $P : \mathcal{H} \to \mathcal{H}$ is a self-adjoint and idempotent linear operator, and if $\mathcal{S} = \{Pf : f \in \mathcal{H}\}$, then $\mathcal{S}$ is necessarily a linear subspace of $\mathcal{H}$ and, for any $f \in \mathcal{H}$, Pf is the orthogonal projection of f on to the subspace $\mathcal{S}$. The proofs of this theorem and the above statements can be found in Conway (1990). Equation (7.18) provides a way to find the projection, as illustrated by the next example.

Example 7.5 (Projection on to a finite dimensional subspace) Let $\mathcal{H}$ be a Hilbert space and $f \in \mathcal{H}$. Let $\mathcal{M}$ be the subspace of $\mathcal{H}$ spanned by the vectors $f_1, \ldots, f_p$ in $\mathcal{H}$. We want to find the projection of f on to $\mathcal{M}$. Let f_0 be this projection; that is, $f_0 = P_{\mathcal{M}} f$. Because $f_0 \in \mathcal{M}$, it is a linear combination of $f_1, \ldots, f_p$, say $f_0 = a_1 f_1 + \cdots + a_p f_p$. By the projection formula (7.18), we have $\langle f - f_0, g \rangle = 0$ for all $g \in \mathcal{M}$. In particular, this holds for $f_1, \ldots, f_p$:

$$\langle f - f_0, f_i \rangle = 0, \quad i = 1, \ldots, p.$$

Since the left-hand side is

$$\langle f, f_i \rangle - \left\langle \sum_{j=1}^{p} a_j f_j, f_i \right\rangle = \langle f, f_i \rangle - \sum_{j=1}^{p} \langle f_j, f_i \rangle a_j,$$

the coefficients $a_1, \ldots, a_p$ satisfy the following p equations

$$\sum_{j=1}^{p} \langle f_j, f_i \rangle a_j = \langle f, f_i \rangle.$$

In matrix notation,

$$\begin{pmatrix} a_1 \\ \vdots \\ a_p \end{pmatrix} = \begin{pmatrix} \langle f_1, f_1 \rangle & \cdots & \langle f_1, f_p \rangle \\ \vdots & \cdots & \vdots \\ \langle f_p, f_1 \rangle & \cdots & \langle f_p, f_p \rangle \end{pmatrix}^{-1} \begin{pmatrix} \langle f, f_1 \rangle \\ \vdots \\ \langle f, f_p \rangle \end{pmatrix}$$

The projection of f onto $\mathcal{M}$ is therefore

$$P_{\mathcal{M}} f = (\langle f, f_1 \rangle, \cdots, \langle f, f_p \rangle) \begin{pmatrix} \langle f_1, f_1 \rangle & \cdots & \langle f_1, f_p \rangle \\ \vdots & \cdots & \vdots \\ \langle f_p, f_1 \rangle & \cdots & \langle f_p, f_p \rangle \end{pmatrix}^{-1} \begin{pmatrix} f_1 \\ \vdots \\ f_p \end{pmatrix}, \tag{7.19}$$

provided that the matrix $\{\langle f_i, f_j\rangle\}_{i,j=1}^{p}$ is invertible. This matrix is called the Gram matrix with respect to the set $\{f_1, \ldots, f_n\}$, and will be written as $G(f_1, \ldots, f_p)$. □

Example 7.6 (Ordinary Least Squares) This is a specialization of Example 7.5 to $\mathcal{H} = \mathbb{R}^n$, which gives the formula for the Ordinary Least Squares. Let x_1, ..., x_p, $p \leq n$, be a set of linear independent vectors in $\mathbb{R}^n$, and let $\mathcal{L}$ be the linear subspace spanned by x_1,..., x_p. Let y be another vector in $\mathbb{R}^n$. Define the inner product in $\mathbb{R}^n$ by $\langle x, y\rangle = x^T y$. In this case, the Gram matrix $G(x_1, \ldots, x_p)$ can be expressed as $X^T \Sigma X$, where X is the n by p matrix whose ith column is x_i^T. The vector $\{\langle f, f_i\rangle\}$ can be expressed as $X^T y$. Hence, the projection of y onto $\mathcal{L}$ can be expressed in matrix form as

$$\hat{y} = X(X^T X)^{-1} X^T y \equiv X\hat{\beta},$$

where $\hat{\beta} = (X^T X)^{-1} X^T y$ is just the Ordinary Least Squares estimate and $\hat{y}$ is the prediction vector of y. □

Example 7.7 (Conditional Expectation) Let (X, Y) be a random element. Suppose P_{XY} and P_Y denote the probability measures induced by (X, Y) and Y respectively. Then, it can be shown that $L_2(P_Y)$ is a subspace of $L_2(P_{XY})$. Let f be a member of $L_2(P_{XY})$. We now show that $f_0(Y) = E[f(X, Y)|Y]$ is the projection of f on to $L_2(P_Y)$ using the projection formula (7.18). If g is an arbitrary member of $L_2(P_Y)$, then

$$\begin{aligned}\langle f - f_0, g\rangle &= \int (f - f_0) g \, dP_{XY} \\ &= E\{[f(X, Y) - f_0(Y)]g(Y)\} \\ &= E\{[f(X, Y) - E(f(X, Y)|Y)]g(Y)\} \\ &= E[f(X, Y)g(Y)] - E[E(f(X, Y)|Y)g(Y)].\end{aligned}$$

Since the second term in the last line is

$$E[E(f(X, Y)|Y)g(Y)] = E[E(g(Y)f(X, Y)|Y)] = E[g(Y)f(X, Y)],$$

we have

$$\langle f - f_0, g\rangle = 0,$$

which means f_0 is the projection of f on to $L_2(P_Y)$. □

Before proceeding to the next example, we first introduce the concept of the Moore-Penrose inverse (See, for example, Kollo and von Rosen, 2005).

Definition 7.16 *Let A be a matrix. The Moore-Penrose inverse A^+ is defined to be the (unique) matrix that satisfies*

$$AA^+A = A, \quad A^+AA^+ = A^+, \quad (AA^+)^T = AA^+, \quad (A^+A)^T = A^+A.$$

Example 7.8 Let Σ be a positive definite matrix in $\mathbb{R}^{p\times p}$ and consider the Hilbert space consisting of the linear space $\mathbb{R}^p$ and the inner product defined by $\langle x, y\rangle = x^T\Sigma y$. Let $\mathcal{G}$ be a linear subspace of $\mathbb{R}^p$ of dimension $q \le p$ and let $\{v_1, \ldots, v_r\}$, $r \ge q$, be a set of vectors in $\mathbb{R}^p$ that span $\mathcal{G}$. Let V be the $p \times r$ matrix $(v_1, \ldots, v_r)$. Note that $v_1, \ldots, v_r$ need not be linearly independent, and hence $V^T\Sigma V$ need not be invertible. Let

$$P_{\mathcal{G}}(\Sigma) = V(V^T\Sigma V)^+V^T\Sigma.$$

We now show that this matrix is the projection operator with range $\mathcal{G}$.

We first note that

$$\begin{aligned} P_{\mathcal{G}}(\Sigma)P_{\mathcal{G}}(\Sigma) &= V(V^T\Sigma V)^+V^T\Sigma V(V^T\Sigma V)^+V^T\Sigma \\ &= V(V^T\Sigma V)^+V^T\Sigma = P_{\mathcal{G}}(\Sigma). \end{aligned}$$

Thus $P_{\mathcal{G}}(\Sigma)$ is idempotent. Moreover, for any $x, y \in \mathbb{R}^p$,

$$\langle x, P_{\mathcal{G}}(\Sigma)y\rangle = x^T\Sigma V(V^T\Sigma V)^+V^T\Sigma y = \langle P_{\mathcal{G}}(\Sigma)x, y\rangle.$$

Thus $P_{\mathcal{G}}(\Sigma)$ is self-adjoint. Since, for any $x \in \mathbb{R}^p$, $P_{\mathcal{G}}(\Sigma)x = Vz$ for some $z \in \mathbb{R}^p$, the range of $P_{\mathcal{G}}(\Sigma)$ is contained in $\mathrm{span}(V) = \mathcal{G}$. Since $\mathrm{span}(V)$ is a q-dimensional subspace, V has rank q. Because Σ is nonsingular, $V(V^T\Sigma V)^+V^T\Sigma$ also has rank q. Thus the range of $P_{\mathcal{G}}(\Sigma)$ is not a proper subset of $\mathcal{G}$. □

Problems

7.1. Suppose $X_n = (Y_{n1}, \ldots, Y_{nk})^T$, $n = 1, 2, \ldots$, are k-dimensional random vectors. Show that X_n converges almost everywhere to a random vector $X = (Y_1, \ldots, Y_k)^T$ if and only if Y_{ni} converges almost everywhere to Y_i for each $i = 1, \ldots, k$.

7.2. Suppose $X_n = (Y_{n1}, \ldots, Y_{nk})^T$, $n = 1, 2, \ldots$, are k-dimensional random vectors. Show that X_n converges in probability to a random vector $X = (Y_1, \ldots, Y_k)^T$ if and only if Y_{ni} converge to Y_i in probability for each $i = 1, \ldots, k$.

7.3. Show that, if X_n is a sequence of random vectors that converges almost everywhere to a random vector, then the sequence converges to X in probability.

7.4. Show that a random vector X is degenerate at a — that is, $P(X = a) = 1$ — if and only if $P \circ X^{-1} = \delta_a$. Furthermore, use the Portmanteau theorem to show that $X_n \xrightarrow{\mathcal{D}} \delta_a$ if and only if $X_n \xrightarrow{P} a$.

7.5. Suppose X_n is a sequence of random vectors that converges in probability to a random vector X; that is, $P(\|X_n - X\| > \epsilon) \to 0$ for any $\epsilon > 0$. Show that, for any open set $G \in \mathcal{F}_X$, $\liminf P(X_n \in G) \geq P(X \in G)$. Use the Portmanteau theorem to conclude that $X_n \xrightarrow{\mathcal{D}} X$.

7.6. Prove part 3 of Proposition 7.1.

7.7. Suppose that $X_1, X_2, \ldots$ are independent and uniformly bounded random variables with mean $E(X_n) = 0$ for all n. Let $S_n = X_1 + \cdots + X_n$ and let $s_n^2 = \mathrm{var}(S_n)$. Show that, if $s_n \to \infty$, then $S_n / s_n \xrightarrow{\mathcal{D}} N(0, 1)$.

7.8. Let $X_1, \cdots, X_n$ be uncorrelated random variables with means $\mu_1, \cdots, \mu_n$ and variances $\sigma_1^2, \cdots, \sigma_n^2$. Suppose that $(\sigma_1^2 + \cdots + \sigma_n^2)/n^2 \to 0$ as $n \to \infty$. Prove that

$$n^{-1} \sum_{i=1}^{n} (X_i - \mu_i) \xrightarrow{P} 0.$$

7.9. Let $X_1, X_2, \ldots$ be a sequence of i.i.d. random variables with finite first four moments. Denote $E(X_i)$ by μ and $\mathrm{var}(X_i)$ by σ^2. Let $S_n = \sum_{i=1}^n (X_i - \overline{X})^2/(n-1)$ be the unbiased estimate of σ^2.

(a) Show that $S_n \xrightarrow{P} \sigma^2$.
(b) Show that

$$\frac{\sum_{i=1}^n (X_i - \mu)}{\sqrt{n}\, S_n} \xrightarrow{\mathcal{D}} N(0, 1).$$

7.10. Let $X_1, \cdots, X_n$ be an i.i.d. sequence with $E(X_i) = \mu \neq 0$ and $\mathrm{var}(X_i) = \sigma^2 < \infty$. Find the asymptotic distribution of

(a) $\sqrt{n}\left(\overline{X}^{-1} - \mu^{-1}\right)$,
(b) $\sqrt{n}\left(\overline{X}^2 - \mu^2\right)$,
(c) $\sqrt{n} \log(\overline{X}/\mu)$,
(d) $\sqrt{n}\left(e^{\overline{X}} - e^{\mu}\right)$.

7.11. Suppose that $\sqrt{n}(X_n - \mu) \xrightarrow{\mathcal{D}} N(0, \sigma^2)$, and that g is a function of X_n with continuous second derivative such that $g'(\mu) = 0$ and $g''(\mu) \neq 0$. Find the asymptotic distribution of $n[g(X_n) - g(\mu)]$. (Hint: a version of the Taylor's theorem states that if f has continuous kth derivative, then

$$f(x) = f(x_0) + f'(x_0)(x - x_0) + \cdots + f^{(k)}(\xi)(x - x_0)^k / k!$$

for some ξ satisfying $|\xi - x_0| \leq |x - x_0|$.)

7.12. Show that Definition 7.6 is equivalent to boundedness in probability.

7.13. Use the Dominated Convergence Theorem to show that X is P-integrable (i.e. $E_P(|X|) < \infty$) if and only if

$$\lim_{\alpha \to \infty} E_P[|X| I(|X| \geq \alpha)] = 0.$$

7.14. Show that the following rules hold for O_P and o_P. For any positive sequences a_n and b_n, we have

$$\begin{aligned} O_P(a_n) + O_P(b_n) &= O_P(\max(a_n, b_n)), \\ o_P(a_n) + o_P(b_n) &= o_P(\max(a_n, b_n)), \\ o_P(a_n) + O_P(a_n) &= O_P(a_n). \end{aligned}$$

7.15. Show that, if a sequence of random vectors X_n converges in distribution, or in probability, or almost everywhere, to a random vector X, then $X_n = O_P(1)$.

7.16. Show that, if X_n is a sequence of integrable random vectors with $E\|X_n\|$ being a bounded sequence, then $X_n = O_P(1)$.

7.17. Suppose that X_n is a sequence of random vectors taking values in $A \subseteq \mathbb{R}^k$, and $f : A \to \mathbb{R}^m$ satisfies the following Lipschitz condition:

$$\frac{\|f(x_1) - f(x_2)\|}{\|x_1 - x_2\|} < K$$

for some $K > 0$ and for all $x_1, x_2 \in A, x_1 \neq x_2$. Show that, if $X_n = O_P(1)$, then $f(X_n) = O_P(1)$.

7.18. Show that, if $f_1, \ldots, f_n$ are orthogonal elements of a Hilbert space $\mathcal{H}$, then

$$\|f_1 + \cdots + f_n\|^2 = \|f_1\|^2 + \cdots + \|f_n\|^2.$$

7.19. Show that a finite-dimensional Hilbert space is complete using the fact that the real number system is complete. That is, if $\{x_n\}$ is a sequence of numbers such that, for any $\epsilon > 0$, there exists n_ϵ such that

$$|x_n - x_m| < \epsilon, \text{ for all } n, m \geq n_\epsilon,$$

then x_n converges to a real number.

7.20. Show that, if f and g are members of a Hilbert space $\mathcal{H}$, then

$$\|f + g\|^2 + \|f - g\|^2 = 2(\|f\|^2 + \|g\|^2).$$

7.21. Let A be a $p \times p$ symmetric and positive semidefinite matrix. Let G and B be $p \times q$ matrices. Suppose that $B^T AB$ is non-singular. Show that the following inequality holds

$$G^T AG \succeq (G^T AB)(B^T AB)^{-1}(B^T AG).$$

7.22. Show that Theorem 7.17 can be generalized to the case where $[G, G]$ is not invertible using the Moore-Penrose inverse. That is, if S and G are member of $\mathcal{H}^p$, then

$$[S, S] \succeq [S, G][G, G]^+[G, S].$$

7.23. Prove Proposition 7.4.

7.24. Let $P_{\mathcal{M}} : \mathcal{H} \to \mathcal{H}$ be the operator defined by (7.19). Show that $P_{\mathcal{M}}$ is an idempotent and self adjoint linear operator.

7.25. In the setting of Example 7.7, show that $L_2(P_Y)$ is a linear subspace of $L_2(P_{XY})$.

7.26. In the setting of Example 7.7, define P to be the operator

$$L_2(P_{XY}) \to L_2(P_{XY}), \quad f \mapsto E[f(X, Y)|Y].$$

Show that P is an idempotent, and self adjoint linear operator.

References

Billingsley, P. (1995). *Probablity and Measure*. Third Edition. Wiley.

Billingsley, P. (1999). *Convergence of Probablity Measures*. Second Edition. Wiley.

Conway, J. B. (1990). *A course in functional analysis*. Second edition. Springer, New York.

Cramér, H. and Wold, H. (1936). Some theorems on distriutions functions. *Journal of the London Mathematical Society*, **s1-11**, 290–294.

Kelley, J. L. (1955). *General Topology*. D. Van Nostrand Company, Inc. Princeton.

Kollo, T. and von Rosen, D. (2005). *Advanced Multivariate Statistics with Matrices*. Springer

Mann, H. B. and Wald, A. (1943). On stochastic limit and order relationships. *The Annalso of Mathematical Statistics*, **14**, 217–226.

Serfling, R. J. (1980). *Approximation Theorems of Mathematical Statistics*. Wiley.

8

Asymptotic theory for Maximum Likelihood Estimation

The theoretical properties of the Maximum Likelihood Estimate introduced in Section 2.6 will be discussed in this chapter. This is one of the most commonly used estimators. Suppose X is a random element whose distribution belongs to a parametric family $\{P_\theta : \theta \in \Theta \subseteq \mathbb{R}^p\}$. Intuitively, if the data $X(\omega) = x$ is observed, then a reasonable estimate of true parameter θ_0 would be the $\theta \in \Theta$ that makes the observed data x most likely to be detected, because, after all, it is x, and not some other values x' of X, that has occurred. Thus, the Maximum Likelihood Estimate seems to be derived from the following dictum: "only the most likely to occur, occurs". This is, of course, not true in a literal sense: sometimes the least likely does occur. However, as a general tendency this seems plausible. Indeed, without any prior knowledge about θ, we seem to have no reason to think otherwise. In this chapter we systematically develop the theoretical properties for the Maximum Likelihood Estimate: its consistency, its asymptotic normality, and its optimality.

8.1 Maximum Likelihood Estimation

Let $X_{1:n} = (X_1, \ldots, X_n)$ be a sample of random vectors of dimension m that take values in a measurable space $(\Omega_n, \mathcal{F}_n)$, with a joint distribution belonging to a parametric family, say $\{P_\theta : \theta \in \Theta \subseteq \mathbb{R}^p\}$. Suppose that P_θ is dominated by a σ-finite measure μ, and let $f_\theta = dP_\theta/d\mu$ be the density of $X_{1:n}$.

Definition 8.1 *Suppose that, for each $x_{1:n} \in \Omega_n$, $\sup_{\theta\in\Theta} f_\theta(x_{1:n})$ can be reached within Θ. Then the Maximum Likelihood Estimate is defined as*

$$\hat{\theta}_n = \underset{\theta\in\Theta}{\operatorname{argmax}}\, [f_\theta(X_{1:n})] = \underset{\theta\in\Theta}{\operatorname{argmax}}\, [\log f_\theta(X_{1:n})].$$

In the above definition, the two argmax are the same because logarithm is a strictly increasing function, and strictly increasing transformations do not affect the maximizer of a function. Taking logarithm brings great convenience,

B. Li and G. J. Babu, *A Graduate Course on Statistical Inference*,
Springer Texts in Statistics, https://doi.org/10.1007/978-1-4939-9761-9_8

as it transforms a product into a sum, to which the Law of Large Numbers and the Central Limit Theorem can be applied. The function $\theta \mapsto f_\theta(X_{1:n})$ is called the likelihood; the function $\theta \mapsto \log f_\theta(X_{1:n})$ is called the log likelihood.

Because the MLE $\hat{\theta}$ is the maximizer of the log likelihood, it satisfies the equation

$$\partial \log f_\theta(X_{1:n})/\partial\theta = 0, \tag{8.1}$$

provided that the function $\theta \mapsto f_\theta(X_{1:n})$ is differentiable. This equation is called the likelihood equation. The function on the left is called the score function, and will be denoted by

$$s(\theta, X_{1:n}) = \partial \log f_\theta(X_{1:n})/\partial\theta.$$

Sometimes equation (8.1) is also called the score equation. The score function has some interesting properties, as described in the next proposition.

Proposition 8.1 *If $f_\theta(x_{1:n})$ and $s(\theta, x_{1:n})f_\theta(x_{1:n})$ satisfy $DUI^+(\theta, \mu)$, then for all $\theta \in \Theta$,*

$$E_\theta\left[s(\theta, X_{1:n})\right] = 0, \tag{8.2}$$

$$E_\theta\left[s(\theta, X_{1:n})s(\theta, X_{1:n})^T\right] = -E_\theta\left[\partial s(\theta, X_{1:n})/\partial\theta^T\right]. \tag{8.3}$$

Proof. Because f_θ is the density of $X_{1:n}$, we have

$$\int_{\Omega_n} f_\theta d\mu = 1.$$

Differentiating both sides of the equation and evoking the first DUI$^+$ condition, we have

$$\int_{\Omega_n} \frac{\partial f_\theta}{\partial\theta}\, d\mu = 0. \tag{8.4}$$

Because $f_\theta(x_{1:n}) > 0$ on Ω_n, we have

$$\frac{\partial f_\theta(x_{1:n})}{\partial\theta} = \frac{\partial \log f_\theta(x_{1:n})}{\partial\theta} f_\theta(x_{1:n}) = s(\theta, x_{1:n})f_\theta(x_{1:n}). \tag{8.5}$$

Hence the left-hand side of (8.4) is simply $E_\theta[s(\theta, X_{1:n})]$. This proves the first identity.

Differentiating the equation

$$\int_{\Omega_n} s(\theta, x_{1:n})f_\theta(x_{1:n})d\mu(x_{1:n}) = 0$$

with respect to θ^T and evoking the second DUI$^+$ condition, we have

$$\int_{\Omega_n} \frac{\partial s(\theta, x_{1:n})}{\partial\theta^T} f_\theta(x_{1:n})d\mu(x_{1:n}) + \int_{\Omega_n} s(\theta, x_{1:n})\frac{f_\theta(x_{1:n})}{\partial\theta^T}d\mu(x_{1:n}) = 0.$$

The first term on the left-hand side is simply $E_\theta[\partial s(\theta, X_{1:n})/\partial \theta^T]$. The second term, by relation (8.5) again, can be rewritten as

$$E_\theta[s(\theta, X_{1:n})s(\theta, X_{1:n})^T].$$

□

The matrix on the left-hand side of (8.3) is called the Fisher information contained in $X_{1:n}$, denoted by $I_{1:n}(\theta)$; the identity (8.3) is called the information identity. The relation (8.2) is known as the unbiasedness of the score (see for example, Godambe, 1960).

We will focus on the case where $X_1, \ldots, X_n$ are i.i.d. with a density function $h_\theta(x)$, where $\theta \in \Theta \subseteq \mathbb{R}^p$. In this case,

$$f_\theta(X_{1:n}) = \prod_{i=1}^n h_\theta(X_i).$$

The log-likelihood is

$$\log f_\theta(X_{1:n}) = \sum_{i=1}^n \log h_\theta(X_i) \propto E_n[\log h_\theta(X)].$$

Obviously, for the purpose of maximizing $\log f_\theta(X_{1:n})$, it is equivalent to use the sum version $\sum_{i=1}^n \log h_\theta(X_i)$ or average version $E_n \log h_\theta(X)$, as they only differ by a proportionality constant n^{-1}. As we will see, the latter is in many ways more convenient to use than the former.

By an abuse of notation, let $s(\theta, X_i)$ be the score for a single observation; that is, $s(\theta, X_i) = \partial h_\theta(X_i)/\partial\theta$. Then Proposition 8.1 holds for a single observation as well. That is,

$$\begin{aligned} &E_\theta\left[s(\theta, X_i\right] = 0, \\ &E_\theta\left[s(\theta, X_i)s(\theta, X_i)^T\right] = -\left[\partial s(\theta, X_i)/\partial\theta^T\right]. \end{aligned}$$

The matrix $E_\theta\left[s(\theta, X_i)s(\theta, X_i)^T\right]$ is called the Fisher information contained in a single observation, denoted by $I(\theta)$. It can be easily shown that

$$I_{1:n}(\theta) = nI(\theta), \quad s(\theta, X_{1:n}) = \sum_{i=1}^n s(\theta, X_i) = nE_n[s(\theta, X)].$$

The assumption in Proposition 8.1 that the support of P_θ does not depend on θ is quite important. Without it, even the definition of score function is problematic. For example, consider the density $h_\theta(x)$ for the uniform distribution on $(0, \theta)$:

$$h_\theta(x) = \theta^{-1} I(x < \theta).$$

This function is not differentiable at $\theta = x$. The log likelihood is only defined for $\theta > x$. If we define the derivatives of $\log h_\theta(x)$ only in the region $\theta > x$, then

$$\partial \log h_\theta(x)/\partial\theta = -\theta^{-1}, \; \partial^2 \log h_\theta(x)/\partial\theta^2 = \theta^{-2}, \; [\partial \log h_\theta(x)/\partial\theta]^2 = \theta^{-2}.$$

Thus none of the identities in Proposition 8.1 is satisfied.

Even in the case where the support of P_θ does not depend on θ, it is still possible that the rest of the conditions in Proposition 8.1 are violated — see Problem 8.3. Nevertheless, conditions such as those in Proposition 8.1 are satisfied for a wide variety of commonly encountered statistical problems, and are often evoked in statistical inference.

8.2 Cramér's approach to consistency

In Section 2.6 we proved the Fisher consistency of the MLE. There are two other types of consistency, strong consistency and consistency, which we define below. Let $T = T(X_{1:n})$ be a statistic.

Definition 8.2 *A statistic $T(X_{1:n})$ is a weakly consistent estimate (or simply consistent estimate) of the parameter θ_0 if, under P_{θ_0}, $T(X_{1:n}) \xrightarrow{P} \theta_0$. It is a strongly consistent estimate of θ_0 if, under P_{θ_0}, $T(X_{1:n}) \to \theta_0$ almost everywhere.*

It turns out that the MLE is consistent in both senses, under different sets of conditions. The methods for proving these two types of consistency are also quite different. The weak consistency was proved in Cramér (1946); the strong consistency was proved by Wald (1949). Each result reveals a different nature of the MLE: the first one reveals the property of the score function; the second reveals the property of the likelihood function. Both approaches are used widely in statistical research. In this section we focus on Cramér's approach.

Suppose $X_1, \ldots, X_n$ are i.i.d. with the common density belonging to a parametric family $\{f_\theta(x) : \theta \in \Theta \subseteq \mathbb{R}^p\}$. As before, let $s(\theta, x)$ be the score of a single observation. Cramér's statement of consistency does not directly state "the MLE is consistent". Rather, it states, roughly, that there is a sequence of consistent solutions to the likelihood equation. The existence part of the statement is established by a fixed point theorem (see Conway, 1990), which is stated below.

Proposition 8.2 (Brouwer's Fixed Point Theorem) *Let B be a closed unit ball in $\mathbb{R}^p$. Suppose that $h : B \mapsto B$ is a continuous function. Then there is an $x \in B$ such that $h(x) = x$*

The next theorem is Cramer's version of consistency of the MLE.

Theorem 8.1 *Suppose $X_1, \ldots, X_n$ are i.i.d. random variables or vectors having a density f_{θ_0} (with respect to a σ-finite measure μ) belonging to a parametric family $\{f_\theta : \theta \in \Theta \in \mathbb{R}^p\}$. Suppose, furthermore,*

1. *$f_\theta(x)$ and $s(\theta, x) f_\theta(x)$ satisfy $DUI^+(\theta, \mu)$;*
2. *$s(\theta, x)$ satisfies $DUI(\theta, A, P_{\theta_0})$, where A is the (common) support of f_θ;*
3. *for all $\theta \in \Theta$, the entries of $I(\theta)$ are finite, and the matrix is positive definite;*
4. *in a neighborhood of θ_0, $E_n[s(\theta, X)]$ converges in probability uniformly to $E[s(\theta, X)]$.*

Then, there is a sequence of estimators $\{\hat{\theta}_n\}$ such that

i. with probability tending to 1, $\hat{\theta}_n$ is a solution to $E_n[s(\theta, X)] = 0$,
ii. $\hat{\theta}_n \xrightarrow{P} \theta_0$.

Proof. Henceforth for notational simplicity, the subscript θ_0 will be dropped from E_{θ_0}. Let R_n be the set of solutions of $E_n[s(\theta, X)] = 0$. If $R_n \neq \varnothing$, define

$$\delta_n = \inf_{\theta \in R_n} \|\theta - \theta_0\|.$$

Then there is a sequence $\{\theta_{n,k} : k = 1, 2, \ldots\}$ in R_n such that $\|\theta_{n,k} - \theta_0\| \to \delta_n$ as $k \to \infty$. Since the sequence $\{\theta_{n,k} : k = 1, 2, \ldots\}$ is bounded, it contains a subsequence $\{\theta_{n,k_\ell} : \ell = 1, 2, \ldots\}$ that converges to some $\hat{\theta}_{n,0}$. Note that $\|\hat{\theta}_{n,0} - \theta_0\| = \delta_n$ because otherwise $\|\theta_{n,k} - \theta_0\| \to \delta_n$ would have been impossible.

Because $E_n[s(\theta, X)]$ is continuous, $E_n[s(\theta_{n,k_\ell}, X)] \to E_n[s(\hat{\theta}_{n,0}, X)]$ as $\ell \to \infty$. Therefore, $E_n[s(\hat{\theta}_{n,0}, X)] = 0$; that is, $\hat{\theta}_{n,0} \in R_n$. Now define

$$\hat{\theta}_n = \begin{cases} \hat{\theta}_{n,0} & \text{if } R_n \neq \varnothing \\ 0 & \text{if } R_n = \varnothing \end{cases}$$

The value 0 for $\hat{\theta}_n$ doesn't affect the asymptotic result in any way. It can be replaced by any constant in the parameter space.

To show that $\{\hat{\theta}_n\}$ satisfies *i* and *ii* in the theorem, expand $(\theta - \theta_0)^T E_n[s(\theta, X)]$, for θ sufficiently close to θ_0, as

$$\begin{aligned}(\theta - \theta_0)^T E_n[s(\theta, X)] &= (\theta - \theta_0)^T E[s(\theta, X)] \\ &\quad + \left\{(\theta - \theta_0)^T E_n[s(\theta, X)] - (\theta - \theta_0)^T E[s(\theta, X)]\right\}.\end{aligned}$$

Because, by condition 2, $E[s(\theta, X)] = \int s(\theta, X) dP_{\theta_0}$ is differentiable with respect to θ under the integral sign, we have

$$\begin{aligned}(\theta - \theta_0)^T E[s(\theta, X)] = (\theta - \theta_0)^T E[s(\theta_0, X)] + (\theta - \theta_0)^T E\left[\frac{\partial s(\theta_0, X)}{\partial \theta^T}\right](\theta - \theta_0) \\ + o(\|\theta - \theta_0\|^2),\end{aligned}$$

where $\|\cdot\|$ denotes the Euclidean norm.

By condition 1 and Proposition 8.1 (as applied to a single observation), the first term on the right-hand side is 0, and the second term on the right-hand side is $-(\theta-\theta_0)^T I(\theta_0)(\theta-\theta_0)$. Hence,

$$(\theta-\theta_0)^T E[s(\theta,X)] = -(\theta-\theta_0)^T I(\theta_0)(\theta-\theta_0) + o(\|\theta-\theta_0\|^2). \tag{8.6}$$

By condition 3, $I(\theta_0)$ is a positive definite matrix. Consequently for sufficiently small $\epsilon > 0$, $(\theta-\theta_0)^T E[s(\theta,X)] < 0$ whenever $\|\theta-\theta_0\| \le \epsilon$. Because $(\theta-\theta_0)^T E[s(\theta,X)]$ is continuous on the compact set $B(\theta_0,\epsilon)$, it attains its maximum within the closed ball. Hence for some $\delta > 0$,

$$-\delta > \sup_{\theta\in B(\theta_0,\epsilon)} \{(\theta-\theta_0)^T E s(\theta,X)\} \ge \sup_{\theta\in \partial B(\theta_0,\epsilon)} \{(\theta-\theta_0)^T E s(\theta,X)\} \tag{8.7}$$

where $\partial B(\theta_0,\epsilon)$ is the boundary $\{\theta : \|\theta-\theta_0\| = \epsilon\}$.

Next, note that

$$\begin{aligned}\sup_{\theta\in \partial B(\theta_0,\epsilon)} \{(\theta-\theta_0)^T E_n s(\theta,X)\} \le & \sup_{\theta\in \partial B(\theta_0,\epsilon)} \{(\theta-\theta_0)^T E s(\theta,X)\} \\ & + \sup_{\theta\in \partial B(\theta_0,\epsilon)} \left\{(\theta-\theta_0)^T \left[E_n s(\theta,X) - E s(\theta,X)\right]\right\}.\end{aligned}$$

By (8.7), the first term on the right is smaller than $-\delta$. By Cauchy-Schwarz inequality, the second term is no greater than

$$\begin{aligned}&\sup_{\theta\in \partial B(\theta_0,\epsilon)} \{\|\theta-\theta_0\| \, \|E_n s(\theta,X) - E s(\theta,X)\|\} \\ &\qquad \le \epsilon \sup_{\theta\in \partial B(\theta_0,\epsilon)} \|E_n s(\theta,X) - E s(\theta,X)\|,\end{aligned}$$

where, by condition 4, the right-hand side converges to 0 in probability. Therefore,

$$P(B_n) \to 1, \text{ where } B_n = \left\{ \sup_{\theta\in \partial B(\theta_0,\epsilon)} \{(\theta-\theta_0) E_n[s(\theta,X)]\} \le 0 \right\}. \tag{8.8}$$

Let $A_n = \{\omega : R_n \cap B(\theta_0,\epsilon) \ne \varnothing\}$. Then, on the event A_n^c, $E_n[s(\theta,X)]{=}0$ has no solution on $B(\theta_0,\epsilon)$. In other words, $E_n[s(\theta,X)] \ne 0$ for all $\theta \in B(\theta_0,\epsilon)$. Consequently the mapping h defined on the unit ball by

$$h(\eta) = E_n[s(\theta_0+\epsilon\eta, X)]/\|E_n s(\theta_0+\epsilon\eta, X)\|$$

is continuous. By Brouwer's Fixed Point Theorem, there is an η^* such that $\|\eta^*\| \le 1$ and $h(\eta^*) = \eta^*$. That is,

$$E_n s(\theta_0+\epsilon\eta^*, X)/\|E_n s(\theta_0+\epsilon\eta^*, X)\| = \eta^*. \tag{8.9}$$

From this equality we also see that $\|\eta^*\| = 1$, which implies $\eta^{*T}h(\eta^*) = \eta^{*T}\eta^* = 1$.

Now let $\theta^* = \theta_0 + \epsilon\eta^*$. Then $\theta^* \in \partial B(\theta_0, \epsilon)$ and

$$(\theta^* - \theta_0)^T E_n s(\theta^*, X) = \epsilon\eta^{*T} E_n s(\theta^*, X) = \epsilon\|E_n s(\theta_0 + \epsilon\eta^*, X)\| > 0,$$

where, for the second equality, we have used the relation (8.9). Hence, on A_n^c,

$$\sup_{\theta \in \partial B(\theta_0, \epsilon)} \{(\theta - \theta_0)^T E_n s(\theta, X)\} > 0.$$

Consequently, $B_n \subseteq A_n$. So by (8.8), $P(A_n) \to 1$. But since

$$R_n \cap B(\theta_0, \epsilon) \neq \varnothing \Rightarrow R_n \neq \varnothing \text{ and } \|\hat{\theta}_{0,n} - \theta_0\| \leq \epsilon,$$

we have

$$P(E_n[s(\hat{\theta}_n, X)] = 0) \to 1, \quad P(\|\hat{\theta}_n - \theta_0\| \leq \epsilon) \to 0.$$

□

Cramér's consistency result does not guarantee any specific solution to be consistent when there are multiple solutions. It merely asserts that consistent solution or solutions exist with probability tending to 1. Nevertheless, if the likelihood equation only has one solution, then Cramér's consistency statement can guarantee that solution to be consistent.

The sufficient conditions for the uniform convergence condition 4 will be further discussed in the next section. In the special case where $p = 1$, the uniform convergence is unnecessary, because the boundary of the closed ball $B(\theta_0, \epsilon)$ is simply the set of two points $\{\theta_0 - \epsilon, \theta_0 + \epsilon\}$. The convergence of $E_n[s(\theta, X)]$ to $E[s(\theta, X)]$ is guaranteed by the law of large numbers. In the next example we verify the sufficient conditions of Theorem 8.1 for $p = 1$ in a Poisson regression problem.

Example 8.1 Suppose X is a random variable with density f_X, and the conditional distribution Y given $X = x$ is Poisson$(e^{\theta x})$. For simplicity, we assume $\Theta = \mathbb{R}$, and $\Omega_X = \mathbb{R}$. Suppose, for all $\theta \in \Theta$,

$$0 < \int_{-\infty}^{\infty} x^2 e^{\theta x} f_X(x)dx < \infty. \tag{8.10}$$

The goal here is to verify the first three conditions in Theorem 8.1.

Since the joint density of (X, Y) is

$$f_\theta(x, y) = (e^{\theta xy}/y!)\, e^{-e^{\theta x}} f_X(x),$$

the log likelihood function is $\log f_\theta(x, y) = \theta xy - e^{\theta x} + \text{constant}$, and the score function for a single observation is

$$s(\theta, x, y) = x(y - e^{\theta x}). \tag{8.11}$$

Conditions 1 and 2 can now be verified by straighforward calculations. Condition 3 is simply assumption (8.10). □

The Cramér's type consistency for all generalized linear models (McCullagh and Nelder, 1989) can be verified in a similar way, where the predictor X can be a vector. Here, we have made the simplifying assumption that the predictor X is random. This is not an unreasonable assumption since the estimation is based on the conditional distribution of $Y|X$, and the marginal density f_X plays no role. The proof of the case where $X_1, \ldots, X_n$ are fixed can be carried out in the same spirit, but requires more careful treatment of details.

8.3 Almost everywhere uniform convergence

In this section we further explore the condition in Theorem 8.1 that requires $E_n[s(\theta, X)]$ to converge to $Es[(\theta, X)]$ uniformly over the boundary set $\partial B(\theta_0, \epsilon)$, that is

$$\sup_{\theta \in \partial B(\theta_0, \epsilon)} |E_n s(\theta, X) - E_n s(\theta, X)| \xrightarrow{P} 0.$$

For a set $\mathcal{F}$ of integrable functions, we are interested in whether the convergence

$$\sup_{f \in \mathcal{F}} |E_n f(X) - Ef(X)| \to 0 \quad [P], \tag{8.12}$$

holds. In the case of Theorem 8.1, the set of functions is $\{s(\theta, X) : \theta \in \partial B(\theta_0, \epsilon)\}$. This type of uniform convergence is also important for the Wald-type consistency that will be discussed in Section 8.4.

If $\mathcal{F}$ consists of a single function, then convergence (8.12) reduces to the strong law of large numbers. But if $\mathcal{F}$ contains too many functions, then uniform convergence over $\mathcal{F}$ will not hold. Then, what is the "appropriate size" for $\mathcal{F}$ to ensure uniform convergence? To answer this question we first have to define the "size of a family of functions". Let $\mathcal{S}$ denote the class of functions f on Ω_X such that $\|f\|_1 = E(|f(X)|) < \infty$.

Definition 8.3 *Given two members ℓ, u of $\mathcal{S}$, the bracket $[\ell, u]$ is the set*

$$\{f \in \mathcal{S} : \ell(x) \le f(x) \le u(x) \text{ for all } x \in \Omega_X\}.$$

The next theorem provides sufficient conditions under which a class $\mathcal{F}$ of functions satisfies (8.12).

Theorem 8.2 (Glivenko-Cantelli Theorem) *Let $X_1, X_2, \ldots$ be an i.i.d. sequence with distribution P. Suppose, for every $\epsilon > 0$, there exists an integer $m_\epsilon \geq 1$ and a set of functions $\{g_{ij} \in \mathcal{S} : i = 1, \ldots, m_\epsilon,\ j = 1, 2\}$ such that $\|g_{i2} - g_{i1}\|_1 < \epsilon$ and $\mathcal{F} \subset \cup_{i=1}^{m_\epsilon}[g_{i1}, g_{i2}]$. Then*

$$\sup_{f \in \mathcal{F}} |E_n f(X) - Ef(X)| \to 0 \quad [P].$$

Proof. If $f \in [\ell, u]$ and $\|u - \ell\| < \epsilon$, then

$$E_n \ell(X) \leq E_n f(X) \leq E_n u(X), \quad E\ell(X) \leq Ef(X) \leq Eu(X).$$

It follows that

$$\begin{aligned} E_n f(X) - Ef(X) &\leq E_n u(X) - E\ell(X) \\ &\leq (E_n u(X) - Eu(X)) + (Eu(X) - E\ell(X)) \\ &\leq (E_n u(X) - Eu(X)) + \epsilon, \\ E_n f(X) - Ef(X) &\geq E_n \ell(X) - Eu(X) \\ &\geq (E_n \ell(X) - E\ell(X)) - (Eu(X) - E\ell(X)) \\ &\geq (E_n \ell(X) - E\ell(X)) - \epsilon. \end{aligned}$$

Hence

$$|E_n f(X) - Ef(X)| \leq \max\{|E_n \ell(X) - E\ell(X)|, |E_n u(X) - Eu(X)|\} + \epsilon.$$

Therefore,

$$\sup_{f \in \mathcal{F}} |E_n f(X) - Ef(X)| \leq \max_{i=1,\ldots,m_\epsilon;\, j=1,2} |E_n g_{ij}(X) - Eg_{ji}(X)| + \epsilon.$$

By the strong law of large numbers each term inside the maximum on the right converges to 0 almost everywhere. By Problem 8.6, the maximum itself converges to 0 almost everywhere. It follows that, for each $\epsilon > 0$,

$$P\left(\limsup_{n \to \infty} \sup_{f \in \mathcal{F}} |E_n f(X) - Ef(X)| \leq \epsilon\right) = 1.$$

By Problem 8.7, the above implies

$$P\left(\limsup_{n \to \infty} \sup_{f \in \mathcal{F}} |E_n f(X) - Ef(X)| = 0\right) = 1.$$

□

Thus, a sufficient condition for almost everywhere uniform convergence over a class $\mathcal{F}$ is that for each $\epsilon > 0$, $\mathcal{F}$ is covered by finite number of $[\ell, u]$ with $\|\ell - u\| < \epsilon$. The next proposition gives a sufficient condition for this.

Proposition 8.3 *Let $\mathcal{F} = \{g(\theta, x) : \theta \in A\}$. Suppose that A is a compact set in $\mathbb{R}^p$, that $g(\theta, x)$ is continuous in θ for every x, and that, for all $\theta \in A$, $|g(\theta, x)|$ is dominated by an integrable function $G(x)$. Then, for any $\epsilon > 0$, there exists an integer $m_\epsilon \geq 1$ and a set of functions $\{g_{ij} \in \mathcal{S} : i = 1, \ldots, m_\epsilon, j = 1, 2\}$ such that $\|g_{i2} - g_{i1}\|_1 < \epsilon$ and $\mathcal{F} \subseteq \cup_{i=1}^{m_\epsilon} [g_{i1}, g_{i2}]$.*

Proof. Let $\epsilon > 0$. For any $\theta \in A$, let $B^\circ(\theta, \delta)$ be the open ball centered at θ with radius δ; that is, $B^\circ(\theta, \delta) = \{\theta' : \|\theta' - \theta\| < \delta\}$. Let

$$u(\theta, x, \delta) = \sup_{\theta' \in B^\circ(\theta, \delta)} g(\theta', x) \quad \text{and} \quad \ell(\theta, x, \delta) = \inf_{\theta' \in B^\circ(\theta, \delta)} g(\theta', x).$$

Because for each x, $g(\theta, x)$ is continuous in θ, for any $\epsilon > 0$ there is a $\delta > 0$ such that $\theta' \in B^\circ(\theta, \delta) \Rightarrow |g(\theta', x) - g(\theta, x)| < \epsilon$. Hence, for such a θ',

$$\begin{aligned} g(\theta', x) &= g(\theta, x) + g(\theta', x) - g(\theta, x) \\ &\leq g(\theta, x) + \sup_{\theta' \in B^\circ(\theta, \delta)} |g(\theta', x) - g(\theta, x)| \\ &\leq g(\theta, x) + \epsilon. \end{aligned}$$

Taking supremum, we have

$$u(\theta, x, \delta) = \sup_{\theta' \in B^\circ(\theta, \delta)} g(\theta', x) \leq g(\theta, x) + \epsilon.$$

Therefore $\limsup_{\delta \to 0} u(\theta, x, \delta) \leq g(\theta, x)$. Similarly, $\liminf_{\delta \to 0} \ell(\theta, x, \delta) \geq g(\theta, x)$. Thus we have shown that

$$\lim_{\delta \to 0} u(\theta, x, \delta) = \lim_{\delta \to 0} \ell(\theta, x, \delta) = g(\theta, x).$$

We next use this result to construct a finite collection of $[\ell, u]$ to cover $\mathcal{F}$. Note that

$$\begin{aligned} &E\{u(\theta, X, \delta) - \ell(\theta, X, \delta)\} \\ &= E\{u(\theta, X, \delta) - g(\theta, X\} + E\{g(\theta, X) - \ell(\theta, X, \delta)\}. \end{aligned}$$

By assumption $g(\theta, x)$ is dominated by $G(x)$, and hence $|u(\theta, x, \delta)|$ and $|\ell(\theta, x, \delta)|$ are both dominated by $G(x)$. By the Dominated Convergence Theorem, both of the two terms on the right-hand side converge to 0 as $\delta \to 0$. Consequently,

$$\lim_{\delta \to 0} E\{u(\theta, X, \delta) - \ell(\theta, X, \delta)\} = 0.$$

Hence, for any $\epsilon > 0$, there is a $\delta_\theta > 0$ (which may depend on θ), ℓ_θ, u_θ such that $\|\ell_\theta - u_\theta\|_1 < \epsilon$, where $\ell_\theta(x) = \ell(\theta, x, \delta_\theta)$, and $u_\theta(x) = u(\theta, x, \delta_\theta)$. Now consider the class of open balls

$$\mathcal{O} = \{B^\circ(\theta, \delta_\theta) : \theta \in A\}.$$

This is an open cover of A. Because A is compact, there is a finite subcover, say, $\{B^\circ(\theta_i, \delta_{\theta_i}) : i = 1, \ldots, m\}$ of A. Then, the collection of $\{[\ell_{\theta_i}, u_{\theta_i}] : i = 1, \ldots, m\}$ must cover $\mathcal{F}$ because, if $f \in \mathcal{F}$, then $f = f(\theta, x)$ for some $\theta \in A$, which must belong to one of the open balls, say $\theta \in B^\circ(\theta_i, \delta_{\theta_i})$. Then,

$$\ell(\theta_i, x, \delta_{\theta_i}) \leq \sup_{\theta' \in B^\circ(\theta_i, \delta_{\theta_i})} f(\theta', x) \leq f(\theta, x) \leq \sup_{\theta' \in B^\circ(\theta_i, \delta_{\theta_i})} f(\theta', x) = u(\theta_i, x, \delta_{\theta_i}).$$

Thus f must be in the bracket $[\ell_{\theta_i}, u_{\theta_i}]$. □

8.4 Wald's approach to consistency

The Wald approach to consistency (Wald, 1949) is quite different from Cramer's approach in that it relies on the properties of the log likelihood, rather than the score function. Also, it asserts that the MLE is consistent, rather than the existence of a solution to the score equation.

The structure of the proof is similar to those given in Wong (1986) and van der Vaart (1998). The intuition is the following. Suppose again $X_1, \ldots, X_n$ are an i.i.d. sample from a density $f_\theta(X)$. The log likelihood is proportional to $E_n \log f_\theta(X)$. As shown in Theorem 2.1, if the family $\{P_\theta : \theta \in \Theta\}$ is identifiable, then $E[\log f_\theta(X)]$ is uniquely maximized at the true parameter θ_0. If

$$\sup_{\theta \in \Theta} |E_n \log f_\theta(X) - E \log f_\theta(X)| \to 0 \quad [P]$$

then we would expect that the maximizer $\hat{\theta}_n$ of $E_n[\log f_\theta(X)]$ to be close to θ_0, which is the maximizer of $E[\log f_\theta(X)]$. Let $R_n(\theta) = E_n[\log f_\theta(X)]$ and $R(\theta) = E[\log f_\theta(X)]$. The next theorem makes the above intuition rigorous.

Theorem 8.3 *Suppose R has a unique maximizer θ_0 and R_n has a unique maximizer $\hat{\theta}_n \in \Theta$ modulo P. Suppose, furthermore,*

1. $\lim_{n \to \infty} \sup_{\theta \in \Theta} |R_n(\theta) - R(\theta)| = 0 \quad [P]$*;*
2. *For every* $\epsilon > 0$, $\sup_{\|\theta - \theta_0\| > \epsilon} R(\theta) < R(\theta_0)$.

Then $\hat{\theta}_n \to \theta_0$ $[P]$.

Assumption *1* says that R_n converges uniformly to R almost everywhere, which would be true if Θ is a compact set and $\log f_\theta(X)$ satisfies the conditions in Proposition 8.3. Condition *2* is called "$R(\theta)$ has a well-separated maximum", which rules out the situations where, although $R(\theta)$ is less than $R(\theta_0)$, the former can get arbitrarily close to the latter (van der Vaart, 1998).

Proof of Theorem 8.3. Since a probability 0 set does not affect our argument, we can assume R_n has a unique maximizer without loss of generality. We first show $R(\hat{\theta}_n) \to R(\theta_0)$ $[P]$. Since $R(\hat{\theta}_n) \le R(\theta_0)$, it suffices to show that

$$\liminf_{n\to\infty} R(\hat{\theta}_n) - R(\theta_0) \ge 0 \quad [P].$$

Since

$$R(\hat{\theta}_n) - R(\theta_0) = R(\hat{\theta}_n) - R_n(\hat{\theta}_n) + R_n(\hat{\theta}_n) - R_n(\theta_0) + R_n(\theta_0) - R(\theta_0),$$

we have

$$\begin{aligned}\liminf_{n\to\infty}[R(\hat{\theta}_n) - R(\theta_0)] \ge &\liminf_{n\to\infty}[R(\hat{\theta}_n) - R_n(\hat{\theta}_n)] \\ &+ \liminf_{n\to\infty}[R_n(\hat{\theta}_n) - R_n(\theta_0)] + \liminf_{n\to\infty}[R_n(\theta_0) - R(\theta_0)].\end{aligned}$$

By condition *1*, the first term and last term on the right-hand side are 0. Therefore,

$$\liminf_{n\to\infty}[R(\hat{\theta}_n) - R(\theta_0)] \ge \liminf_{n\to\infty}[R_n(\hat{\theta}_n) - R_n(\theta_0)],$$

where the right-hand side is nonnegative because $\hat{\theta}_n$ is the maximizer of $R_n(\theta)$.

Next, we show the desired relation

$$P(\lim_{n\to\infty} \|\hat{\theta}_n - \theta_0\| = 0) = P(\limsup_{n\to\infty} \|\hat{\theta}_n - \theta_0\| = 0) = 1.$$

By Problem 8.7, it suffices to show that, for any $\epsilon > 0$,

$$P(\limsup_{n\to\infty} \|\hat{\theta}_n - \theta_0\| < \epsilon) = 1. \tag{8.13}$$

If $\limsup_{n\to\infty} \|\hat{\theta}_n - \theta_0\| \ge \epsilon$, then there a subsequence $\{\hat{\theta}_{n_k} : k = 1, 2, \ldots\}$ such that $\|\hat{\theta}_{n_k} - \theta_0\| > \epsilon/2$. By condition *2*, there is a $\delta > 0$ such that

$$\sup_{\|\theta-\theta_0\|>\epsilon/2} R(\theta) \le R(\theta_0) - \delta.$$

Thus, along the subsequence $\{\hat{\theta}_{n_k} : k = 1, 2, \ldots\}$, $R(\hat{\theta}_{n_k}) \le R(\theta_0) - \delta$, making it impossible for $R(\hat{\theta}_n)$ to converge to $R(\theta_0)$. Hence

$$P\left(\limsup_{n\to\infty} \|\hat{\theta}_n - \theta_0\| \ge \epsilon\right) \le P\left(\lim_{n\to\infty} R(\hat{\theta}_n) \ne R(\theta_0)\right),$$

where the right-hand side is 0 because we already established

$$R(\hat{\theta}_n) \to R(\theta_0) \ [P].$$

This proves (8.13). □

With slightly more effort the uniform convergence condition (condition *1*) in the above Theorem can be relaxed. Instead of assuming $R_n(\theta)$ converges uniformly to $R(\theta)$ over Θ, we can assume $R_n(\theta)$ converges uniformly to $R(\theta)$ over a subset K of Θ, and assume the maximizer of $R_n(\theta)$ is always in K. This condition is known as essential compactness (Wong, 1986), and would be satisfied if, for example, $R_n(\theta)$ concave.

Corollary 8.1 *Suppose R has a unique maximizer θ_0 and, with probability 1, R_n has a unique maximizer $\hat{\theta}_n$. Suppose, furthermore,*

1. *there is a subset $K \subseteq \Theta$, whose interior contains θ_0, such that*
$$\sup_{\theta \in K} |R_n(\theta) - R(\theta)| \to 0 \quad [P];$$
2. *with probability 1, R_n has a unique maximizer $\tilde{\theta}_n$ over K;*
3. $P\left(\liminf_{n\to\infty}\left\{\sup_{\theta' \notin K} R_n(\theta) < \sup_{\theta\in\Theta} R_n(\theta)\right\}\right) = 1;$
4. *For every $\epsilon > 0$,*
$$\sup\{R(\theta) : \|\theta - \theta_0\| \geq \epsilon, \theta \in K\} < R(\theta_0).$$

Then $\hat{\theta}_n \to \theta_0 \quad [P]$.

Proof. Again, we can assume, without loss of generality, that R_n has a unique maximizer $\hat{\theta}_n$ over Θ and a unique maximizer $\tilde{\theta}_n$ over K. By condition *3*,

$$\begin{aligned}
&P\left(\lim_{n\to\infty} \|\hat{\theta}_n - \theta_0\| = 0\right) \\
&= P\left(\left\{\lim_{n\to\infty} \|\hat{\theta}_n - \theta_0\| = 0\right\} \cap \liminf_{n\to\infty}\left\{\sup_{\theta'\notin K} R_n(\theta) < \sup_{\theta\in\Theta} R_n(\theta)\right\}\right).
\end{aligned}$$

The event

$$\liminf_{n\to\infty}\left\{\sup_{\theta'\notin K} R_n(\theta) < \sup_{\theta\in\Theta} R_n(\theta)\right\}$$

happens if and only if, for all sufficiently large n, $\hat{\theta}_n = \tilde{\theta}_n$. Hence

$$\begin{aligned}
&P\left(\lim_{n\to\infty} \|\hat{\theta}_n - \theta_0\| = 0\right) \\
&= P\left(\lim_{n\to\infty} \|\hat{\theta}_n - \theta_0\| = 0,\ \hat{\theta}_n = \tilde{\theta}_n \text{ for sufficiently large } n\right) \\
&= P\left(\lim_{n\to\infty} \|\tilde{\theta}_n - \theta_0\| = 0,\ \hat{\theta}_n = \tilde{\theta}_n \text{ for sufficiently large } n\right) \\
&= P\left(\lim_{n\to\infty} \|\tilde{\theta}_n - \theta_0\| = 0\right).
\end{aligned}$$

From the proof of Theorem 8.3, we see that the probability on the right is 1. □

The well-separated maximum condition (condition *1*) of Theorem 8.3 is guaranteed if K is a compact set, f is upper semi-continuous, and f has a unique maximizer over K. Recall from Chapter 2 that, in the maximum likelihood estimation context, the function $R(\theta) = E[\log f_\theta(X)]$ has a unique maximizer if the family $\{f_\theta : \theta \in \Theta\}$ is identifiable.

The next proposition gives a set of sufficient conditions for a function f to have a well separated maximum on a compact set.

Proposition 8.4 *Suppose K is a compact set and $f(\theta)$ is an upper semi-continuous function on K. Suppose f has a unique maximum over K and the maximizer is an interior point of K. Then f has a well-separated maximum on K.*

Proof. If f is not well-separated on K then there is an $\epsilon > 0$ such that

$$\sup\{f(\theta) : \|\theta - \theta_0\| > \epsilon, \theta \in K\} = f(\theta_0).$$

Then there is a sequence $\{\theta_n\} \subseteq K \setminus B^\circ(\theta_0, \epsilon)$ such that $f(\theta_n) \to f(\theta_0)$. But because the set $K \setminus B^\circ(\theta_0, \epsilon)$ is compact this sequence has a subsequence say $\{\theta_{n_k}\}$ such that $\theta_{n_k} \to \theta^*$ with θ^* in K. By upper semi-continuity of f we have $\limsup_{k\to\infty} f(\theta_{n_k}) \leq f(\theta^*)$. However, by the uniqueness of maximum $f(\theta^*) < f(\theta_0)$, contradicting to $f(\theta_n) \to f(\theta_0)$. □

The next proposition gives a set of sufficient conditions for essential compactness.

Proposition 8.5 *Suppose that R has a unique maximuizer θ_0. Suppose*

1. *There is a subset $K \subseteq \Theta$ whose interior contains θ_0 such that*

$$\lim_{n\to\infty} \sup_{\theta\in K} |R_n(\theta) - R(\theta)| = 0 \quad [P];$$

2. *R has a well-separated maximum over K;*
3. *$R_n(\theta)$ is concave with probability 1.*

Then,

$$P\left(\liminf_{n\to\infty}\left\{\sup_{\theta\notin K} R_n(\theta) < \sup_{\theta\in\Theta} R_n(\theta)\right\}\right) = 1. \tag{8.14}$$

Note that, in condition *3*, we do not need R_n to be strictly concave with probability 1.

Proof of Proposition 8.5. Let $\epsilon > 0$ be such that $B^\circ(\theta_0, \epsilon)$ is contained in the interior of K. By assumptions *1* and *2* there is a $\delta > 0$ such that

$$P\left(\limsup_{n\to\infty} \sup_{K\setminus B^\circ(\theta_0,\epsilon)} R_n(\theta) < R(\theta_0) - \delta\right) = 1.$$

By condition *1*, we have $R_n(\theta_0) \to R(\theta_0)$ $[P]$, which implies

$$P\left(\liminf_{n\to\infty} R_n(\theta_0) > R(\theta_0) - \delta/2\right) = 1$$
$$\Rightarrow P\left(\liminf_{n\to\infty} \sup_{B^\circ(\theta_0,\epsilon)} R_n(\theta) > R(\theta_0) - \delta/2\right) = 1.$$

Let A_n, B_n and C_n be the sets

$$A_n = \left\{\limsup_{n\to\infty} \sup_{K\setminus B^\circ(\theta_0,\epsilon)} R_n(\theta) < R(\theta_0) - \delta\right\},$$
$$B_n = \left\{\liminf_{n\to\infty} \sup_{B^\circ(\theta_0,\epsilon)} R_n(\theta) > R(\theta_0) - \delta/2\right\},$$
$$C_n = \{R_n \text{ is concave }\}.$$

Then,
$$P(A_nB_nC_n) = 1.$$

However, $A_nB_nC_n$ implies that, for sufficiently large n the supremum of R_n outside K cannot be greater than $R(\theta_0) - \delta$, which implies (8.14). □

Problems 8.8 through 8.11 contain step-to-step verifications of all the conditions in Theorem 8.3 in the Poisson regression setting.

8.5 Asymptotic normality

In this section we show that, if $\hat{\theta}$ is the MLE, then $\sqrt{n}(\hat{\theta} - \theta_0)$ converges in distribution to a Normal random vector. Here, we have omitted the subscript n of $\hat{\theta}_n$ for simplicity. As before, let $X_1, \ldots, X_n, \ldots$ be an i.i.d. sequence of random variables or vectors with probability density function $f_\theta(x)$, where $\theta \in \Theta \subseteq \mathbb{R}^p$. Let $s(\theta, X)$ be the score function and $I(\theta)$ be the Fisher information matrix as defined previously. Let $J(\theta)$ and $K(\theta)$ denote the matrices

$$E_\theta\left[\partial s(\theta, X)/\partial\theta^T\right], \quad E_\theta[s(\theta, X)s^T(\theta, X)], \tag{8.15}$$

respectively. Although under the mild conditions in Proposition 8.1 we have $K(\theta) = -J(\theta) = I(\theta)$, it makes the proof clearer if we use two separate symbols.

For finite-dimensional spaces, all matrix norms are equivalent so long as convergence is concerned. For definiteness, we use the Frobenius matrix norm:

$$\|A\|_{\mathrm{F}} = \sqrt{\textstyle\sum_{i,j} a_{ij}^2}.$$

We will also use the matrix version of the rules of O_P and o_P in Theorem 7.15. In particular, we say a sequence of random matrices A_n is of the order $O_P(a_n)$

or $o_P(a_n)$ if $\|A_n\|_F = O_P(a_n)$ or $o_P(a_n)$. The first relation in Theorem 7.15, for example, is to be understood as: if $A_n = O_P(a_n)$, $B_n = O_p(b_n)$, then $A_n B_n = O_p(a_n b_n)$, where $A_n B_n$ is the matrix product. The proof of this extension is left as an exercise (Problem 8.14).

The idea for developing the asymptotic distribution of the MLE is to use Taylor expansion of $E_n s(\hat{\theta}, X)$ at θ_0, and show that the remainder of the Taylor approximation is stochastically small. A concise way to bound the remainder is via the notion of stochastic equicontinuity.

Definition 8.4 *Let $\{g_n(\theta, X_{1:n}) : n = 1, 2, \ldots\}$ be a sequence of random functions that map into $\mathbb{R}^q$. The sequence is said to be stochastically equicontinuous if, for any $\eta > 0$, $\epsilon > 0$, there is a $\delta > 0$ such that*

$$\limsup_{n\to\infty} P\left(\sup_{\|\theta_1-\theta_2\|<\delta} \|g_n(\theta_1 X_{1:n}) - g_n(\theta_2, X_{1:n})\| > \epsilon\right) < \eta. \tag{8.16}$$

This definition can be made more general. For example, both the domain and the range of $g_n(\cdot, X_{1:n})$ can be metric spaces. But this version is sufficient for our discussion. Also, in our application of this concept $g_n(\theta, X_{1:n})$ is actually a matrix rather than a vector. But a matrix can be viewed as a vector by stacking its columns.

Recall that, if $g(\theta)$ is continuous at θ_0 and $\hat{\theta} \xrightarrow{P} \theta_0$, then $g(\hat{\theta}) \xrightarrow{P} g(\theta_0)$. Stochastic equicontinuity plays a similar role as continuity, but in the context when $g(\theta)$ is replaced by a sequence of random functions. Specifically, stochastic equicontinuity guarantees that, if $\hat{\theta}$ is a consistent estimate of θ_0, then the distance between $g_n(\hat{\theta}, X_{1:n})$ and $g_n(\theta_0, X_{1:n})$ converges in probability to 0.

Proposition 8.6 *If $\{g_n(\theta, X_{1:n}) : n = 1, 2, \ldots\}$ is stochastically equicontinuous and $\hat{\theta} \xrightarrow{P} \theta_0$, then*

$$\|g_n(\hat{\theta}, X_{1:n}) - g_n(\theta_0, X_{1:n})\| \xrightarrow{P} 0.$$

Proof. Let $\epsilon > 0, \eta > 0$, and let $\delta > 0$ be a number such that (8.16) is satisfied. Because $\hat{\theta} \xrightarrow{P} \theta_0$, we have $P(\|\hat{\theta} - \theta_0\| < \delta) \to 1$. Hence

$$\begin{aligned}
0 &\le \limsup_{n\to\infty} P\left(\|g_n(\hat{\theta}, X_{1:n}) - g_n(\theta_0, X_{1:n})\| > \epsilon\right) \\
&= \limsup_{n\to\infty} P\left(\|g_n(\hat{\theta}, X_{1:n}) - g_n(\theta_0, X_{1:n})\| > \epsilon, \|\hat{\theta} - \theta_0\| < \delta\right) \\
&\le \limsup_{n\to\infty} P\left(\sup_{\|\theta_1-\theta_2\|<\delta} \|g_n(\theta_1, X_{1:n}) - g_n(\theta_2, X_{1:n})\| > \epsilon\right) < \eta.
\end{aligned}$$

Since $\eta > 0$ is arbitrary, it follows that

$$\limsup_{n\to\infty} P\left(\|g_n(\hat{\theta}, X_{1:n}) - g_n(\theta_0, X_{1:n})\| > \epsilon\right) = 0.$$

□

We now give a further consequence of stochastic equicontinuity when $g_n(\theta, X_{1:n})$ takes the special form $E_n[g(\theta, X)]$. It follows directly from the above proposition and the law of large numbers.

Corollary 8.2 *If $g(\theta, X)$ is P_{θ_0}-integrable, $\{E_n g(\theta, X) : n \in \mathbb{N}\}$ is stochastically equicontinuous, and $\hat{\theta} \xrightarrow{P} \theta_0$, then $E_n[g(\hat{\theta}, X)] \xrightarrow{P} E[g(\theta_0, X)]$.*

The next proposition gives a sufficient condition for stochastic equicontinuity when $g_n(\theta, X_{1:n})$ takes the form $E_n[g(\theta, X)]$.

Proposition 8.7 *Suppose $X_1, \cdots, X_n$ are i.i.d. random vectors with distribution P_{θ_0}. If $g(\theta, X)$ is differentiable with respect to θ, and there is a P_{θ_0}-integrable function $M(X)$ such that*

$$\sup_{\theta \in \Theta} \|\partial g(\theta, X)/\partial \theta^T\|_{\mathrm{F}} \leq M(X),$$

then the sequence $\{E_n[g(\theta, X)] : n \in \mathbb{N}\}$ is stochastically equicontinuous.

Proof. By Taylor's mean value theorem, for any $\theta_1, \theta_2 \in \Theta$, there is a ξ on the line joining θ_1 and θ_2 such that

$$g(\theta_2, X) = g(\theta_1, X) + [\partial g(\xi, X)/\partial \theta^T] (\theta_2 - \theta_1).$$

Hence

$$\begin{aligned} \|g(\theta_2, X) - g(\theta_1, X)\| &\leq \|[\partial g(\xi, X)/\partial \theta^T] (\theta_2 - \theta_1)\| \\ &\leq \left\|\partial g(\xi, X)/\partial \theta^T\right\|_{\mathrm{F}} \|\theta_2 - \theta_1\| \\ &\leq M(X) \|\theta_2 - \theta_1\|, \end{aligned}$$

where the second equality follows from Problem 8.15. It follows that

$$\|E_n[g(\theta_2, X)] - E_n[g(\theta_1, X)]\| \leq E_n[M(X)] \|\theta_2 - \theta_1\|.$$

Then, for any $\delta > 0$,

$$\sup_{\|\theta_1 - \theta_2\| < \delta} \|E_n[g(\theta_2, X)] - E_n[g(\theta_1, X)]\| \leq E_n[M(X)] \, \delta.$$

Let $\epsilon > 0$ and $\eta > 0$. Because $E_n[M(X)] \xrightarrow{P} E[M(X)]$, $E_n[M(X)]$ is bounded in probability (Problem 8.12). Hence there is a $K > 0$ such that

$$\limsup_{n \to \infty} P(E_n M(X) > K) < \eta.$$

Let $\delta > 0$ be so small that $\epsilon/\delta > K$. Then

$$\limsup_{n\to\infty} P\left(E_n[M(X)] > \epsilon/\delta\right) < \eta.$$

Hence

$$\limsup_{n\to\infty} P\left(\sup_{\|\theta_1-\theta_2\|<\delta} \|E_n[g(\theta_2, X)] - E_n[g(\theta_1, X)]\| > \epsilon\right)$$
$$\leq \limsup_{n\to\infty} P\left(E_n M(X)\delta > \epsilon\right) < \eta.$$

□

We are now ready to derive the asymptotic distribution of the maximum likelihood estimate.

Theorem 8.4 *Suppose that $\hat{\theta}$ is a consistent solution to likelihood equation $E_n[s(\theta, X)] = 0$ and that θ_0 is an interior point of Θ. Suppose, furthermore,*

1. *$f_\theta(x)$ satisfies $DUI^+(\theta, \mu)$;*
2. *$s(\theta, x)f_\theta(x)$ satisfies $DUI^+(\theta, \mu)$;*
3. *the sequence $\{E_n[\partial s(\theta, X)/\partial\theta^T] : n \in \mathbb{N}\}$ is stochastically equicontinuous over $B(\theta_0, \epsilon)$ for some $\epsilon > 0$.*

Then $\sqrt{n}(\hat{\theta} - \theta_0) \xrightarrow{\mathcal{D}} N(0, I^{-1}(\theta_0))$, provided that $I(\theta_0)$ is positive definite.

Note that by the first two conditions and Proposition 8.1, the information identities (8.2) and (8.3) hold — particularly for $X_{1:n} = X_1$. The assumption that $K(\theta_0) = I(\theta_0)$ is positive definite amounts to assuming the score $s(\theta_0, X)$ is non-degenerate under P_{θ_0}, which is also quite mild. Condition 3 is local in nature. Although we do not know θ_0, in a particular problem we can check whether this condition is satisfied for every interior point of Θ.

Proof. By Taylor's theorem,

$$0 = E_n[s(\hat{\theta}, X)] = E_n[s(\theta_0, X)] + [E_n\dot{s}(\xi, X)](\hat{\theta} - \theta_0),$$

for some ξ on the line joining θ_0 and $\hat{\theta}$. Because $\xi \xrightarrow{P} \theta_0$ and $\{E_n\dot{s}(\theta, X) : n = 1, 2, \ldots\}$ is stochastically equicontinuous, we have, by Corollary 8.2,

$$E_n\dot{s}(\xi, X) = J(\theta_0) + o_P(1),$$

where $o_P(1)$ means a matrix sequence whose Frobenius norm tends to 0. Hence

$$0 = E_n[s(\theta_0, X)] + J(\theta_0)(\hat{\theta} - \theta_0) + o_P(1)(\hat{\theta} - \theta_0). \tag{8.17}$$

Since $J(\theta_0)$ is invertible, we have

$$(\hat{\theta} - \theta_0) = -J(\theta_0)^{-1}E_n[s(\theta_0, X)] + J(\theta_0)^{-1}o_P(1)(\hat{\theta} - \theta_0).$$

Rearranging the above equality, we have

$$\sqrt{n}[I_p + o_P(1)](\hat{\theta} - \theta_0) = -J^{-1}(\theta_0)\{\sqrt{n}E_n[s(\theta_0, X)]\}.$$

By the central limit theorem,

$$\sqrt{n}E_n[s(\theta_0, X)] \xrightarrow{\mathcal{D}} N(0, K(\theta_0)),$$

where $K(\theta_0) = E[s(\theta_0, X)s^T(\theta_0, X)]$. Hence, by Slutsky's theorem,

$$-[J^{-1}(\theta_0) + o_P(1)][\sqrt{n}E_n s(\theta_0, X)] \xrightarrow{\mathcal{D}} N(0, J^{-1}(\theta_0)K(\theta_0)J^{-1}(\theta_0)).$$

Since, by condition *2*, $K(\theta_0) = -J(\theta_0) = I(\theta_0)$, the right-hand side is simply $N(0, I^{-1}(\theta_0))$, leading to

$$\sqrt{n}[I_p + o_P(1)](\hat{\theta} - \theta_0) \xrightarrow{\mathcal{D}} N(0, I^{-1}(\theta_0)),$$

which implies $\sqrt{n}(\hat{\theta} - \theta_0) \xrightarrow{\mathcal{D}} N(0, I^{-1}(\theta_0))$ by Slutsky's Theorem (Problem 8.13). □

Problems

8.1. Suppose that X is a random variable that takes values in $\Omega_X \subseteq \mathbb{R}$ having an exponential family density (with respect to a σ-finite measure μ) of the form

$$f_\theta(x) = \frac{e^{\theta t(x)}}{\int_{\Omega_X} e^{\theta t(x)} d\mu}, \quad \theta \subseteq \Theta \in \mathbb{R}. \tag{8.18}$$

1. Show that the identities in Proposition 8.1 hold.
2. Show that the Fisher information $I(\theta)$ is of the form $\text{var}_\theta[t(X)]$.
3. Repeat 1 and 2 if θ in (8.18) is replaced by a monotone increasing function of $\psi(\theta)$. That is,

$$f_\theta(x) = \frac{e^{\psi(\theta) t(x)}}{\int_{\Omega_X} e^{\psi(\theta) t(x)} d\mu}.$$

8.2. Suppose that X is a random variable that takes values in $\Omega_X \subseteq \mathbb{R}$ whose density with respect to a σ-finite measure μ is given by (8.18), where $\text{var}_\theta[t(X)] > 0$.

1. Show that the function $\mu(\theta) = E_\theta[t(X)]$ is strictly increasing.
2. Show that the maximum likelihood estimate is given by

$$\hat{\theta} = \mu^{-1}(E_n[t(X)]).$$

8.3. Suppose that X is a random variable in $(0,1)$ whose density is of the form

$$f_\theta(x) = \frac{\theta}{\theta - 1} x^{-1/\theta}, \quad \theta > 1.$$

1. Show that $f_\theta(x)$ does not satisfy $\mathrm{DUI}^+(\theta, \lambda)$, λ being the Lebesgue measure.
2. Show that $E_\theta[s(\theta, X)] \neq 0$.

8.4. Suppose X is a random variable whose density belongs to a parametric family $\{f_\theta(X) : \theta \in \Theta \subseteq \mathbb{R}^p\}$. Suppose the support of f_θ does not depend on θ. The Kullback-Leibler divergence between f_θ and f_{θ_0} is defined as

$$K(\theta) = E_\theta \left\{ \log \left[\frac{f_\theta(X)}{f_{\theta_0}(X)} \right] \right\}.$$

Derive the second-order Taylor polynomial for $K(\theta)$ and express it in terms of the Fisher information $I(\theta_0)$.

8.5. Suppose that $\{A_{1,n}\}, \ldots, \{A_{k,n}\}$ are k sequences of events and that, for each $i = 1, \ldots, k$, $P(A_{i,n}) \to 1$ as n goes to infinity. Show that

$$\lim_{n \to \infty} P\left(\bigcap_{i=1}^{k} A_{i,n} \right) = 1.$$

8.6. Suppose that $\{X_{1,n}\}, \ldots, \{X_{k,n}\}$ are k sequences of random variables.

1. Show that if $X_{i,n} \xrightarrow{P} 0$ for each $i = 1, \ldots, k$, then
$$\max_{i=1,\ldots,k} (|X_{1,n}|, \ldots, |X_{k,n}|) \xrightarrow{P} 0.$$
2. Show that if $X_{i,n} \to 0$ $[P]$ for each $i = 1, \ldots, k$, then
$$\max_{i=1,\ldots,k} (|X_{1,n}|, \ldots, |X_{k,n}|) \to 0 \ [P].$$

8.7. Let $X \geq 0$ be a random variable. Show that if $P(X \leq \epsilon) = 1$ for all $\epsilon > 0$, then $P(X = 0) = 1$.

8.8. Suppose Y conditioning on $X = x$ is distributed as Poisson$(e^{\theta^T x})$, $\theta \in \Theta = \mathbb{R}^p$, and X has a density h with respect to the Lebesgue measure. Let $\Omega_X = \{x \in \mathbb{R}^p : f_X(x) > 0\}$ be the support of X, and suppose that Ω_X contains a nonempty open set in $\mathbb{R}^p$. Let $f_\theta(x, y)$ be the joint density of (X, Y) under θ. Prove that $\{f_\theta(x, y) : \theta \in \Theta\}$ is identifiable.

8.9. Suppose the joint distribution of (X, Y) is as defined in Problem 8.8. Furthermore, suppose that

1. $E\left|\log f_X(X)\right| < \infty$;
2. for any $C > 0$, $E(e^{C\|X\|}) < \infty$, $E(\|X\| e^{C\|X\|}) < \infty$.

Let A be a compact subset of $\mathbb{R}^p$, and let $\mathcal{F} = \{\log f_\theta(x, y) : \theta \in A\}$. Use Proposition 8.3 to show that for all $\epsilon > 0$, $\mathcal{F}$ can be covered by finite number of bracket functions $[\ell, u]$ such that $\|\ell - u\| < \epsilon$.

8.10. Under the conditions of Problem 8.9, use Proposition 8.4 to show that $E_\theta f(X, Y)$ has a well-separated maximum on any compact set A that contains θ_0 as an interior point.

8.11. Suppose the conditions of Problem 8.9 are satisfied. Suppose, further more, that $E_{\theta_0}(e^{\theta^T X} X X^T)$ has finite entries for all $\theta \in \mathbb{R}^p$. Use Proposition 8.5 to show that any compact set A in $\mathbb{R}^p$ whose interior contains θ_0 is essentially compact. That is, condition (8.14) is satisfied.

8.12. Suppose X_n is a sequence of random vectors that converges in probability to a fixed vector a. Show that, $X_n = O_P(1)$.

8.13. Suppose A_n is a sequence of random matrices that converges in probability to A, and X_n a sequence of random vectors. Use Slutsky's Theorem to show that, if $A_n X_n \xrightarrow{\mathcal{D}} U$, then $A X_n \xrightarrow{\mathcal{D}} U$.

8.14. Suppose A_n and B_n are random matrices, and a_n, b_n are sequences of positive numbers. Prove the following statements:

1. if $A_n = O_P(a_n)$ and $B_n = O_P(b_n)$, then $A_n B_n = O_P(a_n b_n)$;
2. if $A_n = O_P(a_n)$ and $B_n = o_P(b_n)$, then $A_n B_n = o_P(a_n b_n)$;
3. if $A_n = o_P(a_n)$ and $B_n = o_P(b_n)$, then $A_n B_n = o_P(a_n b_n)$.

8.15. Suppose A is a $p \times q$ dimensional matrix and b is a q dimensional vector. Show that

$$\|Ab\| \le \|A\|_{\mathrm{F}} \, \|b\|.$$

8.16. Suppose A is an invertible square matrix and Δ is a matrix of the same dimensions. Show that, there exists an $\epsilon > 0$ such that, for all $\|\Delta\|_{\mathrm{F}} \le \epsilon$, $A + \Delta$ is invertible. Let $\bar{B}(A, \epsilon)$ be the closed ball $\{A + \Delta : \|\Delta\|_{\mathrm{F}} \le \epsilon\}$. Let $f : \bar{B}(A, \epsilon) \to \mathbb{R}^{p \times p}$ be defined via

$$f(\Delta) = (A + \Delta)^{-1}.$$

Show that $f(\Delta)$ is continuous at $\Delta = 0$.

8.17. Suppose A_n is a sequence of random matrices that converges in probability to a nonrandom, invertible matrix A. Use the result of Problem 8.16 to show that A_n is invertible with probability tending to 1.

8.18. Show that the condition "$\hat{\theta}$ is a consistent solution to $E_n[s(\theta, X)] = 0$" in Theorem 8.4 can be replaced by "$\hat{\theta}$ is consistent and with probability tending to 1 is a solution to $E_n[s(\theta, X)] = 0$." Notice that the latter is the conclusion of the Cramer-type consistency in Theorem 8.1.

8.19. Suppose $X_1, \ldots, X_n$ are i.i.d. random variables whose common distribution is P_{θ_0}, where θ_0 is an interior point of a p-dimensional parameter space Θ. Let $\hat{\theta}$ be a p-dimensional statistic, and let f be a function from $\Theta \times \Omega_X$ to $\mathbb{R}$. Suppose

1. $\sqrt{n}(\hat{\theta} - \theta_0) \xrightarrow{\mathcal{D}} N(0, \Sigma)$ for some positive semidefinite matrix Σ;
2. for each $x \in \Omega_X$, the function $\theta \mapsto f(\theta, x)$ is differentiable;
3. for any $\theta \in \Theta$, the components of $\partial f(\theta, X)/\partial\theta$ are P_{θ_0}-integrable;
4. the sequence of random functions $\{E_n[\partial f(\theta, X)/\partial\theta] : n = 1, 2, \ldots\}$ is stochastically equicontinuous over Θ.

Derive the asymptotic distribution of $\sqrt{n}[E_n f(\hat{\theta}, X) - Ef(\theta_0, X)]$.

8.20. Suppose X is a random variable whose distribution belongs to a multi-parameter exponential family:

$$f_\theta(x) = \frac{e^{\theta^T t(x)}}{\int e^{\theta^T t(x)} d\mu(x)}, \quad \theta \in \Theta \subseteq \mathbb{R}^p, \ x \in \Omega_X \subseteq \mathbb{R},$$

where μ is a σ-finite measure on Ω_X and t is a function from Ω_X to $\mathbb{R}^p$. Suppose $X_1, \ldots, X_n$ are an i.i.d. sample from X. Let $\hat{\theta}$ be the maximum likelihood estimate.

1. Derive the asymptotic distribution of $\sqrt{n}(\hat{\theta} - \theta_0)$.
2. Suppose we estimate the moment generating function of $t(X)$, $\varphi(u) = E_{\theta_0}[e^{u^T t(X)}]$, by $\hat{\varphi}(u) = E_{\hat{\theta}}[e^{u^T t(X)}]$. Prove that

$$\sqrt{n}[\hat{\varphi}(u) - \varphi(u)] \xrightarrow{\mathcal{D}} N(0, \Lambda),$$

where Λ is

$$\varphi^2(u)[E_{\theta_0+u}t(X) - E_{\theta_0}t(X)]^T \{\text{var}_{\theta_0}[t(X)]\}^{-1}[E_{\theta_0+u}t(X) - E_{\theta_0}t(X)].$$

8.21. Suppose $X_1, \ldots, X_n$ are an i.i.d. sample from the uniform distribution $U[0, \theta]$; that is,

$$f_\theta(x) = \theta^{-1}, \quad 0 \le x \le \theta.$$

1. Derive the maximum likelihood estimate $\hat{\theta}$ of θ_0.
2. Derive the asymptotic distribution of $n(\hat{\theta} - \theta_0)$.
3. Explain the discrepancy between this asymptotic distribution and that given by Theorem 8.4.

References

Conway, J. B. (1990). *A course in functional analysis.* Second edition. Springer, New York.

Cramér, H. (1946). *Mathematical Methods of Statistics.* Princeton University Press.

Godambe, V. P. (1960). An optimum property of regular maximum likelihood estimation. *The Annals of Mathematical Statistics*, **31**, 1208–1211.

McCullagh, P. and Nelder, J. A. (1989). *Generalized Linear Models, Second Edition*, Chapman & Hall.

van der Vaart, A. W. (1998). *Asymptotic Statistics.* Cambridge University Press.

Wald, A. (1949). Note on the consistency of maximum likelihood estimate. *Annals of Mathematical Statistics*, **20**, 595–601.

Wong, W. H. (1986). Theory of partial likelihood. *The Annals of Statistics.* **14**, 88–123.

9

Estimating equations

In this chapter we develop the theory of estimating equations. Estimating equations are a generalization of the maximum likelihood method, but they do not require a fully specified probability model. Instead, they only require the functional forms of their first two moments. Estimating equations have wide range of applications: for example, the quasi likelihood method (Wedderburn, 1974; McCullagh, 1983; Godambe and Thompson, 1989; Heyde, 1997), and the Generalized Estimating Equations (Liang and Zeger, 1986; Zeger and Liang, 1986) are two important types of estimating equations that are widely used in Generalized Linear Models and longitudinal data analysis. (Yes, a Generalized Estimating Equation is indeed a special type of estimating equation due to the commonly adopted convention, even though this sounds like an oxymoronic statement). In addition, estimating equations are related to the Generalized Method of Moments (Hansen, 1982), which is popular in econometrics. Estimating equations have also been developed in conjunction with the Martingale theory, and are a powerful method for statistical inference for stochastic processes (Heyde, 1997). As a natural extension of Maximum Likelihood Estimation and Method of Moments, estimating equations have their combined flavors and advantages.

In addition to their wide applications in data analysis, estimating equations are also a convenient theoretical framework to develop many aspects of statistical inference, such as conditional inference (Godambe, 1976; Lindsay, 1982), nuisance parameters, efficient estimator, and the information bound. They make some optimal results in statistical inference transparent via projections in Hilbert spaces.

9.1 Optimal Estimating Equations

The basic theory of optimal estimating equations are introduced and developed by Godambe (1960, 1976), Durbin (1960), and Crowder (1987). See also Morton (1981) and Jarrett (1984). An estimating equation is any measurable

B. Li and G. J. Babu, *A Graduate Course on Statistical Inference*,
Springer Texts in Statistics, https://doi.org/10.1007/978-1-4939-9761-9_9

and P_θ-integrable function of θ and X. We usually assume its expectation under θ is either 0 or statistically ignorable, so that under mild conditions its solution is asymptotically Normal with a variance matrix depending on the estimating equation. The optimal estimating equation in a class of estimating equations is the one whose solution has the smallest asymptotic variance among the class in terms of Louwner's ordering. The explicit form of the optimal estimating equation can be derived by the multivariate Cauchy-Schwarz inequality developed in Section 7.9.

As before, let $X_1, \ldots, X_n$ be an i.i.d. sample with probability distribution P_θ, where θ is a p-dimensional parameter in a parametric space $\Theta \subseteq \mathbb{R}^p$. Throughout this chapter, we assume that the parametric family $\{P_\theta : \theta \in \Theta\}$ is dominated by a σ-finite measure μ, and denote the density of P_θ with respect to μ by f_θ.

Definition 9.1 *A function $g : \Theta \times \Omega_X \to \mathbb{R}^p$ is called an unbiased estimating equation if*

$$E_\theta[g(\theta, X)] = 0$$

for all $\theta \in \Theta$.

An unbiased estimating equation is a generalization of the score function $s(\theta, X)$, which satisfies $E_\theta[s(\theta, X)] = 0$ under the assumptions of Proposition 8.1. It also includes many other estimates, such as the method of moments, the least squares estimate, and the quasi likelihood estimate (Wedderburn, 1974; McCullagh, 1983). Given an estimating equation $g(\theta, X)$, the parameter θ is estimated by solving the equation

$$E_n[g(\theta, X)] = 0.$$

Let $L_2(P_\theta)$ be the class of all P_θ-square-integrable random variables, and let

$$L_2^p(P_\theta) = L_2(P_\theta) \times \cdots \times L_2(P_\theta)$$

be the p-fold Cartesian product of $L_2(P_\theta)$. We often assume that, for each θ, $g(\theta, X)$ is a member of $L_2^p(P_\theta)$. This is stated as the following formal definition for easy reference.

Definition 9.2 *An estimating equation $g(\theta, X)$ is P_θ-square-integrable if $g(\theta, X) \in L_2^p(P_\theta)$ for each $\theta \in \Theta$. Furthermore, a class of estimating equations $\mathcal{G}$ is said to be P_θ-square-integrable if its members are P_θ-square-integrable.*

The optimal estimating equation is defined in terms of the amount of information contained in an estimating equation, as introduced by Godambe (1960). We now give a general definition of this information.

Definition 9.3 *Suppose that $g(\theta, X)$ is an unbiased and P_θ-square-integrable estimating equation, and $g(\theta, x) f_\theta(x)$ satisfies $DUI^+(\theta, \mu)$. Then the following matrix*

$$I_g(\theta) = E_\theta[\partial g^T(\theta, X)/\partial\theta]\{E_\theta[g(\theta, X)g^T(\theta, X)]\}^+ E_\theta[\partial g(\theta, X)/\partial\theta^T].$$

is called the information contained in $g(\theta, X)$.

To understand the intuition behind this definition, consider the special case where θ is a scalar, and the information $I_g(\theta)$ takes the form

$$I_g(\theta) = \frac{\{E_\theta[\dot{g}(\theta, X)]\}^2}{E_\theta[g^2(\theta, X)]}.$$

Intuitively, since $\hat{\theta}$ is solved from $E_n[g(\theta, X)] = 0$, a larger slope of $E_{\theta_0}[g(\theta, X)]$ at θ_0 would benefit estimation, because a slight departure from θ_0 would cause a large change in $E_n[g(\theta, X)]$, forcing its root to be close to θ_0. On the other hand, a smaller variance $\text{var}_{\theta_0}[g(\theta_0, X)]$ would benefit estimation, because it will make the random function $E_n[g(\theta, X)]$ packed tightly around 0 when θ is near θ_0, which makes the variation of the solution small. Thus the ratio of these two quantities characterizes the tendency of an estimating equation having a root close to the true parameter value. Also, as we shall see in Section 9.5, $I_g^{-1}(\theta)$ is the asymptotic variance of $\sqrt{n}(\hat{\theta}_g - \theta_0)$, where $\hat{\theta}_g$ is any consistent solution to the estimating equation $E_n[g(\theta, X)] = 0$. Thus maximizing the information of an estimating equation amounts to minimizing the asymptotic variance of its solution.

We now define the optimal estimating equation in a class $\mathcal{G}$ of estimating equations.

Definition 9.4 *Suppose* $\mathcal{G}$ *is a class of unbiased and* P_θ*-square-integrable estimating equations such that, for each* $g \in \mathcal{G}$, $g(\theta, x)f_\theta(x)$ *satisfies* $DUI^+(\theta, \mu)$. *If there is a member* g^* *of* $\mathcal{G}$ *such that* $I_{g^*} \succeq I_g$ *for all* $g \in \mathcal{G}$, *then* g^* *is called the optimal estimating equation in* $\mathcal{G}$.

We next develop a general method for constructing the optimal estimating equation. For all practical purposes a class $\mathcal{G}$ of estimating equations may be assumed to be a linear manifold. That is, for each θ, the set $\{g(\theta, X) : g \in \mathcal{G}\}$ is a linear manifold in $L_2^p(P_\theta)$. As we will see in the subsequent development, if this linear manifold is closed, then the optimal estimating equation can be obtained by projecting the score function $s(\theta, X)$ on to the linear subspace. However, the set $\{g(\theta, X) : g \in \mathcal{G}\}$ may not be closed: for example the DUI assumption may not be preserved after taking an $L_2(P_\theta)$-limit. While it might be possible to get around this problem by using a more general definition of a derivative, we take the simpler approach to assume that $\mathcal{G}$ contains the optimal estimating equations, which is sufficient for the discussions here. The method for constructing optimal estimating equations by projecting the score function on to a class of estimating equations is referred to as the projected score method by Small and McLeish (1989).

The next lemma is a generalization of the information identity (8.3) in Proposition 8.1.

Lemma 9.1 *If $g(\theta, x)$ is an unbiased estimating equation such that $g(\theta, x) f_\theta(x)$ satisfies DUI$^+(\theta, \mu)$, then*

$$E_\theta \left[\frac{\partial g(\theta, X)}{\partial \theta^T} \right] = -E_\theta[g(\theta, X) s^T(\theta, X)]. \tag{9.1}$$

Proof. By unbiasedness of $g(\theta, X)$, we have

$$\int_A g(\theta, X) f_\theta(X) d\mu = 0,$$

where A is the common support of $f_\theta(x)$. Because $g(\theta, X) f_\theta(X)$ satisfies DUI$^+(\theta, \mu)$, by differentiating both sides of the equation we have

$$\int_A \frac{\partial g(\theta, X)}{\partial \theta^T} f_\theta(X) d\mu + \int_A g(\theta, X) \frac{\partial f_\theta(X)/\partial \theta^T}{f_\theta(X)} f_\theta(X) d\mu = 0.$$

The asserted result follows because first term on the left is $E_\theta[\partial g(\theta, X)/\partial \theta^T]$, and the second term on the left is $E_\theta[g(\theta, X) s^T(\theta, X)]$. □

Note that, if we take $g(\theta, X)$ to be $s(\theta, X)$, then the identity (9.1) reduces to the information identity (8.3) in Proposition 8.1. Before stating the next theorem, we define the inner product matrix of two members of $L_2^p(P_\theta)$, $h_1(\theta, X)$ and $h_2(\theta, X)$, to be

$$[h_1, h_2] = E_\theta[h_1(\theta, X) h_2^T(\theta, X)].$$

Theorem 9.1 *Suppose $\mathcal{G}$ is a class of unbiased and P_θ-square-integrable-estimating equations such that for each $g \in \mathcal{G}$, $g(\theta, x) f_\theta(x)$ satisfies DUI$^+(\theta, \mu)$. If there is a member g^* of $\mathcal{G}$ satisfying $[s - g^*, g] = 0$ for all $g \in \mathcal{G}$, then $I_{g^*} \succeq I_g$ for all $g \in \mathcal{G}$.*

Proof. By the multivariate Cauchy-Schwarz inequality (the more general version in Problem 7.22), we have, for any $g \in \mathcal{G}$,

$$[g^*, g^*] \succeq [g^*, g][g, g]^+[g, g^*].$$

By assumption, $[g^*, g] = [s, g]$. As shown in Lemma 9.1, under the DUI and the unbiasedness assumption, we have

$$[g, s] = -E_\theta[\partial g(\theta, X)/\partial \theta^T]. \tag{9.2}$$

Hence

$$[g^*, g][g, g]^+[g, g^*] = E[\partial g^T/\partial \theta][E(g g^T)]^+ E(\partial g/\partial \theta^T) = I_g.$$

Applying the above relation to $g = g^*$, we have

$$[g^*, g^*] = [g^*, g^*][g^*, g^*]^+[g^*, g^*] = I_{g^*}.$$

Therefore $I_{g^*} \succeq I_g$. □

It follows immediately from the above theorem that the score function $s(\theta, x)$ is the optimal estimating equation among a class of estimating equations $\mathcal{G}$, as long as it belongs to $\mathcal{G}$. In particular, we can take $\mathcal{G}$ to be the class of all unbiased, P_θ-square-integrable estimating equations that satisfy the DUI condition. This optimal property is stated in the next theorem, whose simple proof is omitted. The core idea of this result is due to Godambe (1960) and Durbin (1960).

Corollary 9.1 *Suppose $\mathcal{G}$ is the class of all unbiased, P_θ-square-integrable estimating equations such that, for each $g \in \mathcal{G}$, $f_\theta(x)$ and $g(\theta, x) f_\theta(x)$ satisfy $DUI^+(\theta, \mu)$. If $s(\theta, x)$ is a member of $\mathcal{G}$, then, for any $g \in \mathcal{G}$, we have $I_s(\theta) \succeq I_g(\theta)$ for all $\theta \in \Theta$.*

The optimality of maximum likelihood estimation can be stated at several levels. The above result stated in terms of estimating equations is an intuitive and relatively simple optimal property. We will revisit this issue to give a more general form of this optimality in Chapter 10.

By itself, Corollary 9.1 does not lead to any new method; it merely states that the maximum likelihood estimate is optimal in this sense. A more interesting case is when $s(\theta, X)$ is not among the estimating equations considered, in which case it leads to new methods such as the quasi likelihood method and the generalized estimating equations.

The next proposition shows that the condition $[s - g^*, g] = 0$ for all $g \in \mathcal{G}$ uniquely determines an optimal estimating equation in $\mathcal{G}$ provided that $\mathcal{G}$ is a linear space.

Proposition 9.1 *Suppose that $\mathcal{G}$ is a linear manifold of unbiased and P_θ-square-integrable estimating equations such that, for each $g \in \mathcal{G}$, $g(\theta, x) f_\theta(x)$ satisfies $DUI^+(\theta, \mu)$. If there exist g_1^* and g_2^* in $\mathcal{G}$ satisfying $[s - g_i^*, g] = 0$, for $i = 1, 2$ and for all $g \in \mathcal{G}$, then $g_1^* = g_2^*$.*

Proof. By assumption,

$$[s - g_1^*, g] = 0, \quad [s - g_2^*, g] = 0$$

for all $g \in \mathcal{G}$. Since $\mathcal{G}$ is a linear manifold, $g_1^* - g_2^*$ is a member of $\mathcal{G}$. Hence

$$[s - g_1^*, g_1^* - g_2^*] = 0, \quad [s - g_2^*, g_1^* - g_2^*] = 0.$$

Subtracting the first equation from the second, we have

$$[g_1^* - g_1^*, g_1^* - g_2^*] = 0,$$

which implies $g_1^* - g_2^* = 0$ by Proposition 7.4. □

In the next three sections we develop some special optimal estimating equations, including the quasi likelihood estimating equation, and the Generalized Estimating Equation. They are the foundations of some important statistical methodologies, and also serve to illustrate how to construct optimal estimating equations using the general method introduced in this section under different settings. Ideally all the results in the next three sections can be written rigorously as lemmas, theorems, and corollaries, but doing so would involve lengthy regularity conditions that may obscure otherwise simple ideas. We will instead develop them somewhat informally without stating all the regularity conditions.

9.2 Quasi likelihood estimation

Suppose that $(X_1, Y_1), \ldots, (X_n, Y_n)$ are i.i.d. with a joint distribution from a parametric family $\{P_\theta : \theta \in \Theta \subseteq \mathbb{R}^p\}$. We assume X_i is a p-dimensional random vector, and Y_i is a random variable. Again, assuming X_i to be random is purely for convenience. The following estimating equation is based entirely on conditional moments given X. That is, we are in effect treating X as fixed. The optimal estimating equation we get is the same as that of assuming X_i to be fixed.

Here, we do not assume the full parametric form of P_θ, but instead, we only assume the forms of the first two conditional moments of Y given X:

$$E_\theta(Y|X) = \mu(\theta^T X), \quad \text{var}_\theta(Y|X) = V(\theta^T X),$$

where μ and V are known functions. For example, in a log linear model for the Poisson data,

$$\mu(\theta^T X) = e^{\theta^T X}, \quad V(\theta^T X) = e^{\theta^T X}.$$

Denote the conditional density of $Y|X$ by $f_\theta(y|x)$ and the marginal density of X by $f_X(x)$. Neither $f_\theta(y|x)$ nor f_X needs to be specified except the conditional moments $\mu(\cdot)$ and $V(\cdot)$.

Consider the class of unbiased estimating equations of the form

$$\mathcal{G} = \{a_\theta(X)(Y - \mu(\theta^T X)) : a \text{ is a function from } \Theta \times \Omega_X \text{ to } \mathbb{R}^p\}.$$

Our goal is to find the optimal estimating equation in $\mathcal{G}$ that satisfies $[s - g^*, g] = 0$ for all $g \in \mathcal{G}$, where, as before, $s = s(\theta, x, y)$ represents the true score function. Despite its appearance, this process does not require the form of s. Our optimal estimating equation is derived from the defining relation

$$E_\theta(g^* g^T) = E_\theta(s g^T) = -E_\theta(\partial g^T / \partial \theta).$$

In our setting, $g^* = a_\theta^*(X)(Y - \mu(\theta^T X))$, $g = a_\theta(X)(Y - \mu(\theta^T X))$. For convenience, we abbreviate $a_\theta^*(X)$, $a_\theta(X)$, and $\mu(\theta^T X)$ by a^*, a, and μ, respectively. The above relation specializes to

$$E_\theta[a^* a^T (Y - \mu)^2] = -E_\theta \left[\frac{\partial a^T}{\partial \theta} (Y - \mu) \right] + E_\theta \left(\frac{\partial \mu}{\partial \theta} a^T \right). \tag{9.3}$$

The left-hand side is

$$E_\theta\{a^* a^T E[(Y - \mu)^2 | X]\} = E_\theta(a^* a^T V).$$

The first term on the right-hand side of (9.3) is

$$-E_\theta \left[\frac{\partial a^T}{\partial \theta} E(Y - \mu | X) \right] = 0.$$

Thus, a^* satisfies the following functional equation

$$E_\theta(a^* a^T V) = E_\theta \left(\frac{\partial \mu}{\partial \theta} a^T \right).$$

By inspection, we see that if we take $a^* = (\partial \mu / \partial \theta)/V$, then the above equation is satisfied for all a. Thus we arrive at the following optimal estimating equation

$$g^*(\theta, X, Y) = [\partial \mu(\theta^T X)/\partial \theta] \, [Y - \mu(\theta^T X)]/V(\theta^T X).$$

This optimal estimating equation is called the quasi score function. See Wedderburn (1974), McCullagh (1983), and Li and McCullagh (1994). The maximum quasi likelihood estimate is defined as the solution to the estimating equation

$$E_n[g^*(\theta, X, Y)] = 0.$$

The information contained in the quasi score is

$$I_{g^*} = E[(\partial \mu / \partial \theta)(\partial \mu / \partial \theta)^T / V].$$

The above construction can be easily extended to vector-valued Y_i, say, of dimension q. In this case, $\mu(\theta^T X)$ is a q-dimensional vector and $V(\theta^T X)$ is a $q \times q$ matrix. Consider the class $\mathcal{G}$ of estimating equations of the form

$$A(\theta, X)[Y - \mu(\theta^T X)],$$

where $A(\theta, X)$ is a $p \times q$ random matrix that may depend on θ and X but does not depend on Y. By Theorem 9.1, if there is an $A^*(\theta, X)$ such that $g^* = A^*(\theta, X)[Y - \mu(\theta^T X)]$ satisfies

$$\begin{aligned} &[A(\theta, X)(Y - \mu(\theta^T X)), s] \\ &\qquad = [A(\theta, X)(Y - \mu(\theta^T X)), A^*(\theta, X)(Y - \mu(\theta^T X))], \end{aligned} \tag{9.4}$$

for all $A(\theta, X)$, then g^* is an optimal estimating equation. The left-hand side is

$$\begin{aligned}[A(\theta, X)(Y - \mu(\theta^T X)), s] &= E_\theta\{\partial[A(\theta, X)(Y - \mu(\theta^T X))]/\partial\theta^T\} \\ &= -\, E_\theta[A(\theta, X)\partial\mu(\theta^T X)/\partial\theta^T].\end{aligned}$$

The right-hand side of (9.4) is

$$E_\theta[A(\theta, X)V(\theta^T X)A^{*T}(\theta, X)]$$

So, if we let

$$A^{*T}(\theta, X) = V^{-1}(\theta^T X)\partial\mu(\theta^T X)/\partial\theta^T,$$

then (9.4) is satisfied for all $A(\theta, X)$. Thus, the quasi score function in this case is

$$[\partial\mu^T(\theta^T X)/\partial\theta]V^{-1}(\theta^T X)[Y - \mu(\theta^T X)].$$

This is the form given in McCullagh (1983).

9.3 Generalized Estimating Equations

Generalized Estimating Equations (GEE) were introduced by Liang and Zeger (1986), and Zeger and Liang (1986) to deal with the situations where each subject has multiple observations that may be dependent. They are a flexible and effective method for handling longitudinal data, and are widely used in that area. Suppose we have observations

$$\{(X_{ik}, Y_{ik}) : k = 1, \ldots, n_i, \; i = 1, \ldots, n\},$$

where X_{ik} are p-dimensional predictors, and Y_{ik} are 1-dimensional response. We write

$$Y_i = (Y_{i1}, \ldots, Y_{in_i})^T, \quad X_i = (X_{i1}, \ldots, X_{in_i}).$$

Note that Y_i (and also X_i) may have different dimensions for different i. The responses within the same subject, $\{Y_{i1}, \ldots, Y_{in_i}\}$, may be dependent, but the responses for different subjects are assumed to be independent.

As in quasi likelihood, we assume the functional forms of the conditional mean and variance:

$$E(Y_{ik}|X_{ik}) = \mu(X_{ik}^T\beta), \quad \mathrm{var}(Y_{ik}|X_{ik}) = V(X_{ik}^T\beta),$$

where μ and V are known functions. We write

$$\begin{aligned}\mu_i(X_i, \beta) &= (\mu(\beta^T X_{i1}), \ldots, \mu(\beta^T X_{in_i}))^T, \\ V_i(X_i, \beta) &= \mathrm{diag}\left(V(\beta^T X_{i1}), \ldots, V(\beta^T X_{in_i})\right).\end{aligned}$$

Within the same subject, we assume

$$\mathrm{var}(Y_i|X_i) = V_i(X_i, \beta)^{1/2}R_i(\alpha)V_i(X_i, \beta)^{1/2},$$

where, for each i, $R_i(\alpha)$ is a known matrix-valued function of α. These $R_i(\alpha)$ are called the working correlation matrices. An important feature of the GEE is that $R_i(\alpha)$ need not be correctly specified for GEE to yield $\sqrt{n}$-consistent estimates, but if $R_i(\alpha)$ are correctly specified, then GEE is the optimal estimating equation, as described below.

Consider the following class of linear (in Y) and unbiased estimating equations

$$\mathcal{G} = \left\{ \sum_{i=1}^{n} W_i(X_i, \alpha, \beta)(Y_i - \mu_i(X_i, \beta)) : W_i(X_i, \beta) \in \mathbb{R}^{p \times n_i} \right\}.$$

It is natural to consider this class of estimating equations because their inner products can be completely specified by the form of the first two moments that we assume. We seek the optimal estimating equation within this class. Let

$$S(\alpha, \beta, X_{1:n}, Y_{1:n}) = \sum_{i=1}^{n} s(\alpha, \beta, X_i, Y_i)$$

be the score function. Let G be an arbitrary member of $\mathcal{G}$, and G^* be the optimal estimating equation in $\mathcal{G}$. That is,

$$\begin{aligned} G(\alpha, \beta, X_{1:n}, Y_{1:n}) &= \sum_{i=1}^{n} W_i(X_i, \alpha, \beta)(Y_i - \mu_i(X_i, \beta)), \\ G^*(\alpha, \beta, X_{1:n}, Y_{1:n}) &= \sum_{i=1}^{n} W_i^*(X_i, \alpha, \beta)(Y_i - \mu_i(X_i, \beta)). \end{aligned}$$

For convenience, we omit the arguments of the functions and write

$$\begin{aligned} &S = S(\alpha, \beta, X_{1:n}, Y_{1:n}), \; s_i = s(\alpha, \beta, X_i, Y_i), \\ &G = G(\alpha, \beta, X_{1:n}, Y_{1:n}), \\ &W_i = W_i(X_i, \alpha, \beta), \; R_i = R_i(\alpha), \; \mu_i = \mu_i(\beta, X_i), \; V_i = V_i(\beta, X_i). \end{aligned}$$

The optimal estimating equation G^* is determined by the equation

$$[S, G] = [G^*, G] \tag{9.5}$$

for all $G \in \mathcal{G}$. The right-hand side is

$$\begin{aligned} E[G^* G^T] &= \sum_{i=1}^{n} E\left[W_i^*(Y_i - \mu_i)(Y_i - \mu_i)^T W_i^T\right] \\ &= \sum_{i=1}^{n} E\left\{W_i^* E[(Y_i - \mu_i)(Y_i - \mu_i)^T | X_i] W_i^T\right\} \qquad (9.6) \\ &= \sum_{i=1}^{n} E\left(W_i^* V_i^{1/2} R_i V_i^{1/2} W_i^T\right). \end{aligned}$$

The left-hand side of (9.5) is

$$E(SG^T) = \sum_{i=1}^{n} E\left[s_i(Y_i - \mu_i)^T W_i^T\right].$$

By Lemma 9.1, the summand in the right-hand side is

$$\begin{aligned} &- E\left\{(\partial/\partial\beta)[(Y_i - \mu_i)^T W_i^T]\right\} \\ &= -E\left\{[\partial(Y_i - \mu_i)^T/\partial\beta]W_i^T\right\} - E\left\{(Y_i - \mu_i)^T(\partial W_i^T/\partial\beta)\right\}. \end{aligned}$$

In the second term, $(Y_i - \mu_i)^T$ is a row vector, and W_i^T is an $n_i \times p$ matrix. The expression $(Y_i - \mu_i)^T(\partial W_i^T/\partial\beta)$ simply means the $p \times p$ matrix whose jth row is $(Y_i - \mu_i)^T(\partial W_i^T/\beta_j)$. Since

$$\begin{aligned} E\left\{[\partial(Y_i - \mu_i)^T/\partial\beta]W_i^T\right\} &= -E\left[(\partial\mu_i^T/\partial\beta)W_i^T\right] \\ E\left\{(Y_i - \mu_i)^T(\partial W_i^T/\partial\beta)\right\} &= E[E(Y_i - \mu_i|X_i)^T(\partial W_i^T/\partial\beta)] = 0, \end{aligned}$$

we have

$$[S, G] = E\left[(\partial\mu_i^T/\partial\beta)W_i^T\right]. \tag{9.7}$$

By (9.6) and (9.7), the defining relation (9.5) for an optimal estimating equation specializes to the following form in the GEE setting:

$$\sum_{i=1}^{n} E\left(W_i^* V_i^{1/2} R_i V_i^{1/2} W_i^T\right) = \sum_{i=1}^{n} E\left[(\partial\mu_i^T/\partial\beta)W_i^T\right]. \tag{9.8}$$

If we let

$$W_i^* = (\partial\mu_i^T/\partial\beta)V_i^{-1/2}R_i^{-1}V_i^{-1/2},$$

then (9.8) holds for all W_i. Thus we arrive at the following optimal estimating equation

$$\sum_{i=1}^{n}[\partial\mu_i(X_i,\beta)^T/\partial\beta][V_i^{1/2}(X_i,\beta)R_i(\alpha)V_i^{1/2}(X_i,\beta)]^{-1}[Y_i - \mu_i(X_i,\beta)] = 0.$$

At the sample level, the parameters α and β are estimated by the following iterative regime. For a fixed $\hat{\alpha}$, we estimate β by solving the above equation. For a fixed $\hat{\beta}$, we define the residuals as

$$\hat{r}_{ik} = \frac{Y_{ik} - \mu_i(X_i,\hat{\beta})}{V_i^{1/2}(X_i,\hat{\beta})}.$$

We then estimate (k,k')th component of R_i as

$$(\hat{R}_i)_{kk'} = \frac{1}{N-p}\sum_{i=1}^{n} \hat{r}_{ik}\hat{r}_{ik'}.$$

In this case, we do not assume any further structure in the correlation matrix R_i, so that the distinct entries of R_i themselves constitute α. But it is possible to build more structure into the $R_i(\alpha)$, such as the autoregressive model, the exchangeable correlation model. See Liang and Zeger (1986) for further information.

9.4 Other optimal estimating equations

The above two sections are concerned with the quasi score function and the generalized estimating equation, both of which can be regarded as the optimal estimating equations among the linear estimating equations of the form

$$A(\theta, X)[Y - \mu(\theta^T X)],$$

where $A(\theta, X)$ is a scalar or a vector. In this section we make a brief exposition of several other types of optimal estimating equations to get a broader view of the methology of estimating equations.

Crowder (1987) considered the following general class of estimating equations

$$g(\theta, X, Y) = W(\theta, X)u(\theta, X, Y),$$

where u is a function on $\Omega \times \Theta$ to $\mathbb{R}^r$, whose components are unbiased and P_θ-square-integrable estimating equations. The components of u may be any fixed set of functions of X, Y, θ. In particular, Crowder (1987) considered the special case where u is the vector of linear and quadratic polynomials of Y, as a modification of the quasi score function when the fourth moment of Y is available.

Another type of optimal estimating equations are the ones used for conditional inference, which were developed by Waterman and Lindsay (1996). Suppose $(X_1, \ldots, X_n)$ is distributed as P_θ, where θ consists of a parameter of interest ψ and a nuisance parameter λ. For simplicity, we will focus on the case where both ψ and λ are scalars. But the following development can be extended in a straightforward (albeit quite tedious) manner to the vector-valued ψ and λ. The vector-valued ψ and λ will be considered in Section 9.8, but there we will only consider the first-order projection, a special case of the mth-order projection developed below.

We are interested in statistical inference about ψ. The parameter λ is not of interest, but it is needed to specify a meaningful statistical model. Ideally, if we have a statistic T that is sufficient for λ, then the conditional distribution of $(X_1, \ldots, X_n)|T$ does not depend on the nuisance parameter λ, and we can infer about ψ using this conditional distribution. This is, in fact, the approach we took in Chapter 4 to develop the UMPU test for a parameter of interest for an exponential family distribution, for which such a statistic T does exit. However, existence of such a T is a very stringent requirement, which a majority of distributions outside the exponential family do not meet. Waterman and Lindsay (1996) proposed a generalization of the conditional score $s^{(c)}(\psi, X_1, \ldots, X_n) = \partial \log f_{X_1,\ldots,X_n|T}(x_1, \ldots, x_n|t;\psi)/\partial\psi$ using the idea of optimal estimating equations for the situations where the sufficient statistic T for λ does not exist. For convenience, only for the rest of this section, we will use X to abbreviate $X_1, \ldots, X_n$, so that, for example, $s^{(c)}(\psi, X_1, \ldots, X_n)$ is abbreviated by $s^{(c)}(\psi, X)$. After this section, we will resume the convention that $X_1, \ldots, X_n$ are random copies of X — a convention that has been used throughout the rest of the book.

To develop the intuition, let us first derive an alternative representation of the conditional score $s^{(c)}(\psi, X)$ when T exists. Since

$$f_X(x;\theta) = f_{X|T}(x|t;\psi) f_T(t;\theta), \tag{9.9}$$

we have

$$s_\psi(\theta, X) = s^{(c)}(\psi, X) - s^{(m)}(\theta, T),$$

where $s_\psi(\theta, x) = \partial \log f_X(x;\theta)/\partial\psi$ is the score function for ψ, and $s^{(m)}(\theta, T)$ is the marginal score $\partial \log f_T(t;\theta)/\partial\psi$. Next, consider the linear subspace $\mathcal{B}$ of $L_2(P_\theta)$ spanned by

$$b_k(x) = \frac{\partial^k f_X(x;\theta)/\partial\lambda^k}{f_X(x;\theta)}, \quad k = 1, 2, \ldots$$

These functions are called the Bhattacharyya basis (Bhattacharyya, 1946) (and hence the notation $\mathcal{B}$). By (9.9), it is easy to check that

$$b_k(x) = \frac{\partial^k f_T(t;\theta)/\partial\lambda^k}{f_T(t;\theta)} \equiv c_k(t).$$

If the set of functions $\{c_k(t) : k = 1, 2, \ldots\}$ is rich enough so that $s^{(m)}(\theta, t)$ is contained in $\mathcal{B}$, then $s^{(m)}(\theta, T)$ is, in fact, the projection of $s_\psi(\theta, X)$ on to $\mathcal{B}$ in terms of the $L_2(P_\theta)$ inner product. To see this, take any basis function $c_k(t)$ from $\mathcal{B}$, and we have

$$\begin{aligned} E_\theta\{c_k(T)[s_\psi(\theta, X) - s^{(m)}(\theta, T)]\} &= E_\theta[c_k(T)s^{(c)}(\psi, X)] \\ &= E_\theta\{c_k(T)E_\psi[s^{(c)}(\psi, X)|T]\} = 0. \end{aligned}$$

Consequently, $s^{(c)}(\psi, X) = s_\psi(\theta, X) - s^{(m)}(\theta, T)$ is the projection of $s_\psi(\theta, X)$ on to $\mathcal{B}^\perp$. In other words, $s^{(c)}(\psi, X)$ is the optimal estimating equation in $\mathcal{B}^\perp$.

The point of the above derivation is that, while the conditional score $s^{(c)}(\psi, X)$ is defined only when there is a sufficient statistic T for λ, the projection $s_\psi(\theta, X)$ on to $\mathcal{B}^\perp$ is not restricted by the existence of T. Motivated by this, Waterman and Lindsay (1996) proposed to use the projection of $s_\psi(\theta, X)$

on to $\mathcal{B}_m^\perp$ as the generalized conditional score to conduct statistical inference about ψ, where $\mathcal{B}_m = \text{span}\{b_1(x), \ldots, b_m(x)\}$.

We can now derive the explicit form of this projection. Let $P_{\mathcal{B}_m}$ be the projection operator on to the subspace of $L_2(P_\theta)$ spanned by the set $\mathcal{B}_m$. By Example 7.5, the projection of $s_\psi = s_\psi(\theta, X)$ on to $\mathcal{B}_m$ is

$$P_{\mathcal{B}_m} s_\psi = (\langle s_\psi, b_1\rangle, \ldots, \langle s_\psi, b_m\rangle) \begin{pmatrix} \langle b_1, b_1\rangle & \cdots & \langle b_1, b_m\rangle \\ \vdots & \ddots & \vdots \\ \langle b_m, b_1\rangle & \cdots & \langle b_m, b_m\rangle \end{pmatrix} \begin{pmatrix} b_1 \\ \vdots \\ b_m \end{pmatrix}.$$

Thus, the projection of s_ψ on to $\mathcal{B}_m^\perp$ is $s_\psi - P_{\mathcal{B}_m} s_\psi$.

An important — and most commonly used — special case is $m = 1$, where the above projection becomes

$$s_\psi - \frac{E_\theta(s_\psi s_\lambda)}{E_\theta(s_\lambda s_\lambda)} s_\lambda = s_\psi - \frac{I_{\psi\lambda}}{I_{\lambda\lambda}} s_\lambda,$$

where $s_\lambda = \partial \log f_X(x;\theta)/\partial\lambda$ is the score function for λ; $I_{\psi\lambda}$ and $I_{\lambda\lambda}$ are the (ψ,ψ)- and (ψ,λ)-components of the Fisher information. This estimating equation is commonly known as the efficient score, to which we will return in Section 9.8.

9.5 Asymptotic properties

Let $X_1, \ldots, X_n$ be independent copies of X, where X is distributed as P_θ. Let g be an unbiased estimating equation. We estimate the parameter θ by solving the estimating equation

$$E_n[g(\theta, X)] = 0.$$

The existence of consistent solutions, as stated in the next theorem, can be proved using a similar method as that used in Theorem 8.1.

Theorem 9.2 *Suppose $X_1, \ldots, X_n$ are i.i.d. random variables or vectors having a density f_{θ_0} belonging to a parametric family $\{f_\theta : \theta \in \Theta \subseteq \mathbb{R}^p\}$, and the support B of f_θ does not depend on θ. Suppose $g : \Theta \times \Omega_X \to \mathbb{R}^p$ is an unbiased estimating equation such that*

1. *for all $\theta \in \Theta$, the matrix $A(\theta_0) = E[\partial g(\theta, X)/\partial\theta^T]$ is negative definite;*
2. *$g(\theta, x)$ satisfies $DUI(\theta, B, P_{\theta_0})$;*
3. *in a neighborhood of θ_0, $E_n[g(\theta, X)]$ converges in probability uniformly to $E[g(\theta, X)]$.*

Then, there is a sequence of estimators $\{\hat\theta_n\}$ such that

i. with probability tending to 1, $\hat\theta_n$ is a solution to $E_n[g(\theta, X)] = 0$, *ii.* $\hat{\theta}_n \xrightarrow{P} \theta_0$.

The assumption that $A(\theta_0)$ is negative definite is not unreasonable. For example, if $g = g^*$ is the optimal estimating equation, then we have

$$E[\partial g^*(\theta, X)/\partial \theta^T] = -E[g^*(\theta, X) g^{*T}(\theta, X)],$$

which is negative definite if $g^*(\theta, X)$ is not degenerate for any θ. Another example is when $E_n[g(\theta, X)]$ is derived from the derivative of a concave objective function, in which case, $\partial g(\theta, X)/\partial \theta^T$ is the Hessian matrix of a concave function, which is negative semi-definite. The proof of this theorem is similar to the proof of Theorem 8.1, and is left as an exercise.

Since the estimating equation method does not start with an objective function to maximize or minimize, Wald's approach to consistency cannot be directly applied. So a commonly used consistency statement is existence of a consistent solution to an estimating equation as in Theorem 9.2, which is not as specific as the Wald-type consistency statement. However, it is possible to construct an objective function such that a Wald-type consistency statement can be obtained. For example, Li (1993, 1996) introduced a deviance function for the quasi-likelihood method as a function of the first two moments. It was shown that the minimax of this deviance function is consistent under mild assumptions.

Next, we derive the asymptotic distribution of a consistent solution to an estimating equation. Let $J_g(\theta)$ and $K_g(\theta)$ denote the matrices

$$J_g(\theta) = E_\theta[\partial g(\theta, X)/\partial \theta^T], \quad K_g(\theta) = E_\theta[g(\theta, X) g^T(\theta, X)]. \tag{9.10}$$

The information contained in the estimating equation g is simply

$$I_g(\theta) = J_g^T(\theta) K_g^{-1}(\theta) J_g(\theta). \tag{9.11}$$

As before, let $B^\circ(\theta_0, \epsilon)$ denote the open ball centered at θ_0 with radius ϵ.

Theorem 9.3 *Suppose*

1. *$g(\theta, x)$ is an unbiased, P_θ-square-integrable estimating equation such that $g(\theta, x) f_\theta(x)$ satisfies $DUI^+(\theta, \mu)$;*
2. *the sequence $\{E_n[\partial g(\theta, X)/\partial \theta^T] : n = 1, 2, \ldots\}$ is stochastically equicontinuous over $B^\circ(\theta_0, \epsilon)$;*
3. *the true parameter θ_0 is an interior point of Θ.*

If $\hat{\theta}$ is a consistent solution to likelihood equation $E_n[g(\theta, X)] = 0$, then $\sqrt{n}(\hat{\theta} - \theta_0) \xrightarrow{\mathcal{D}} N(0, I_g^{-1}(\theta_0))$.

Proof. The similar arguments in the proof of Theorem 8.4 lead to

$$\sqrt{n}(\hat{\theta} - \theta_0) \xrightarrow{\mathcal{D}} N(0, J_g^{-1}(\theta_0) K_g(\theta_0) J_g^{-T}(\theta_0))$$

But here, we no longer have

$$K_g(\theta) = -J_g(\theta) = I_g(\theta).$$

So, by (9.11), the asymptotic distribution of $\sqrt{n}(\hat{\theta} - \theta)$ is $N(0, I_g(\theta_0))$. □

The finite-sample optimality in Corollary 9.1, combined with the asymptotic normality in Theorem 9.3, shows that the maximum likelihood estimate has the smallest asymptotic variance (in terms of Louwner's ordering) among the solutions to estimating equations $E_n[g(\theta, X)] = 0$, where g satisfies the conditions in Theorem 9.3. This property is a form of asymptotic efficiency of the maximum likelihood estimate.

Similarly, in the situations where $\mathcal{G}$ does not include the true score function, if g^* is the optimal estimating equation in $\mathcal{G}$, and if $\hat{\theta}_{g^*}$ is the solution to the estimating equation $E_n[g^*(\theta, X)] = 0$, then $\sqrt{n}(\hat{\theta}_{g^*} - \theta_0)$ converges in distribution to $N(0, I_{g^*}^{-1}(\theta_0))$. Since $I_{g^*} \succeq I_g$ for any $g \in \mathcal{G}$, $\sqrt{n}(\hat{\theta}_{g^*} - \theta_0)$ has the the smallest asymptotic variance among the solutions to the estimating equations in $\mathcal{G}$.

9.6 One-step Newton-Raphson estimate

The maximum likelihood estimate is not the unique estimate that achieves asymptotic efficiency. In fact, if we are given a $\sqrt{n}$-consistent estimator, then it is quite easy to construct an asymptotically efficient estimator. We first give a formal definition of $\sqrt{n}$-consistency.

Definition 9.5 *We say that $\tilde{\theta}$ is a $\sqrt{n}$-consistent estimate of θ_0 if $\tilde{\theta} - \theta_0 = O_P(n^{-1/2})$.*

Before introducing the one-step Newton-Raphson estimate, we now briefly review the Newton-Raphson algorithm. Consider a generic equation

$$F(\theta) = 0,$$

where $F : \Theta \to \mathbb{R}^p$ is a differentiable function. Suppose we have an initial value $\theta^{(0)}$. Then, near $\theta^{(0)}$, we can approximate $F(\theta)$ by its first-order Taylor polynomial $\hat{F}(\theta) = \theta^{(0)} + [\partial F(\theta^{(0)})/\partial \theta^T](\theta - \theta^{(0)})$. Instead of solving $F(\theta) = 0$, we solve the linear equation $\hat{F}(\theta) = 0$, which gives

$$\theta = \theta^{(0)} - [\partial F(\theta^{(0)})/\partial \theta^T]^{-1} F(\theta^{(0)}).$$

Motivated by this approximation, the Newton-Raphson algorithm consists of the iterations

$$\theta^{(k+1)} = \theta^{(k)} - [\partial F(\theta^{(k)})/\partial \theta^T]^{-1} F(\theta^{(k)})$$

until convergence.

In our context, the equation to solve is $E_n[s(\theta, X)] = 0$. So the Newton-Raphson algorithm for computing the maximum likelihood estimate is

$$\theta^{(k+1)} = \theta^{(k)} - [E_n \partial s(\theta^{(k)}, X)/\partial\theta^T]^{-1} E_n s(\theta^{(k)}, X).$$

By the central limit theorem, $E_n[\partial s(\theta_0, X)/\partial\theta^T]$ is a $\sqrt{n}$-consistent estimate of $-I(\theta_0)$. So it is reasonable to replace $E_n[\partial s(\theta^{(k)}, X)/\partial\theta^T]$ by $-I(\theta^{(k)})$ in the above formula, resulting in

$$\theta^{(k+1)} = \theta^{(k)} + I^{-1}(\theta^{(k)}) E_n[s(\theta^{(k)}, X)].$$

This is called the Fisher scoring algorithm (Cox and Hinkley, 1974).

The one-step Newton-Raphson estimate is derived from this, except that the initial value is a $\sqrt{n}$-consistent estimate, and that we only need to apply the formula (9.12) once. That is, if $\tilde{\theta}$ is a $\sqrt{n}$-consistent estimate of θ_0, then the one-step Newton-Raphson estimate is

$$\hat{\theta} = \tilde{\theta} + I^{-1}(\tilde{\theta}) E_n[s(\tilde{\theta}, X)]. \tag{9.12}$$

The next theorem implies that even though the updated θ has not converged yet after one iteration, this estimate is fully efficient: it has the same asymptotic distribution as the maximum likelihood estimate.

In fact, we shall prove a more general result. Let g be an unbiased, P_θ-square-integrable estimating equation such that $g(\theta, x) f_\theta(x)$ satisfies $\text{DUI}^+(\theta, \mu)$. If we define $\tilde{\theta}_g$ as the one-step Newton-Raphson estimator

$$\tilde{\theta}_g = \tilde{\theta} - J_g^{-1}(\tilde{\theta}) E_n[g(\tilde{\theta}, X)], \tag{9.13}$$

then $\sqrt{n}(\tilde{\theta}_g - \theta_0)$ has the same asymptotic distribution as $\sqrt{n}(\hat{\theta}_g - \theta_0)$, where $\hat{\theta}_g$ is a consistent solution to $E_n[g(\theta, X)] = 0$. Consequently, if g^* is the optimal estimating equation in $\mathcal{G}$, then $\tilde{\theta}_{g^*}$ has the smallest asymptotic variance among the solutions to estimating equations in $\mathcal{G}$. In particular, for $g(\theta, X) = s(\theta, X)$, the estimate in (9.12) is an asymptotically efficient estimate.

Theorem 9.4 *Suppose*

1. *$\tilde{\theta}$ is a $\sqrt{n}$-consistent estimator of θ_0;*
2. *g is an unbiased, P_θ-square-integrable estimating equation such that $g(\theta, x) f_\theta(x)$ satisfies $DUI^+(\theta, \mu)$;*
3. *the sequence of random functions $\{E_n[\partial g(\theta, X)/\partial\theta^T] : n = 1, 2, \ldots\}$ is stochastic equicontinuous in a neighborhood of θ_0;*
4. *$J_g(\theta)$ is continuous in θ.*

If $\tilde{\theta}_g$ is defined as (9.13), then

$$\sqrt{n}(\tilde{\theta}_g - \theta_0) \xrightarrow{\mathcal{D}} N\left(0, I_g^{-1}(\theta_0)\right).$$

Proof. By Taylor's mean value theorem,

$$E_n[g(\tilde{\theta}, X)] = E_n[g(\theta_0, X)] + \frac{\partial E_n[g(\xi, X)]}{\partial \theta^T}(\tilde{\theta} - \theta_0), \tag{9.14}$$

for some ξ. By the stochastic equicontinuity condition and Corollary 8.2,

$$\frac{\partial E_n[g(\xi, X)]}{\partial \theta^T} = J_g(\theta_0) + o_P(1). \tag{9.15}$$

Substituting (9.15) into the right-hand side of (9.14), we have

$$E_n[g(\tilde{\theta}, X)] = E_n[g(\theta_0, X)] + J_g(\theta_0)(\tilde{\theta} - \theta_0) + o_P(1)(\tilde{\theta} - \theta_0).$$

Because $\tilde{\theta}$ is $\sqrt{n}$-consistent, the term $o_P(1)(\tilde{\theta} - \theta_0)$ is of the order $o_P(n^{-1/2})$, resulting in

$$E_n[g(\tilde{\theta}, X)] = E_n[g(\theta_0, X)] + J_g(\theta_0)(\tilde{\theta} - \theta_0) + o_P(n^{-1/2}). \tag{9.16}$$

Moreover, since $J_g(\theta)$ is continuous, by the Continuous Mapping Theorem (see Theorem 7.1), $J_g(\tilde{\theta}) = J_g(\theta_0) + o_P(1)$. Hence, by the definition of $\tilde{\theta}_g$,

$$\tilde{\theta}_g = \tilde{\theta} + [J_g^{-1}(\theta_0) + o_P(1)]E_n[g(\tilde{\theta}, X)].$$

Now substituting (9.16) into the right-hand side of the above equation, we have

$$\tilde{\theta}_g = \tilde{\theta} - [J_g^{-1}(\theta_0) + o_P(1)]\left\{E_n[g(\theta_0, X)] + J_g(\theta_0)(\tilde{\theta} - \theta_0) + o_P(n^{-1/2})\right\}.$$

Since $E[g(\theta_0, X)] = 0$, by the Central Limit Theorem, $\sqrt{n}E_n[g(\theta_0, X)]$ converges in distribution to a Normal random vector. Hence $E_n[g(\theta_0, X)]$ is of the order $O_P(n^{-1/2})$. Because $\tilde{\theta}$ is $\sqrt{n}$-consistent, the term $J_g(\theta_0)(\tilde{\theta} - \theta_0)$ is also of the order $O_P(n^{-1/2})$. Therefore,

$$o_P(1)\left\{E_n[g(\theta_0, X)] + J_g(\theta_0)(\tilde{\theta} - \theta_0) + o_P(n^{-1/2})\right\} = o_P(n^{-1/2}).$$

Consequently,

$$\begin{aligned}\tilde{\theta}_g &= \tilde{\theta} - J_g^{-1}(\theta_0)\left\{E_n[g(\theta_0, X)] + J_g(\theta_0)(\tilde{\theta} - \theta_0)\right\} + o_P(n^{-1/2}) \\ &= \theta_0 - J_g^{-1}(\theta_0)E_n[g(\theta_0, X)] + o_P(n^{-1/2}).\end{aligned}$$

Hence

$$\sqrt{n}(\tilde{\theta}_g - \theta_0) = -J_g^{-1}(\theta_0)\sqrt{n}E_n[g(\theta_0, X)] + o_P(1).$$

By the Central Limit Theorem, the leading term on the right-hand side converges in distribution to

$$N\left(0, J_g^{-1}(\theta_0)K_g(\theta_0)J_g^{-T}(\theta_0)\right) = N(0, I_g^{-1}(\theta_0)).$$

The desired result now follows from Slutsky's theorem. □

9.7 Asymptotic linear form

In Section 9.5 we have shown that, if $\hat{\theta}_g$ is consistent, then $\sqrt{n}(\hat{\theta}_g - \theta_0)$ converges in distribution to a Normal random vector, which, in turn, implies that $\hat{\theta}_g$ is $\sqrt{n}$-consistent. In this section we develop a stronger asymptotic property for $\hat{\theta}_g$ than the asymptotic normality.

Theorem 9.5 *If the conditions in Theorem 9.3 hold, then*

$$\hat{\theta}_g = \theta_0 - J_g^{-1}(\theta_0)\, E_n[g(\theta_0, X)] + o_P(n^{-1/2}). \tag{9.17}$$

Note that, by straightforward applications of the Central Limit Theorem and Slutsky's theorem, we can deduce from (9.17) the asymptotic Normality

$$\sqrt{n}(\hat{\theta}_g - \theta_0) \xrightarrow{\mathcal{D}} N(0, I_g^{-1}(\theta_0)).$$

Proof of Theorem 9.5. Since, by Theorem 9.3, $\hat{\theta}_g$ is $\sqrt{n}$-consistent, and by the proof of Theorem 9.4, the relation (9.16) holds for $\hat{\theta}_g$. In this case, $E_n[g(\hat{\theta}_g, X)] = 0$ because $\hat{\theta}_g$ is a solution to the equation $E_n[g(\theta, X)] = 0$. Hence

$$0 = E_n[g(\theta_0, X)] + J_g(\theta_0)(\hat{\theta}_g - \theta_0) + o_P(n^{-1/2}),$$

which implies the desired result. □

An estimate that satisfies a relation such as (9.17) is said to be asymptotically linear. See, for example, Bickel, Klaassen, Ritov, and Wellner (1993). More generally, we have the following definition.

Definition 9.6 *Suppose $X_1, \ldots, X_n$ are independent copies of a random vector X, whose P_{θ_0} belongs to a parametric family $\{P_\theta : \theta \in \Theta \subseteq \mathbb{R}^p\}$. An estimate $\hat{\theta}$ of θ_0 is asymptotically linear if*

$$\hat{\theta} = \theta_0 + E_n[\psi(\theta_0, X)] + o_P(n^{-1/2}), \tag{9.18}$$

where

1. *$E[\psi(\theta_0, X)] = 0$;*
2. *the covariance matrix $var[\psi(\theta_0, X)]$ has finite elements.*

Note that, because $\psi(\theta_0, X)$ is a mean 0 function, the term $E_n[\psi(\theta_0, X)]$ in (9.18) is of the order $O_P(n^{-1/2})$ by the Central Limit Theorem. This condition is satisfied by a large number of statistics. The term "asymptotic linear" refers to the fact that, after ignoring the term $o_P(n^{-1/2})$, the leading term is a linear functional of the empirical distribution F_n. A typical estimate is asymptotically linear; whereas a typical test statistic is asymptotically quadratic. The function ψ in (9.18) is called the influence function. So, for example, the influence function of $\hat{\theta}_g$ is $-J_g^{-1}(\theta_0)g(\theta_0, X)$. By the Central Limit Theorem, an asymptotically linear estimate $\hat{\theta}$ has the following asymptotic Normal distribution:

$$\sqrt{n}(\hat{\theta} - \theta_0) \xrightarrow{\mathcal{D}} N\left(0, \text{var}[\psi(\theta_0, X)]\right).$$

Two important special cases are when g is the score function s or the optimal estimating equation $g^* \in \mathcal{G}$. In both cases,

$$I_{g^*}(\theta) = K_{g^*}(\theta) = -J_{g^*}(\theta), \quad I(\theta) = K(\theta) = -J(\theta).$$

Thus, the maximum likelihood estimate $\hat{\theta}$ and a consistent solution $\hat{\theta}_{g^*}$ of an optimal estimating equation g^* have the following asymptotic linear forms:

$$\begin{aligned}\hat{\theta} &= \theta_0 + I^{-1}(\theta_0)E_n[s(\theta_0, X)] + o_P(n^{-1/2}),\\ \hat{\theta}_{g^*} &= \theta_0 + I_{g^*}^{-1}(\theta_0)E_n[g^*(\theta_0, X)] + o_P(n^{-1/2}).\end{aligned}$$

The asymptotic linear form provides more information about an estimate than the asymptotic distribution. For example, if we know the asymptotic linear forms of two estimates are

$$\begin{aligned}\hat{\theta} &= \theta_0 + E_n\psi_1(\theta_0, X) + o_P(n^{-1/2}),\\ \tilde{\theta} &= \theta_0 + E_n\psi_2(\theta_0, X) + o_P(n^{-1/2}),\end{aligned}$$

then we know the joint asymptotic distribution of $[\sqrt{n}(\hat{\theta} - \theta_0), \sqrt{n}(\tilde{\theta} - \theta_0)]$ is

$$N\left(0, \begin{pmatrix} E[\psi_1(\theta_0, X)\psi_1^T(\theta_0, X)] & E[\psi_1(\theta_0, X)\psi_2^T(\theta_0, X)] \\ E[\psi_2(\theta_0, X)\psi_1^T(\theta_0, X)] & E[\psi_2(\theta_0, X)\psi_2^T(\theta_0, X)] \end{pmatrix}\right).$$

However, if we only know the asymptotic distributions of the random vectors $\sqrt{n}(\hat{\theta} - \theta_0)$ and $\sqrt{n}(\tilde{\theta} - \theta_0)$, then we cannot deduce the joint asymptotic distribution of the two random vectors. Fortunately, in most cases where we know a statistic is asymptotically Normal, its asymptotic linear form is readily available.

9.8 Efficient score for parameter of interest

In this section we take a closer look at the efficient score described at the end of Section 9.4. We will consider the more general case where θ, ψ, and λ are vectors. That is, θ is a p-dimensional parameters consisting of an r-dimensional subvector ψ, which is the parameter of interest, and an s-dimensional subvector λ, which is the nuisance parameter.

For the purpose of estimating ψ, we need to extend the notion of the information about the whole parameter θ contained in an estimating equation $g(\theta, x)$, as defined in Definition 9.3, to that of the information about a part (ψ) of the parameter θ contained in the estimating equation $g(\theta, x)$.

Definition 9.7 *Suppose $g : \Theta \times \Omega_X \to \mathbb{R}^r$ is an unbiased, P_θ-square-integrable estimating equation such that $g(\theta, x) f_\theta(x)$ satisfies DUI$^+$ (ψ, μ). Then the matrix*

$$I_g(\psi|\theta) = E_\theta \left(\frac{\partial g^T(\theta, X)}{\partial \psi} \right) \left\{ E_\theta \left[g(\theta, X) g^T(\theta, X) \right] \right\}^+ E_\theta \left(\frac{\partial g(\theta, X)}{\partial \psi^T} \right)$$

is called the information about ψ contained in $g(\theta, X)$.

This is a generalization of the information matrix $I_g(\theta)$ in Definition 9.3 because, when ψ is the entire parameter θ, $I_g(\theta|\theta) = I_g(\theta)$. The optimal estimating equation for ψ can be found in the same way as that for θ. Recall that s_ψ stands for the ψ-component of $s(\theta, X)$. That is, $s_\psi(\theta, x) = \partial \log f_\theta(x)/\partial \psi$. The proof of the next theorem is similar to that of Theorem 9.1 and is left as an exercise.

Theorem 9.6 *Suppose $\mathcal{G}$ is a class of unbiased, P_θ-square-integrable estimating equations of dimension r such that, for each $g \in \mathcal{G}$, $g(\theta, x) f_\theta(x)$ satisfies DUI$^+$ (θ, μ). If there is a member g^* of $\mathcal{G}$ such that $[s_\psi - g^*, g] = 0$ for all $g \in \mathcal{G}$, then $I_{g^*}(\psi|\theta) \succeq I_g(\psi|\theta)$ for all $g \in \mathcal{G}$.*

When estimating ψ, it is natural to consider a class of estimating equations that are, in some sense, insensitive to the nuisance parameter λ. We now give a formal definition for such estimating equations.

Definition 9.8 *An unbiased and P_θ-square-integrable estimating equation $g : \Theta \times \Omega_X \to \mathbb{R}^r$ is said to be insensitive to λ to the first order if $g(\theta, x) f_\theta(x)$ satisfies DUI$^+$ (θ, μ), and*

$$E\left(\frac{\partial g(\theta, X)}{\partial \lambda^T} \right) = 0. \tag{9.19}$$

Let $\mathcal{G}_{\psi \cdot \lambda}$ denote the class of such estimating equations.

The next proposition gives a characterization of the space $\mathcal{G}_{\psi \cdot \lambda}$.

Proposition 9.2 *Let $g : \Theta \times \Omega_X \to \mathbb{R}^r$ be an unbiased, P_θ-square-integrable estimating equation such that $g(\theta, x) f_\theta(x)$ satisfies DUI$^+$$(\theta, \mu)$. Then g belongs to $\mathcal{G}_{\psi \cdot \lambda}$ if and only if $[g, s_\lambda] = 0$.*

Proof. By the DUI$^+$ assumption,

$$\int (\partial g/\partial \lambda^T) f_\theta d\mu + \int g s_\lambda f_\theta d\mu = 0.$$

Hence $[g, s_\lambda] = 0$ if and only if $g \in \mathcal{G}_{\psi \cdot \lambda}$. □

By Problem 9.14 we see that the optimal estimating equation in $\mathcal{G}_{\psi \cdot \lambda}$ that maximizes $I_g(\psi|\theta)$ is

$$s_\psi - [s_\psi, s_\lambda][s_\lambda, s_\lambda]^{-1} s_\lambda.$$

Note that the Fisher information for θ can be written as the matrix

$$I(\theta) = \begin{pmatrix} [s_\psi, s_\psi] & [s_\psi, s_\lambda] \\ [s_\lambda, s_\psi] & [s_\lambda, s_\lambda] \end{pmatrix} \equiv \begin{pmatrix} I_{\psi\psi} & I_{\psi\lambda} \\ I_{\lambda\psi} & I_{\lambda\lambda} \end{pmatrix}.$$

Thus the optimal estimating equation in $\mathcal{G}_{\psi\cdot\lambda}$ can be written as

$$g^* = s_\psi - I_{\psi\lambda} I_{\lambda\lambda}^{-1} s_\lambda. \tag{9.20}$$

Note that the information for ψ contained in $g^*(\theta, X)$ is

$$I_{g^*}(\psi|\theta) = I_{\psi\psi} - I_{\psi\lambda} I_{\lambda\lambda}^{-1} I_{\lambda\psi}. \tag{9.21}$$

Because of the special importance of the optimal estimating equation g^* in (9.20) and the information $I_{g^*}(\psi|\theta)$ in statistical inference, we give them special names and notations.

Definition 9.9 *The optimal estimating equation $g^*(\theta, X)$ in (9.20) is called the efficient score for ψ, and is written as $s_{\psi\cdot\lambda}(\theta, X)$; the information $I_{g^*}(\psi|\theta)$ in (9.21) is called the efficient information for ψ, and is written as $I_{\psi\cdot\lambda}(\theta)$.*

Before proceeding further, let us review some formulas about inversion of a block matrix. The next proposition can be proved by straightforward matrix multiplication.

Proposition 9.3 (Inversion of a block matrix) *Let B be a p by p nonsingular and symmetric matrix, which is partitioned into*

$$B = \begin{pmatrix} B_{11} & B_{12} \\ B_{21} & B_{22} \end{pmatrix},$$

where B_{11} is an $r \times r$ matrix, $B_{12} = B_{21}^T$ is an $r \times s$ matrix, and B_{22} is an $s \times s$ matrix, with $r + s = p$. Let the inverse B^{-1} of B be partitioned, in accordance with the above dimensions, as

$$B^{-1} = \begin{pmatrix} (B^{-1})_{11} & (B^{-1})_{12} \\ (B^{-1})_{21} & (B^{-1})_{22} \end{pmatrix}.$$

Then

$$\begin{aligned}
(B^{-1})_{11} &= (B_{11} - B_{12} B_{22}^{-1} B_{21})^{-1} \\
(B^{-1})_{22} &= (B_{22} - B_{21} B_{11}^{-1} B_{12})^{-1} \\
(B^{-1})_{12} &= -(B^{-1})_{11} B_{12} B_{22}^{-1} = -B_{11}^{-1} B_{12} (B^{-1})_{22} = (B^{-1})_{21}^T.
\end{aligned}$$

Partition the inverse Fisher information $I^{-1}(\theta)$ into block matrix

$$I^{-1} = \begin{pmatrix} (I^{-1})_{\psi\psi} & (I^{-1})_{\psi\lambda} \\ (I^{-1})_{\lambda\psi} & (I^{-1})_{\lambda\lambda} \end{pmatrix}.$$

By Proposition 9.3, we have

$$[(I^{-1})_{\psi\psi}]^{-1} = I_{\psi\psi} - I_{\psi\lambda} I_{\lambda\lambda}^{-1} I_{\lambda\psi} = I_{\psi\cdot\lambda}.$$

Thus, the efficient information is nothing but the inverse of the (ψ, ψ)-block of the inversed Fisher information matrix. Because the matrix $I_{\psi\lambda} I_{\lambda\lambda}^{-1} I_{\lambda\psi}$ is positive semi-definite, the above equality also implies that

$$I_{\psi\psi}(\theta) \succeq I_{\psi\cdot\lambda}(\theta) \tag{9.22}$$

for all $\theta \in \Theta$ in terms of Louwner's ordering. The interpretation of this inequality is the following. The matrix $I_{\psi\psi}$ is the information contained in $s(\psi, \lambda, X)$ when λ is treated as known; the matrix $I_{\psi\cdot\lambda}$ is the information in $s(\theta, X)$ about ψ, but λ is not treated as known. So $I_{\psi\psi}$ must be larger than $I_{\psi\cdot\lambda}$ because the former assumes more information. Also note that the equality in (9.22) holds if and only if $I_{\psi\lambda} = 0$, which means s_ψ and s_λ are orthogonal in the $L_2(P_\theta)$ geometry. Thus, under the orthogonality of s_ψ and s_λ, estimation accuracy of ψ is not increased by knowing λ, or decreased by not knowing λ.

We now use the efficient score and efficient information to express asymptotic linear form of the ψ-component of the maximum likelihood estimate.

Theorem 9.7 *Suppose the conditions in Theorem 9.7 are satisfied with $g(\theta, X)$ being the score function $s(\theta, X)$. Denote by $\hat{\psi}$ and $\hat{\lambda}$ the ψ-component and the λ-component of the MLE $\hat{\theta}$. Then $\hat{\psi}$ has the following asymptotic linear form*

$$\hat{\psi} = \psi_0 + I_{\psi\cdot\lambda}^{-1} E_n[s_{\psi\cdot\lambda}(\theta_0, X)] + o_P(n^{-1/2}).$$

Proof. Rewrite the expansion (9.17) in terms of $\hat{\psi}$ and $\hat{\lambda}$ to obtain

$$\begin{pmatrix} \hat{\psi} - \psi_0 \\ \hat{\lambda} - \lambda_0 \end{pmatrix} = \begin{pmatrix} (I^{-1})_{\psi\psi} & (I^{-1})_{\psi\lambda} \\ (I^{-1})_{\lambda\psi} & (I^{-1})_{\lambda\lambda} \end{pmatrix} \begin{pmatrix} E_n s_\psi(\theta_0, X) \\ E_n s_\lambda(\theta_0, X) \end{pmatrix} + o_P(n^{-1/2}).$$

From this we can read off the expansion of $\hat{\psi} - \psi_0$, as follows

$$\begin{aligned} \hat{\psi} - \psi_0 &= (I^{-1})_{\psi\psi} E_n s_\psi(\theta_0, X) + (I^{-1})_{\psi\lambda} E_n s_\lambda(\theta_0, X) + o_P(n^{-1/2}) \\ &= (I^{-1})_{\psi\psi} \left[E_n s_\psi(\theta_0, X) + (I^{-1})_{\psi\psi}^{-1} (I^{-1})_{\psi\lambda} E_n s_\lambda(\theta_0, X) \right] + o_P(n^{-1/2}). \end{aligned}$$

From Proposition 9.3 we see that

$$\begin{aligned} (I^{-1})_{\psi\psi} &= I_{\psi\cdot\lambda}^{-1} \\ (I^{-1})_{\psi\psi}^{-1} (I^{-1})_{\psi\lambda} &= (I^{-1})_{\psi\psi}^{-1} \left[-(I^{-1})_{\psi\psi} I_{\psi\lambda} I_{\lambda\lambda}^{-1} \right] = -I_{\psi\lambda} I_{\lambda\lambda}. \end{aligned}$$

It follows that

$$\begin{aligned}\hat{\psi} - \psi_0 =& I_{\psi\cdot\lambda}^{-1} E_n \left[s_\psi(\theta_0, X) - I_{\psi\lambda} I_{\lambda\lambda}^{-1} s_\lambda(\theta_0, X) \right] + o_P(n^{-1/2}) \\ =& I_{\psi\cdot\lambda}^{-1} E_n s_{\psi\cdot\lambda}(\theta_0, X) + o_P(n^{-1/2}),\end{aligned}$$

as desired. □

From the above expansion we can easily write down the asymptotic distribution of the MLE for the parameter of interest.

Corollary 9.2 *Under the assumptions in Theorem 9.7, we have*

$$\sqrt{n}(\hat{\psi} - \psi_0) \xrightarrow{\mathcal{D}} N(0, I_{\psi\cdot\lambda}^{-1}(\theta_0)).$$

Recall that $s_{\psi\cdot\lambda}$ is the optimal estimating equation in $\mathcal{G}_{\psi\cdot\lambda}$. Thus $I_{\psi\cdot\lambda}(\theta)$ is the upper bound of the information $I_g(\psi|\theta)$ for any estimating equation in $\mathcal{G}_{\psi\cdot\lambda}$. Meanwhile, if we pretend λ_0 to be known, then Theorem 9.3 implies that any consistent solution $\hat{\psi}_g$ to the estimating equation

$$E_n[g(\psi, \lambda_0, X)] = 0$$

has asymptotic distribution $\sqrt{n}(\hat{\psi}_g - \psi_0) \xrightarrow{\mathcal{D}} N(0, I_g^{-1}(\psi_0|\theta_0))$. Thus, intuitively, we can say that $\hat{\psi}$ has the smallest asymptotic variance among the solutions to all estimating equations in $\mathcal{G}_{\psi\cdot\lambda}$. Of course, this statement is not rigorous as we pretend λ_0 to be known. This statement will be made rigorous by Theorem 9.9.

The efficient score and efficient information share some similarities with the score $s(\theta, X)$ and the information $I(\theta)$ when there is no nuisance parameter. For example, the information identity has its analogy for the efficient score.

Theorem 9.8 *Suppose $s_{\psi\cdot\lambda}(\theta, X)$ is P_θ-square-integrable, and $f_\theta(x)$ and $s_{\psi\cdot\lambda}(\theta, x) f_\theta(x)$ satisfy DUI$^+$ (θ, μ). Then*

$$E_\theta[s_{\psi\cdot\lambda}(\theta, X)] = 0, \tag{9.23}$$

$$E_\theta\left[\frac{\partial s_{\psi\cdot\lambda}(\theta, X)}{\partial \lambda^T}\right] = 0, \tag{9.24}$$

$$E_\theta\left[\frac{\partial s_{\psi\cdot\lambda}(\theta, X)}{\partial \psi^T}\right] = -E\left[s_{\psi\cdot\lambda}(\theta, X) s_{\psi\cdot\lambda}^T(\theta, X)\right] = -I_{\psi\cdot\lambda}(\theta). \tag{9.25}$$

Proof. The equality (9.23) follows from $E_\theta[s(\theta, X)] = 0$, as can be verified by differentiating the equation $\int f_\theta(x) d\mu(x) = 1$ with respect to θ.

To establish (9.24), first observe that by the definition of $s_{\psi\cdot\lambda}(\theta, X)$, we have

$$E_\theta\left[\frac{\partial s_{\psi\cdot\lambda}(\theta, X)}{\partial \lambda^T}\right] = E_\theta\left[\frac{\partial s_\psi(\theta, X)}{\partial \lambda^T}\right] - E_\theta\left[\frac{\partial (I_{\psi\lambda} I_{\lambda\lambda}^{-1} s_\lambda(\theta, X))}{\partial \lambda^T}\right]. \tag{9.26}$$

The second term on the right is

$$-E_\theta\left[\frac{\partial(I_{\psi\lambda}I_{\lambda\lambda}^{-1})}{\partial\lambda^T}s_\lambda(\theta,X)+I_{\psi\lambda}I_{\lambda\lambda}^{-1}\frac{\partial s_\lambda(\theta,X)}{\partial\lambda^T}\right],$$

where the notation $[\partial(I_{\psi\lambda}I_{\lambda\lambda}^{-1})/\partial\lambda^T]s_\lambda(\theta,X)$ simply means the matrix

$$\sum_{i=1}^{s}\frac{\partial(I_{\psi\lambda}I_{\lambda\lambda}^{-1})}{\partial\lambda_i}s_{\lambda_i}(\theta,X),$$

$s_{\lambda_i}(\theta,X)$ being the score function for λ_i, $\partial\log f_\theta(X)/\partial\lambda_i$. Since the above term has expectation 0,

$$E_\theta\left[\frac{\partial(I_{\psi\lambda}I_{\lambda\lambda}^{-1}s_\lambda(\theta,X))}{\partial\lambda^T}\right]=E_\theta\left[I_{\psi\lambda}I_{\lambda\lambda}^{-1}\frac{\partial s_\lambda(\theta,X)}{\partial\lambda^T}\right]=-I_{\psi\lambda}I_{\lambda\lambda}^{-1}I_{\lambda\lambda}=-I_{\psi\lambda}.$$

Hence the right-hand side of (9.26) reduces to $-I_{\psi\lambda}+I_{\psi\lambda}=0$, proving (9.24).

Finally, to establish (9.25), first differentiate the equation

$$\int s_{\psi\cdot\lambda}(\theta,x)f_\theta(x)d\mu(x)=0, \tag{9.27}$$

with respect to λ to obtain

$$E_\theta\left[\frac{\partial s_{\psi\cdot\lambda}(\theta,X)}{\partial\lambda^T}\right]=-\,E\left[s_{\psi\cdot\lambda}(\theta,X)s_\lambda^T(\theta,X)\right].$$

By (9.24), the left-hand side is 0, and hence $E\left[s_{\psi\cdot\lambda}(\theta,X)s_\lambda^T(\theta,X)\right]=0$, leading to

$$E\left[s_{\psi\cdot\lambda}(\theta,X)s_\psi^T(\theta,X)\right]=E\left[s_{\psi\cdot\lambda}(\theta,X)s_{\psi\cdot\lambda}^T(\theta,X)\right].$$

Next, differentiate the equation (9.27) with respect to ψ to obtain

$$E_\theta\left[\frac{\partial s_{\psi\cdot\lambda}(\theta,X)}{\partial\psi^T}\right]=-E\left[s_{\psi\cdot\lambda}(\theta,X)s_\psi^T(\theta,X)\right]=-E\left[s_{\psi\cdot\lambda}(\theta,X)s_{\psi\cdot\lambda}^T(\theta,X)\right],$$

proving (9.25). □

One way to use estimating equations in $\mathcal{G}_{\psi\cdot\lambda}$, such as the efficient score $s_{\psi\cdot\lambda}(\theta,X)$, is to estimate ψ by solving the equation

$$E_n[g(\psi,\tilde{\lambda},X)]=0, \tag{9.28}$$

where $\tilde{\lambda}$ is some estimate of the nuisance parameter λ_0. Since an estimating equation in $\mathcal{G}_{\psi\cdot\lambda}$ is insensitive to the nuisance parameter λ_0, intuitively, it should be able to tolerate a relatively poor estimate of λ_0 while still producing an accurate estimate of ψ. Indeed, the next theorem shows that even if $\tilde{\lambda}-\lambda_0=o_P(n^{-1/4})$, which can be much slower than the parametric rate $O_P(n^{-1/2})$, the solution to (9.28) produces an estimate for ψ that is asymptotically equivalent to the solution to $E_n[g(\psi, \lambda_0, X)] = 0$, where λ_0 is treated as known.

We need to introduce some additional notations. Suppose $h(\lambda)$ is a function from $\mathbb{R}^s$ to $\mathbb{R}^r$ with components $h_1(\lambda), \ldots, h_r(\lambda)$. We use $\partial^2 h(\lambda)/\partial\lambda\partial\lambda^T$ denote the $r \times s \times s$ array

$$\left\{ \frac{\partial^2 h_i(\lambda)}{\partial\lambda_j \partial\lambda_k} : i = 1, \ldots, r, j, k = 1, \ldots, s \right\}.$$

Furthermore, if $a, b \in \mathbb{R}^s$, then the notation $a^T \left[\partial^2 h(\lambda)/\partial\lambda\partial\lambda^T\right] b$ represents the r-dimensional vector whose ith component is

$$\sum_{j=1}^{s} \sum_{k=1}^{s} a_j b_k \frac{\partial^2 h_i(\lambda)}{\partial\lambda_j \partial\lambda_k}.$$

For an estimating equation $g \in \mathcal{G}_{\psi\cdot\lambda}$, let

$$J_g(\psi|\theta) = E_\theta\left[\frac{\partial g(\theta, X)}{\partial\psi^T}\right], \quad K_g(\psi|\theta) = E_\theta\left[g(\theta, X) g^T(\theta, X)\right],$$

so that we have

$$I_g(\psi|\theta) = J_g(\psi|\theta)^T K_g^{-1}(\psi|\theta) J_g(\psi|\theta).$$

As before, we use $\mathbb{N}$ to denote the set of natural numbers $\{1, 2, \ldots\}$. To our knowledge, the following result has not been recorded in the statistical literature.

Theorem 9.9 *Suppose that g is an estimating equation in $\mathcal{G}_{\psi\cdot\lambda}$ satisfying the following additional conditions:*

1. *$g(\theta, X)$ is twice differentiable with respect to λ and the entries of the $r \times s \times s$ array;*
2. *in a neighborhood of θ_0, the sequences of random elements*

$$\left\{ E_n\left[\frac{\partial g(\psi, \lambda, X)}{\partial\psi^T}\right] : n \in \mathbb{N} \right\}, \quad \left\{ E_n\left[\frac{\partial^2 g(\psi, \lambda, X)}{\partial\lambda\partial\lambda^T}\right] : n \in \mathbb{N} \right\}$$

   are stochastically equicontinuous;
3. *the matrices $J_g(\psi|\theta)$ and $K_g(\psi|\theta)$ are nonsingular.*

If $\tilde{\lambda}$ is an estimate of λ_0 such that $\tilde{\lambda} - \lambda_0 = o_P(n^{-1/4})$, and $\hat{\psi}$ is a consistent solution to the estimating equation $E_n[g(\psi, \tilde{\lambda}, X)] = 0$, then

$$\sqrt{n}(\hat{\psi} - \psi_0) \xrightarrow{\mathcal{D}} N\left(0, I_g^{-1}(\psi_0|\theta_0)\right).$$

Proof. Applying Taylor's mean value theorem to the function $\psi \mapsto E_n[g(\psi, \tilde{\lambda}, X)]$, we have

$$0 = E_n[g(\hat{\psi}, \tilde{\lambda}, X)] = E_n[g(\psi_0, \tilde{\lambda}, X)] + E_n\left[\frac{g(\xi, \tilde{\lambda}, X)}{\partial \psi^T}\right](\hat{\psi} - \psi_0)$$

for some ξ between ψ_0 and $\hat{\psi}$. Because $(\xi, \tilde{\lambda})$ converges in probability to (ψ_0, λ_0), by the first equicontinuity assumption in *2* and Corollary 8.2, we see that $E_n[g(\xi, \tilde{\lambda}, X)/\partial \psi^T]$ differs from $J(\psi_0|\theta_0)$ by $o_P(1)$. Hence

$$0 = E_n[g(\psi_0, \tilde{\lambda}, X)] + [J_g(\psi_0|\theta_0) + o_P(1)](\hat{\psi} - \psi_0). \tag{9.29}$$

Next, applying (the second-order) Taylor's mean-value theorem to the function $\lambda \mapsto E_n[g(\psi_0, \lambda, X)]$, we have

$$\begin{aligned} E_n[g(\psi_0, \tilde{\lambda}, X)] = & E_n[g(\psi_0, \lambda_0, X)] + E_n\left[\frac{\partial g(\psi_0, \lambda_0, X)}{\partial \lambda^T}\right](\tilde{\lambda} - \lambda_0) \\ & + \frac{1}{2}(\tilde{\lambda} - \lambda_0)^T E_n\left[\frac{\partial^2 g(\psi_0, \lambda_1, X)}{\partial \lambda \partial \lambda^T}\right](\tilde{\lambda} - \lambda_0), \end{aligned} \tag{9.30}$$

for some λ_1 on the line joining $\tilde{\lambda}$ and λ_0. By the second equicontinuity assumption in *2* and Corollary 8.2,

$$E_n\left[\frac{\partial^2 g(\psi_0, \lambda^\dagger, X)}{\partial \lambda \partial \lambda^T}\right] \xrightarrow{P} E\left[\frac{\partial^2 g(\psi_0, \lambda_0, X)}{\partial \lambda \partial \lambda^T}\right].$$

Therefore, the term on the left-hand side above is $O_P(1)$, and hence the third term on the right-hand side of (9.30) is of the order $o_P(n^{-1/2})$. Moreover, because $g \in \mathcal{G}_{\psi \cdot \lambda}$, we have $E[\partial g(\theta_0, X)/\partial \lambda^T] = 0$. Hence, by the central limit theorem,

$$E_n\left[\frac{\partial g(\psi_0, \lambda_0, X)}{\partial \lambda^T}\right] = O_P(n^{-1/2}),$$

which implies that the second term in (9.30) is of the order $o_P(n^{-3/4})$. So the following approximation holds:

$$E_n[g(\psi_0, \tilde{\lambda}, X)] = E_n[g(\psi_0, \lambda_0, X)] + o_P(n^{-1/2}).$$

Substituting this into the right-hand side of (9.29) results in

$$0 = E_n[g(\psi_0, \lambda_0, X)] + [J_g(\psi_0|\theta_0) + o_P(1)](\hat{\psi} - \psi_0) + o_P(n^{-1/2}).$$

Multiplying both sides of the above equation by the matrix $J_g^{-1}(\psi_0|\theta_0)$ from the left, we obtain

$$[I_r + o_P(1)](\hat{\psi} - \psi_0) = -J_g^{-1}(\psi_0|\theta_0)E_n[g(\psi_0, \lambda_0, X)] + o_P(n^{-1/2}).$$

Hence, by the central limit theorem and Slutsky's theorem,

$$\sqrt{n}[I_r + o_P(1)](\hat{\psi} - \psi_0) \xrightarrow{\mathcal{D}} N(0, I_g(\psi_0|\theta_0)),$$

which implies the asserted result by Problem 8.13. □

An important special case of Theorem 9.9 is when g is the efficient score, which says that for any $\tilde{\lambda}$ that converges to λ_0 at a rate faster than $n^{-1/4}$, a consistent solution to

$$E_n[s_{\psi\cdot\lambda}(\psi, \tilde{\lambda}, X)] = 0$$

has the asymptotic distribution $N(0, I_{\psi\cdot\lambda}(\theta_0)^{-1})$. This result also makes precise the optimal statement following Corollary 9.2. That is, for any estimating equation $g \in \mathcal{G}_{\psi\cdot\lambda}$, and any estimate $\tilde{\lambda}$ satisfying $\tilde{\lambda} = \lambda_0 + o_P(n^{-1/4})$, the asymptotic variance of the solution to

$$E_n[g(\psi, \tilde{\lambda}, X)] = 0$$

reaches its lower bound (in terms of Louwner's ordering) when g is the efficient score $s_{\psi\cdot\lambda}$.

Problems

9.1. Prove Theorem 9.2 by following the proof of Theorem 8.1.

9.2. Let $X_1, \cdots, X_n$ be i.i.d. with density $f(x; \mu) = x^{\mu-1}e^{-x}/\Gamma(\mu)$.

1. Find all the solutions to the estimating equation

$$\sum_{i=1}^{n}(X_i^2 - \mu - \mu^2) = 0, \tag{9.31}$$

   and decide which one is consistent. Derive the asymptotic distribution of this solution.
2. Find the maximum likelihood estimate of μ and derive its asymptotic distribution.
3. Let $\hat{\mu}^{(1)} = \bar{X}$, $\hat{\mu}^{(2)}$ be the consistent solution to the estimating equation (9.31), and $\hat{\mu}^{(3)}$ be the maximum likelihood estimate of μ. Let $V_1(\mu)$, $V_2(\mu)$, and $V_3(\mu)$ be the asymptotic variances of the above three estimators of μ. Using a computer to plot them against μ. What is your conclusion?

9.3. Suppose that $(X_1, Y_1), \ldots, (X_n, Y_n)$ are an i.i.d. sample from (X, Y), where X is a random vector in $\mathbb{R}^p$, and Y is a random variable. Suppose that conditional distribution of $Y|X$ is given by $N(e^{\beta^T x}, e^{\beta^T x})$ for some $\beta \in \mathbb{R}^p$, and that the marginal distribution of X does not depend on β.

1. Derive the score function $s(\beta, X, Y)$.

2. Derive the Fisher information matrix $I(\beta)$.
3. Derive the quasi score function $g^*(\beta, X, Y)$.
4. Derive the information contained in g^*, and show that this information is smaller (in terms of Louwner's ordering) than obtained in part 2.
5. Write down Fisher scoring iterative algorithms for estimating the maximum likelihood estimate and maximum quasi likelihood estimate.

9.4. In the setting of Problem 9.3, suppose we estimate β by the Least Squares method — that is, by minimizing

$$E_n(Y - e^{\beta^T X})^2$$

over $\beta \in \mathbb{R}^p$ to estimate β.

1. Derive the estimating equation for β.
2. Compute the information contained in this estimating equation, and show that it is smaller (in terms of Louwner's ordering) than that obtained in part 4 of Problem 9.3.

9.5. Suppose $\theta \in \Theta \subseteq \mathbb{R}^p$ is a p-dimensional parameter and $g_1, \ldots, g_m$ are p-dimensional unbiased, P_θ-square-integrable estimating equations such that $g_i(x) f_\theta(x)$, $i = 1, \ldots, m$, satisfy $\mathrm{DUI}^+(\theta, \mu)$. Consider the following class of unbiased estimating equations

$$\mathcal{G} = \{A_1(\theta) g_1(\theta, X) + \cdots + A_m(\theta) g_m(\theta, X) : A_1(\theta), \ldots, A_m(\theta) \in \mathbb{R}^{p \times p}\},$$

where $A_i(\theta)$ are $p \times p$ nonrandom matrices that may depend on θ. Derive the optimal equation g^* in $\mathcal{G}$.

9.6. Consider the generalized estimating equation described in Section 9.3 with the following simplifications: $n_i = m$ are the same for all $i = 1, \ldots, n$, and

$$\mu_i(X_i, \beta) = \mu(X_i^T \beta), \quad V_i(X_i, \beta) = V(X_i^T \beta), \quad R_i(\alpha) = R(\alpha).$$

Furthermore, assume that $(X_1, Y_1), \ldots, (X_n, Y_n)$ are an i.i.d. sample from (X, Y). These lead the following generalized estimating equation

$$E_n \left\{ [\partial \mu(\beta^T X)^T / \partial \beta][V^{1/2}(\beta^T X) R(\alpha) V^{1/2}(\beta^T X)]^{-1} [Y - \mu(\beta^T X)] \right\} = 0.$$

Denote the term in $E_n\{\cdots\}$ by $g(\beta, \alpha, X, Y)$. Assume that $\hat{\alpha}$ converges in probability to a fixed vector α_0 with the rate $o_P(n^{-1/4})$. Let $\hat{\beta}$ be a consistent solution to the equation

$$E_n[g(\beta, \hat{\alpha}, X, Y)] = 0.$$

Let β_0 be the true parameter.

1. Derive the asymptotic distribution of $\sqrt{n}(\hat{\beta} - \beta_0)$ when the form of $R(\alpha)$ might be misspecified.
2. Derive the asymptotic distribution of $\sqrt{n}(\hat{\beta} - \beta_0)$ when the form of $R(\alpha)$ is correctly specified.
3. Show that, when R is correctly specified, $g(\beta, \alpha, X, Y)$ is the optimal estimating equation among estimating equations of the form

$$A(X, \alpha, \beta)[Y - \mu(\beta^T X)],$$

where $A(X, \alpha, \beta)$ is an $m \times m$ random matrix that may depend on X, α, β but does not depend on Y.

9.7. Prove Theorem 9.6.

9.8. Let $X = (X_1, \ldots, X_n)$ be a sample with joint distribution P_θ, where θ is a two dimensional vector with a parameter of interest ψ and a nuisance parameter λ. Suppose P_θ has a density f_θ with respect to a σ-finite measure μ. Suppose there is a sufficient statistic $T = T(X_{1:n})$ for the nuisance parameter λ — that is, the conditional density $f_{X|T}(x|t; \psi)$ does not depend on λ. Let $\mathcal{G}_\psi$ be a class of estimating equations for ψ satisfying the following conditions:

(i). each $g(\psi, X)$ in $\mathcal{G}_\psi$ is a function of X and ψ;
(ii). $E_{\psi,\lambda}[g(\psi, X)] = 0$ for all values of λ;
(iii). $g(\psi, X)$ is P_θ-square-integrable, and $g(\psi, x) f_\theta(x)$ is differentiable with respect to ψ under the integral with respect to μ.

Let $s_{X|T}(\psi, X)$ be the conditional score function $\partial \log f_{X|T}(x|t; \psi)/\partial \psi$ and assume that it satisfies condition (iii). Note that this conditional score depends on both X and T, but since T is a function of X, we can write it as $s_{X|T}(\psi, X)$.

1. Show that the conditional score satisfies (ii).
2. Assuming $s_{X|T}(\psi; X)$ belongs to $\mathcal{G}_\psi$, show that it is the optimal estimating equation in that class in terms of the information $I_g(\psi|\theta)$.
3. Let $s_\psi(\theta; X)$ be the unconditional score for $(\partial/\partial\psi) \log f_\theta(X)$. Show that

$$s_{X|T}(\psi, X) = s_\psi(\theta; X) - E(s_\psi(\theta; X)|T).$$

(This problem is inspired by Godambe (1976)).

9.9. Suppose $X \sim \text{Poisson}(\lambda_1)$, $Y \sim \text{Poisson}(\lambda_2)$, and X, Y are independent. We are interested in estimating $\psi = \lambda_1/\lambda_2$, treating λ_2 as the nuisance parameter. For simplicity, we denote λ_2 by λ. Let $T = X + Y$.

1. Show that, for each fixed ψ, T is sufficient for λ.
2. Derive the conditional score function $s_{X|T}(\psi, X)$.
3. Derive the information about ψ contained in the conditional score. Express it in terms of both the original parameter (λ_1, λ_2) and the transformed parameter (ψ, λ).

9.10. In the setting of Problem 9.9, show that the efficient score $s_{\psi\cdot\lambda}$ has the same form as the conditional score in Problem 9.9.

9.11. Suppose X and Y are i.i.d. $N(\lambda,\psi)$. We are interested in estimating ψ in the presence of the nuisance parameter λ. Let $T = X + Y$.

1. Show that, for each fixed ψ, T is sufficient for λ.
2. Find the conditional distribution of X given T.
3. Derive the conditional score $\partial \log f_{X|T}(x|t;\psi)/\partial\psi$.
4. Compute the information about ψ contained in the conditional score.

9.12. In the setting of Problem 9.11, compute the efficient score $s_{\psi\cdot\lambda}(\psi,\lambda,X)$. Compute the information about ψ contained in the efficient score. Compare this information with the information contained in the conditional score as derived in Problem 9.11, and explain the discrepancy.

9.13. In the setting of Problem 9.11, compute the estimating equation $s_\psi - P_{\mathcal{B}_m} s_\psi$ in Section 9.4 with $m = 2$. Show that this estimating equation has the same form as the conditional score obtained in Problem 9.11.

9.14. Suppose θ is a p-dimension parameter consisting of an r-dimensional parameter of interest ψ and an s-dimensional nuisance parameter λ, where $p = r + s$. Let $\mathcal{G}_{\psi\cdot\lambda}$ be the class of estimating equations defined in Definition 9.8. Show that $s_\psi - [s_\psi, s_\lambda][s_\lambda, s_\lambda]^{-1} s_\lambda$ is the optimal estimating equation in $\mathcal{G}_{\psi\cdot\lambda}$ in the sense that it maximizes the information $I_g(\psi|\theta)$ among $g \in \mathcal{G}_{\psi\cdot\lambda}$.

9.15. Prove the following extension of Lemma 9.1. Suppose θ is a p-dimension parameter consisting of an r-dimensional parameter of interest ψ and an s-dimensional nuisance parameter λ, where $p = r + s$. Let $g : \Theta \times \Omega_X \to \mathbb{R}^r$ be an unbiased, P_θ-square-integrable estimating equation for ψ such that $g(\theta,x) f_\theta(x)$ satisfies $\mathrm{DUI}^+(\psi,\mu)$. Let $s_\psi(\theta,X) = \partial \log f_\theta(X)/\partial\psi$. Show that

$$E_\theta\left[\frac{\partial g(\theta,X)}{\partial\psi^T}\right] = -E_\theta[g(\theta,X)s_\psi^T(\theta,X)].$$

9.16. Let $X_1,\cdots,X_n$ be an i.i.d. sample from a distribution with density of the form $f(x;\theta)$, where $\theta \in \mathbb{R}^p$ consists of a parameter of interest $\psi \in \mathbb{R}^r$, and a nuisance parameter $\lambda \in \mathbb{R}^s$, with $r + s = p$. Let $g(\lambda,X)$ be an unbiased estimating equation for λ that satisfies the conditions in Theorems 9.2 and 9.3, and suppose $\tilde\lambda$ is a consistent solution to

$$E_n[g(\lambda,X)] = 0.$$

Let $s_\psi(\psi,\lambda,X)$ be the score for ψ, and let $s_{\psi\cdot\lambda}(\psi,\lambda,X)$ be the efficient score.

1. Let $\tilde\psi$ be a consistent solution to the estimating equation

$$E_n[s_\psi(\psi,\tilde\lambda,X)] = 0.$$

Derive the asymptotic distribution of $\sqrt{n}(\tilde\psi - \psi_0)$.

2. Let $\hat{\psi}$ be a consistent solution to the estimating equation

$$E_n s_{\psi\cdot\lambda}(\psi, \tilde{\lambda}, X) = 0.$$

   Derive the asymptotic distribution of $\sqrt{n}(\hat{\psi} - \psi_0)$.
3. Compare the asymptotic variances of $\sqrt{n}(\tilde{\psi} - \psi_0)$ and $\sqrt{n}(\hat{\psi} - \psi_0)$.

9.17. Suppose θ is a p-dimensional parameter, consisting of a parameter of interest $\psi \in \mathbb{R}^r$, and a nuisance parameter $\lambda \in \mathbb{R}^s$, where $r + s = p$. Let $s_{\psi\cdot\lambda}(\psi, \lambda, X)$ be the efficient score, and let $I_{\psi\cdot\lambda}(\psi, \lambda)$ be the efficient information. Suppose $\tilde{\psi}$ and $\tilde{\lambda}$ are estimates of ψ_0 and λ_0 such that

$$\tilde{\psi} - \psi_0 = O_P(n^{-1/2}), \quad \tilde{\lambda} - \lambda_0 = o_P(n^{-1/4}).$$

Moreover, suppose:

1. $s_{\psi\cdot\lambda}(\psi, \lambda, X)$ is differentiable with respect to ψ and twice differentiable with respect to λ;
2. the entries of the arrays

$$A(\psi, \lambda) = \frac{\partial s_{\psi\cdot\lambda}(\psi, \lambda, X)}{\partial \psi^T}, \quad B(\psi, \lambda) = \frac{\partial^2 s_{\psi\cdot\lambda}(\psi, \lambda, X)}{\partial \lambda \partial \lambda^T}$$

   are P_θ-integrable;
3. the sequences of random arrays

$$\{E_n[A(\psi, \lambda)] : n = 1, 2, \ldots\}, \quad \{E_n[B(\psi, \lambda)] : n = 1, 2, \ldots\}$$

   are stochastic equicontinuous in an open ball centered at $\theta_0 = (\psi_0^T, \lambda_0^T)^T$.

Let $\hat{\psi}$ be the one-step Newton-Raphson estimate for the parameter of interest ψ_0, defined as

$$\hat{\psi} = \tilde{\psi} - I_{\psi\cdot\lambda}^{-1}(\tilde{\psi}, \tilde{\lambda}) E_n[s_{\psi\cdot\lambda}(\tilde{\psi}, \tilde{\lambda}, X)].$$

1. Show that

$$\hat{\psi} = \psi_0 - I_{\psi\cdot\lambda}^{-1}(\psi_0, \lambda_0) E_n[s_{\psi\cdot\lambda}(\psi_0, \lambda_0, X)] + o_P(n^{-1/2}).$$

2. Derive the asymptotic distribution of $\sqrt{n}(\hat{\psi} - \psi_0)$.

9.18. Suppose that the parameter $\theta \in \mathbb{R}^p$ consists of a parameter of interest $\psi \in \mathbb{R}^r$ and a nuisance parameter $\lambda \in \mathbb{R}^s$, with $r + s = p$. Let $g(\theta, X)$ be a p-dimensional unbiased and P_θ-square-integrable estimating equation such that $g(\theta, x) f_\theta(x)$ satisfies $\mathrm{DUI}^+(\theta, \mu)$. Let g_ψ be the first r components of g and g_λ be the last s components of g. Let $J_g(\theta)$ and $K_g(\theta)$ be the matrices defined in (9.10). Decompose J_g and J_g^{-1} as block matrices according to the dimensions of ψ and λ, as follows:

$$J_g = \begin{pmatrix} (J_g)_{\psi\psi} & (J_g)_{\psi\lambda} \\ (J_g)_{\lambda\psi} & (J_g)_{\lambda\lambda} \end{pmatrix}, \quad J_g^{-1} = \begin{pmatrix} (J_g^{-1})_{\psi\psi} & (J_g^{-1})_{\psi\lambda} \\ (J_g^{-1})_{\lambda\psi} & (J_g^{-1})_{\lambda\lambda} \end{pmatrix}.$$

Let K_g and K_g^{-1} be decomposed similarly. Let

$$g_{\psi\cdot\lambda}(\theta, X) = g_\psi(\theta, X) - (J_g)_{\psi\lambda}[(J_g)_{\lambda\lambda}]^{-1} g_\lambda(\theta, X).$$

Let $\hat{\theta}$ be a consistent solution to $E_n[g_{\psi\cdot\lambda}(\theta, X)] = 0$, and let $\hat{\psi}$ be its first r components. You may impose further regularity conditions such as stochastic equicontinuity.

1. Show that

$$\hat{\psi} = \psi_0 + (J_g^{-1})_{\psi\psi} E_n[g_{\psi\cdot\lambda}(\theta, X)] + o_P(n^{-1/2}).$$

2. Derive the asymptotic distribution of $\sqrt{n}(\hat{\psi} - \psi_0)$, and express the asymptotic variance in terms of the sub-matrices of J_g and K_g.

9.19. In the setting of Problem 9.18. Suppose $\tilde{\lambda}$ is an estimate of λ_0 satisfying $\tilde{\lambda} - \lambda_0 = o_P(n^{-1/4})$. Let $\hat{\psi}$ be a consistent solution to the estimating equation

$$E_n[g_{\psi\cdot\lambda}(\psi, \tilde{\lambda}, X)] = 0.$$

Show that $\sqrt{n}(\hat{\psi} - \psi_0)$ has the same asymptotic distribution as the one obtained in Problem 9.18.

References

Bhattacharyya, A. (1946). On some analogues of the amount of information and their use in statistical estimation. *Sankhya: The Indian Journal of Statistics.* **8**, 1–14.

Bickel, P. J., Klaassen, C. A. J., Ritov, Y. Wellner, J. A. (1993). *Efficient and Adaptive Estimation for Semiparametric Models.* The Johns Hopkins University Press.

Cox, D. R. and Hinkley, D. (1974). *Theoretical Statistics.* Chapman & Hall.

Crowder, M. (1987). On linear and quadratic estimating functions. *Biometrika*, **74**, 591–597.

Durbin, J. (1960). Estimation of parameters in time-series regression models. *Journal of the Royal Statistical Society, Series B*, **22**, 139–153.

Godambe, V. P. (1960). An optimum property of regular maximum likelihood estimation. *The Annals of Mathematical Statistics*, **31**, 1208–1211.

Godambe, V. P. (1976). Conditional likelihood and unconditional optimum estimating equations. *Biometrikca*, **63**, 277–284.

Godambe, V. P. and Thompson, M. E. (1989). An extension of quasi-likelihood estimation. *Journal of Statistical Planning and Inference*, **22**, 137–152.

Hansen, L. P. (1982). Large sample properties of Generalized Method of Moments Estimators. *Econometrica*, **50**, 1029–1054.

Heyde, C. C. (1997). *Quasi-Likelihood and its Application: a General Approach to Optimal Parameter Estimation.* Springer.

Jarrett, R. G. (1984). Bounds and expansions for Fisher information when the moments are known. *Biometrika*, **71**, 101–113.

Li, B. (1993). A deviance function for the quasi-likelihood method. *Biometrika*, **80**, 741–753.

Li, B. (1996). A minimax approach to consistency and efficiency for estimating equations. *The Annals of Statistics*, **24**, 1283–1297.

Li, B. and McCullagh, P. (1994). Potential functions and conservative estimating functions. *The Annals of Statistics*, **22**, 340–356.

Liang, K. Y. and Zeger, S. L. (1986). Longitutinal data analysis using generalized linear models. *Biometrika*, **73**, 13–22.

Lindsay, B. (1982). Conditional score functions: some optimality results. *Biometrika*, **69**, 503–512.

McCullagh, P. (1983). Quasi-likelihood functions. *The Annals of Statistics*, **11**, 59–67.

Morton, R. (1981). Efficiency of estimating equations and the use of pivots. *Biometrika*, **68**, 227–233.

Small, C. G. and McLeish, D. L. (1989). Projection as a method for increasing sensitivity and eliminating nuisance parameters. *Biometrika*, **76**, 693–703.

Waterman, R. P and Lindsay, B. G. (1996). Projected score methods for approximating conditional scores. *Biometrika*, **83**, 1–13.

Wedderburn, R. W. M. (1974). Quasi-likelihood functions, Generalized Linear Models, and the Gauss-Newton method. *Biometrika*, **61**, 439–447.

Zeger, S. L. and Liang, K. Y. (1986). Longitudinal data analysis for discrete and continuous outcomes. *Biometrics*, **42**, 121–130.

10

Convolution Theorem and Asymptotic Efficiency

In Chapters 8 and 9 we have developed optimality of the maximum likelihood estimate among the class of solutions to estimating equations in terms of the information the asymptotic variance. The optimality of the maximum likelihood estimate, in fact, goes much deeper. In this chapter we show that the maximum likelihood estimate, as well as estimates that are asymptotically equivalent to it, have the smallest asymptotic variance among all regular estimates. This is a wide class of estimates that includes not only the asymptotically linear estimates such as the solutions to estimating equations, but also asymptotically nonlinear (and therefore asymptotically non-Gaussian) estimates. We systematically develop the theory underlying this general result: the framework of Local Asymptotic Normality and the Convolution Theorem (Le Cam 1953, 1960; Hájek 1970). This is an amazingly logical system that leads to far-reaching results with a small set of assumptions. Some techniques introduced in this chapter, such as Le Cam's third lemma and the convolution theorem, will also be useful for developing local alternative distributions for asymptotic hypothesis tests in the next chapter.

As a historical note, it had been known since Fisher (1922, 1925) that the maximum likelihood estimate has the smallest asymptotic variance among, roughly, all estimates that are asymptotically normal. But Fisher did not give a rigorous proof and counterexamples were found, the first of which by J. L. Hodges, Jr. in an unpublished paper, which was cited by Le Cam (1953). This led to intensive research in the ensuing years on what kind of estimates can reach Fisher's asymptotic variance lower bound, how wide a class of estimates the lower bound applies to, as well as how meaningful the counterexamples are. Hall and Mathiason (1990), van der Vaart (1997, 1998), and Le Cam and Yang (2000) are excellent references on this topic. Some of the developments here echo the logic lines in these works.

B. Li and G. J. Babu, *A Graduate Course on Statistical Inference*,
Springer Texts in Statistics, https://doi.org/10.1007/978-1-4939-9761-9_10

10.1 Contiguity

Recall that a probability measure P is absolutely continuous with respect to a probability measure Q if, for any measurable set A, $Q(A) = 0$ implies $P(A) = 0$. Contiguity is an analogue of this condition for two sequences of probability measures.

Definition 10.1 *Let $\{P_n\}$ and $\{Q_n\}$ be two sequences of probability measures. P_n is said to be contiguous with respect to Q_n if, for any sequence A_n, $Q_n(A_n) \to 0$ implies $P_n(A_n) \to 0$. This property is written as $P_n \lhd Q_n$. If $P_n \lhd Q_n$ and $Q_n \lhd P_n$, then P_n and Q_n are said to be mutually contiguous, and this property is expressed as $P_n \lhd\rhd Q_n$.*

Even though contiguity between sequences of probability measures is analogous to absolute continuity between two probability measures, the latter does not imply the former logically. In particular, even if $P_n \ll Q_n$ for every n, that does not imply $P_n \lhd Q_n$. For example, let P_n be the distribution of $N(0, 1/n)$ and Q_n be the distribution of $N(1, 1/n)$. In this case $P_n \ll Q_n$ for each n. The limiting distribution of P_n is a point mass at 0 and the limiting distribution of Q_n is a point mass at 1. So if, for each n, A_n is the set $(-1/2, -1/2)$, then $Q_n(A_n) \to 0$ and yet $P_n(A_n) \to 1$. Proof of the next Proposition is left as an exercise.

Proposition 10.1 *The following statements are equivalent:*

1. *$P_n \lhd Q_n$.*
2. *Whenever $Q_n(A_n) \to 1$, we have $P_n(A_n) \to 1$.*
3. *If T_n is any sequence of random variables with $Q_n(|T_n| \geq \epsilon) \to 0$, then $P_n(|T_n| \geq \epsilon) \to 0$.*

We now focus on two results known as Le Cam's first and third lemmas.

10.2 Le Cam's first lemma

Le Cam's first lemma is concerned with a set of sufficient and necessary conditions for contiguity. Recall that, for two probability measures P and Q, if $P \ll Q$, then

$$E_Q\left(\frac{dP}{dQ}\right) = \int \frac{dP}{dQ} dQ = \int dP = 1, \tag{10.1}$$

where E_Q denotes the expectation with respect to the probability measure Q. Le Cam's first Lemma is analogous to this result when P and Q are replaced by sequences of probability measures $\{P_n\}$ and $\{Q_n\}$ and absolute continuity $P \ll Q$ is replaced by contiguity $P_n \lhd Q_n$.

The following technical lemma is established first.

Lemma 10.1 *Suppose that $g_n : \mathbb{R} \to \mathbb{R}$ is a sequence of functions such that, for any $\epsilon > 0$, $\liminf_{n\to\infty} g_n(\epsilon) \geq 0$. Then there is a sequence $\epsilon_n \downarrow 0$ such that $\liminf_{n\to\infty} g_n(\epsilon_n) \geq 0$.*

Proof. Since for each integer $k \geq 1$, $\liminf_{n\to\infty} g_n(1/k) \geq 0$, there is a positive integer n_k such that, for all $n \geq n_k$, $g_n(1/k) > -1/k$. Without loss of generality, we can assume that $n_{k+1} > n_k$ for all $k = 1, 2, \ldots$. Let $\epsilon_n = 1/k$ for $n_k \leq n < n_{k+1}$. Then $g_n(\epsilon_n) > -\epsilon_n$ for all $n \geq n_1$. Clearly $\epsilon_n \downarrow 0$ and $\liminf_{n\to\infty} g_n(\epsilon_n) \geq 0$. □

In the following discussion, multiple probability measures are considered on single measurable spaces. So for clarity of the exposition, we revamp the notation for convergence in distribution and convergence in probability. For $n \in \mathbb{N}$, let X_n and X be random variables defined, respectively, on the probability spaces $(\Omega_n, \mathcal{F}_n, P_n)$ and $(\Omega, \mathcal{F}, P)$ taking values in $(\mathbb{R}^k, \mathcal{R}^k)$.

We write

$$X_n \xrightarrow[P_n]{\mathcal{D}} X,$$

if X_n converges in distribution to X under the sequence $\{P_n\}$; that is, for every bounded and continuous f on $\mathbb{R}^k$,

$$\int f(X_n)dP_n \to \int f(X)dP.$$

Similarly, if X_n converges in P_n-probability to a constant a; that is,

$$P_n(\|X_n - a\| > \epsilon) \to 0,$$

for every $\epsilon > 0$, then we write $X_n \xrightarrow{P_n} a$.

The first statement of the next theorem is analogous to (10.1), and the second statement is analogous to $P \ll Q$.

Theorem 10.1 (Le Cam's first lemma) *Let P_n and Q_n be sequences of probability measures on measurable spaces $(\Omega_n, \mathcal{F}_n)$, and assume $P_n \equiv Q_n$. Then the following statements are equivalent:*

1. *If $dP_n/dQ_n \xrightarrow[Q_n]{\mathcal{D}} V$ along a subsequence, then $E(V) = 1$.*
2. *$P_n \lhd Q_n$.*
3. *If $dQ_n/dP_n \xrightarrow[P_n]{\mathcal{D}} U$ along a subsequence, then $P(U > 0) = 1$.*

Proof. $1 \Rightarrow 2$. Suppose $Q_n(A_n) \to 0$ for a sequence of measurable sets $\{A_n\}$. Then we shall show that $P_n(A_n) \to 0$, proving statement 2. First note that, since $P_n \ll Q_n$, the sequence of densities $\{dP_n/dQ_n\}$ is tight under the sequence of measures $\{Q_n\}$. By Prohorov's Theorem (Theorem 7.10), every subsequence $\{n'\}$ has a further subsequence $\{n''\}$, and a random variable V, such that $dP_{n''}/dQ_{n''} \xrightarrow[Q_{n''}]{\mathcal{D}} V$.

Since $Q_n(A_n) \to 0 \Rightarrow 1 - I_{A_n} \xrightarrow{Q_n} 1$, by a version of Slutsky's theorem (Corollary 7.3),

$$0 \leq (dP_{n''}/dQ_{n''})(1 - I_{A_{n''}}) \underset{Q_{n''}}{\xrightarrow{\mathcal{D}}} V.$$

Since $P_n(A_n) = \int I_{A_n}(dP_n/dQ_n)dQ_n$, we have by Fatou's lemma (Lemma 7.4), and equation (10.1),

$$E(V) \leq \liminf_{n''\to\infty} E_{Q_{n''}}((dP_{n''}/dQ_{n''})(1 - I_{A_{n''}})) = 1 - \limsup_{n''\to\infty} P_{n''}(A_{n''}).$$

By statement 1, $E(V) = 1$, so $P_{n''}(A_{n''}) \to 0$. Thus we have shown that every subsequence of the sequence $\{P_n(A_n)\}$ of real numbers, contains a further subsequence $P_{n''}(A_{n''}) \to 0$. Hence $P_n(A_n) \to 0$.

$2 \Rightarrow 3$. Suppose $dQ_{n'}/dP_{n'} \underset{P_{n'}}{\xrightarrow{\mathcal{D}}} U$ for some subsequence $\{n'\}$. Clearly, $P(U \geq 0) = 1$, as $dQ_{n'}/dP_{n'} \geq 0$. Thus it suffices to show that $P(U = 0) = 0$. By the Portmanteau Theorem (Theorem 7.6), for any $\epsilon > 0$,

$$\liminf_{n'\to\infty} P_{n'}(dQ_{n'}/dP_{n'} < \epsilon) - P(U < \epsilon) \geq 0. \tag{10.2}$$

By Lemma 10.1, there exists a sequence $\epsilon_{n'} \downarrow 0$ such that

$$\liminf_{n'\to\infty} \{P_{n'}(dQ_{n'}/dP_{n'} < \epsilon_{n'}) - P(U < \epsilon_{n'})\} \geq 0.$$

Since $P(U < \epsilon) \geq P(U = 0)$ for all $\epsilon > 0$, we have

$$\liminf_{n'\to\infty} P_{n'}(dQ_{n'}/dP_{n'} < \epsilon_{n'}) \geq \limsup_{n'\to\infty} P(U < \epsilon_{n'}) \geq P(U = 0). \tag{10.3}$$

It remains to show that the left-hand side of (10.3) is 0. Since $P_n \equiv Q_n$, by Radon-Nikodym Theorem (see Theorem 1.11)

$$\begin{aligned} Q_{n'}(dQ_{n'}/dP_{n'} < \epsilon_{n'}) &= \int_{dQ_{n'}/dP_{n'}<\epsilon_{n'}} (dQ_{n'}/dP_{n'})dP_{n'} \\ &\leq \epsilon_{n'} \int dP_{n'} = \epsilon_{n'} \to 0. \end{aligned} \tag{10.4}$$

Since $P_{n'} \lhd Q_{n'}$, (10.4) implies

$$P_{n'}(dQ_{n'}/dP_{n'} < \epsilon_{n'}) \to 0.$$

Therefore the left-hand side of (10.3) is 0.

$3 \Rightarrow 1$. Let μ_n be the probability measure $(P_n + Q_n)/2$. Then $P_n \ll \mu_n$ and $Q_n \ll \mu_n$. Since $P_n \equiv Q_n$, by Theorem 1.13, the probability densities p_n and q_n of P_n and Q_n with respect to μ_n satisfy

$$0 \leq p_n, q_n \leq 2,\ \mu_n\{p_n = 0\} = \mu_n\{q_n = 0\} = 0,\ p_n + q_n = 2,$$
$$\mu_n\left\{\frac{p_n}{q_n} = \frac{dP_n}{dQ_n}\right\} = 1, \text{ and } \mu_n\left\{\frac{q_n}{p_n} = \frac{dQ_n}{dP_n}\right\} = 1. \tag{10.5}$$

Clearly for any $c > 0$,

$$\{(p_n/q_n) < c\} \Leftrightarrow \{(2 - q_n)/q_n < c\} \Leftrightarrow \{q_n > 2/(1+c)\},$$

and hence

$$\begin{aligned} E_{Q_n}\left(\frac{p_n}{q_n} I_{\{p_n/q_n \leq c\}}\right) &= E_{\mu_n}\left(p_n I_{\{p_n/q_n \leq c\}}\right) \\ &\geq E_{\mu_n}\left(p_n I_{\{p_n/q_n < c\}}\right) \\ &= E_{\mu_n}\left((2 - q_n) I_{\{q_n > 2/(1+c)\}}\right). \end{aligned} \tag{10.6}$$

Now suppose $dP_n/dQ_n \xrightarrow[Q_n]{\mathcal{D}} V$ along a subsequence $\{n'\}$. By (10.5), we have $(p_{n'}/q_{n'}) \xrightarrow[Q_{n'}]{\mathcal{D}} V$. Since, for any $K > 0$,

$$\mu_n(q_n > K) \leq (1/K) E_{\mu_n}(dQ_n/d\mu_n) = 1/K,$$
$$P_n(q_n/p_n > K) = P_n(dQ_n/dP_n > K) \leq (1/K) E_{P_n}(dQ_n/dP_n) = 1/K,$$

the sequence $\{q_n\}$ is tight under $\{\mu_n\}$, and the sequence $\{(q_n/p_n)\}$ is tight under $\{P_n\}$. So, by Lemma 7.6, there is a further subsequence $\{n''\}$ of $\{n'\}$ such that

$$\frac{q_{n''}}{p_{n''}} \xrightarrow[P_{n''}]{\mathcal{D}} U, \tag{10.7}$$

$$\frac{p_{n''}}{q_{n''}} \xrightarrow[Q_{n''}]{\mathcal{D}} V, \text{ and } q_{n''} \xrightarrow[\mu_{n''}]{\mathcal{D}} W, \tag{10.8}$$

for some random variables U, W. Hence by Fatou's lemma (Lemma 7.4), (10.5), and bounded convergence theorem (Theorem 1.8),

$$E(V) \leq 1, \quad E(W) = 1. \tag{10.9}$$

In view of (10.6) and (10.8), by applying Portmanteau Theorem (Theorem 7.6) to the upper bounded upper semi-continuous function $f(x) = xI_{\{x \leq c\}}$ and the lower bounded lower semi-continuous function $g(x) = (2 - x)I_{\{x > 2/(1+c)\}}$, we get for any $c > 0$,

$$\begin{aligned} E(V) \geq E\left(VI_{\{V \leq c\}}\right) &\geq \limsup_{n'' \to \infty} E_{Q_{n''}}\left(\frac{p_{n''}}{q_{n''}} I_{\{p_{n''}/q_{n''} \leq c\}}\right) \\ &\geq \liminf_{n'' \to \infty} E_{\mu_{n''}}\left((2 - q_{n''}) I_{\{q_{n''} > 2/(1+c)\}}\right) \\ &\geq E\left((2 - W) I_{\{W > (2/(1+c)\}}\right). \end{aligned} \tag{10.10}$$

Now let $c \to \infty$ in (10.10), and use (10.9) to get

$$\begin{aligned} 1 \geq E(V) \geq & E\left((2-W)I_{\{W>0\}}\right) \\ =& 2P(W>0) - E(W) \\ =& 1 - 2P(W=0). \end{aligned} \tag{10.11}$$

To complete the proof it is sufficient to prove $P(W=0)=0$. Toward this end, let $0<\epsilon<1$, and apply Portmanteau Theorem 7.6 to (10.7) and (10.8) to conclude

$$\begin{aligned} P(W=0) \leq P(W<\epsilon) \leq & \liminf_{n''\to\infty} \mu_{n''}(q_{n''}<\epsilon) \\ \leq & \limsup_{n''\to\infty} P_{n''}((q_{n''}/p_{n''}) \leq \epsilon) \\ \leq & P(U \leq \epsilon). \end{aligned}$$

By the right continuity of probability distribution functions and Statement 3, it follows that $P(W=0) \leq P(U \leq \epsilon) \downarrow P(U=0)=0$, as $\epsilon \downarrow 0$. □

10.3 Le Cam's third lemma

Le Cam's third lemma establishes the limit form of the local alternative distributions P_n based on the limit form of the null distributions Q_n. This is useful for proving the convolution theorem and for developing the asymptotic power of a test statistic under the local alternative distributions. To understand the intuition behind Le Cam's third lemma, it is again helpful to make an analogy with the situation involving two probability measures P, Q. If $P \ll Q$ and U is a random vector, then for any measurable set B,

$$P(U \in B) = E_P[I_B(U)] = E_Q\left[I_B(U)\frac{dP}{dQ}\right].$$

Le Cam's third lemma is an analogous statement with probability measures P and Q replaced by sequences of probability measures P_n and Q_n, and the absolute continuity $P \ll Q$ with contiguity $P_n \lhd Q_n$.

Theorem 10.2 (Le Cam's third lemma) *Suppose that P_n, Q_n are probability measures defined on $(\Omega_n, \mathcal{F}_n)$ such that $P_n \equiv Q_n$ and $P_n \lhd Q_n$. If $\{U_n\}$ is a sequence of random vectors in $\mathbb{R}^k$ such that*

$$(U_n, dP_n/dQ_n) \xrightarrow[Q_n]{\mathcal{D}} (U, V), \tag{10.12}$$

then $L(B) = E(I_B(U)V)$ defines a probability measure on the Borel sets of $\mathbb{R}^k$, and $U_n \xrightarrow[P_n]{\mathcal{D}} L$.

Proof. Because $P_n \triangleleft Q_n$ and $dP_n/dQ_n \xrightarrow[Q_n]{\mathcal{D}} V$, by Le Cam's first lemma $L(\mathbb{R}^k) = E(V) = 1$. As $V \geq 0$, clearly $L(B) \geq 0$ for all Borel sets B. By Problem 1.26, L is a probability measure on $\mathbb{R}^k$. By the definition of L, $E[f(U)V] = \int f dL$ holds for any measurable indicator function f. Thus, by Theorem 7.11,

$$E[f(U)V] = \int f dL, \tag{10.13}$$

for all non-negative measurable functions f, and hence clearly (10.13) holds for all measurable functions that are bounded below.

We shall now show that $U_n \xrightarrow[P_n]{\mathcal{D}} L$. Since $P_n \equiv Q_n$, for any measurable function f that is bounded below, we have

$$E_{P_n}(f(U_n)) = E_{Q_n}\left(f(U_n)\frac{dP_n}{dQ_n}\right).$$

Thus for any lower semi-continuous function f that is bounded from below, the function that maps (u, v) to $f(u)v$ is a also lower semi-continuous and is bounded from below. Thus, by the Portmanteau Theorem and the convergence $(U_n, dP_n/dQ_n) \xrightarrow[Q_n]{\mathcal{D}} (U, V)$, we have

$$\liminf_{n\to\infty} \int f(U_n) dP_n = \liminf_{n\to\infty} \int f(U_n)\frac{dP_n}{dQ_n} dQ_n \geq E(f(U)V). \tag{10.14}$$

Hence by (10.13) and (10.14),

$$\liminf_{n\to\infty} \int f(U_n) dP_n \geq \int f dL.$$

Now another application of Portmanteau Theorem yields $U_n \xrightarrow[P_n]{\mathcal{D}} L$. □

In the above proof we have used $\int f dL = E[f(U)V]$ for any measurable function bounded from below. We now expand this result somewhat and state it as a corollary for future reference. This corollary is a direct consequence of Theorem 7.11.

Corollary 10.1 *Suppose U is a random vector and V is a nonnegative random variable with $E(V) = 1$. Let L be the set function defined by $L(B) = E[I_B(U)V]$ for all Borel sets B of $\mathbb{R}^k$. Then L defines a probability measure and the equality*

$$E[f(U)V] = \int f(U) dL$$

holds (i) for any nonnegative measurable function f; (ii) for any measurable function $f(u)$ such that $E|f(U)V| < \infty$ and $\int |f(U)| dL < \infty$.

In particular, the moment generating function (if it exists) and the characteristic function of L are, respectively,

$$\phi_L(t) = E(e^{t^T U} V), \quad \kappa_L(t) = E(e^{it^T U} V).$$

The next corollary gives the local alternative distribution of (U_n, L_n) under the conditions in Le Cam's third lemma.

Corollary 10.2 *Suppose that the assumptions in Theorem 10.2 hold. Then $L(B) = E[I_B(U, \log V)V]$ defines a probability measure and*

$$(U_n, L_n) \xrightarrow[P_n]{\mathcal{D}} L.$$

Proof. By (10.12) and the continuous mapping theorem,

$$(U_n, \log(dP_n/dQ_n), dP_n/dQ_n) \xrightarrow[Q_n]{\mathcal{D}} (U, \log V, V).$$

By Theorem 10.2 (with U_n replaced by the random vector $(U_n, \log(dP_n/dQ_n))$), we have

$$(U_n, \log(dP_n/dQ_n)) \xrightarrow[P_n]{\mathcal{D}} L,$$

as desired. □

10.4 Local asymptotic Normality

We now set up the assumptions and notations for a framework known as the local asymptotic Normality. See Le Cam (1960), Hall and Mathiason (1990), van der Vaart (1998), and Le Cam and Yang (2000). This framework will be important for the development of both the convolution theorem in Section 10.5 and the local alternative distributions of hypothesis testing in Chapter 11. We first make some assumptions about the parametric family that is the basis of our discussions.

Assumption 10.1 (parametric model) *1. $(\Omega_n, \mathcal{F}_n)$, $n \in \mathbb{N}$, is a sequence of measurable spaces;*
2. for each n, $\{P_{n\theta} : \theta \in \Theta \subseteq \mathbb{R}^p\}$ is a homogeneous parametric family of probability measures on $(\Omega_n, \mathcal{F}_n)$ dominated by a σ-finite measure μ_n.
3. The density $f_{n\theta} = dP_{n\theta}/d\mu_n$ is differentiable with respect to θ.

Recall from Section 2.1.1, a parametric family $\{P_{n\theta} : \theta \in \Theta\}$ being homogeneous means that $P_{n\theta'} \equiv P_{n\theta''}$ for all $\theta', \theta'' \in \Theta$. This assumption is stronger than necessary and is introduced to simplify the subsequent technical developments. Note that above definition makes no reference to the random vectors underlying the distributions $P_{n\theta}$.

To facilitate the asymptotic analysis, we localize the above parametric family around an interior point θ_0 in the parameter space, as described by the next definition.

Definition 10.2 (local parametric model) *Suppose Assumption 10.1 holds. Let θ_0 be an interior point of Θ, and let $\theta_n(\delta) = \theta_0 + n^{-1/2}\delta \in \Theta$.*

1. *The set $\{P_{n\theta_n(\delta)} : \delta \in \mathbb{R}^p\}$ is called the local parametric family around θ_0.*
2. *The sequence $\{P_{n\theta_0} : n \in \mathbb{N}\}$ is called the null sequence and is denoted by $\{Q_n : n \in \mathbb{N}\}$.*
3. *The sequence $\{P_{n\theta_n(\delta)} : n \in \mathbb{N}\}$ is called the local alternative sequence and is denoted by $\{P_n(\delta) : n \in \mathbb{N}\}$.*
4. *$L_n(\delta) = \log(dP_n(\delta)/dQ_n)$ is called the local log likelihood ratio.*
5. *$S_n = n^{-1/2}\partial(\log f_{n\theta})/\partial\theta|_{\theta=\theta_0} = n^{-1/2}\partial L_n(\delta)/\partial\delta|_{\delta=0}$ is called the standardized score.*

When it causes no ambiguity, we will write $\theta_n(\delta)$, $P_n(\delta)$, and $L_n(\delta)$ simply as θ_n, P_n, and L_n. The vector δ serves as the localized parameter — localized within a neighborhood with size of the order $n^{-1/2}$. In a hypothesis test setting, the sequence Q_n can be regarded as the null hypothesis, and the sequence $P_n(\delta)$ the local alternative hypothesis. In the estimation setting, $\{P_n(\delta) : \delta \in \mathbb{R}^p\}$ is simply a family of distributions indexed by the local parameter δ. The localization scale $n^{-1/2}$ is chosen to guarantee contiguity of $P_n(\delta)$ with respect to Q_n.

The next assumption is the local asymptotic normal assumption (LAN). It assumes that the standard score function S_n is asymptotically Normal, and the second-order Taylor expansion of L_n has a desired remainder term. The conditions for these are quite mild, which are satisfied not only for the independent case but also for some stochastic processes.

Assumption 10.2 (LAN) *Under Assumption 10.1, we further assume*

$$S_n \xrightarrow[Q_n]{\mathcal{D}} N(0, I(\theta_0)) \tag{10.15}$$

$$L_n(\delta) \stackrel{Q_n}{=} \delta^T S_n - \delta^T I(\theta_0)\delta/2 + o_P(1), \tag{10.16}$$

where $I(\theta_0)$ is a positive definite matrix, called the Fisher information. If these conditions hold, then we say (S_n, L_n) satisfies LAN.

The notation $\stackrel{Q_n}{=}$ in (10.16) indicates the probability underlying $o_P(1)$ is Q_n: thus, $X_n \stackrel{Q_n}{=} Y_n + o_P(1)$ means that for any $\epsilon > 0$,

$$Q_n(\|X_n - Y_n\| > \epsilon) \to 0.$$

This notation will be used repeatedly throughout the rest of the Chapter. The next lemma shows that $P_n(\delta)$ is contiguous with respect to Q_n under LAN.

Lemma 10.2 *If* $(S_n, L_n(\delta))$ *satisfies LAN, then* $P_n(\delta) \lhd Q_n$.

Proof. By Slutsky's theorem, $L_n(\delta) \xrightarrow[Q_n]{\mathcal{D}} L(\delta)$, where

$$L(\delta) \sim N(-\delta^T I(\theta_0)\delta/2, \delta^T I(\theta_0)\delta).$$

By the continuous mapping theorem,

$$dP_n/dQ_n = e^{L_n(\delta)} \xrightarrow[Q_n]{\mathcal{D}} e^{L(\delta)}.$$

The expectation of $e^{L(\delta)}$ is simply the moment generating function of $L(\delta)$ evaluated at 1, which is

$$E(e^{L(\delta)}) = \phi_{L(\delta)}(1) = \exp[(-\delta^T I\delta/2) + (\delta^T I\delta)1^2/2] = 1.$$

Here, we used the fact that the moment generating function of normal distribution with mean μ and variance σ^2 is $\exp(\mu t + \sigma^2 t^2/2)$. Hence, by Theorem 10.1, $P_n(\delta) \lhd Q_n$. □

We next develop a set of sufficient conditions for LAN under the i.i.d. parametric model. To do so, we first give a rigorous definition of the i.i.d. parametric model and the regularity conditions needed.

Assumption 10.3 (i.i.d. parametric model) *1.* $X_1, X_2, \ldots$ *are i.i.d. random vectors with distribution belonging to a homogeneous parametric family* $\{P_\theta : \theta \in \Theta\}$ *defined on a measurable space* $(\Omega, \mathcal{F})$.
2. P_θ *is dominated by a* σ*-finite measure* μ, *with its density denoted by* $f_\theta = dP_\theta/d\mu$.
3. f_θ *is differentiable with respect to* θ.

Under the i.i.d. model, the various quantities in Assumption 10.1 and Definition 10.2 reduce to the following:

- $(\Omega_n, \mathcal{F}_n, \mu_n) = (\Omega \times \cdots \times \Omega, \mathcal{F} \times \cdots \times \mathcal{F}, \mu \times \cdots \times \mu)$;
- $P_{n\theta} = P_\theta \times \cdots \times P_\theta$;
- $L_n(\delta) = nE_n[\ell(\theta_n, X) - \ell(\theta_0, X)]$, where $\ell(\theta, X) = \log[f_\theta(X)]$;
- $S_n = n^{1/2}E_n[s(\theta, X)]$, where $s(\theta, X) = \partial[\log f_\theta(X)]/\partial\theta$.

The next proposition gives the sufficient conditions for LAN. Following the notations in Chapter 9 (see equation (9.10)), let

$$J(\theta) = E_\theta[\partial s(\theta, X)/\partial\theta^T], \quad K(\theta) = E_\theta[s(\theta, X)s^T(\theta, X)].$$

Proposition 10.2 *Suppose Assumption 10.3 holds and*

1. *$\ell(\theta, x)$ is twice differentiable;*
2. *$f_\theta(X)$ and $s(\theta, X)f_\theta(X)$ satisfy DUI$^+(\theta, \mu)$;*
3. *$s(\theta, X)$ is P_θ-square integrable;*
4. *the sequence of random matrices $\{E[\partial^2 \ell(\theta, X)/\partial\theta\partial\theta^T] : n \in \mathbb{N}\}$ is stochastically equicontinuous in a neighborhood of θ_0.*

Then $J(\theta) = -K(\theta)$, and LAN is satisfied with $I(\theta) = -J(\theta) = K(\theta)$.

Proof. Under Assumption 10.3, $S_n = \sqrt{n} E_n[s(\theta_0, X)]$. Because $f_\theta(X)$ satisfies DUI$^+(\theta, \mu)$, $s(\theta, X)$ is an unbiased estimating equation. Because $s(\theta, X)$ is P_θ-square-integrable, by the Central Limit Theorem,

$$S_n \xrightarrow[Q_n]{\mathcal{D}} N(0, K(\theta_0)), \tag{10.17}$$

By Taylor's mean value theorem,

$$\begin{aligned} E_n[\ell(\theta_n, X)] =& E_n[\ell(\theta_0, X)] + n^{-1/2} E_n[\partial \ell(\theta_0, X)/\partial\theta^T]\delta \\ &+ n^{-1}\delta^T E_n[\partial \ell(\theta^\dagger, X)/\partial\theta\partial\theta^T]\delta \end{aligned}$$

for some $\theta^\dagger$ between θ_0 and θ_n. By $\theta_n \to \theta_0$, the stochastic equicontinuity condition *4*, and Corollary 8.2, we have

$$E_n[\partial \ell(\theta^\dagger, X)/\partial\theta\partial\theta^T] \stackrel{Q_n}{=} J(\theta_0) + o_P(1).$$

Hence

$$L_n \stackrel{Q_n}{=} \delta^T S_n - \delta^T (J + o_P(1))\delta/2 = \delta^T S_n - \delta^T J\delta/2 + o_P(1), \tag{10.18}$$

where J is the abbreviation of $J(\theta_0)$. Finally, because $s(\theta, X)f_\theta(X)$ satisfies DUI$^+(\theta, \mu)$, we have

$$-J(\theta) = K(\theta) = I(\theta). \tag{10.19}$$

Now the proposition follows from (10.17), (10.18), and (10.19). □

10.5 The convolution theorem

Now we are ready to prove the Le Cam-Hajek convolution theorem (see, for example, Bickel, Klaassen, Ritov, and Wellner (1993)). This theorem asserts, roughly, if $\tilde{\theta}$ is a regular estimate (see below for a definition) of θ_0, then $\sqrt{n}(\tilde{\theta} - \theta_0)$ can be written as the sum of two asymptotically independent random vectors, the first of which has asymptotic distribution $N(0, I^{-1}(\theta_0))$. This is significant because it implies that the asymptotic variance of any regular estimate is greater than or equal to the asymptotic variance of the maximum likelihood estimate. This form of optimality of the MLE is much stronger than the one stated in Section 9.5 following the proof of Theorem 9.3, because a regular estimate need not be asymptotically linear or asymptotically normal.

We first introduce the concept of a regular estimate. Let $\theta_0 \in \Theta$ be the true parameter. Let $h : \Theta \to \mathbb{R}^r$, $r \leq p$, be a differentiable function.

Definition 10.3 *Under the local parametric model in Definition 10.2, we say that $\hat{\psi}$ is a regular estimate of $h(\theta_0)$ if*

$$\sqrt{n}\{\hat{\psi} - h[\theta_n(\delta)]\} \xrightarrow[P_n(\delta)]{\mathcal{D}} Z,$$

where Z is a random vector whose distribution does not depend on δ.

Of special importance are the following two scenarios:

1. $h(\theta) = \theta$ for all $\theta \in \Theta$. In this case $\hat{\theta}$ is a regular estimate of θ_0 if and only if

$$\sqrt{n}[\hat{\theta} - \theta_n(\delta)] \xrightarrow[P_n(\delta)]{\mathcal{D}} Z,$$

   where the distribution of Z does not depend on δ.
2. $\theta = (\psi^T, \lambda^T)^T$, where $\psi \in \mathbb{R}^r$ is the parameter of interest, and $\lambda \in \mathbb{R}^s$ is the nuisance parameter. In this case $\hat{\psi}$ is a regular estimate of ψ_0 if and only if

$$\sqrt{n}[\hat{\psi} - \psi_n(\delta)] \xrightarrow[P_n(\delta)]{\mathcal{D}} Z,$$

   where $\psi(\delta)$ is the first r components of $\theta_n(\delta)$, and the distribution of Z does not depend on δ.

Intuitively, regularity means that in the vicinity of θ_0, the asymptotic distribution of $\sqrt{n}[\hat{\psi} - h(\theta)]$ under $P_{n\theta}$ is essentially the same. Thus it is a type of smoothness of the limiting distribution. Technically, since regularity concerns the asymptotic distribution under the local alternative distribution $P_n(\delta)$, it links $\sqrt{n}(\hat{\psi} - h(\theta_0))$ with the likelihood ratio L_n. Indeed, a main point of the convolution theorem is that, if $\hat{\psi}$ is regular and (S_n, L_n) satisfies LAN, then the joint distribution of $\sqrt{n}(\hat{\psi} - h(\theta_0))$ and L_n converges weakly. We are now ready to prove the convolution theorem.

Theorem 10.3 (Convolution Theorem) *If $\hat{\psi}$ is a regular estimate of $h(\theta_0)$ and (S_n, L_n) satisfies LAN, then*

$$\sqrt{n}\{\hat{\psi} - h[\theta_n(\delta)]\} \xrightarrow[P_n(\delta)]{\mathcal{D}} \dot{h}^T I^{-1} S + R, \tag{10.20}$$

where $S \perp\!\!\!\perp R$, and S, I are as defined in LAN – that is, $S_n \xrightarrow[Q_n]{\mathcal{D}} S$ and $E(SS^T) = I$.

Proof. Let $U_n = \sqrt{n}[\hat{\psi} - h(\theta_0)]$. Since $\hat{\psi}$ is regular,

$$\sqrt{n}\{\hat{\psi} - h[\theta_n(\delta)]\} \xrightarrow[P_n(\delta)]{\mathcal{D}} U$$

where the distribution of U does not depend on δ. Taking $\delta = 0$, we have $U_n \xrightarrow[Q_n]{\mathcal{D}} U$. By the LAN assumption, we also have $S_n \xrightarrow[Q_n]{\mathcal{D}} N(0, I)$. Therefore both sequences $\{U_n\}$ and $\{S_n\}$ are tight under Q_n, implying that (U_n, S_n) is jointly tight under Q_n. By Prohorov's theorem, for any subsequence $\{n'\}$ there is a further subsequence $\{n''\}$ such that $(U_{n''}, S_{n''}) \xrightarrow[Q_n]{\mathcal{D}} W$, where the first r arguments of W has the same distribution as U and the last p arguments of W has the same distribution as S. Naturally, we denote this fact by $(U_{n''}, S_{n''}) \xrightarrow[Q_n]{\mathcal{D}} (U, S)$.

Now fix a subsequence $\{n'\}$ and for convenience write the further subsequence $\{n''\}$ as $\{k\}$. By the LAN assumption and Slutsky's theorem,

$$(U_k, L_k(\delta)) \overset{Q_k}{=} (U_k, \delta^T S_k - \delta^T I\delta/2) + o_P(1) \xrightarrow[Q_k]{\mathcal{D}} (U, \delta^T S - \delta^T I\delta/2).$$

By Continuous Mapping Theorem,

$$(U_k, dP_k(\delta)/dQ_k) \xrightarrow[Q_k]{\mathcal{D}} (U, e^{\delta^T S - \delta^T I\delta/2}).$$

By Le Cam's third Lemma (Theorem 10.2), if L is the set function $L(B) = E[I_B(U)e^{\delta^T S - \delta^T I\delta/2}]$, then L is a probability measure, and $U_k \xrightarrow[P_k(\delta)]{\mathcal{D}} L$. By Corollary 10.1, the characteristic function of L is

$$\kappa_L(t) = E\left(e^{it^T U + \delta^T S - \delta^T I\delta/2}\right). \tag{10.21}$$

The same characteristic function can be deduced from the regularity of $\hat{\psi}$. Because h is differentiable, U_k can be expressed as

$$U_k = \sqrt{k}[\hat{\psi} - h(\theta_k(\delta))] + \dot{h}^T\delta + o(1).$$

Hence $U_k \xrightarrow[P_k(\delta)]{\mathcal{D}} U + \dot{h}^T\delta$, and an alternative expression for the characteristic function of the limit law of U_k under $P_k(\delta)$ is

$$\kappa_L(t) = E\left(e^{it^T U + it^T \dot{h}^T\delta}\right). \tag{10.22}$$

From (10.21) and (10.22) we have

$$E\left(e^{it^T U + \delta^T S - \delta^T I\delta/2}\right) = E\left(e^{it^T U + it^T \dot{h}^T\delta}\right).$$

By Lemma 2.2, both sides of the above equality are analytic functions of $\delta \in \mathbb{R}^p$. Hence, by the analytic continuation theorem (Theorem 2.7), the equality holds for all $\delta \in \mathbb{C}^p$, the p-fold Cartesian product of the complex plane $\mathbb{C}$. Take $\delta = iu$, where $u \in \mathbb{R}^p$. Then

$$\begin{aligned} &E(e^{it^TU+iu^TS+u^TIu/2}) = E(e^{it^TU-t^T\dot{h}^Tu}) \\ \Rightarrow &E(e^{it^TU+iu^TS}) = e^{-t^T\dot{h}^Tu-u^TIu/2}E(e^{it^TU}). \end{aligned} \tag{10.23}$$

Since the right hand side depends only on U, the joint distribution of (U, S) does not depend on the subsequence k. Therefore, $(U_n, S_n) \xrightarrow[Q_n]{\mathcal{D}} (U, S)$ along the whole sequence. By continuous mapping theorem

$$(U_n - \dot{h}^TI^{-1}S_n, S_n) \xrightarrow[Q_n]{\mathcal{D}} (U - \dot{h}^TI^{-1}S, S) \equiv (R, S).$$

Next, let us show that R and S are independent. The characteristic function of (R, S) is

$$\begin{aligned} \kappa_{(R,S)}(t, u) &= E\left(e^{it^T(U-\dot{h}^TI^{-1}S)+iu^TS}\right) \\ &= E\left(e^{it^TU+i(-I^{-1}\dot{h}t+u)^TS}\right) = \kappa_{(U,S)}(t, -\dot{h}^TI^{-1}t + u) \end{aligned}$$

By (10.23),

$$\begin{aligned} &\kappa_{(U,S)}(t, -\dot{h}^TI^{-1}t + u) \\ &= e^{-t^T\dot{h}^T(-I^{-1}\dot{h}t+u)-(-I^{-1}\dot{h}t+u)^TI_\theta(-I^{-1}\dot{h}t+u)/2}E(e^{it^TU}) \\ &= e^{t^T\dot{h}^TI^{-1}\dot{h}t/2-u^TIu/2}E(e^{it^TU}). \end{aligned}$$

Therefore,

$$\kappa_{(R,S)}(t, u) = \left(e^{-u^TIu/2}\right)\left(e^{t^T\dot{h}^TI^{-1}\dot{h}t/2}Ee^{it^TU}\right).$$

Since the characteristic function of (R, S) factorizes into the product of a function of t and a function of u, R and S are independent. Therefore, $U - \dot{h}^TI^{-1}S$ and $\dot{h}^TI^{-1}S$ are independent. So

$$U_n \xrightarrow[Q_n]{\mathcal{D}} U = (U - \dot{h}^TI^{-1}S) + \dot{h}^TI^{-1}S,$$

where $U - \dot{h}^TI^{-1}S$ and $\dot{h}^TI^{-1}S$ are independent.

Finally, because $\hat{\psi}$ is regular, the asymptotic distribution of $\sqrt{n}\{\hat{\psi} - h[\theta_n(\delta)]\}$ under $P_n(\delta)$ is the same as the asymptotic distribution of U_n under Q_n. Thus we have (10.20). □

The next corollary gives an alternative form of the convolution theorem.

Corollary 10.3 *Suppose (S_n, L_n) satisfies LAN and $\hat{\psi}$ is a regular estimate of $h(\theta_0)$. Then*

$$\sqrt{n}[\hat{\psi} - h(\theta_0)] = \dot{h}^T I^{-1} S_n + R_n,$$

where $(S_n, R_n) \xrightarrow[Q_n]{\mathcal{D}} (S, R)$ and $S \perp\!\!\!\perp R$.

Proof. By the proof of Theorem 10.3, $(S_n, R_n) \xrightarrow[Q_n]{\mathcal{D}} (S, R)$, where $S \perp\!\!\!\perp R$ and $R_n = \sqrt{n}[\hat{\psi} - h(\theta_0)] - \dot{h}^T I^{-1} S_n$. □

The name "convolution" is motivated by the fact that, because of the independence between S and R, the distribution of U is the convolution of the distribution of R and the distribution of $\dot{h}^T I^{-1} S$. An important fact emerged in the proof of the above theorem — that is, (U_n, S_n) converges in distribution to a random vector (U, S). This is not automatically implied by $U_n \xrightarrow[Q_n]{\mathcal{D}} U$, $S_n \xrightarrow[Q_n]{\mathcal{D}} S$, and $L_n \xrightarrow[Q_n]{\mathcal{D}} \log(V)$. Instead it was deduced from the regularity of $\hat{\psi}$ and the LAN condition using the argument via subsequences. This result is of importance in its own and we record it below as a corollary.

Corollary 10.4 *If (S_n, L_n) satisfies LAN and $\hat{\psi}$ is a regular estimate of $h(\theta_0)$, then $(U_n, S_n) \xrightarrow[Q_n]{\mathcal{D}} (U, S)$ for some random vector in $\mathbb{R}^{k+p}$, where $U_n = \sqrt{n}[\hat{\psi} - h(\theta_0)]$. The characteristic function of (U, S) is*

$$\kappa_{U,S}(t, u) = e^{-t^T \dot{h}^T u - u^T I u/2} E(e^{it^T U}).$$

To complete the picture of regular estimate and convolution theorem, we show that the convolution form, in fact, characterizes a regular estimate; that is, an estimate that can be written as the convolution form must be regular. We first prove a lemma.

Lemma 10.3 *If $X \sim N(\mu, \Sigma)$, then, for any $s \in \mathbb{C}^p$,*

$$E(e^{s^T X}) = \exp(\mu^T s + s^T \Sigma s/2). \tag{10.24}$$

Proof. Since the moment generating function of X is the right-hand side of (10.24) with $s \in \mathbb{R}^p$, the equality (10.24) holds for all $s \in \mathbb{R}^p$. Because the functions on both sides of (10.24) are analytic functions of s, by the analytic continuation theorem (Theorem 2.7), the equality holds for all $s \in \mathbb{C}^p$. □

Theorem 10.4 *Suppose that (S_n, L_n) satisfies LAN. Then the following statements are equivalent:*

1. $\hat{\psi}$ is a regular estimate of $h(\theta_0)$;

2. $\sqrt{n}[\hat{\psi} - h(\theta_0)] = \dot{h}^T I^{-1} S_n + R_n$ *where* $(S_n, R_n) \xrightarrow[Q_n]{\mathcal{D}} (S, R)$ *with* $S \perp\!\!\!\perp R$.

Proof. 1 ⇒ 2. This is Corollary 10.3.
2 ⇒ 1. By the differentiability of h and the LAN assumption,

$$\begin{pmatrix} \sqrt{n}\{\hat{\psi} - h[\theta_n(\delta)]\} \\ L_n \end{pmatrix} \overset{Q_n}{=} \begin{pmatrix} \dot{h}^T I^{-1} S_n + R_n - \dot{h}^T \delta \\ \delta^T S_n - \delta^T I \delta / 2 \end{pmatrix} + o_P(1).$$

Because $(S_n, R_n) \xrightarrow[Q_n]{\mathcal{D}} (S, R)$, the above implies

$$\begin{pmatrix} \sqrt{n}\{\hat{\psi} - h[\theta_n(\delta)]\} \\ L_n \end{pmatrix} \xrightarrow[Q_n]{\mathcal{D}} \begin{pmatrix} \dot{h}^T I^{-1} S + R - \dot{h}^T \delta \\ \delta^T S - \delta^T I \delta / 2 \end{pmatrix}.$$

By Le Cam's third lemma and Corollary 10.1, $\sqrt{n}\{\hat{\psi} - h[\theta_n(\delta)]\} \xrightarrow[P_n(\delta)]{\mathcal{D}} L$, where the characteristic function of L is

$$\begin{aligned} \kappa_L(t) &= E\left\{\exp[i(\dot{h}^T I^{-1} S + R - \dot{h}^T \delta)^T t] \exp(\delta^T S - \delta^T I \delta / 2)\right\} \\ &= E\left\{\exp[i t^T \dot{h}^T I^{-1} S - i t^T \dot{h}^T \delta + \delta^T S - \delta^T I \delta / 2]\right\} \kappa_R(t) \qquad (10.25) \\ &= E\left\{\exp[(i I^{-1} \dot{h} t + \delta)^T S]\right\} \exp(-i t^T \dot{h}^T \delta - \delta^T I \delta / 2) \kappa_R(t), \end{aligned}$$

where the second equality holds because $S \perp\!\!\!\perp R$, and κ_R is the characteristic function of R. Because $S \sim N(0, I)$, by Lemma 10.3,

$$\begin{aligned} E\left\{\exp[(i I^{-1} \dot{h} t + \delta)^T S]\right\} &= \exp\left[(i I^{-1} \dot{h} t + \delta)^T I (i I^{-1} \dot{h} t + \delta)/2\right] \\ &= \exp\left(-t^T \dot{h}^T I^{-1} \dot{h} t / 2 + i t^T \dot{T}^T \delta + \delta^T I \delta / 2\right). \end{aligned}$$

Substituting this into the right-hand side of (10.25), we have

$$\kappa_L(t) = \exp\left(-t^T \dot{h}^T I^{-1} \dot{h} t / 2\right) \kappa_R(t).$$

Since this characteristic function is independent of δ, the probability measure L does not depend on δ. Hence $\hat{\psi}$ is a regular estimate of $h(\theta_0)$. □

10.6 Asymptotically efficient estimates

Equipped with the convolution theorem, we can now answer the question raised at the beginning of this chapter: the maximum likelihood estimate — or any estimate that is asymptotically equivalent to it — is optimal among how large a class of estimates? The convolution theorem implies that if $\hat{\psi}$ is a regular estimate of $h(\theta_0)$ and (S_n, L_n) satisfies LAN, then

$$\sqrt{n}[\hat{\psi} - h(\theta_0)] \xrightarrow[Q_n]{\mathcal{D}} \dot{h}^T I^{-1} S + R, \quad S \perp\!\!\!\perp R.$$

Thus the asymptotic variance of $\sqrt{n}[\hat{\psi} - h(\theta_0)]$ is bounded from below by $\dot{h}^T I^{-1} \dot{h}$ in terms of Louwner's ordering. In other words, any regular estimate of $h(\theta_0)$ that is asymptotically normal with asymptotic variance $\dot{h}^T I^{-1} \dot{h}$ is optimal among all regular estimates. This leads to the following formal definition.

Definition 10.4 *Under the LAN assumption, an estimate $\hat{\psi}$ of $h(\theta_0)$ is asymptotically efficient if it is regular with asymptotic distribution*

$$\sqrt{n}[\hat{\psi} - h(\theta_0)] \xrightarrow[Q_n]{\mathcal{D}} N(0, \dot{h}^T I^{-1} \dot{h}), \tag{10.26}$$

where $I = E(SS^T)$, and S is as defined in LAN.

Note that, because $\hat{\psi}$ is a regular estimate of $h(\theta_0)$, expression (10.26) is equivalent to

$$\sqrt{n}[\hat{\psi} - h(\theta_n(\delta))] \xrightarrow[P_n(\delta)]{\mathcal{D}} N(0, \dot{h}^T I^{-1} \dot{h}) \quad \text{for all } \delta \in \mathbb{R}^p.$$

Moreover, the above convergence implies the regularity of $\hat{\psi}$ and the convergence (10.26). Hence we have the following equivalent definition of an asymptotically efficient estimate.

Definition 10.5 *Under the LAN assumption, an estimate $\hat{\psi}$ of $h(\theta_0)$ is asymptotically efficient if*

$$\sqrt{n}[\hat{\psi} - h(\theta_n(\theta))] \xrightarrow[P_n(\delta)]{\mathcal{D}} N(0, \dot{h}^T I^{-1} \dot{h}) \quad \textit{for all } \delta \in \mathbb{R}^p. \tag{10.27}$$

In the case of $h(\theta) = \theta$, the right-hand side of (10.26) is $N(0, I^{-1})$. As shown in Chapter 8, this is the asymptotic distribution of the maximum likelihood estimate in the i.i.d. case. Hence the maximum likelihood estimate is asymptotically efficient estimate of θ_0 in that case. In the case of $h(\theta) = \psi$, the right-hand side of (10.26) is $N(0, \dot{h}^T I^{-1} \dot{h})$, where $\dot{h}(\theta_0) = (I_r, 0)$. Following the notations in Section 9.8, let $I_{\psi\psi}$ be the $r \times r$ upper left block I and $I_{\psi\lambda}$ be the upper right block of dimension $r \times s$, and so on, and make the similar partition to I^{-1}. Then

$$\dot{h}^T I^{-1} \dot{h} = (U_r, 0) I^{-1} (U_r, 0)^T = (I^{-1})_{\psi\psi} = (I_{\psi\psi} - I_{\psi\lambda} I_{\lambda\lambda}^{-1} I_{\lambda\psi})^{-1} \equiv I_{\psi\cdot\lambda}^{-1}.$$

As in Section 9.8, we call $I_{\psi\cdot\lambda}$ the efficient information. Note that, here, I has a much more general meaning than the Fisher information in the i.i.d. case, as the LAN assumption accommodates far more probability models than the i.i.d. model. In this i.i.d. case, this definition reduces to the definition in Section 9.8 because $E(SS^T)$ is precisely $E[s(\theta_0, X) s(\theta_0, X)^T]$. We have shown in Chapter 9 that, if $\hat{\psi}$ is the ψ-component of the maximum likelihood estimate $\hat{\theta}$, then $\sqrt{n}(\hat{\psi} - \psi)$ has asymptotic normal with variance $I_{\psi\cdot\lambda}^{-1}$. Thus, $\hat{\psi}$ is an asymptotically efficient estimate of ψ_0 among all regular estimates of ψ_0.

More generally, if $\hat{\theta}$ is the maximum likelihood estimate, then, by the δ-method, $\sqrt{n}(h(\hat{\theta}) - h(\theta_0))$ has asymptotic distribution $N(0, \dot{h}^T I^{-1} \dot{h})$. Thus $\hat{\psi} = h(\hat{\theta})$ is an asymptotically efficient estimate of $h(\theta_0)$. Furthermore, as we have shown in Chapter 9, there is a wide class of estimates, such as the one-step Newton-Raphson estimates, that have the same asymptotic distribution as the maximum likelihood estimate. All these estimates are asymptotically efficient.

The next theorem gives a sufficient and necessary condition for an estimate to be asymptotically efficient.

Theorem 10.5 *If (S_n, L_n) satisfies LAN, then the following conditions are equivalent:*

1. *$\hat{\psi}$ is an asymptotically efficient estimate of $h(\theta_0)$;*
2. *$\sqrt{n}[\hat{\psi} - h(\theta_0)] \stackrel{Q_n}{=} \dot{h}^T I^{-1} S_n + o_P(1)$.*

Proof. 2 $\Rightarrow$ 1. Obviously, statement 2 implies $\sqrt{n}[\hat{\psi} - h(\theta_0)] \xrightarrow[Q_n]{\mathcal{D}} N(0, \dot{h}^T I \dot{h})$. To see that $\hat{\psi}$ is regular, note that statement 2 means

$$\sqrt{n}[\hat{\psi} - h(\theta_0)] = \dot{h}^T I^{-1} S_n + R_n$$

where $R_n \xrightarrow[Q_n]{\mathcal{D}} 0$. Because $S_n \xrightarrow[Q_n]{\mathcal{D}} S$ and 0 and S are independent, by Theorem 10.4, $\hat{\psi}$ is a regular estimate of $h(\theta_0)$.

1 $\Rightarrow$ 2. Let $U_n = \sqrt{n}[\hat{\psi} - h(\theta_0)]$. Because $\hat{\psi}$ is a regular estimate of $h(\theta_0)$,

$$U_n = \dot{h}^T I^{-1} S_n + R_n,$$

where R_n and S_n are asymptotically independent. Hence the asymptotic variance of U_n is the sum of the asymptotic variance of $\dot{h}^T I^{-1} S_n$ and the asymptotic variance of R_n. By assumption, the asymptotic variance of U_n is $\dot{h}^T I^{-1} \dot{h}$. Since $S_n \xrightarrow[Q_n]{\mathcal{D}} N(0, I)$, the asymptotic variance of $\dot{h}^T I^{-1} S_n$ is also $\dot{h}^T I^{-1} \dot{h}$. So the asymptotic variance of R_n is 0, implying $R_n \stackrel{Q_n}{=} o_P(1)$. □

10.7 Augmented LAN

An important special case of Le Cam's third lemma is when (U_n, S_n) has a joint asymptotic normal distribution in addition to the LAN assumption on (S_n, L_n). This not only provides us concrete examples and verification criteria for regular estimates, but also plays a critical role in the development of local alternative distribution for hypothesis testing that will be done in the next chapter. Following Hall and Mathiason (1990), we refer to the LAN assumption together with the joint asymptotic normal assumption on (U_n, S_n) as the augmented local asymptotic normal assumption, or ALAN (Hall and Mathiason abbreviated this assumption as LAN$^{\#}$).

Assumption 10.4 (ALAN) *Let $\{U_n : n \in \mathbb{N}\}$ be a sequence random vectors on $(\Omega_n, \mathcal{F}_n)$. We say that (U_n, S_n, L_n) satisfies ALAN if (10.16) holds and*

$$\begin{pmatrix} U_n \\ S_n \end{pmatrix} \xrightarrow[Q_n]{\mathcal{D}} N\left[\begin{pmatrix} 0 \\ 0 \end{pmatrix}, \begin{pmatrix} \Sigma_U & \Sigma_{US} \\ \Sigma_{SU} & I \end{pmatrix}\right]. \tag{10.28}$$

From Assumption 10.4, we can easily derive the asymptotic joint distribution of (U_n, L_n) under Q_n.

Proposition 10.3 *If (U_n, S_n, L_n) satisfies ALAN, then*

$$\begin{pmatrix} U_n \\ L_n(\delta) \end{pmatrix} \xrightarrow[Q_n]{\mathcal{D}} N\left[\begin{pmatrix} 0 \\ -\delta^T I\delta/2 \end{pmatrix}, \begin{pmatrix} \Sigma_U & \Sigma_{US}\delta \\ \delta^T\Sigma_{SU} & \delta^T I\delta \end{pmatrix}\right]. \tag{10.29}$$

Proof. Let (U, S) be the random vector whose joint distribution is the multivariate Normal distribution on the right-hand side of (10.28). By the continuous mapping theorem,

$$\begin{pmatrix} U_n \\ \delta^T S_n - \delta^T I\delta/2 \end{pmatrix} \xrightarrow[Q_n]{\mathcal{D}} \begin{pmatrix} U \\ \delta^T S - \delta^T I\delta/2 \end{pmatrix}.$$

By (10.16) and Slutsky's Theorem,

$$\begin{pmatrix} U_n \\ L_n(\delta) \end{pmatrix} \xrightarrow[Q_n]{\mathcal{D}} \begin{pmatrix} U \\ \delta^T S - \delta^T I\delta/2 \end{pmatrix}.$$

Because

$$\begin{aligned} &E(U) = 0, \quad E(\delta^T S - \delta^T I\delta/2) = -\delta^T I\delta/2 \\ &\mathrm{var}(U) = \Sigma_U, \quad \mathrm{cov}(U, \delta^T S - \delta^T I\delta/2) = \Sigma_{US}\delta, \\ &\mathrm{var}(\delta^T S - \delta^T I\delta/2) = \delta^T I\delta, \end{aligned}$$

the distribution of $(U^T, \delta^T S - \delta^T I\delta/2)^T$ is the right-hand side of (10.29). □

Recall from Corollary 10.4 that regularity of $\hat{\psi}$ and LAN together implies the joint convergence in distribution of (U_n, S_n). From this and the above proposition we can easily deduce the following sufficient and necessary condition for ALAN.

Proposition 10.4 *Suppose $\hat{\psi}$ is a regular estimate of $h(\theta_0)$, and let $U_n = \sqrt{n}[\hat{\psi} - h(\theta_0)]$. Then the following conditions are equivalent:*

1. (S_n, L_n) *satisfies LAN and* $U_n \xrightarrow[Q_n]{\mathcal{D}} N(0, \Sigma_U)$*;*

2. (U_n, S_n, L_n) *satisfies ALAN with* $\Sigma_{US} = \dot{h}^T$*.*

Proof. 2 ⇒ 1. If 2 holds, then (U_n, S_n) converges to a multivariate normal random vector. Hence S_n converges marginally to a multivariate normal vector, which, together with the assumption on L_n in ALAN, implies that (S_n, L_n) satisfies LAN. Since U_n also converges marginally to a multivariate normal vector, $U_n \xrightarrow[Q_n]{\mathcal{D}} N(0, \Sigma_U)$ holds.

1 ⇒ 2. Since $\hat{\psi}$ is regular and (S_n, L_n) satisfies LAN, by Corollary 10.4, $(U_n, S_n) \xrightarrow[Q_n]{\mathcal{D}} (U, S)$ with characteristic function

$$\kappa_{U,S}(t, u) = e^{-t^T \dot{h}^T u - u^T I u/2} E(e^{it^T U}).$$

Because $U \sim N(0, \Sigma_U)$,

$$\begin{aligned}\kappa_{U,S}(t, u) &= e^{-t^T \dot{h}^T u - u^T I u/2} e^{-t^T \Sigma_U t/2} \\ &= e^{-\frac{1}{2}(u^T I u - 2t^T \dot{h}^T u + t^T \Sigma_U t)},\end{aligned}$$

which is the characteristic function of a multivariate normal distribution with mean 0 and variance matrix

$$\begin{pmatrix} \Sigma_U & \dot{h}^T \\ \dot{h} & I \end{pmatrix}.$$

Thus statement 2 holds. □

We now develop sufficient conditions for ALAN under the i.i.d. parametric model. Let $\hat{\theta}$ be an estimate of θ_0 and let $U_n = \sqrt{n}(\hat{\theta} - \theta_0)$. The next proposition gives the sufficient conditions for (10.28).

Proposition 10.5 *Suppose Assumption 10.3 holds and*

1. $\hat{\theta}$ *is an asymptotically linear estimate of* θ_0 *with influence function* ψ*;*
2. $s(\theta, X)$ *is* P_θ*-square integrable and* $s(\theta, x) f_\theta(x)$ *satisfies* $DUI^+(\theta, \mu)$*;*

Then condition (10.28) is satisfied with $U_n = \sqrt{n}(\hat{\theta} - \theta_0)$*.*

Proof. By Definition 9.6, $\hat{\theta}$ being an asymptotically linear estimate of θ_0 means

$$\sqrt{n}(\hat{\theta} - \theta_0) = \sqrt{n} E_n[\psi(\theta_0, X)] + o_P(1),$$

where $\psi(\theta_0, X)$ is P_θ-square integrable and $E[\psi(\theta_0, X)] = 0$. Also, under Assumption 10.3, $S_n = \sqrt{n} E_n[s(\theta_0, X)]$. By the Central Limit Theorem and Slutsky's Theorem,

$$\begin{pmatrix} U_n \\ S_n \end{pmatrix} \xrightarrow[Q_n]{\mathcal{D}} N\left[\begin{pmatrix} 0 \\ 0 \end{pmatrix}, \begin{pmatrix} E[\psi(\theta_0, X)\psi^T(\theta_0, X)] & E[\psi(\theta_0, X)s^T(\theta_0, X)] \\ E[s(\theta_0, X)\psi^T(\theta_0, X)] & E[s(\theta_0, X)s^T(\theta_0, X)] \end{pmatrix}\right]$$

By condition 2, the Fisher information $I(\theta_0)$ is well defined and it is the matrix $E[s(\theta_0, X)s^T(\theta_0, X)]$. □

10.8 Le Cam's third lemma under ALAN

Le Cam's third lemma reduces to a particularly convenient form under ALAN, which will be used heavily in the next chapter. We first develop the specific forms of the measure $L(B)$ in Theorem 10.2 and Corollary 10.2.

Lemma 10.4 *Suppose U is a p-dimensional random vector and V is a positive random variable and their joint distribution is determined by*

$$\begin{pmatrix} U \\ \log V \end{pmatrix} \sim N\left[\begin{pmatrix} \mu_U \\ -\sigma^2/2 \end{pmatrix}, \begin{pmatrix} \Sigma_U & \beta \\ \beta^T & \sigma^2 \end{pmatrix}\right]. \tag{10.30}$$

If L is the probability measure on $(\mathbb{R}^p, \mathcal{R}^p)$ defined by

$$L(B) = E[I_B(U)V],$$

then the c.d.f. of L is $N(\mu_U + \beta, \Sigma_U)$.

Proof. Let $W = \log V$. Then $L(B)$ can be rewritten as $E[I_B(U)e^W]$. By Corollary 10.1, the characteristic function of L is

$$\phi_L(t) = E(e^{it^T U} e^W) = \exp\left[\begin{pmatrix} it \\ 1 \end{pmatrix}^T \begin{pmatrix} U \\ W \end{pmatrix}\right]$$

By (10.30) and Lemma 10.3, the right-hand side is

$$\begin{aligned} &\exp\left[\begin{pmatrix} it \\ 1 \end{pmatrix}^T \begin{pmatrix} \mu_U \\ -\sigma^2/2 \end{pmatrix}^T + \frac{1}{2}\begin{pmatrix} it \\ 1 \end{pmatrix}^T \begin{pmatrix} \Sigma_U & \beta \\ \beta^T & \sigma^2 \end{pmatrix} \begin{pmatrix} it \\ 1 \end{pmatrix}\right] \\ &= \exp\left(it^T(\mu_U + \beta) - t^T \Sigma_U t/2\right), \end{aligned}$$

which is the characteristic function of $N(\mu_U + \beta, \Sigma_U)$. □

If we replace U in the lemma by $(U^T, \log V)^T$ then we get the following result.

Corollary 10.5 *Suppose U is a p-dimensional random vector and V is a positive random variable with their joint distribution determined by (10.30). If L is the probability measure on $(\mathbb{R}^{p+1}, \mathcal{R}^{p+1})$ defined by*

$$L(B) = E[I_B(U, \log V)V] \tag{10.31}$$

then the c.d.f. of L is

$$N\left[\begin{pmatrix} \mu_U + \beta \\ \sigma^2/2 \end{pmatrix}, \begin{pmatrix} \Sigma_U & \beta \\ \beta^T & \sigma^2 \end{pmatrix}\right]. \tag{10.32}$$

Proof. If (10.30) holds then

$$\begin{pmatrix} U \\ \log V \\ \log V \end{pmatrix} \sim N\left[\begin{pmatrix} \mu_U \\ -\sigma^2/2 \\ -\sigma^2/2 \end{pmatrix}, \begin{pmatrix} \Sigma_U & \beta & \beta \\ \beta^T & \sigma^2 & \sigma^2 \\ \beta^T & \sigma^2 & \sigma^2 \end{pmatrix}\right]. \tag{10.33}$$

By Lemma 10.4 (with U replaced by $(U^T, \log V)^T$), the probability measure L defined by (10.31) corresponds to the distribution

$$N\left[\begin{pmatrix} \mu_U + \beta \\ -\sigma^2/2 + \sigma^2 \end{pmatrix}, \begin{pmatrix} \Sigma_U & \beta \\ \beta^T & \sigma^2 \end{pmatrix}\right] = N\left[\begin{pmatrix} \mu_U + \beta \\ \sigma^2/2 \end{pmatrix}, \begin{pmatrix} \Sigma_U & \beta \\ \beta^T & \sigma^2 \end{pmatrix}\right],$$

as desired. □

The next theorem gives the special form of Le Cam's third lemma under the ALAN assumption.

Theorem 10.6 *If (U_n, S_n, L_n) satisfies ALAN, then*

$$\begin{pmatrix} U_n \\ L_n \end{pmatrix} \xrightarrow[P_n(\delta)]{\mathcal{D}} N\left[\begin{pmatrix} \Sigma_{US}\delta \\ \delta^T I \delta/2 \end{pmatrix}, \begin{pmatrix} \Sigma_U & \Sigma_{US}\delta \\ \delta^T \Sigma_{SU} & \delta^T I \delta \end{pmatrix}\right]. \tag{10.34}$$

In particular, $U_n \xrightarrow[P_n(\delta)]{\mathcal{D}} N(\Sigma_{US}\delta, \Sigma_U)$.

Proof. By Proposition 10.3, $(U_n^T, L_n)^T \xrightarrow[Q_n]{\mathcal{D}} (U^T, W)^T$, where $(U^T, W)^T$ is the random vector whose joint distribution is the right-hand side of (10.29). By the continuous mapping theorem,

$$\begin{pmatrix} U_n \\ dP_n/dQ_n \end{pmatrix} \xrightarrow[Q_n]{\mathcal{D}} \begin{pmatrix} U \\ e^W \end{pmatrix}.$$

By Corollary 10.2, $(U_n^T, L_n)^T \xrightarrow[P_n]{\mathcal{D}} L$, where L is defined by (10.31). By Corollary 10.5 this measure is, in fact, the distribution on the right-hand side of (10.34). □

A variation of Le Cam's third lemma under the ALAN assumption concerns the joint distribution of U_n and the standardized score function S_n, which is given by the next corollary.

Corollary 10.6 *If (U_n, S_n, L_n) satisfies ALAN, then*

$$\begin{pmatrix} U_n \\ S_n \end{pmatrix} \xrightarrow[P_n(\delta)]{\mathcal{D}} N\left[\begin{pmatrix} \Sigma_{US}\delta \\ I\delta \end{pmatrix}, \begin{pmatrix} \Sigma_U & \Sigma_{US} \\ \Sigma_{SU} & I \end{pmatrix}\right]. \tag{10.35}$$

Proof. Let $(U^T, S^T)^T$ represent the random vector whose joint distribution is (10.28). By the continuous mapping theorem,

$$\begin{pmatrix} U_n \\ S_n \\ \delta^T S_n - \delta^T I\delta/2 \end{pmatrix} \xrightarrow[Q_n]{\mathcal{D}} \begin{pmatrix} U \\ S \\ \delta^T S - \delta^T I\delta/2 \end{pmatrix}.$$

Hence, by (10.16) and Slutsky's theorem,

$$\begin{pmatrix} U_n \\ S_n \\ L_n \end{pmatrix} \xrightarrow[Q_n]{\mathcal{D}} \begin{pmatrix} U \\ S \\ \delta^T S - \delta^T I\delta/2 \end{pmatrix}.$$

Because

$$\text{cov}\left[\begin{pmatrix} U \\ S \end{pmatrix}, \delta^T S - \delta^T I\delta/2\right] = \begin{pmatrix} \Sigma_{US}\delta \\ I\delta \end{pmatrix},$$

by Theorem 10.6, (10.35) holds. □

10.9 Superefficiency

In the last two sections we have shown that $\dot{h}^T I^{-1} \dot{h}$ is the lower bound of the asymptotic variances of all regular estimates. An estimate whose asymptotic variance reaches this lower bound is asymptotically efficient. In this section we use an example to show that it is possible for an estimate that is not regular to have a smaller asymptotic variance than an asymptotically efficient estimate. We call such estimates *superefficient estimates.* More specifically, let $\hat{\theta}$ be an estimate such that $\sqrt{n}(\hat{\theta} - \theta)$ converges in distribution under $P_{n\theta}$. Let $\text{AV}_{\hat{\theta}}(\theta)$ be the asymptotic variance of $\sqrt{n}(\hat{\theta} - \theta)$ under $P_{n\theta}$. Let $I(\theta)$ be the Fisher information.

Definition 10.6 *An estimate $\hat{\theta}$ is superefficient if $\text{AV}_{\hat{\theta}}(\theta) \leq I^{-1}(\theta)$ for all $\theta \in \Theta$, and $\text{AV}_{\hat{\theta}}(\theta) < I^{-1}(\theta)$ for some $\theta \in \Theta$. Here, when θ is a vector, the inequality is in terms of Louwner's ordering.*

For convenience, let us refer to an estimate that is not regular as an *irregular estimate.* We first prove a lemma.

Lemma 10.5 *Suppose $X_n = O_P(1)$. Then for any sequences $\{a_n\}$ and $\{b_n\}$ such that $a_n \to \infty$ and $b_n > 0$, we have $I(X_n > a_n) = o_P(b_n)$.*

The point of this lemma is that if $I(X_n > a_n) = o_P(1)$, then its order of magnitude is arbitrarily small — for example $I(X_n > a_n) = o_P(n^{-100})$. This fact will prove convenient for the discussions in this section.

Proof. Since $X_n \le |X_n|$, it suffices to show that $I(|X_n| > a_n) = o_P(b_n)$. This means, for any $\epsilon > 0$, $P(b_n^{-1} I(|X_n| > a_n) > \epsilon) \to 0$. Because $b_n > 0$, $b_n^{-1} I(|X_n| > a_n) > \epsilon$ if and only if $I(|X_n| > a_n) = 1$. Hence we only need to show $P(|X_n| > a_n) \to 0$. For any fixed $\epsilon > 0$, let $K > 0$ be such that $P(|X_n| > K) < \epsilon$ for all n. Because $a_n \to \infty$, $a_n > K$ for all sufficiently large n. Therefore, for sufficiently large n, $P(|X_n| > a_n) < \epsilon$. Because ϵ is arbitrary we have $\limsup_{n\to\infty} P(|X_n| > a_n) = 0$, as desired. □

The next example describes an estimate, called Hodges-Lehmann estimate, that is irregular and superefficient.

Example 10.1 Suppose that $X_1, X_2, \ldots$ are i.i.d. $N(\theta, 1)$ where $\theta \in \mathbb{R}$. Let $\hat\theta$ be the estimator

$$\hat\theta = \begin{cases} \bar X & \text{if } |\bar X| > n^{-\frac14} \\ a\bar X & \text{if } |\bar X| \le n^{-\frac14} \end{cases}$$

where $0 \le a < 1$. We first show that the above estimate is superefficient. Let $\mathrm{AV}_{\hat\theta}(\theta)$ denote the asymptotic variance of $\sqrt{n}(\hat\theta - \theta)$. We will show that

$$\mathrm{AV}_{\hat\theta}(\theta) \begin{cases} \le I^{-1}(\theta) & \text{for all } \theta \ne 0 \\ < I^{-1}(\theta) & \text{for all } \theta = 0. \end{cases} \tag{10.36}$$

Note that $\sqrt{n}\,(\bar X - \theta) \overset{\mathcal{D}}{=} Z$, where Z has normal distribution with mean zero and unit variance.

Case I: $\theta = 0$. In this case,

$$\begin{aligned} \sqrt{n}(\hat\theta - 0) &= \sqrt{n}\,\bar X I(|\bar X| > n^{-\frac14}) + \sqrt{n}\,a\bar X I(|\bar X| \le n^{-\frac14}). \\ &\overset{\mathcal{D}}{=} ZI(|Z| > n^{\frac14}) + aZI(|Z| \le n^{\frac14}) \\ &\overset{\mathcal{D}}{\to} aZ \sim N(0, a^2). \end{aligned}$$

Case II: $\theta \ne 0$. In this case

$$\sqrt{n}(\hat\theta - \theta) = \sqrt{n}\,\bar X I(|\bar X| > n^{-\frac14}) + \sqrt{n}\,a\bar X I(|\bar X| \le n^{-\frac14}) - \sqrt{n}\,\theta.$$

Without loss of generality, assume $\theta > 0$. Note that

$$\begin{aligned} |\bar X| \le n^{-\frac14} &\Leftrightarrow -n^{-\frac14} \le \bar X \le n^{-\frac14} \\ &\Leftrightarrow \sqrt{n}(-n^{-\frac14} - \theta) \le \sqrt{n}(\bar X - \theta) \le \sqrt{n}(n^{-\frac14} - \theta) \\ &\Rightarrow \sqrt{n}(\theta - \bar X) \ge \sqrt{n}(\theta - n^{-\frac14}). \end{aligned}$$

Because $\theta > 0$, the right-hand side of the last line goes to ∞. But we also know that, under P_θ, the term $\sqrt{n}(\theta - \bar X) = O_P(1)$ (in fact, $\sqrt{n}(\theta - \bar X) \sim N(0,1)$). Therefore, by Lemma 10.5, for any sequence $b_n > 0$, we have

$$I\left(\sqrt{n}(\theta - \bar{X}) \geq \sqrt{n}(\theta - n^{-\frac{1}{4}})\right) = o(b_n),$$

which implies $I(|\bar{X}| \leq n^{-\frac{1}{4}}) = o_P(b_n)$. Take $b_n = n^{-1/2}$, then we have $I(|\bar{X}| \leq n^{-\frac{1}{4}}) = o_P(n^{-1/2})$. Hence

$$\sqrt{n}(\hat{\theta} - \theta) = \sqrt{n}\bar{X}(1 + o_P(n^{-1/2})) + \sqrt{n}a\bar{X}\ o_P(n^{-1/2}) - \sqrt{n}\theta.$$

Because $\bar{X} \xrightarrow{P} \theta$, $\sqrt{n}\bar{X} = O_P(n^{1/2})$. Therefore

$$\sqrt{n}(\hat{\theta} - \theta) = \sqrt{n}(\bar{X} - \theta) + o_P(1) \xrightarrow{\mathcal{D}} N(0, 1).$$

Meanwhile, it is easy to see that the Fisher information in this case is $I(\theta) \equiv 1$. Therefore (10.36) holds.

Next, we show that $\hat{\theta}$ is irregular at $\theta = 0$. Since $\theta_n(\delta) = \delta/\sqrt{n}$, under $P_{\theta_n(\delta)}$, $\sqrt{n}(\bar{X} - \delta/\sqrt{n})$ is distributed as $N(0, 1)$ and therefore has order of magnitude $O_P(1)$. In the meantime,

$$\begin{aligned} |\bar{X}| > n^{-\frac{1}{4}} &\Leftrightarrow \bar{X} > n^{-\frac{1}{4}} \text{ or } \bar{X} < -n^{-\frac{1}{4}} \\ &\Leftrightarrow \bar{X} - n^{-\frac{1}{2}}\delta > n^{-\frac{1}{4}} - n^{-\frac{1}{2}}\delta \text{ or } \bar{X} - n^{-\frac{1}{2}}\delta < -n^{-\frac{1}{4}} - n^{-\frac{1}{2}}\delta \\ &\Leftrightarrow \sqrt{n}(\bar{X} - n^{-\frac{1}{2}}\delta) > n^{\frac{1}{4}} - \delta \text{ or } -\sqrt{n}(\bar{X} - n^{-\frac{1}{2}}\delta) > n^{\frac{1}{4}} + \delta. \end{aligned}$$

Hence

$$\begin{aligned} &I(|\bar{X}| > n^{-\frac{1}{4}}) \\ &\leq I\left(\sqrt{n}(\bar{X} - n^{-\frac{1}{2}}\delta) > n^{\frac{1}{4}} - \delta\right) + I\left(-\sqrt{n}(\bar{X} - n^{-\frac{1}{2}}\delta) > n^{\frac{1}{4}} + \delta\right). \end{aligned}$$

Because

$$\begin{aligned} \sqrt{n}(\bar{X} - n^{-\frac{1}{2}}\delta) &= O_P(1), \quad n^{\frac{1}{4}} - \delta \to \infty, \\ -\sqrt{n}(\bar{X} - n^{-\frac{1}{2}}\delta) &= O_P(1), \quad n^{\frac{1}{4}} + \delta \to \infty, \end{aligned}$$

we have, by Lemma 10.5, for any $b_n > 0$,

$$I(|\bar{X}| > n^{-\frac{1}{4}}) = o_P(b_n), \quad \text{and in particular,} \quad I(|\bar{X}| > n^{-\frac{1}{4}}) = o_P(1).$$

Hence, under $P_{\theta_n(\delta)}$,

$$\hat{\theta} = \bar{X}I(|\bar{X}| > n^{-\frac{1}{4}}) + a\bar{X}I(|\bar{X}| \leq n^{-\frac{1}{4}}) = a\bar{X} + o_P(n^{-1/2}).$$

It follows that

$$\begin{aligned} \sqrt{n}(\hat{\theta} - n^{-\frac{1}{2}}\delta) &= \sqrt{n}(a\bar{X} - n^{-\frac{1}{2}}\delta) + o_P(1) \\ &= \sqrt{n}(a(\bar{X} - n^{-\frac{1}{2}}\delta + n^{-\frac{1}{2}}\delta) - n^{-\frac{1}{2}}\delta) + o_P(1) \\ &= \sqrt{n}a(\bar{X} - n^{-\frac{1}{2}}\delta) + \sqrt{n}(an^{-\frac{1}{2}}\delta - n^{-\frac{1}{2}}\delta) + o_P(1) \\ &= \sqrt{n}a(\bar{X} - n^{-\frac{1}{2}}\delta) + (a-1)\delta + o_P(1) \xrightarrow{\mathcal{D}} N((a-1)\delta, a^2). \end{aligned}$$

Since this distribution depends on δ, $\hat{\theta}$ is irregular at $\theta = 0$. □

The existence of superefficient estimates does not diminish the importance of asymptotically efficient estimates, as superefficient estimates are somewhat pathological. Suppose $\hat{\theta}$ is a superefficient estimate. Let us say that $\theta \in \Theta$ is a superefficient point if $\text{AV}_{\hat{\theta}}(\theta) < I^{-1}(\theta)$. Let S be the set of all superefficient points. Then it can be shown that S has Lebesgue measure 0. See Le Cam (1953, 1960); Bahadur (1964); van der Vaart (1997). This issue will be further explored in a series problems in the Problems section.

Problems

10.1. Let P_n and Q_n probability measures defined by the following distributions:

1. $Q_n = N(0, 1/n)$, $P_n = N(0, 1/n^2)$;
2. $Q_n = N(0, 1/n)$, $P_n = N(1/\sqrt{n}, 1/n)$;
3. $Q_n = U(0, 2/n)$, $P_n = U(0, 1/n)$;
4. $Q_n = U(0, 1/n)$, $P_n = U(0, 1/n^2)$.

In each of the above scenarios,

1. Show that $dP_n/dQ_n \xrightarrow[Q_n]{\mathcal{D}} V$ for some V, and find the distribution of V;
2. Compute $E(V)$;
3. Prove or disprove $P_n \lhd Q_n$.

10.2. Show that, if $c > 0$, then

1. $f(x) = xI_{\{x \le c\}}$ is an upper semi-continuous function bounded from above;
2. $g(x) = (2 - x)I_{\{x > 2/(1+c)\}}$ is a lower semi-continuous function bounded from below.

10.3. Under the assumptions of Theorem 10.2, show that the following statements are equivalent:

1. $P_n \lhd\rhd Q_n$;
2. If $dP_n/dQ_n \xrightarrow[Q_n]{\mathcal{D}} V$ along a subsequence, then $E(V) = 1$, $P(V > 0) = 1$.

10.4. Suppose that $\hat{\theta}$ is a regular estimator of θ_0 and h is a differentiable function. Show that $h(\hat{\theta})$ is a regular estimator of $h(\theta_0)$.

10.5. Suppose that $X_1, \ldots, X_n$ are i.i.d. with $E_\mu(X) = \mu$ and $\text{var}_\mu(X) = 1$. We are interested in estimating μ^2 by $\bar{X}^2$. Let us say $\{n^a\}$ is normalizing sequence if a is so chosen that $n^a(\bar{X}^2 - \mu^2) = O_P(1)$ and $n^a(\bar{X}^2 - \mu^2) \ne o_P(1)$.

1. If $\mu \ne 0$, find the normalizing sequence n^a and the asymptotic distribution of $n^a(\bar{X}^2 - \mu^2)$.
2. If $\mu = 0$, find the normalizing sequence n^a and the asymptotic distribution of $n^a(\bar{X}^2 - \mu^2)$. What is the asymptotic variance of this distribution?

3. If $\mu_n = n^{-1/2}\delta$ where $\delta \neq 0$, find the normalizing sequence n^a and the asymptotic distribution of $n^a(\bar{X}^2 - \mu_n^2)$. What is the asymptotic variance of this distribution?
4. Is $\bar{X}^2$ a regular estimator of μ^2 if at $\mu = 0$?

10.6. Suppose that $\hat{\theta}$ is a regular estimate of a scalar parameter θ_0, and that $(\sqrt{n}(\hat{\theta} - \theta_0), S_n, L_n)$ satisfies ALAN. Denote the limiting distribution of $\sqrt{n}(\hat{\theta} - \theta_n(\delta))$ under $P_n(\delta)$ by $N(0, \sigma^2)$. Let h be a differentiable function of θ. Define the notion of a normalizing sequence as in the last problem.

1. Suppose that $\dot{h}(\theta_0) \neq 0$. Find the normalizing sequence n^a in $n^a[h(\hat{\theta}) - h(\theta_0)]$ and derive the asymptotic distribution of $n^a[h(\hat{\theta}) - h(\theta_n(\delta))]$ under $P_n(\delta)$. Is $h(\hat{\theta})$ regular at θ_0?
2. Suppose that h is twice differential at θ_0 and that $\dot{h}(\theta_0) = 0$. Find the normalizing sequence n^a in $n^a[h(\hat{\theta}) - h(\theta_0)]$ and derive the asymptotic distribution of $n^a[h(\hat{\theta}) - h(\theta_n)]$ under $P_n(\delta)$. Is $h(\hat{\theta})$ regular at θ_0?

10.7. Let θ be a p-dimensional vector, and suppose that $\hat{\theta}$ is a regular estimate of θ, and that $(\sqrt{n}(\hat{\theta} - \theta_0), S_n, L_n)$ satisfies ALAN. Let

$$\tilde{\theta} = \begin{cases} \hat{\theta} & \text{if } \|\hat{\theta} - \theta_0\| > n^{-\frac{1}{4}} \\ \theta_0 + a(\hat{\theta} - \theta_0) & \text{if } \|\hat{\theta} - \theta_0\| \leq n^{-\frac{1}{4}} \end{cases}$$

where $0 < a < 1$. Denote $U_n = \sqrt{n}(\tilde{\theta} - \theta_0)$. Show that U_n satisfies ALAN at under θ_0, and derive the asymptotic covariance Σ_{US} between U_n and S_n under θ_0. Is $\tilde{\theta}$ a regular estimate of θ_0?

10.8. Suppose T_n is a regular estimate of $\theta_0 \in \mathbb{R}^p$. Assuming $p \geq 0$, the James-Stein-type estimator of θ based on T_n can be defined as follows

$$U_n = T_n - (p - 2)\frac{T_n}{\|\sqrt{n}T_n\|^2},$$

where $\|\cdot\|$ is the Euclidean norm. In the following, let Z represent the standard Normal random variable.

1. Write down the local alternative asymptotic distribution of $\sqrt{n}(U_n - \theta_n(\delta))$ at $\theta_0 \neq 0$ in terms of Z. Is U_n regular at $\theta \neq 0$?
2. Write down the local alternative asymptotic distribution of $\sqrt{n}(U_n - \theta_n(\delta))$ at $\theta_0 = 0$ in terms of Z. Is U_n regular at $\theta_0 = 0$?

10.9. Suppose Q_n is the probability measure $N(0, \sigma_n^2)$ and P_n is the probability measure $N(\theta_n, \sigma_n^2)$. Prove the following statements:

1. If $\lim_{n\to\infty} \mu_n/\sigma_n \to \rho$ for some $\rho \neq 0$, then $P_n \lhd Q_n$;
2. If $\lim_{n\to\infty} \mu_n/\sigma_n = 0$, then $P_n \lhd Q_n$;
3. If $\lim_{n\to\infty} \mu_n/\sigma_n = \infty$, then P_n is not contiguous with respect to Q_n.

10.10. A distance between two probability measures, say P and Q, is defined by

$$\|P - Q\| = \sup_A |P(A) - Q(A)|,$$

where the supremum is taken all measurable sets. Let $\{P_n\}$ and $\{Q_n\}$ be two sequences of probability measures. Show that if $\|P_n - Q_n\| \to 0$ as $n \to \infty$ then $P_n \lhd\rhd Q_n$.

10.11. Let $K : R^1 \mapsto R^1$ be a function such that (a) $0 < K(0) < 1$, (b) $\lim_{|u|\to\infty} K(u) = 1$, and (c) K is differentiable and has bounded derivative; that is, $\dot{K}(u) < C$ for some $C > 0$. Let $\hat{\theta}$ be a regular estimator of θ.

1. Show that $K(n^{1/4}\hat{\theta}) \xrightarrow{P} K(0)$ under $\theta = 0$ and $K(n^{1/4}\hat{\theta}) \xrightarrow{P} 1$ under $\theta \neq 0$.
2. Show that $K(n^{1/4}\hat{\theta}) \xrightarrow{P} K(0)$ under $\theta_n = n^{-1/2}\delta$ and $K(n^{1/4}\hat{\theta}) \xrightarrow{P} 1$ under $\theta_n = \theta + n^{-1/2}\delta$, with $\theta_0 \neq 0$.
3. Let $\tilde{\theta} = K(n^{1/4}\hat{\theta})\hat{\theta}$. Derive the asymptotic distribution of $\sqrt{n}(\tilde{\theta} - \theta_n(\delta))$ under $\theta_n(\delta)$, where $\theta_n(\delta) = n^{-1/2}\delta$. Is $\tilde{\theta}$ regular at 0?

Remarks on Problems 10.12 through 10.15

An alternative definition of a regular estimate is the following: $\hat{\psi}$ is a regular estimate of $h(\theta_0)$ if, for any sequence of parameters θ_n such that $\sqrt{n}(\theta_n - \theta_0)$ is bounded, we have $\sqrt{n}[\hat{\psi} - h(\theta_0)] \xrightarrow[P_{n\theta_n}]{\mathcal{D}} Z$, where the distribution of Z does not depend on the sequence θ_n chosen. This alternative definition obviously implies Definition 10.3; it is equivalent to Definition 10.3 under the following assumption: if $\delta_n \to \delta$, then

$$L_n(\delta_n) \overset{Q_n}{=} L_n(\delta) + o_P(1). \tag{10.37}$$

Problems 10.12 through 10.15 provide a proof of the equivalence, and also give a sufficient condition for (10.37). These problems touch on many aspects of this chapter.

10.12. Suppose that $(S_n, L_n(\delta))$ satisfies LAN, and $\hat{\psi}$ is a regular estimate of $h(\theta_0)$ in the sense of Definition 10.3. Show that

$$\begin{pmatrix} U_n \\ L_n(\delta) \end{pmatrix} \xrightarrow[Q_n]{\mathcal{D}} \begin{pmatrix} U \\ \delta^T S - \delta^T I \delta/2 \end{pmatrix}.$$

10.13. Suppose the conditions in Problem 10.12 are satisfied. Consider a sequence of local alternative parameters θ_n such that

$$\sqrt{n}(\theta_n - \theta_0) \equiv \delta_n \to \delta$$

for some $\delta \in \mathbb{R}^p$. Note that θ_n can be written as $\theta_0 + n^{-1/2}\delta_n$, and $P_{n\theta_n}$ can be written as $P_n(\delta_n)$. Let

$$L_n(\delta_n) = \log[dP_{n\theta_n}/dQ_n] = \log[dP_n(\delta_n)/dQ_n].$$

Suppose that $L_n(\delta_n) \overset{Q_n}{=} L_n(\delta) + o_P(1)$. Show that

$$\sqrt{n}[\hat{\psi} - h(\theta_n(\delta_n))] \xrightarrow[P_n(\delta_n)]{\mathcal{D}} Z,$$

where the distribution of Z is the same as the limiting distribution of $\sqrt{n}[\hat{\psi} - h(\theta_n(\delta))]$ under $P_n(\delta)$. Hence conclude that the distribution of Z doesn't depend on the sequence $\{\delta_n\}$ chosen.

10.14. Suppose the conditions in Problem 10.12 are satisfied and, whenever $\delta_n \to \delta$

$$L_n(\delta_n) = L_n(\delta) + o_P(1).$$

Use the argument via subsequences to show that, for any θ_n such that $\sqrt{n}(\theta_n - \theta_0)$ is bounded, $\sqrt{n}[\hat{\psi} - h(\theta_n)] \xrightarrow[P_{n\theta_n}]{\mathcal{D}} Z$, where the distribution of Z is the same as the limiting distribution of $\sqrt{n}(\hat{\psi} - h(\theta_n(\delta))]$ under $P_n(\delta)$. Hence conclude that the distribution of Z does not depend on the sequence $\{\theta_n\}$ chosen.

10.15. Under the assumptions of Proposition 10.2, prove that condition (10.37) is satisfied for all $\delta \in \mathbb{R}^p$.

10.16. Suppose that $U \in \mathbb{R}^r$ and $S \in \mathbb{R}^p$, where $p \geq r$, are random vectors with joint distribution F and marginal distribution F_U and F_S. Suppose H is a $p \times r$ matrix, I is a $p \times p$ positive definite matrix. Suppose F_S is $N(0, I)$. For each $\delta \in \mathbb{R}^p$, let $G_{U,\delta}$ be a measure defined by

$$G_{U,\delta}(B) = E_F[I_B(U)e^{\delta^T S - \delta^T I \delta/2}].$$

Show that $G_{U,\delta}$ is a probability measure. Moreover, suppose that, for any $\delta \in \mathbb{R}^p$, $U + H^T\delta$ has distribution $G_{U,\delta}$. Show that

$$U - H^T I^{-1} S \perp\!\!\!\perp S,$$

and derive the characteristic function of $R = U - H^T I^{-1} S$.

10.17. Suppose $\hat{\psi}$ is a regular estimate of $h(\theta_0)$ and (S_n, L_n) satisfies LAN. Show that $(\sqrt{n}(\hat{\psi} - h(\theta_0)), L_n) \xrightarrow[Q_n]{\mathcal{D}} (U, W)$ with characteristic function

$$\kappa_{U,W}(t, v) = e^{-t^T \dot{h}^T \delta v - v^2 \delta^T I \delta/2} E(e^{it^T U}).$$

10.18. Under the i.i.d. model (Assumption 10.3), suppose that $\hat{\psi}$ is an asymptotically linear estimate of $h(\theta_0)$ with influence function $\rho(\theta, X)$. That is,

$$\sqrt{n}[\hat{\psi} - h(\theta_0)] \overset{Q_n}{=} E_n[\rho(\theta_0, X)] + o_P(n^{-1/2}).$$

where $E[\rho(\theta_0, X)] = 0$ and $\rho(\theta_0, X)$ is P_{θ_0}-square-integrable. Suppose that $\rho(\theta, X) f_\theta(X)$ satisfies $\text{DUI}^+(\theta, \mu)$. Use Theorem 10.4 to show that $\hat{\psi}$ is a regular estimate of $h(\theta_0)$ if and only if

$$E\left[\frac{\partial \rho(\theta_0, X)}{\partial \theta^T}\right] = -\dot{h}(\theta_0).$$

10.19. Recall from Theorem 9.7 that, under the assumptions of Theorem 9.5, a consistent solution $\hat{\theta}$ to the estimating equation $E_n[g(\theta, X)] = 0$ can be expanded as the form

$$\hat{\theta} = \theta_0 - J_g^{-1}(\theta_0) E[g(\theta_0, X)] + o_P(1),$$

where

$$J_g(\theta_0) = E\left[\frac{\partial g(\theta_0, X)}{\partial \theta^T}\right].$$

Show that, under the conditions in Theorem 9.5, $\hat{\theta}$ is a regular estimate of θ_0.

10.20. Suppose that (S_n, L_n) satisfies LAN, and $\theta = (\psi^T, \lambda^T)$, where $\psi \in \mathbb{R}^r$ is the parameter of interest, and $\lambda \in \mathbb{R}^s$ is the nuisance parameter. Suppose $\hat{\psi}$ is a regular estimate of ψ. Show that $\sqrt{n}(\hat{\psi} - \psi_0) \perp\!\!\!\perp S_\lambda$, where S_λ is the last s components of S, and S is as defined in the LAN assumption.

10.21. Suppose that $X_1, \ldots, X_n$ are i.i.d. with density $f(x - \theta)$ where f is a known symmetric p.d.f. defined on $\mathbb{R}$ satisfying

$$\int_{-\infty}^{\infty} [\dot{f}(t)/f(t)]^2 f(t) dt < \infty, \quad \lim_{t \to \infty} f(t) = 0,$$

where $\dot{f}$ denote the derivative of f. Let $T_1 = \bar{X}$ and T_2 be the sample median of X. It can be shown that T_2 satisfies

$$T_2 \stackrel{Q_n}{=} \theta_0 - \frac{1}{f(0)} E_n[I(X \le \theta_0) - 1/2] + o_P(1).$$

1. Derive the asymptotic distribution of $\sqrt{n}(T_2 - \theta_0)$ under Q_n.
2. Show that T_2 satisfies LAN and is a regular estimate of θ_0.
3. Derive the asymptotic distribution of $\sqrt{n}(T_2 - \theta_0)$ under $P_n(\delta)$.
4. Derive the asymptotic distribution of $\sqrt{n}(T_1 - \theta_0, T_2 - \theta_0)$ under Q_n.
5. Derive the asymptotic distribution of $\sqrt{n}(T_1 - \theta_0, T_2 - \theta_0)$ under $P_n(\delta)$.

Remarks on Problems 10.22 through 10.28

Bahadur (1964) gave a relatively simple proof that the collection of superefficient points have Lebesgue measure 0. This method uses the Neyman-Pearson Lemma to derive an inequality concerning the null and local alternative distributions. The next few problems walk through his proof (with adaptation to our context, assumptions, and notations). Proving the various steps of this result turns out to be excellent exercises, as it involves many techniques developed in this chapter. In the context of 1-dimensional θ, it suffices to consider one δ value, say $\delta = 1$. Thus in the following we use $\theta_n(1)$, $L_n(1)$, and $P_n(1)$ and so on. Also, note that $P_n(1)$ and $P_{n\theta_{n(1)}}$ mean the same probability measure. Throughout these problems, we will always make the following assumptions

1. $(S_n, L_n(1))$ satisfies LAN;
2. $\hat{\theta}$ is an estimate of θ_0 such that $\sqrt{n}(\hat{\theta} - \theta_0) \xrightarrow[Q_n]{\mathcal{D}} N(0, v(\theta_0))$.

Our goal is to show that the collection of superefficient points $\{\theta_0 : v(\theta_0) < I^{-1}(\theta_0)\}$ has Lebesgue measure 0.

10.22. Use Corollary 10.1 to show that $L_n(1) \xrightarrow[P_n(1)]{\mathcal{D}} N(I/2, I)$.

10.23. Let $K_n = [L_n(1) + I/2]/\sqrt{I}$. Show that

1. $K_n \xrightarrow[Q_n]{\mathcal{D}} N(0,1), \quad K_n \xrightarrow[P_n(1)]{\mathcal{D}} N(\sqrt{I}, 1)$;
2. for any $k > \sqrt{I}$, $\lim_{n\to\infty} P_{n\theta_n(1)}(K_n \geq k) < 1/2$.

10.24. Use the Neyman-Pearson lemma to show that, if C is the event $K_n \geq k$, and D is any other event such that

$$P_{n\theta_n(1)}(C) < P_{n\theta_n(1)}(D),$$

then $P_{n\theta_0}(C) < P_{n\theta_0}(D)$.

10.25. Show that, if

$$\limsup_n P_{n\theta_n(1)}(\hat{\theta} \geq \theta_n(1)) \geq 1/2,$$

then, for any $k > \sqrt{I}$, there is a subsequence n' such that

$$P_{n'\theta_{n'}(1)}(K_{n'} \geq k) < P_{n'\theta_{n'}(1)}(\hat{\theta} \geq \theta_{n'}(1)),$$

whence use the result of Problem 10.24 to conclude that

$$P_{n'\theta_0}(K_{n'} \geq k) < P_{n'\theta_0}(\hat{\theta} \geq \theta_0). \tag{10.38}$$

10.26. By taking limits on both sides of (10.38), show that $v(\theta_0) \geq k^{-2}$ for any $k > I^{1/2}(\theta_0)$. Conclude that $v(\theta_0) \geq I^{-1}(\theta_0)$, and whence that

$$\{\theta_0 : \limsup_n P_{n\theta_n(1)}(\hat{\theta} \geq \theta_n(1)) \geq 1/2\} \subseteq \{\theta_0 : v(\theta_0) \geq I^{-1}(\theta_0)\}.$$

10.27. Let $\Delta_n(\theta) = |P_\theta(\hat{\theta} < \theta) - 1/2|$, and Φ the c.d.f. of $N(0,1)$. Use the Bounded Convergence Theorem to show that

$$\lim_{n\to\infty} \int \Delta_n(\theta + n^{1/2}) d\Phi(\theta) = 0.$$

Using this to show that $\Delta_n \xrightarrow{\Phi} 0$ (i.e. Δ_n converges in Φ-probability to 0).

10.28. Using the fact that, if $U_n \xrightarrow{P} a$, then $U_n \to a$ almost surely along some subsequence of $\{n\}$, to show that

$$\Phi\left(\{\theta_0 : \liminf_n \Delta_n(\theta_n) = 0\}\right) = 1,$$

whence conclude, in turn,

1. $\Phi(\{\theta_0 : \limsup_n P_{n\theta_n(1)}(\hat{\theta} \geq \theta_n(1)) \geq 1/2\}) = 1$;
2. $\Phi(\{\theta_0 : v(\theta_0) \geq I^{-1}(\theta_0)\}) = 1$.
3. the set $\{\theta_0 : v(\theta_0) < I^{-1}(\theta_0)\}$ has Lebesgue measure 0.

References

Bahadur, R. R. (1964). On Fisher's bound for asymptotic variances. *The Annals of Mathematical Statistics.* **35**, 1545–1552.

Bickel, P. J., Klaassen, C. A. J., Ritov, Y. Wellner, J. A. (1993). *Efficient and Adaptive Estimation for Semiparametric Models.* The Johns Hopkins University Press.

Fisher, R. A. (1922). On the mathematical foundations of theoretical statistics. *Philosophical Transactions of the Royal Society A*, **222**, 594–604.

Fisher, R. A. (1925). Theory of statistical estimation. *Proc. Cambridge Phil. Soc.*, **22**, 700–725.

Hájek, J. (1970). A characterization of limiting distributions of regular estimates. *Zeitschrift für Wahrscheinlichkeitstheorie und Verwandte Gebiete.* **14**, 323–330.

Hall, W. J. and Mathiason, D. J. (1990). On large-sample estimation and testing in parametric models. *Int. Statist. Rev.*, **58**, 77–97.

Le Cam, L. (1953). On some asymptotic Properties of maximum likelihood estimates and related Bayes estimates. *Univ. California Publ. Statistic.* **1**, 277–330.

Le Cam, L. (1960). Locally asymptotically normal families of distributions. *Univ. California Publ. Statistic.* **3**, 370–98.

Le Cam, L. and Yang, G. L. (2000). *Asymptotics in Statistics: Some Basic Concepts.* Second Edition. Springer.

van der Vaart, A. W. (1997). Superefficiency. *Festschrift for Lucien Le Cam: Research Papers in Probability and Statistics,* 397–410. Springer.

van der Vaart, A. W. (1998). *Asymptotic Statistics.* Cambridge University Press.

11

Asymptotic Hypothesis Test

In this chapter we develop various asymptotic methods for testing statistical hypotheses under the general framework of Quadratic Form tests (QF test, Hall and Mathiason 1990), a class of statistics that are asymptotically equivalent to quadratic forms in statistics that satisfy the ALAN assumption in Chapter 10. The asymptotic null and local alternative distributions of a QF test can be easily derived from Le Cam's third lemma. Several commonly used test statistics will be shown to be special cases of QF tests, including Wilks's likelihood ratio test, Wald's test, Rao's score test, Neyman's $C(\alpha)$ test, the Lagrangian multiplier test, as well as tests based on estimating equations. We first consider the testing problem that involves an explicit parameter of interest and an explicit nuisance parameter, and then the more general testing problem where the null hypothesis is specified by an arbitrary nonlinear equation of parameters. We will also introduce the concept of asymptotically efficient QF test whose local power is the greatest among the collection of all QF tests, and Pitman's efficiency that can be used to numerically compare the powers of two tests.

11.1 Quadratic Form test

Consider the setting where θ is a p-dimensional parameter consisting of an r-dimensional parameter of interest ψ and an s-dimensional nuisance parameter λ; that is, $\theta = (\psi, \lambda)$. We allow $r = p$, so as to accommodate the special case $\psi = \theta$ – that is, the entire parameter θ is of interest. Thus, the null hypothesis is

$$H_0 : \psi = \psi_0. \tag{11.1}$$

Here, in the asymptotic approach to hypothesis testing, we consider the following sequence of local alternative hypotheses:

$$H_1^{(n)} : \theta = \theta_n(\delta), \text{ where } \theta_n(\delta) = \theta_0 + n^{-1/2}\delta.$$

B. Li and G. J. Babu, *A Graduate Course on Statistical Inference*,
Springer Texts in Statistics, https://doi.org/10.1007/978-1-4939-9761-9_11

Note that the local alternative hypothesis involves the nuisance parameter λ_0, which does not appear in the null hypothesis. However, λ_0 will always remain offstage, and will not affect the further development in anyway. As before, let Q_n denote the null probability measure $P_{n\theta_0}$ and $P_n(\delta)$ the local alternative probability measure $P_{n\theta_n(\delta)}$.

A more general testing problem than (11.1) is

$$H_0 : h(\theta) = 0, \tag{11.2}$$

where h is an arbitrary differentiable function. This will be taken up in a later section. Although the setting (11.1) is a special case of (11.2), and all the related theories under (11.1) are special cases of their counterparts under (11.2), we will nevertheless first develop the special case and then move on to the more general case. We choose this somewhat inefficient way of presentation because (11.1) is the most commonly used form of hypothesis, and also because this special case helps to develop intuition, especially that related to the efficient score and efficient information. We now give the definition of a Quadratic Form test for the hypothesis (11.1). Let L_n and S_n be the local log likelihood ratio and the standardized score as defined in Definition 10.2 and the note immediately following it. For random vectors V_n and W_n, recall that we write $V_n \overset{Q_n}{=} W_n + o_P(1)$ to mean $Q_n(\|V_n - W_n\| > \epsilon) \to 0$ for every $\epsilon > 0$.

Definition 11.1 *A test statistic $T_n \in \mathbb{R}$ is a Quadratic Form (QF) test if there is an r-dimensional random vector U_n such that (U_n, S_n, L_n) satisfies ALAN with $\Sigma_U \succ 0$, and such that*

$$T_n \overset{Q_n}{=} U_n^T \Sigma_U^{-1} U_n + o_P(1). \tag{11.3}$$

Using Le Cam's third lemma under ALAN (Corollary 10.6), we can easily derive the asymptotic null and alternative distribution of a QF test.

Theorem 11.1 *If T_n is a QF test of the form (11.3), then, for any $\delta \in \mathbb{R}^p$,*

$$T_n \xrightarrow[P_n(\delta)]{\mathcal{D}} \chi^2_r\left(\delta^T \Sigma_{SU} \Sigma_U^{-1} \Sigma_{US} \delta\right).$$

In particular, when $\delta = 0$, we have $T_n \xrightarrow[Q_n]{\mathcal{D}} \chi^2_r$.

Proof. Because T_n is a QF test, it can be written as (11.3) where

$$\begin{pmatrix} U_n \\ S_n \end{pmatrix} \xrightarrow[Q_n]{\mathcal{D}} N\left[\begin{pmatrix} 0 \\ 0 \end{pmatrix}, \begin{pmatrix} \Sigma_U & \Sigma_{US} \\ \Sigma_{SU} & I \end{pmatrix}\right].$$

By Corollary 10.6, $U_n \xrightarrow[P_n(\delta)]{\mathcal{D}} N(\Sigma_{US}\delta, \Sigma_U)$. Hence

$$\Sigma_U^{-1/2} U_n \xrightarrow[P_n(\delta)]{\mathcal{D}} N(\Sigma_U^{-1/2} \Sigma_{US}\delta, I_r),$$

which implies

$$U_n^T \Sigma_U^{-1} U_n \xrightarrow[P_n(\delta)]{\mathcal{D}} \chi_r^2(\delta^T \Sigma_{SU} \Sigma_U^{-1} \Sigma_{US} \delta).$$

Also, because $P_n(\delta) \lhd Q_n$, by Proposition 10.3,

$$T_n \overset{P_n(\delta)}{=} U_n^T \Sigma_U^{-1} U_n + o_P(1).$$

Now apply Slutsky's theorem to prove the asserted convergence. □

The above theorem applies to all QF tests, which include many well known test statistics. Thus, as long as we can show that a particular statistic is a QF test, then we can automatically write down their null and local alternative distributions using the the above theorem. In the following few sections we shall show that a set of most commonly used test statistics are QF tests.

11.2 Wilks's likelihood ratio test

First, consider the parametric model as specified by Assumption 10.1. Suppose $\hat{\theta}$ is the maximum likelihood estimate, and $\tilde{\theta} = (\psi_0^T, \tilde{\lambda}^T)^T$ is the maximum likelihood estimate under the constraint $\psi = \psi_0$. Then the likelihood ratio test is defined as

$$T_n = 2\log(dP_{n\hat{\theta}}/dP_{n\tilde{\theta}}),$$

which is also known as Wilks's test (Wilks, 1938). Under the i.i.d. parametric model (Assumption 10.3), the above reduces to

$$T_n = 2nE_n[\ell(\hat{\theta}, X) - \ell(\tilde{\theta}, X)].$$

The next theorem shows that T_n is a QF test under the i.i.d. model. Recall the notations

$$J(\theta) = E_\theta[\partial s(\theta, X)/\partial \theta^T], \quad K(\theta) = E_\theta[s(\theta, X)s^T(\theta, X)],$$

where $s(\theta, X)$ is the score function $\partial \log f_\theta(X)/\partial \theta$ for an individual observation X. Also recall that, if $f_\theta(X)$ and $s(\theta, X)f_\theta(X)$ satisfy $\text{DUI}^+(\theta, \mu)$, and $s(\theta, X)$ is P_θ-square integrable, then $K(\theta) = -J(\theta)$, and the common matrix $I(\theta)$ is known as the Fisher information. Let $I_{\psi \cdot \lambda}(\theta)$ and $s_{\psi \cdot \lambda}(\theta, X)$ be the efficient information and efficient score defined in Section 9.8. Let

$$J_n(\theta) = E_n[\partial s(\theta, X)/\partial \theta^T]. \tag{11.4}$$

Theorem 11.2 *Suppose Assumption 10.3 holds and*

1. *$\ell(\theta, x)$ is twice differentiable;*
2. *$f_\theta(X)$ and $s(\theta, X)f_\theta(X)$ satisfy $DUI^+(\theta, \mu)$;*

3. $s(\theta, X)$ *is* P_θ*-square integrable and* $K(\theta)$ *is positive definite;*
4. the sequence of random matrices $\{J_n(\theta) : n \in \mathbb{N}\}$ *in (11.4) is stochastically equicontinuous in a neighborhood of* θ_0*;*
5. $\hat{\theta}$ *and* $\tilde{\lambda}$ *are consistent estimates of* θ_0 *and* λ_0*, respectively.*

Then T_n *is a QF test of the form* $T_n \stackrel{Q_n}{=} U_n^T \Sigma_U^{-1} U_n + o_P(1)$*, where*

$$U_n = n^{1/2} E_n[s_{\psi\cdot\lambda}(\theta_0, X)], \quad \Sigma_U = I_{\psi\cdot\lambda}(\theta_0), \quad \Sigma_{US} = (I_{\psi\cdot\lambda}(\theta_0), 0).$$

The above theorem is intended to cover the case $r = p$ as well. In this case, $s_{\psi\cdot\lambda}$ and $I_{\psi\cdot\lambda}$ are to be understood as s and I, and the theorem asserts that $T_n \stackrel{Q_n}{=} U_n^T \Sigma_U^{-1} U_n + o_P(1)$, where

$$U_n = n^{1/2} E_n[s(\theta_0, X)], \quad \Sigma_U = I(\theta_0), \quad \Sigma_{US} = I(\theta_0).$$

Proof of Theorem 11.2. First, consider the case $r < p$. By conditions 2 and 3,

$$K(\theta) = -J(\theta) = I(\theta), \tag{11.5}$$

which implies, in particular, $J(\theta)$ is invertible. Rewrite T_n as $T_n^{(1)} - T_n^{(2)}$, where

$$T_n^{(1)} = 2nE_n[\ell(\hat{\theta}, X) - \ell(\theta_0, X)], \quad T_n^{(2)} = 2nE_n[\ell(\tilde{\theta}, X) - \ell(\theta_0, X)].$$

By Taylor's theorem,

$$\begin{aligned} T_n^{(1)}/(2n) &= E_n[\ell(\hat{\theta}, X)] - E_n[\ell(\theta_0, X)] \\ &= E_n[s^T(\theta_0, X)](\hat{\theta} - \theta_0) + \frac{1}{2}(\hat{\theta} - \theta_0)^T J_n(\theta^\dagger)(\hat{\theta} - \theta_0), \end{aligned}$$

for some $\theta^\dagger$ between θ_0 and $\hat{\theta}$. Because $\theta^\dagger \xrightarrow{Q_n} \theta_0$, by the stochastic equicontinuity assumption and Corollary 8.2,

$$J_n(\theta^\dagger) \stackrel{Q_n}{=} -I(\theta_0) + o_P(1).$$

Hence

$$T_n^{(1)}/(2n) \stackrel{Q_n}{=} E_n[s^T(\theta_0, X)](\hat{\theta} - \theta_0) + \frac{1}{2}(\hat{\theta} - \theta_0)^T[-I(\theta_0) + o_P(1)](\hat{\theta} - \theta_0).$$

By Theorem 9.5 (as applied to $g(\theta, x) = s(\theta, x)$), and the assumption that the MLE $\hat{\theta}$ is consistent, we have

$$\hat{\theta} - \theta_0 \stackrel{Q_n}{=} I^{-1}(\theta_0) E_n[s(\theta_0, X)] + o_P(1). \tag{11.6}$$

Hence

$$T_n^{(1)}/(2n) \stackrel{Q_n}{=} [(\hat{\theta}-\theta_0)^T I(\theta_0) + o_P(n^{-1/2})](\hat{\theta}-\theta_0) + \frac{1}{2}(\hat{\theta}-\theta_0)^T[-I(\theta_0)+o_P(1)](\hat{\theta}-\theta_0).$$

Because (11.6) also implies $\hat{\theta}-\theta_0 \stackrel{Q_n}{=} O_P(n^{-1/2})$, we have

$$T_n^{(1)}/(2n) \stackrel{Q_n}{=} \frac{1}{2}(\hat{\theta}-\theta_0)^T I(\theta_0)(\hat{\theta}-\theta_0) + o_P(n^{-1}). \tag{11.7}$$

By a similar argument we can show that

$$T_n^{(2)}/(2n) \stackrel{Q_n}{=} \frac{1}{2}(\hat{\lambda}-\lambda_0)^T I_{\lambda\lambda}(\theta_0)(\hat{\lambda}-\lambda_0) + o_P(n^{-1}). \tag{11.8}$$

Next, we establish an asymptotic linear relation between $\hat{\theta}-\theta_0$ and $\tilde{\lambda}-\lambda_0$. Applying Theorem 9.5 to the estimating equation $E_n[s_\lambda(\psi_0,\lambda,X)]=0$ where ψ_0 is fixed and λ alone is the argument, under the assumption that $\tilde{\lambda}$ is a consistent solution to this estimating equation (condition 5), we have

$$\begin{aligned}\tilde{\lambda}-\lambda_0 &\stackrel{Q_n}{=} I_{\lambda\lambda}^{-1}(\theta_0)E_n[s_\lambda(\theta_0,X)] + o_P(1)\\ &\stackrel{Q_n}{=} [0, I_{\lambda\lambda}^{-1}(\theta_0)]E_n[s(\theta_0,X)] + o_P(1)\\ &\stackrel{Q_n}{=} [0, I_{\lambda\lambda}^{-1}(\theta_0)]I(\theta_0)(\hat{\theta}-\theta_0) + o_P(1),\end{aligned} \tag{11.9}$$

where, for the third equality we used again the relation (11.6). Substituting this into the right-hand side of (11.8), we have

$$\begin{aligned}T_n^{(2)}/(2n) &\stackrel{Q_n}{=} \frac{1}{2}(\hat{\theta}-\theta_0)^T I \begin{pmatrix} 0 \\ I_{\lambda\lambda}^{-1} \end{pmatrix} I_{\lambda\lambda} \begin{pmatrix} 0 & I_{\lambda\lambda}^{-1} \end{pmatrix} I(\hat{\theta}-\theta_0) + o_P(1)\\ &\stackrel{Q_n}{=} \frac{1}{2}(\hat{\theta}-\theta_0)^T \begin{pmatrix} I_{\psi\lambda}I_{\lambda\lambda}^{-1}I_{\lambda\psi} & I_{\psi\lambda} \\ I_{\lambda\psi} & I_{\lambda\lambda} \end{pmatrix}(\hat{\theta}-\theta_0) + o_P(1).\end{aligned}$$

Hence

$$\begin{aligned}T_n/(2n) &= T_n^{(1)}/(2n) - T_n^{(2)}/(2n)\\ &\stackrel{Q_n}{=} \frac{1}{2}(\hat{\theta}-\theta_0)^T \begin{pmatrix} I_{\psi\cdot\lambda} & 0 \\ 0 & 0 \end{pmatrix}(\hat{\theta}-\theta_0) + o_P(n^{-1})\\ &\stackrel{Q_n}{=} \frac{1}{2}(\hat{\psi}-\psi_0)^T I_{\psi\cdot\lambda}(\hat{\psi}-\psi_0) + o_P(n^{-1}).\end{aligned} \tag{11.10}$$

By Theorem 9.7 we have

$$\hat{\psi}-\psi_0 \stackrel{Q_n}{=} I_{\psi\cdot\lambda}^{-1} E_n[s_{\psi\cdot\lambda}(\theta_0,X)] + o_P(n^{-1/2}).$$

Substituting the above into the right-hand side of (11.10), we have

$$\begin{aligned} T_n &\stackrel{Q_n}{=} nE_n[s^T_{\psi\cdot\lambda}(\theta_0, X)]I^{-1}_{\psi\cdot\lambda}(\theta_0)E_n[s_{\psi\cdot\lambda}(\theta_0, X)] + o_P(1) \\ &\equiv U_n^T \Sigma_U^{-1} U_n + o_P(1). \end{aligned} \tag{11.11}$$

It remains to show that (U_n, S_n) converges to a multivariate normal distribution with asserted forms of Σ_U and Σ_{US}. For convenience, we abbreviate $s(\theta_0, X)$, $s_\lambda(\theta_0, X)$, $s_\psi(\theta_0, X)$, and $s_{\psi\cdot\lambda}(\theta_0, X)$ by s, s_ψ, s_λ, and $s_{\psi\cdot\lambda}$. By the central limit theorem

$$\begin{pmatrix} n^{1/2}E_n(s_{\psi\cdot\lambda}) \\ n^{1/2}E_n(s) \end{pmatrix} \xrightarrow[Q_n]{\mathcal{D}} N\left[\begin{pmatrix} 0 \\ 0 \end{pmatrix}, \begin{pmatrix} E(s_{\psi\cdot\lambda}s^T_{\psi\cdot\lambda}) & E(s_{\psi\cdot\lambda}s^T) \\ E(ss^T_{\psi\cdot\lambda}) & E(ss^T) \end{pmatrix}\right],$$

where

$$\begin{aligned} E(s_\lambda s^T_{\psi\cdot\lambda}) &= I_{\lambda\psi} - I_{\lambda\lambda}I^{-1}_{\lambda\lambda}I_{\lambda\psi} = 0, \\ E(s_\psi s^T_{\psi\cdot\lambda}) &= I_{\psi\psi} - I_{\psi\lambda}I^{-1}_{\lambda\lambda}I_{\lambda\psi} = I_{\psi\cdot\lambda}. \end{aligned}$$

Hence $\Sigma_{SU} = (I_{\psi\cdot\lambda}, 0)^T$. Also, by Theorem 9.8,

$$\Sigma_U = E(s_{\psi\cdot\lambda}s^T_{\psi\cdot\lambda}) = I_{\psi\cdot\lambda},$$

thus proving the case of $r < p$.

The case of $r = p$ can be proved by substituting (11.6) into (11.7). □

From Theorems 11.1 and 11.2 we can easily derive the null and local alternative distributions of the likelihood ratio test.

Theorem 11.3 *Under the conditions in Theorem 11.2, we have, for any $\delta \in \mathbb{R}^p$,*

$$T_n \xrightarrow[P_n(\delta)]{\mathcal{D}} \chi^2_r(\delta^T_\psi I_{\psi\cdot\lambda}\delta_\psi),$$

where δ_ψ is the first r components of δ. In particular, when $\delta = 0$, we have $T_n \xrightarrow[Q_n]{\mathcal{D}} \chi^2_r$.

Proof. By Theorem 11.1, $T_n \xrightarrow[P_n(\delta)]{\mathcal{D}} \chi^2_r(\delta^T\Sigma_{SU}\Sigma_U^{-1}\Sigma_{US}\delta)$. Substituting into the noncentrality parameter the forms of Σ_U and Σ_{US} as given in Theorem 11.2, we have

$$\delta^T\Sigma_{SU}\Sigma_U^{-1}\Sigma_{US}\delta = \delta^T\begin{pmatrix} I_{\psi\cdot\lambda} \\ 0 \end{pmatrix} I^{-1}_{\psi\cdot\lambda}\,(I_{\psi\cdot\lambda}\ 0)\delta = \delta^T_\psi I_{\psi\cdot\lambda}\delta_\psi,$$

as desired. □

It is interesting to note that the limiting distribution of T_n under $P_n(\delta)$ only depends on δ_ψ. In the local parametric family $\{P_n(\delta) : \delta \in \mathbb{R}^p\}$, δ_ψ

plays the role of the parameter of interest ψ, and δ_λ plays the role of the nuisance parameter λ. Thus, the limiting distribution of T_n only depends on the parameter of interest in the local parametric family.

In practice, we use the null distribution in Theorem 11.1 to determine the critical value of the rejection region, and use the local alternative distribution to determine the power at an arbitrary $\theta_0 + n^{-1/2}\delta$. Since λ_0 is not specified in the null hypothesis, we can replace it by $\tilde{\lambda}$. The estimated power is $\sqrt{n}$-consistent estimate of the true power at $\theta_0 + n^{-1/2}\delta$.

11.3 Wald's, Rao's, and Neyman's tests

In this section we introduce three more commonly used tests which turn out to be QF-tests having the same form of $U_n^T \Sigma_U^{-1} U_n$ as the likelihood ratio test.

11.3.1 Wald's test

Wald (1943) introduced the following statistic

$$W_n = n(\hat{\theta} - \theta_0)^T I(\theta_0)(\hat{\theta} - \theta_0)$$

for testing the hypothesis $H_0 : \theta = \theta_0$ asymptotically. He showed that this statistic converges in distribution to χ^2_p. For the more general hypothesis (11.1), Wald's test takes the following form.

Definition 11.2 *The Wald test for hypothesis (11.1) takes the form*

$$W_n = n(\hat{\psi} - \psi_0)^T I_{\psi\cdot\lambda}(\bar{\theta})(\hat{\psi} - \psi_0),$$

where $\bar{\theta}$ is any consistent estimate of θ_0.

The often-used $\bar{\theta}$ is the global MLE $\hat{\theta}$, or the constrained MLE $(\psi_0, \tilde{\lambda})$. Since Wald's test itself is already in a quadratic form, it is no surprise that it is a QF test, as shown in the next Theorem.

Theorem 11.4 *If the conditions in Theorem 9.7 hold and $I(\theta)$ is a continuous function of θ, then W_n in Definition 11.2 is a QF test with the same quadratic form as T_n.*

Proof. By Theorem 9.7,

$$\sqrt{n}(\hat{\psi} - \psi_0) \stackrel{Q_n}{=} \sqrt{n} I_{\psi\cdot\lambda}^{-1} E_n[s_{\psi\cdot\lambda}(\theta_0, X)] + o_P(1).$$

Since $I(\theta)$ is continuous, $I(\bar{\theta}) \stackrel{Q_n}{=} I(\theta_0) + o_P(1)$. Hence W_n is of the form (11.11). □

As a consequence, the asymptotic null and local alternative distributions of W_n are the same as those of T_n, as given in Theorem 11.2.

11.3.2 Rao's test

Rao's test, also known as the score test, was introduced by Rao (1948). In the case where $\psi = \theta$, Rao's statistic is of the form

$$R_n = nE_n[s^T(\theta_0, X)]I^{-1}(\theta_0)E_n[s(\theta_0, X)].$$

This, by definition, is in the form of a QF test with $U_n = n^{1/2}E_n[s(\theta_0, X)]$, $\Sigma_U = I(\theta_0)$, and $\Sigma_{US} = I(\theta_0)$. The general form of Rao's test is given by the following definition (Rao, 2001).

Definition 11.3 *Suppose $\tilde{\theta} = (\psi_0^T, \tilde{\lambda}^T)^T$ is the constrained MLE under the null hypothesis $H_0 : \psi = \psi_0$. Then Rao's statistic is*

$$R_n = nE_n[s^T(\tilde{\theta}, X)]I^{-1}(\tilde{\theta})E_n[s(\tilde{\theta}, X)].$$

An interesting feature of Rao's statistic is that it only involves the constrained maximum likelihood estimate $(\psi_0, \tilde{\lambda})$ under the null hypothesis; the global MLE appears nowhere in this statistic. In particular, in the case where $\psi = \theta$, no estimate is needed to perform this test. This gives a numerical advantage to Rao's test, because we only need to perform the maximization over an s-dimensional space. That Rao's test is a QF test will be proved in the next subsection along with the Neyman's $C(\alpha)$ test.

11.3.3 Neyman's $C(\alpha)$ test

This test was introduced by Neyman (1959). One way to motivate Neyman's $C(\alpha)$ test is through its relation with Rao's test, as explained in Kocherlakota and Kocherlakota (1991). Since $\tilde{\lambda}$ is a solution to $E_n[s_\lambda(\psi_0, \lambda, X)] = 0$, we have

$$E_n[s(\tilde{\theta}, X)] = \begin{pmatrix} E_n[s_\psi(\tilde{\theta}, X)] \\ 0 \end{pmatrix}.$$

Hence Rao's test can be equivalently written as

$$\begin{aligned} R_n &= n(E_n s_\psi^T(\tilde{\theta}, X), 0)I^{-1}(\tilde{\theta})(E_n s_\psi^T(\tilde{\theta}, X), 0)^T \\ &= nE_n[s_\psi^T(\tilde{\theta}, X)][I^{-1}(\tilde{\theta})]_{\psi\psi}E_n[s_\psi(\tilde{\theta}, X)] \\ &= nE_n[s_{\psi\cdot\lambda}^T(\psi_0, \tilde{\lambda}, X)]I_{\psi\cdot\lambda}^{-1}(\psi_0, \tilde{\lambda})E_n[s_{\psi\cdot\lambda}(\psi_0, \tilde{\lambda}, X)], \end{aligned} \tag{11.12}$$

where, for the third equality, we used $E_n[s_\psi(\tilde{\theta}, X)] = E_n[s_{\psi\cdot\lambda}(\tilde{\theta}, X)]$, which holds because $E_n[s_\lambda(\tilde{\theta}, X)] = 0$, and $[I^{-1}(\tilde{\theta})]_{\psi\psi} = I_{\psi\cdot\lambda}^{-1}(\tilde{\theta})$, which follows from the formula of the inverse of block matrix given in Proposition 9.3. The right-hand side of (11.12) is precisely the form of Neyman's $C(\alpha)$ test introduced by Neyman (1959) except that the latter does not require $\tilde{\lambda}$ to be the MLE under the null hypothesis $H_0 : \psi = \psi_0$. Instead, Neyman's $C(\alpha)$ test allows $\tilde{\lambda}$ to be any $\sqrt{n}$-consistent estimate of λ_0.

Definition 11.4 *Let $\tilde{\lambda}$ be any $\sqrt{n}$ consistent estimate of λ_0, Neyman's $C(\alpha)$ test is the statistic*

$$N_n = nE_n[s^T_{\psi\cdot\lambda}(\psi_0, \tilde{\lambda}, X)]I^{-1}_{\psi\cdot\lambda}(\psi_0, \tilde{\lambda})E_n[s_{\psi\cdot\lambda}(\psi_0, \tilde{\lambda}, X)].$$

Rao's test is a special case of Neyman's $C(\alpha)$ test when $\tilde{\lambda}$ is the constrained MLE under the null hypothesis. We next prove that N_n — and hence also R_n — is a QF test.

Theorem 11.5 *Suppose Assumption 10.3 holds and*

1. *$\ell(\theta, x)$ is twice differentiable;*
2. *$f_\theta(X)$ and $s(\theta, X)f_\theta(X)$ satisfy $DUI^+(\theta, \mu)$;*
3. *$s(\theta, X)$ is P_θ-square integrable and $K(\theta)$ is positive definite and continuous;*
4. *the sequence of random matrices*

$$\{E_n[\partial s_{\psi\cdot\lambda}(\psi_0, \lambda, X)/\partial\lambda^T] : n \in \mathbb{N}\} \tag{11.13}$$

   is stochastically equicontinuous in a neighborhood of λ_0;
5. *$\tilde{\lambda}$ is a $\sqrt{n}$-consistent estimate of λ_0.*

Then N_n is a QF test with the same quadratic form as T_n.

The assumptions in this theorem are essentially the same as those in Theorem 11.2 except conditions 4 and 5: we only require these conditions for the λ-component. Nevertheless, we do need the derivatives of ℓ with respect to θ (not just with respect to λ) because both the efficient score and the efficient information involve derivatives with respect to θ.

Proof of Theorem 11.5. By Taylor's theorem,

$$\begin{aligned} &E_n[s_{\psi\cdot\lambda}(\psi_0, \tilde{\lambda}, X)] \\ &= E_n[s_{\psi\cdot\lambda}(\theta_0, X)] + E_n\left[\frac{\partial s_{\psi\cdot\lambda}(\psi_0, \lambda^\dagger, X)}{\partial\lambda^T}\right](\tilde{\lambda} - \lambda_0), \end{aligned} \tag{11.14}$$

for some $\lambda^\dagger$ between λ_0 and $\tilde{\lambda}$. By the equicontinuity condition 4, and Corollary 8.2,

$$E_n\left[\frac{\partial s_{\psi\cdot\lambda}(\psi_0, \lambda^\dagger, X)}{\partial\lambda^T}\right] \stackrel{Q_n}{=} E\left[\frac{\partial s_{\psi\cdot\lambda}(\theta_0, X)}{\partial\lambda^T}\right] + o_P(1).$$

However, by Theorem 9.8, the expectation on the right-hand side is 0, leading to

$$E_n\left[\frac{\partial s_{\psi\cdot\lambda}(\psi_0, \lambda^\dagger, X)}{\partial\lambda^T}\right] \stackrel{Q_n}{=} o_P(1).$$

This, together with (11.14) and condition 5, implies

$$\begin{aligned} E_n s_{\psi\cdot\lambda}(\psi_0, \tilde{\lambda}, X) &\overset{Q_n}{=} E_n s_{\psi\cdot\lambda}(\theta_0, X) + o_P(1)\, O_P(n^{-1/2}) \\ &\overset{Q_n}{=} E_n s_{\psi\cdot\lambda}(\theta_0, X) + o_P(n^{-1/2}). \end{aligned} \tag{11.15}$$

Meanwhile, by the continuity and nonsingularity of $I(\theta)$, we have

$$I^{-1}_{\psi\cdot\lambda}(\tilde{\theta}) \overset{Q_n}{=} I^{-1}_{\psi\cdot\lambda}(\theta_0) + o_P(1). \tag{11.16}$$

Now substitute (11.15) and (11.16) into the right-hand side of (11.12) to complete the proof. □

11.4 Asymptotically efficient test

The local asymptotic distribution we have developed for the QF tests allows us to compare the local asymptotic power among these tests. Since, as we will show below, the local asymptotic power of a QF test is an increasing function of the noncentrality parameter of its asymptotic noncentral χ^2 distribution, a QF test with the greatest noncentrality parameter achieves maximum local asymptotic power. We first define the class of regular QF tests, which is the platform for developing the asymptotically efficient QF tests.

Definition 11.5 *A test T_n for $H_0 : \psi = \psi_0$ is regular if $T_n \xrightarrow[P_n(\delta)]{\mathcal{D}} F(\delta)$, where $F(\delta)$ depends on, and only on, δ_ψ in the sense that*

1. *if $\delta_\psi \neq 0$, then $F(\delta) \neq F(0)$;*
2. *if $\delta^{(1)}_\psi = \delta^{(2)}_\psi$, then $F(\delta^{(1)}) = F(\delta^{(2)})$.*

In the case where $\psi = \theta$, the above definition requires that the limiting distribution $F(\delta)$ genuinely depends on δ; that is, $F(\delta) \neq F(0)$ whenever $\delta \neq 0$. This seems to contradict with the definition of a regular estimate, which requires that the limiting distribution of $\sqrt{n}(\hat{\psi} - \psi_0 - n^{-1/2}\delta_\psi)$ under $P_n(\delta)$ to be the same for all δ. This apparent inconsistency of terminologies comes from the fact that T_n is usually of the form

$$T_n \overset{Q_n}{=} n(\hat{\psi} - \psi_0)^T I_{\psi\cdot\lambda}(\theta_0)(\hat{\psi} - \psi_0) + o_P(1),$$

where $\hat{\psi}$ is a regular estimate of ψ_0, and asymptotic distribution of $\sqrt{n}(\hat{\psi}-\psi_0)$ under $P_n(\delta)$, unlike that of $\sqrt{n}(\hat{\psi}-\psi_0-n^{-1/2}\delta_\psi)$ under $P_n(\delta)$, should indeed depend on δ_ψ. The next theorem gives a sufficient and necessary condition for a QF test to be regular. In the following, we use S to denote the limit of S_n; that is, $S_n \xrightarrow[Q_n]{\mathcal{D}} S$, where $S \sim N(0, I)$, I being the Fisher information. This is guaranteed by the ALAN assumption. We use S_ψ to denote the first r components of S, and S_λ the last s components of S.

Theorem 11.6 *A QF test T_n for the hypothesis $H_0 : \psi = \psi_0$ is regular if and only if Σ_{US_ψ} is non-singular and $\Sigma_{US_\lambda} = 0$.*

Proof. By Theorem 11.1, $T_n \xrightarrow[P_n(\delta)]{\mathcal{D}} F(\delta)$, where $F(\delta) = \chi^2_r(\delta^T \Sigma_{SU} \Sigma_U^{-1} \Sigma_{US} \delta)$. The noncentrality parameter can be rewritten as

$$\delta_\psi^T \Sigma_{S_\psi U} \Sigma_U^{-1} \Sigma_{US_\psi} \delta_\psi + 2\delta_\psi^T \Sigma_{S_\psi U} \Sigma_U^{-1} \Sigma_{US_\lambda} \delta_\lambda + \delta_\lambda^T \Sigma_{S_\lambda U} \Sigma_U^{-1} \Sigma_{US_\lambda} \delta_\lambda.$$

If Σ_{US_ψ} is nonsingular and $\Sigma_{US_\lambda} = 0$, then the above reduces to

$$\delta_\psi^T \Sigma_{S_\psi U} \Sigma_U^{-1} \Sigma_{US_\psi} \delta_\psi,$$

where $\Sigma_{S_\psi U} \Sigma_U^{-1} \Sigma_{US_\psi}$ is positive definite. Hence $F(\delta)$ satisfies conditions 1 and 2 in Definition 11.5. Conversely, if $F(\delta)$ satisfies conditions 1 and 2 in Definition 11.5, then, by condition 1, $F((0, \delta_\psi)) \neq F(0)$ for any $\delta_\psi \in \mathbb{R}^r$, implying

$$\delta_\psi^T \Sigma_{S_\psi U} \Sigma_U^{-1} \Sigma_{US_\psi} \delta_\psi > 0.$$

Thus Σ_{US_ψ} is nonsingular. By condition 2 of Definition 11.5, $F(0) = F(0, \delta_\lambda)$ for any $\delta_\lambda \in \mathbb{R}^s$. Hence

$$\delta_\lambda^T \Sigma_{S_\lambda U} \Sigma_U^{-1} \Sigma_{US_\lambda} \delta_\lambda = 0$$

for all $\delta_\lambda \in \mathbb{R}^s$, implying $\Sigma_{US_\lambda} = 0$. □

Note that the quadratic form $U_n^T \Sigma_U^{-1} U_n$ is not unique. In fact, it is invariant under any invertible linear transformation; that is, if we transform U_n to $V_n = AU_n$ for some nonsingular $A \in \mathbb{R}^{r \times r}$, then $V_n \xrightarrow[Q_n]{\mathcal{D}} AU$, with $\Sigma_V = A\Sigma_U A^T$. Hence

$$V_n^T \Sigma_V^{-1} V_n = U_n^T A^T A^{-T} \Sigma_U^{-1} A^{-1} A U_n = U_n^T \Sigma_U^{-1} U_n.$$

Thus, there are infinitely many representations of a QF test. As the next theorem shows, a special choice of U_n shares the same asymptotically independent decomposition as a regular estimate (see the convolution theorem, Theorem 10.3). Let $S_{n,\psi}$ and $S_{n,\lambda}$ be the first r and last s components of S_n, respectively, and let $S_{n,\psi\cdot\lambda}$ be the standardized efficient score $S_{n,\psi} - I_{\psi\lambda} I_{\lambda\lambda}^{-1} S_{n,\lambda}$.

Theorem 11.7 *A QF test T_n for $H_0 : \psi = \psi_0$ is regular if and only if it can be written as*

$$T_n \overset{Q_n}{=} U_n^T \Sigma_U^{-1} U_n + o_P(1),$$

where $U_n = I_{\psi\cdot\lambda}^{-1} S_{n,\psi\cdot\lambda} + R_n$, with (R_n, S_n, L_n) satisfying ALAN and $R \perp\!\!\!\perp S$. Here (R, S) is the limit of (R_n, S_n); that is, $(R_n, S_n) \xrightarrow[Q_n]{\mathcal{D}} (R, S)$.

Proof. Because T_n is a regular QF test, it can be written as $T_n \overset{Q_n}{=} V_n^T \Sigma_V^{-1} V_n + o_P(1)$ for some $V_n \in \mathbb{R}^r$ such that (V_n, S_n, L_n) satisfies ALAN with Σ_{VS_ψ} being a nonsingular matrix and $\Sigma_{VS_\lambda} = 0$. Let $U_n = \Sigma_{VS_\psi}^{-1} V_n$. Then (U_n, S_n, L_n) satisfies ALAN with $\Sigma_{US} = (I_r, 0)$, and

$$T_n \overset{Q_n}{=} U_n^T \Sigma_U^{-1} U_n + o_p(1).$$

The limiting random vector U of U_n can be written as $I_{\psi\cdot\lambda}^{-1} S_{\psi\cdot\lambda} + R$ where $R - U - I_{\psi\cdot\lambda}^{-1} S_{\psi\cdot\lambda}$ and $S_{\psi\cdot\lambda} = S_\psi - I_{\psi\lambda} I_{\lambda\lambda}^{-1} S_\lambda$. To show that S and R are independent, note that

$$\begin{aligned}\mathrm{cov}(R, S) &= \mathrm{cov}(U, S) - \mathrm{cov}(I_{\psi\cdot\lambda}^{-1} S_{\psi\cdot\lambda}, S) \\ &= \Sigma_{US} - I_{\psi\cdot\lambda}^{-1} \mathrm{cov}(S_{\psi\cdot\lambda}, S).\end{aligned}$$

By construction, $\Sigma_{US} = (I_r, 0)$. Also,

$$\mathrm{cov}(S_{\psi\cdot\lambda}, S) = (\mathrm{cov}(S_{\psi\cdot\lambda}, S_\psi), \mathrm{cov}(S_{\psi\cdot\lambda}, S_\lambda)) = (I_{\psi\cdot\lambda}, 0).$$

Hence $\mathrm{cov}(R, S) = 0$. Because R and S are jointly Normal, we have $R \perp\!\!\!\perp S$. The reverse implication is obvious. □

Since $U = I_{\psi\cdot\lambda}^{-1} S_{\psi\cdot\lambda} + R$ with $R \perp\!\!\!\perp S$, the variance of U is always greater than or equal to $I_{\psi\cdot\lambda}^{-1}$ in terms of Louwner's ordering. Since the noncentrality parameter for the distribution of $U^T \Sigma_U^{-1} U$ is

$$\delta^T \begin{pmatrix} I_r \\ 0 \end{pmatrix} \Sigma_U^{-1} (I_r \ 0) \delta = \delta_\psi^T \Sigma_U^{-1} \delta_\psi,$$

$\delta_\psi^T I_{\psi\cdot\lambda} \delta_\psi$ is the upper bound of the noncentrality parameter of any regular QF test. We summarize this optimal result in the next corollary.

Corollary 11.1 *Suppose T_n is a regular QF test. Then the following statements hold:*

1. $T_n \xrightarrow[P_n(\delta)]{\mathcal{D}} \chi_r^2(\delta_\psi^T \Sigma_U^{-1} \delta_\psi)$, *where* $\Sigma_U^{-1} \preceq I_{\psi\cdot\lambda}$;
2. *in the above statement,* $\Sigma_U = I_{\psi\cdot\lambda}^{-1}$ *if and only if T_n can be represented as* $T_n \overset{Q_n}{=} U_n^T \Sigma_U^{-1} U_n + o_P(1)$, *where* $U_n \overset{Q_n}{=} I_{\psi\cdot\lambda}^{-1} S_{n,\psi\cdot\lambda} + o_P(1)$.

Proof. 1. By Theorem 11.7, T_n can be written as

$$T_n \overset{Q_n}{=} U_n^T \Sigma_U^{-1} U_n + o_P(1),$$

where $U_n = I_{\psi\cdot\lambda}^{-1} S_{n,\psi\cdot\lambda} + R_n$ with (R_n, S_n, L_n) satisfying ALAN and $\Sigma_{RS} = 0$. Hence $\Sigma_U = I_{\psi\cdot\lambda}^{-1} + \Sigma_R$, implying $\Sigma_U^{-1} \preceq I_{\psi\cdot\lambda}$.

2. $\Sigma_U^{-1} = I_{\psi\cdot\lambda}$ if and only if $\Sigma_U = I_{\psi\cdot\lambda}^{-1}$ which, by part 1, holds if and only if $\Sigma_R = 0$, which in turn holds if and only if $R_n \stackrel{Q_n}{=} o_P(1)$. □

We now define the the asymptotically efficient QF test. A natural definition would be in terms of the local asymptotic power, which is indeed the approach we adopt. Let $\mathcal{Q}$ denote the collection of all sequences of QF tests. So a member of $\mathcal{Q}$ is $\{T_n : n \in \mathbb{N}\}$ where T_n is a QF test.

Definition 11.6 *An asymptotically efficient QF test is a member $\{T_n^* : n \in \mathbb{N}\}$ of $\mathcal{Q}$ such that for any member $\{T_n : n \in \mathbb{N}\}$ of $\mathcal{Q}$, $\delta \in \mathbb{R}^p$, and any $c > 0$,*

$$P(T^*(\delta) \geq c) \geq P(T(\delta) \geq c),$$

where $T^*(\delta)$ is the limit of T_n^* under $P_n(\delta)$ and $T(\delta)$ is the limit of T_n under $P_n(\delta)$; that is,

$$T_n \xrightarrow[P_n(\delta)]{\mathcal{D}} T(\delta) \text{ and } T_n^* \xrightarrow[P_n(\delta)]{\mathcal{D}} T^*(\delta).$$

This definition is an asymptotic analogue of the UMPU test described in Chapters 3 and 4. Mathew and Nordstrom (1997) showed that a noncentral chi-squared distribution with a larger noncentrality parameter is always stochastically larger than a noncentral chi-squared distribution with the same degrees of freedom and a smaller noncentrality parameter. A rigorous statement of this is given below without proof.

Proposition 11.1 *If $K_1 \sim \chi_r^2(d_1)$ and $K_2 \sim \chi_r^2(d_2)$ and $d_2 \geq d_1$, then, for any $c > 0$,*

$$P(K_2 \geq c) \geq P(K_1 \geq c)$$

Using this proposition we immediately arrive at the following equivalent definition of an asymptotically efficient QF test.

Definition 11.7 *An asymptotically efficient QF test is a member $\{T_n^* : n \in \mathbb{N}\}$ of $\mathcal{Q}$ such that $T_n^* \xrightarrow[P_n(\delta)]{\mathcal{D}} \chi_r^2(\delta_\psi^T I_{\psi\cdot\lambda}\delta_\psi)$ for any $\delta \in \mathbb{R}^p$.*

Since all the four test statistics T_n, W_n, R_n, and N_n developed in the previous sections are QF tests with the same quadratic form

$$S_{n,\psi\cdot\lambda}^T I_{\psi\cdot\lambda}^{-1} S_{n,\psi\cdot\lambda} + o_P(1)$$

under Q_n, where $S_{n,\psi\cdot\lambda} = n^{1/2} E_n[s_{\psi\cdot\lambda}(\theta_0, X)]$, their noncentrality parameters all reach the upper bound $\delta_\psi^T I_{\psi\cdot\lambda}\delta_\psi$. Hence they are all asymptotically efficient.

In the rest of this section we develop a relation between a regular estimate that satisfies ALAN and a regular QF test.

Proposition 11.2 *Suppose that $\hat{\psi}$ is a regular estimate of ψ_0 and (U_n, S_n, L_n) satisfies ALAN with $U_n = \sqrt{n}(\hat{\psi} - \psi_0)$. Then T_n of the form*

$$T_n \stackrel{Q_n}{=} U_n^T \Sigma_U^{-1} U_n + o_P(1) \tag{11.17}$$

is a regular QF test. Furthermore, T_n is asymptotically efficient if and only if $\hat{\psi}$ is asymptotically efficient.

Proof. Because $\hat{\psi}$ is regular and (U_n, S_n, L_n) satisfies ALAN, we have, by Proposition 10.4, $\Sigma_{US} = \dot{h}^T = (I_r, 0)$, where $h(\theta) = \psi = (I_r, 0)\theta$. Hence by Theorem 11.6, T_n is regular.

By Corollary 11.1, $T_n \xrightarrow[P_n(\delta)]{\mathcal{D}} \chi_r^2(\delta_\psi^T \Sigma_U^{-1} \delta_\psi)$. Therefore, T_n is asymptotically efficient QF test if and only if $\Sigma_U^{-1} = I_{\psi\cdot\lambda}$, which holds, in turn, if and only if $\hat{\psi}$ is asymptotically efficient estimate. □

The converse statement of Proposition 11.2 is not true: Problem 11.13 shows that it is possible to construct a regular QF test based on an estimate that is not regular.

11.5 Pitman efficiency

This section is concerned with a numerical measurement of asymptotic relative efficiency between two test statistics introduced by Pitman (1948) in an unpublished lecture notes. See, for example, Neother (1950), Neother (1955), and van Eeden (1963). The general definition given here is from Hall and Mathiason (1990). Recall that each member of $\mathcal{Q}$ converges in distribution to a noncentral χ^2 distribution under $P_n(\delta)$, which characterizes the local asymptotic power in the direction of δ: the larger the noncentrality parameter the greater the asymptotic power in the direction of δ. The relative Pitman efficiency of one member of $\mathcal{Q}$ with respect to another is defined as the ratio of the respective noncentrality parameters. Let $\{T_n\}$ and $\{T_n^*\}$ be two members of $\mathcal{Q}$ with

$$T_n \stackrel{Q_n}{=} U_n^T \Sigma_U^{-1} U_n + o_P(1), \quad T_n^* \stackrel{Q_n}{=} U_n^{*T} \Sigma_{U^*}^{-1} U_n^* + o_P(1),$$

where $(U_n, S_n) \xrightarrow[Q_n]{\mathcal{D}} (U, S)$, $(U_n^*, S_n) \xrightarrow[Q_n]{\mathcal{D}} (U^*, S)$. Let

$$\Sigma_{US} = \text{cov}(U, S), \quad \Sigma_{U^*S} = \text{cov}(U^*, S).$$

Definition 11.8 *The relative Pitman efficiency of $\{T_n\}$ with respect to $\{T_n^*\}$ in the direction of δ is*

$$\mathcal{E}_{T,T^*}(\delta) = \frac{\delta^T \Sigma_{SU} \Sigma_U^{-1} \Sigma_{US}\delta}{\delta^T \Sigma_{SU^*} \Sigma_{U^*}^{-1} \Sigma_{U^*S}\delta}.$$

The Pitman efficiency of $\{T_n\}$ in the direction of δ is its relative Pitman efficiency with respect to any Asymptotically Efficient QF test; that is

$$\mathcal{E}_T(\delta) = \frac{\delta^T \Sigma_{SU} \Sigma_U^{-1} \Sigma_{US}\delta}{\delta_\psi^T I_{\psi\cdot\lambda}\delta_\psi}.$$

Note that if $T_n \stackrel{Q_n}{=} U_n^T \Sigma_U^{-1} U_n + o_P(1)$ is a regular QF test for the parameter of interest ψ, then $\Sigma_{US} = (\Sigma_{US_\psi}, 0)$. Thus $\Sigma_{US}\delta = \Sigma_{US_\psi}\delta_\psi$, and the Pitman efficiency for any regular QF test is of the form

$$\mathcal{E}_T(\delta) = \frac{\delta_\psi^T \Sigma_{S_\psi U} \Sigma_U^{-1} \Sigma_{US_\psi}\delta_\psi}{\delta_\psi^T I_{\psi\cdot\lambda}\delta_\psi}.$$

Pitman efficiency can be equivalently defined using the correlation matrix between two random vectors. If U and V are random vectors with finite and positive definite variance matrices $\Sigma_U = \text{var}(U)$ and $\Sigma_V = \text{var}(V)$, then their correlation matrix is defined to be

$$R_{UV} = \Sigma_U^{-1/2} \Sigma_{UV} \Sigma_V^{-1/2},$$

where $\Sigma_{UV} = \text{cov}(U, V)$. Let

$$\rho^2(U, V) = \text{tr}(R_{UV} R_{VU}) = \text{tr}(R_{VU} R_{UV}).$$

Then, the Pitman efficiency of $\{T_n\}$ can be rewritten as

$$\mathcal{E}_T(\delta) = \rho^2(\delta_\psi^T S_\psi, U).$$

11.6 Hypothesis specified by an arbitrary constraint

We now turn to the general hypothesis testing problem where the null hypothesis is specified by one or a set of equations. Suppose that $h : \Theta \mapsto \mathbb{R}^r$ is a mapping from $\Theta \subseteq \mathbb{R}^p$ to $\mathbb{R}^r$, where $r \leq p$. We are interested in testing

$$H_0 : h(\theta) = 0.$$

The test with an explicit parameter of interest, $H_0 : \psi = \psi_0$, can be regarded as a special case of the above test with $h(\theta) = \psi - \psi_0$. All the tests we developed in the previous sections can be generalized to this setting. The likelihood ratio test and the score test take almost exactly the same form as before, with $\tilde{\theta}$ being the MLE under the constraint $h(\theta) = 0$. However, because there is no explicit parameter of interest in this case, the forms of Wald's test and Neyman's $C(\alpha)$ test have to be modified.

11.6.1 Asymptotic analysis of constrained MLE

Let $\tilde{\theta}$ be the maximizer of the likelihood $E_n[\ell(\theta, X)]$ under the constraint $h(\theta) = 0$. In this subsection we derive the asymptotic linear form of $\tilde{\theta}$ under standard regularity conditions. Let $F(\theta)$ be the Lagrangian

$$F(\theta) = E_n[\ell(\theta, X)] - h^T(\theta)\rho,$$

where $\rho \in \mathbb{R}^r$ is the Lagrangian multiplier. It is well known in Calculus that, if ℓ and h are continuously differentiable with respect to θ, then, for some $\tilde{\rho} \in \mathbb{R}^r$, $(\tilde{\theta}, \tilde{\rho})$ satisfies the system of equations

$$\begin{cases} E_n[s(\theta, X)] - H(\theta)\rho = 0 \\ h(\theta) = 0. \end{cases} \tag{11.18}$$

where $H(\theta)$ is the $p \times r$ matrix $\partial h^T(\theta)/\partial\theta$.

In the following development, we will frequently encounter sequences of random matrices, say A_n, which converges in probability to an invertible matrix A, but which themselves may not be invertible. The next lemma shows that, in this case, the Moore-Penrose inverse of A_n converges in probability to the inverse of A.

We need to use two facts from linear algebra. First, if a matrix $A \in \mathbb{R}^{p\times p}$ is invertible, then there is an open ball

$$B(A, \epsilon) = \{M \in \mathbb{R}^{p\times p} : \|M - A\| < \epsilon\}$$

such that every $M \in B(A, \epsilon)$ is invertible, where the norm is the Frobenius norm. This is because M is invertible if and only if $\det(M)$ is nonzero, and $\det(M)$, being a linear combination of products of entries of M, is continuous in M in the Frobenius norm. This implies there is an open all $\mathrm{B}(A, \epsilon)$ in which every M has $\det(M) \neq 0$.

Second, on any open set G of $\mathbb{R}^{p\times p}$ of invertible matrices, the function $M \mapsto M^{-1}$ is continuous. This is because

$$M^{-1} = [\det(M)]^{-1}\mathrm{adj}(M),$$

where $\mathrm{adj}(A)$ is the adjugate of A (see, for example, Horn and Johnson, 1985, page 22). Since both $[\det(M)]^{-1}$ and $\mathrm{adj}(M)$ are continuous in M on G, M^{-1} is continuous in M on G.

Lemma 11.1 *If $A_n \xrightarrow{P} A$ and A is invertible, then*

1. $P(A_n^+ = A_n^{-1}) \to 1$;
2. $A_n^+ \xrightarrow{P} A^{-1}$.

Proof. 1. As discussed above, since A is invertible, there is an open ball $B(A, \epsilon)$ such that all $M \in B(A, \epsilon)$ are invertible. Because $A_n \xrightarrow{P} A$, $P(A_n \in B(A, \epsilon)) \to 1$. The assertion holds because $A_n \in B(A, \epsilon)$ implies $A_n^+ = A_n^{-1}$.

2. Let $g : \mathbb{R}^{p\times p} \to \mathbb{R}^{p\times p}$ be the function $g(M) = M^+$. Since every M in $B(A,\epsilon)$ is invertible, $g(M) = M^+ = M^{-1}$ on $B(A,\epsilon)$. By the discussion preceding this Lemma, $g(M)$ is continuous on $B(A,\epsilon)$. By the Continuous Mapping Theorem,

$$A_n^+ = g(A_n) \xrightarrow{P} g(A) = A^+ = A^{-1},$$

where the last equality holds because A is invertible. □

The next theorem gives an asymptotic linear form of the Lagrangian multiplier $\tilde{\rho}$ and the constrained MLE $\tilde{\theta}$ under $h(\theta) = 0$. In the following, we abbreviate the gradient matrix $H(\theta_0)$ by H and the Fisher information $I(\theta_0)$ by I. As before, the symbol I should be differentiated from I_k, the $k \times k$ identity matrix.

Theorem 11.8 *Suppose Assumption 10.3 holds and*

1. *$\ell(\theta, x)$ is twice differentiable;*
2. *$f_\theta(X)$ and $s(\theta, X)f_\theta(X)$ satisfy $DUI^+(\theta, \mu)$;*
3. *$s(\theta, X)$ is P_θ-square integrable, and $I(\theta)$ is invertible;*
4. *$h(\theta)$ is continuously differentiable, and $H^T(\theta)I^{-1}(\theta)H(\theta)$ is invertible;*
5. *the sequence of random matrices $\{J_n(\theta) : n \in \mathbb{N}\}$ in (11.4) is stochastically equicontinuous in a neighborhood of θ_0.*

If $\tilde{\theta}$ is a consistent solution to $E_n[s(\theta, X)] = 0$ under the constraint $h(\theta) = 0$, then

$$\begin{aligned}\tilde{\rho} &= (H^T I^{-1} H)^{-1} H^T I^{-1} E_n[s(\theta_0, X)] + o_P(n^{-1/2}),\\ \tilde{\theta} &= \theta_0 + I^{-1} Q_H(I^{-1}) E_n[s(\theta_0, X)] + o_P(n^{-1/2}),\end{aligned} \tag{11.19}$$

where $Q_H(I^{-1}) = I_p - P_H(I^{-1})$, and

$$P_H(I^{-1}) = H(H^T I^{-1} H)^{-1} H^T I^{-1}.$$

Recall from Example 7.8 that the matrix $P_H(I^{-1})$ is simply the projection on to the subspace span(H) of $\mathbb{R}^p$ with respect to the inner product $\langle x, y\rangle_{I^{-1}} = x^T I^{-1} y$. Hence $Q_H(I^{-1})$ is the projection on to span$(H)^\perp$ with respect to the inner product $\langle\cdot,\cdot\rangle_{I^{-1}}$.

Before proving the theorem, we note the following fact: if a_n is a sequence of constants that goes to 0 as $n \to \infty$, and U_n, V_n, W_n are random vectors, then

$$P(U_n = V_n) \to 1,\ V_n = W_n + o_P(a_n) \ \Rightarrow\ U_n = W_n + o_P(a_n). \tag{11.20}$$

The proof of this is left as an exercise.

Proof of Theorem 11.8. To prove the first relation in (11.19), note that, by (11.18),

$$H(\tilde{\theta})\tilde{\rho} = E_n[s(\tilde{\theta}, X)].$$

By the above equality and Taylor's theorem,

$$\begin{aligned} H(\tilde{\theta})\tilde{\rho} &= E_n[s(\theta_0, X)] + J_n(\theta^\dagger)(\tilde{\theta} - \theta_0) \\ h(\tilde{\theta}) &= h(\theta_0) + H^T(\theta^\ddagger)(\tilde{\theta} - \theta_0) \end{aligned} \tag{11.21}$$

for some $\theta^\dagger$ and $\theta^\ddagger$ between θ_0 and $\tilde{\theta}$. Because $h(\tilde{\theta}) = 0$ by definition and $h(\theta_0) = 0$ under the null hypothesis H_0, the second equation in (11.21) reduces to $H^T(\theta^\ddagger)(\tilde{\theta} - \theta_0) = 0$ under H_0. Multiplying the first equation in (11.21) by $H^T(\theta^\ddagger)J_n^+(\theta^\dagger)$ from the left, we have

$$H^T(\theta^\ddagger)J_n^+(\theta^\dagger)H(\tilde{\theta})\tilde{\rho} = H^T(\theta^\ddagger)J_n^+(\theta^\dagger)E_n[s(\theta_0, X)] + R_n, \tag{11.22}$$

where $R_n = H^T(\theta^\ddagger)J_n^+(\theta^\dagger)J_n(\theta^\dagger)(\tilde{\theta} - \theta_0)$.

Because $\{J_n(\theta) : n \in \mathbb{N}\}$ is stocahstically equicontinuous in a neighborhood of θ_0 and $\theta^\dagger \xrightarrow{P} \theta_0$, we have, by Corollary 8.2, $J_n(\theta^\dagger) \xrightarrow{P} -I$, where I is invertible. Hence, by Lemma 11.1, part 1

$$P(J_n^+(\theta^\dagger)J_n(\theta^\dagger) = I_p) \to 1,$$

which, in view of $H^T(\theta^\ddagger)(\tilde{\theta} - \theta_0) = 0$, implies $P(R_n = 0) \to 1$. Since $R_n = 0$ implies

$$H^T(\theta^\ddagger)J_n^+(\theta^\dagger)H(\tilde{\theta})\tilde{\rho} = H^T(\theta^\ddagger)J_n^+(\theta^\dagger)E_n[s(\theta_0, X)],$$

we have

$$\begin{aligned} P\Big(&[H^T(\theta^\ddagger)J_n^+(\theta^\dagger)H(\tilde{\theta})]^+ H^T(\theta^\ddagger)J_n^+(\theta^\dagger)H(\tilde{\theta})\tilde{\rho} \\ &= [H^T(\theta^\ddagger)J_n^+(\theta^\dagger)H(\tilde{\theta})]^+ H^T(\theta^\ddagger)J_n^+(\theta^\dagger)E_n[s(\theta_0, X)]\Big) \to 1. \end{aligned} \tag{11.23}$$

Because $J_n(\theta^\dagger) \xrightarrow{P} -I$ and I is invertible, by Lemma 11.1, part 2, we have

$$J_n^+(\theta^\dagger) \xrightarrow{P} -I^{-1}. \tag{11.24}$$

Because $H(\theta)$ is continuous, we have

$$H(\tilde{\theta}) \xrightarrow{P} H \text{ and } H(\theta^\ddagger) \xrightarrow{P} H. \tag{11.25}$$

Consequently,

$$H^T(\theta^\ddagger)J_n^+(\theta^\dagger)H(\tilde{\theta}) \xrightarrow{P} -H^T I^{-1} H. \tag{11.26}$$

Because $H^T I^{-1} H$ is invertible, by part 1 of Lemma 11.1, we have

$$P\left([H^T(\theta^\ddagger)J_n^+(\theta^\dagger)H(\tilde{\theta})]^+ = [H^T(\theta^\ddagger)J_n^+(\theta^\dagger)H(\tilde{\theta})]^{-1}\right) \to 1. \tag{11.27}$$

By (11.23) and (11.27),

$$P\left(\tilde{\rho} = [H^T(\theta^\ddagger)J_n^+(\theta^\dagger)H(\tilde{\theta})]^+ H^T(\theta^\ddagger)J_n^+(\theta^\dagger)E_n[s(\theta_0, X)]\right) \to 1. \tag{11.28}$$

Now by (11.26) and part 2 of Lemma 11.1,

$$[H^T(\theta^\ddagger)J_n^+(\theta^\dagger)H(\tilde{\theta})]^+ \xrightarrow{P} -(H^T I^{-1} H)^{-1}. \tag{11.29}$$

Hence by (11.24), (11.25), (11.29), and

$$E_n[s(\theta_0, X)] = O_P(n^{-1/2}), \tag{11.30}$$

we have

$$\begin{aligned}
&[H^T(\theta^\ddagger)J_n^+(\theta^\dagger)H(\tilde{\theta})]^+ H^T(\theta^\ddagger)J_n^+(\theta^\dagger)E_n[s(\theta_0, X)] \\
&\qquad = (H^T I^{-1} H)^{-1} H^T I^{-1} E_n[s(\theta_0, X)] + o_P(n^{-1/2}).
\end{aligned}$$

By (11.20), the above relation and (11.28) imply

$$\tilde{\rho} = (H^T I^{-1} H)^{-1} H^T I^{-1} E_n[s(\theta_0, X)] + o_P(n^{-1/2}). \tag{11.31}$$

proving the first expression in (11.19).

Next, substitute (11.31) into the first equation in (11.21) to obtain

$$\begin{aligned}
&J_n(\theta^\dagger)(\tilde{\theta} - \theta_0) \\
&= -E_n[s(\theta_0, X)] + H(\tilde{\theta})\{(H^T I^{-1} H)^{-1} H^T I^{-1} E_n[s(\theta_0, X)] + o_P(n^{-1/2})\} \\
&= -E_n[s(\theta_0, X)] + H(H^T I^{-1} H)^{-1} H^T I^{-1} E_n[s(\theta_0, X)] + o_P(n^{-1/2}) \\
&= -E_n[s(\theta_0, X)] + P_H(I^{-1})E_n[s(\theta_0, X)] + o_P(n^{-1/2}),
\end{aligned}$$

where the second equality follows from (11.25) and (11.30), and the third from the definition of $P_H(I^{-1})$. Hence

$$\begin{aligned}
J_n^+(\theta^\dagger)J_n(\theta^\dagger)(\tilde{\theta} - \theta_0) = &- J_n^+(\theta^\dagger)E_n[s(\theta_0, X)] \\
&+ J_n^+(\theta^\dagger)P_H(I^{-1})E_n[s(\theta_0, X)] \\
&+ J_n^+(\theta^\dagger)o_P(n^{-1/2}),
\end{aligned} \tag{11.32}$$

Because $J_n^+(\theta^\dagger) = O_P(1)$, the last term on the right-hand side of (11.32) is $o_P(n^{-1/2})$. By (11.24) and (11.30), the first and second terms on the right-hand side of (11.32) are $I^{-1}E_n[s(\theta_0, X)] + o_P(n^{-1/2})$ and $-I^{-1}P_H(I^{-1})E_n[s(\theta_0, X)] + o_P(n^{-1/2})$, respectively. Because $P(J_n^+(\theta^\dagger)J_n(\theta^\dagger) = I_p) \to 1$, the left-hand side of (11.32) is $\tilde{\theta} - \theta_0$ with probability tending to 1. Consequently, by (11.20),

$$\tilde{\theta} - \theta_0 = I^{-1}E_n[s(\theta_0, X)] - I^{-1}P_H(I^{-1})E_n[s(\theta_0, X)] + o_P(n^{-1/2}),$$

proving the second expression in (11.19). □

It is informative to investigate the forms of the various quantities in (11.19) in the special case where $h(\theta) = \psi$. In this case, $H^T = (I_r, 0)$ and, as shown in Problem 11.19,

$$I^{-1}Q_H(I^{-1}) = \begin{pmatrix} 0 & 0 \\ 0 & I_{\lambda\lambda}^{-1} \end{pmatrix}, \quad I^{-1}Q_H(I^{-1})E_n[s(\theta, X)] = \begin{pmatrix} 0 \\ E_n[s_\lambda(\theta, X)] \end{pmatrix}.$$

So the second equation in (11.19) reduces to

$$\tilde{\lambda} = \lambda_0 + I_{\lambda\lambda}^{-1} E_n[s_\lambda(\theta_0, X)] + o_P(n^{-1/2}),$$

which is implied by Theorem 9.5 when $g(\theta, X)$ therein is taken to be the estimating equation $(\lambda, x) \mapsto s_\lambda(\psi_0, \lambda, x)$. The Lagrangian multiplier $\tilde{\rho}$ also has an interesting interpretation in the special case of $h(\theta) = \psi$: since $\tilde{\theta} = (0^T, \tilde{\lambda}^T)^T$ and $h(\theta) = \psi$, we have

$$E_n[s(\tilde{\theta}, X)] = \begin{pmatrix} E_n[s(\psi_0, \tilde{\lambda}, X)] \\ 0 \end{pmatrix}, \quad H(\tilde{\theta}) = \begin{pmatrix} I_r \\ 0 \end{pmatrix}.$$

Hence, $\tilde{\rho}$ is simply the plugged-in score $E_n[s_\psi(\psi_0, \tilde{\lambda}, X)]$. Finally, the quantities on the right-hand side of the first equation in (11.19) generalizes the efficient score and efficient information when $h(\theta) = \psi$. As shown in Problem 11.19,

$$(H^T I^{-1} H)^{-1} = I_{\psi\psi} - I_{\psi\lambda} I_{\lambda\lambda}^{-1} I_{\psi\lambda} = I_{\psi\cdot\lambda}. \tag{11.33}$$

Thus $(H^T I^{-1} H)^{-1}$ is the generalization of the efficient information. Also, as shown in Problem 11.19,

$$(H^T I^{-1} H)^{-1} H^T I^{-1} s(\theta, X) = s_{\psi\cdot\lambda}(\theta, X). \tag{11.34}$$

Thus, $(H^T I^{-1} H)^{-1} H^T I^{-1} s(\theta, X)$ is the generalization of the efficient score. This motivates the following definition.

Definition 11.9 *The efficient information and efficient score for the hypothesis $H_0 : h(\theta) = 0$ is*

$$\begin{aligned} I_{(H)}(\theta) &= [H^T(\theta) I^{-1}(\theta) H(\theta)]^{-1} \\ s_{(H)}(\theta, X) &= I_{(H)}(\theta) H^T(\theta) I^{-1}(\theta) s(\theta, X). \end{aligned}$$

In terms of the efficient score, the first equation in (11.19) can be reexpressed as

$$\tilde{\rho} = E_n[s_{(H)}(\theta_0, X)] + o_P(n^{-1/2}). \tag{11.35}$$

Thus, the Lagrangian multiplier is asymptotically equivalent to the efficient score. This point will be useful later.

11.6.2 Likelihood ratio test for general hypotheses

In this subsection we derive the asymptotic distribution of Wilks likelihood test statistic for testing the hypothesis $H_0 : h(\theta) = 0$, which is defined as

$$T_n = 2n[E_n\ell(\hat{\theta}, X) - E_n\ell(\tilde{\theta}, X)], \tag{11.36}$$

where $\hat{\theta}$ is the global MLE and $\tilde{\theta}$ is the MLE under the constraint $h(\theta) = 0$.

Theorem 11.9 *If the assumptions in Theorem 11.8 are satisfied and $\hat{\theta}$ is a consistent solution to the likelihood equation $E_n[s(\theta, X)] = 0$, then T_n in (11.36) is a QF test of the form*

$$T_n \overset{Q_n}{=} U_n^T \Sigma_U^{-1} U_n + o_P(1), \tag{11.37}$$

where

$$U_n = n^{1/2} E_n[s_{(H)}(\theta_0, X)], \quad \Sigma_U = I_{(H)}, \quad \Sigma_{US} = I_{(H)} H^T. \tag{11.38}$$

Proof. From the proof of Theorem 11.2, we have

$$\begin{aligned} 2[E_n\ell(\hat{\theta}, X) - E_n\ell(\theta_0, X)] &= (\hat{\theta} - \theta_0)^T I(\hat{\theta} - \theta_0) + o_P(n^{-1}) \\ &= (E_n s)^T I^{-1} (E_n s) + o_P(n^{-1}), \end{aligned}$$

where s is the abbreviation of $s(\theta_0, X)$. Also, similar to the argument used in that proof, by Taylor's theorem we can show that

$$\begin{aligned} &2[E_n\ell(\tilde{\theta}, X) - E_n\ell(\theta_0, X)] \\ &\qquad = 2\,(E_n s)^T\,(\tilde{\theta} - \theta_0) - (\tilde{\theta} - \theta_0)^T I(\tilde{\theta} - \theta_0) + o_P(n^{-1}). \end{aligned}$$

Substituting the second equation in (11.19) into the right hand side, we have

$$\begin{aligned} &2[E_n\ell(\tilde{\theta}, X) - E_n\ell(\theta_0, X)] \\ &= 2\,(E_n s)^T\, I^{-1} Q_H(I^{-1})\,(E_n s) - (E_n s)^T\, I^{-1}[Q_H(I^{-1})]^2\,(E_n s) + o_P(n^{-1}) \\ &= (E_n s)^T\, I^{-1} Q_H(I^{-1})\,(E_n s) + o_P(n^{-1}), \end{aligned}$$

where, for the second equality, we have used the fact that, as a projection, $Q_H(I^{-1})$ is idempotent. Therefore

$$\begin{aligned} &2[E_n\ell(\hat{\theta}, X) - E_n\ell(\tilde{\theta}_0, X)] \\ &= (E_n s)^T\, I^{-1} P_H(I^{-1})\,(E_n s) + o_P(n^{-1}) \\ &= (E_n s)^T\, I^{-1} H(H^T I^{-1} H)^{-1} H^T I^{-1}\,(E_n s) + o_P(n^{-1}) \\ &= \left(E_n s_{(H)}\right)^T\, I_{(H)}^{-1}\,\left(E_n s_{(H)}\right) + o_P(n^{-1}). \end{aligned}$$

Hence $T_n \stackrel{Q_n}{=} U_n^T \Sigma_U^{-1} U_n + o_P(1)$ with $U_n = n^{1/2} E_n(s_{(H)})$ and $\Sigma_U = I_{(H)}$. By the central limit theorem,

$$\begin{pmatrix} n^{1/2} E_n(s_{(H)}) \\ n^{1/2} E_n(s) \end{pmatrix} \xrightarrow[Q_n]{\mathcal{D}} N\left[\begin{pmatrix} 0 \\ 0 \end{pmatrix}, \begin{pmatrix} E(s_{(H)} s_{(H)}^T) & E(s_{(H)} s^T) \\ E(s s_{(H)}^T) & E(s s^T) \end{pmatrix}\right],$$

It is easy to verify that $E(s_{(H)} s^T) = I_{(H)} H^T$. Hence $\Sigma_{SU} = I_{(H)} H^T$. □

This theorem implies that, for any $\delta \in \mathbb{R}^p$,

$$T_n \xrightarrow[P_n(\delta)]{\mathcal{D}} \chi_r^2(\delta^T H I_{(H)} H \delta).$$

When $h(\theta) = \psi$, the noncentrality parameter reduces to $\delta_\psi^T I_{\psi \cdot \lambda} \delta_\psi$; when $\delta = 0$, $P_n(\delta)$ reduces to Q_n and the noncentral chi-squared distribution reduces to the central chi-squared distribution.

11.6.3 Wald's test and Rao's test for general hypotheses

Wald's and Rao's tests for the general hypothesis $H_0 : h(\theta) = 0$ are defined as follows:

$$W_n = n h^T(\hat{\theta}) I_{(H)}(\hat{\theta}) h(\hat{\theta}) \tag{11.39}$$

$$R_n = n E_n[s^T(\tilde{\theta}, X)] I^{-1}(\tilde{\theta}) E_n[s(\tilde{\theta}, X)]. \tag{11.40}$$

In (11.39), $h(\hat{\theta})$ is a generalization of $\hat{\psi} - \psi_0$, and $I_{(H)}$ is a generalization of $I_{\psi \cdot \lambda}$. In (11.40), R_n takes the same form as Definition 11.3 except that here $\tilde{\theta}$ is the constrained maximizer of $E_n \ell(\theta, X)$ subject to $h(\theta) = 0$ rather than $(\psi_0^T, \tilde{\lambda}^T)^T$. Since, by (11.18), $H(\tilde{\theta})\tilde{\rho} = E_n[s(\tilde{\theta}, X)]$, R_n can be equivalently written as

$$R_n = \tilde{\rho}^T H^T(\tilde{\theta}) I^{-1}(\tilde{\theta}) H(\tilde{\theta}) \tilde{\rho}. \tag{11.41}$$

For this reason, Rao's score test is also known as the Lagrangian multiplier test in the econometrics literature (see Engle, 1984; Bera and Bilias, 2001). The next lemma gives the asymptotic linear form of $h(\hat{\theta})$.

Lemma 11.2 *Suppose Assumption 10.3 and the assumptions 1∼5 in Theorem 11.9 are satisfied. If $\hat{\theta}$ is a consistent solution to the likelihood equation $E_n[s(\theta, X)] = 0$, then*

$$h(\hat{\theta}) = h(\theta_0) + I_{(H)}^{-1} E_n[s_{(H)}(\theta_0, X)] + o_P(n^{-1/2}).$$

Proof. By Taylor's theorem,

$$h(\hat{\theta}) = h(\theta_0) + H^T(\theta^\dagger)(\hat{\theta} - \theta_0)$$

for some $\theta^\dagger$ between θ_0 and $\hat{\theta}$. By Theorem 9.5, as applied to $g(\theta, x) = s(\theta, x)$, we have $\hat{\theta} - \theta_0 = I^{-1} E_n[s(\theta_0, X)] + o_P(n^{-1/2})$. Because $H(\theta)$ is continuous we have $H(\theta^\dagger) = H + o_P(1)$. Hence

$$\begin{aligned} h(\hat{\theta}) &= h(\theta_0) + [H + o_P(1)]^T \{I^{-1} E_n[s(\theta_0, X)] + o_P(n^{-1/2})\} \\ &= h(\theta_0) + H^T I^{-1} E_n[s(\theta_0, X)] + o_P(n^{-1/2}) \\ &= h(\theta_0) + I_{(H)}^{-1} E_n[s_{(H)}(\theta_0, X)] + o_P(n^{-1/2}), \end{aligned}$$

where, for the second equality, we have used $E_n[s(\theta_0, X)] = O_P(n^{-1/2})$. □

We now show that W_n and R_n in (11.39) and (11.40) are QF tests with the same quadratic form $U_n^T \Sigma_U^{-1} U_n$ as in T_n in (11.36).

Theorem 11.10 *Suppose Assumption 10.3 and the assumptions 1∼5 in Theorem 11.9 are satisfied. Suppose, in addition, $I(\theta)$ is continuous.*

1. *If $\hat{\theta}$ is a consistent solution to $E_n[s(\theta, X)] = 0$ then W_n is a QF test of the form specified by (11.37) and (11.38);*
2. *If $\tilde{\theta}$ is a consistent solution to $E_n[s(\theta, X)] = 0$ subject to the constraint $h(\theta) = 0$, then R_n is a QF test of the form specified by (11.37) and (11.38).*

Proof. 1. By Lemma 11.2,

$$W_n \overset{Q_n}{=} n[H^T I^{-1} E_n s + o_P(n^{-1/2})]^T I_{(H)}(\hat{\theta}) [H^T I^{-1} E_n s + o_P(n^{-1/2})],$$

where $E_n s$ is the abbreviation of $E_n[s(\theta_0, X)]$. Because $I(\theta)$ and $H(\theta)$ are continuous, I is invertible, and $H^T I^{-1} H$ is invertible, we have, by Lemma 11.1, $I_{(H)}(\hat{\theta}) \xrightarrow{P} I_{(H)}$. Hence,

$$\begin{aligned} W_n &\overset{Q_n}{=} n[H^T I^{-1} E_n s + o_P(n^{-1/2})]^T [I_{(H)} + o_P(1)][H^T I^{-1} E_n s + o_P(n^{-1/2})] \\ &\overset{Q_n}{=} n(H^T I^{-1} E_n s)^T I_{(H)} (H^T I^{-1} E_n s) + o_P(1) \\ &= n(E_n s_{(H)})^T I_{(H)}^{-1} (E_n s_{(H)}) + o_P(1), \end{aligned}$$

where, for the second equality, we have used $E_n s = O_P(n^{-1/2})$, and in the third line, $s_{(H)}$ is the abbreviation of $s_{(H)}(\theta_0, X)$.
2. By (11.35), (11.41) and the first equation in (11.19),

$$R_n \overset{Q_n}{=} [E_n s_{(H)} + o_P(n^{-1/2})]^T H^T(\tilde{\theta}) I^{-1}(\tilde{\theta}) H(\tilde{\theta}) [E_n s_{(H)} + o_P(n^{-1/2})].$$

Because $\tilde{\theta}$ is consistent, $H(\theta)$ and $I(\theta)$ are continuous and $I(\theta)$ is invertible, we have

$$H^T(\tilde{\theta}) I^{-1}(\tilde{\theta}) H(\tilde{\theta}) \xrightarrow{Q_n} H^T I^{-1} H.$$

Hence

$$R_n \stackrel{Q_n}{=} n[E_n s_{(H)} + o_P(n^{-1/2})]^T [H^T I^{-1} H + o_P(1)][E_n s_{(H)} + o_P(n^{-1/2})]$$
$$= n(E_n s_{(H)})^T I_{(H)}^{-1} (E_n s_{(H)}) + o_P(1),$$

where, for the second equality, we used $E_n(s_{(H)}) = O_P(n^{-1/2})$. □

11.6.4 Neyman's $C(\alpha)$ test for general hypotheses

Next, we extend Neyman's $C(\alpha)$ test in Section 11.3.3 to the general hypothesis $H_0 : h(\theta) = 0$. Replacing the efficient score function and efficient information matrix in Definition 11.4 by $s_{(H)}$ and $I_{(H)}$ leads to the following definition of the extended Neyman's $C(\alpha)$ statistic.

Definition 11.10 *Neyman's $C(\alpha)$ statistic for the general hypothesis H_0 : $h(\theta) = 0$ is defined as*

$$N_n = nE_n[s_{(H)}^T(\tilde{\theta}, X)] I_{(H)}^{-1}(\tilde{\theta}) E_n[s_{(H)}(\tilde{\theta}, X)]. \tag{11.42}$$

where $\tilde{\theta}$ is any $\sqrt{n}$-consistent estimate of θ_0 that satisfies $h(\tilde{\theta}) = 0$.

We next show that the Neyman's $C(\alpha)$ thus defined is a QF-test with the same asymptotic quadratic form as T_n, R_n, and W_n.

Theorem 11.11 *Suppose Assumption 10.3 and the assumptions 1∼5 in Theorem 11.9 are satisfied. Suppose, in addition, $I(\theta)$ is continuous and $\tilde{\theta}$ is a $\sqrt{n}$-consistent estimate of θ_0 satisfying the constraint $h(\tilde{\theta}) = 0$. Then N_n is a QF test of the form specified by (11.37) and (11.38).*

Proof. Abbreviate $I(\tilde{\theta})$ and $H(\tilde{\theta})$ by $\tilde{I}$ and $\tilde{H}$. By definition,

$$E_n[s_{(H)}(\tilde{\theta}, X)] = (\tilde{H}^T \tilde{I}^{-1} \tilde{H})^{-1} \tilde{H}^T \tilde{I}^{-1} E_n[s(\tilde{\theta}, X)].$$

By Taylor's theorem,

$$E_n[s(\tilde{\theta}, X)] = E_n[s(\theta_0, X)] + J_n(\theta^\dagger)(\tilde{\theta} - \theta_0)$$

for some $\theta^\dagger$ between θ_0 and $\tilde{\theta}$. Since $\{J_n(\theta) : n \in \mathbb{N}\}$ is stochastic equicontinuous and $\theta^\dagger \xrightarrow{Q_n} \theta_0$, we have by Corollary 8.2 $J_n(\theta^\dagger) \stackrel{Q_n}{=} -I + o_P(1)$. Hence

$$E_n[s(\tilde{\theta}, X)] \stackrel{Q_n}{=} E_n[s(\theta_0, X)] - I(\tilde{\theta} - \theta_0) + o_P(n^{-1/2}). \tag{11.43}$$

Because both $E_n[s(\theta_0, X)]$ and $\tilde{\theta} - \theta_0$ are of the order $O_P(n^{-1/2})$ under Q_n, we have $E_n[s(\tilde{\theta}, X)] \stackrel{Q_n}{=} O_P(n^{-1/2})$. Furthermore, because $H(\theta)$, $I(\theta)$ are continuous, and $I(\theta)$ and $H^T(\theta)I^{-1}(\theta)H(\theta)$ are invertible, we have, by Lemma 11.1,

$$(\tilde{H}^T \tilde{I}^{-1} \tilde{H})^{-1} \tilde{H}^T \tilde{I}^{-1} \stackrel{Q_n}{=} (H^T I^{-1} H)^{-1} H^T I^{-1} + o_P(1).$$

Therefore,

$$E_n[s_{(H)}(\tilde{\theta}, X)] \stackrel{Q_n}{=} (H^T I^{-1} H)^{-1} H^T I^{-1} E_n[s(\tilde{\theta}, X)] + o_P(n^{-1/2}). \quad (11.44)$$

Now substituting (11.43) into (11.44), we have

$$\begin{aligned} E_n[s_{(H)}(\tilde{\theta}, X)] \stackrel{Q_n}{=} & (H^T I^{-1} H)^{-1} H^T I^{-1} E_n[s(\theta_0, X)] \\ & - (H^T I^{-1} H)^{-1} H^T (\tilde{\theta} - \theta_0) + o_P(n^{-1/2}). \end{aligned} \quad (11.45)$$

By Taylor's theorem

$$\begin{aligned} h(\tilde{\theta}) &= h(\theta_0) + H^T(\theta^\dagger)(\tilde{\theta} - \theta_0) \\ &= h(\theta_0) + H^T(\tilde{\theta} - \theta_0) + o_P(n^{-1/2}), \end{aligned}$$

where the second equality follows from the continuity of $H(\theta)$ and the $\sqrt{n}$-consistency of $\tilde{\theta}$. Because both $\tilde{\theta}$ and θ_0 satisfy the constraint $h(\theta) = 0$, we have $H^T(\tilde{\theta} - \theta_0) = o_P(n^{-1/2})$. Hence the second term on the right hand side of (11.45) is of the order $o_P(n^{-1/2})$, resulting in

$$\begin{aligned} E_n[s_{(H)}(\tilde{\theta}, X)] &= (H^T I^{-1} H)^{-1} H^T I^{-1} E_n[s(\theta_0, X)] + o_P(n^{-1/2}) \\ &= E_n[s_{(H)}(\theta_0, X)] + o_P(n^{-1/2}). \end{aligned}$$

Substituting the above relation as well as $I_{(H)}^{-1}(\tilde{\theta}) \stackrel{Q_n}{=} I_{(H)}^{-1}(\theta_0) + o_P(1)$ into the right hand side of (11.42), we have

$$N_n = n E_n[s_{(H)}(\theta_0, X)]^T I_{(H)}^{-1}(\theta_0) E_n[s_{(H)}(\theta_0, X)] + o_P(1),$$

as desired. □

11.6.5 Asymptotic efficiency

In this subsection we extend the concept of an asymptotically efficient test to the general hypothesis $H_0 : h(\theta) = 0$. We first extend the concept of a regular test for such a hypothesis. Let δ_H be the projection of δ on to span(H) with respect to the I^{-1}-inner product, and $\delta_{H^\perp}$ the projection on to the orthogonal complement of span(H). That is,

$$\delta_H = P_H(I^{-1})\delta, \quad \delta_{H^\perp} = Q_H(I^{-1})\delta.$$

These vectors play the roles of δ_ψ and δ_λ when ψ and λ are explicitly defined.

Definition 11.11 *A statistic T_n for testing $H_0 : h(\theta) = 0$ is regular if $T_n \xrightarrow[P_n(\delta)]{\mathcal{D}} F(\delta)$, where $F(\delta)$ depends on, and only on, δ_H in the sense that*

1. if $\delta_H \neq 0$*, then* $F(\delta) \neq F(0)$*;*
2. if $\delta_H^{(1)} = \delta_H^{(2)}$*, then* $F(\delta^{(1)}) = F(\delta^{(2)})$.

The next theorem, which is a generalization of Theorem 11.6, gives a necessary and sufficient condition for a QF test to be regular. Let S_n, U_n, Σ_{SU} be as defined in Assumption 10.4.

Theorem 11.12 *A QF test* T_n *for the hypothesis* $H_0 : h(\theta) = 0$ *is regular if and only if* $\text{span}(\Sigma_{SU}) = \text{span}(H)$.

Proof. By Theorem 11.1, $T_n \xrightarrow[P_n(\delta)]{\mathcal{D}} \chi_r^2[f(\delta)]$, where $f(\delta) = \delta^T \Sigma_{SU} \Sigma_U^{-1} \Sigma_{US} \delta$. Decomposing Σ_{SU} as $P_H(I^{-1})\Sigma_{SU} + Q_H(I^{-1})\Sigma_{SU}$, we have

$$\begin{aligned} f(\delta) = &\delta_H^T \Sigma_{SU} \Sigma_U^{-1} \Sigma_{US} \delta_H + 2\delta_H^T \Sigma_{SU} \Sigma_U^{-1} \Sigma_{US} \delta_{H^\perp} \\ &+ \delta_{H^\perp}^T \Sigma_{SU} \Sigma_U^{-1} \Sigma_{US} \delta_{H^\perp}. \end{aligned} \tag{11.46}$$

If $\text{span}(\Sigma_{SU}) = \text{span}(H)$, then there is a nonsingular $A \in \mathbb{R}^{r \times r}$ such that $\Sigma_{SU} = HA$ and

$$f(\delta) = \delta_H^T HA\Sigma_U^{-1} A^T H^T \delta_H, \tag{11.47}$$

which depends on and only on δ_H. Conversely, suppose $f(\delta)$ satisfies 1 and 2 of Definition 11.11. Then, for any $\delta \in \text{span}(H)$, $\delta \neq 0$, we have $\delta^T \Sigma_{SU} \Sigma_U^{-1} \Sigma_{US} \delta \neq 0$, implying

$$\text{span}(H) \subseteq \text{span}(\Sigma_{SU} \Sigma_U^{-1} \Sigma_{US}) = \text{span}(\Sigma_{SU}).$$

For any $\delta \perp \text{span}(H)$, $\delta \neq 0$, we have $\delta^T \Sigma_{SU} \Sigma_U^{-1} \Sigma_{US} \delta = f(\delta) = f(0) = 0$, implying

$$\text{span}(H)^\perp \subseteq \text{span}(\Sigma_{SU} \Sigma_U^{-1} \Sigma_{US})^\perp = \text{span}(\Sigma_{SU})^\perp.$$

Hence $\text{span}(\Sigma_{SU}) = \text{span}(H)$. □

The next theorem is a generalization of Theorem 11.7. It says that a QF test for $H_0 : h(\theta) = 0$ is regular if and only if its U_n can be so chosen as to obey the convolution theorem for a regular estimate. We will use $S_{n,(H)}$ and $S_{(H)}$ to represent $I_{(H)} H^T I^{-1} S_n$ and $I_{(H)} H^T I^{-1} S$, respectively, where S is the limit of S_n and $S \sim N(0, I)$. Note also that $S_{n,(H)}$ reduces to the rescaled efficient score $\sqrt{n} E_n[s_{(H)}(\theta_0, X)]$ under the i.i.d. assumption.

Theorem 11.13 *A QF test* T_n *for* $H_0 : h(\theta) = 0$ *is regular if and only if it can be written as the form*

$$T_n \overset{Q_n}{=} U_n^T \Sigma_U^{-1} U_n + o_P(1),$$

where $U_n = I_{(H)}^{-1} S_{n,(H)} + R_n$*, with* (R_n, S_n, L_n) *satisfying ALAN and* $R \perp\!\!\!\perp S$.

Proof. Because T_n is a QF test, it can be written as $T_n \overset{Q_n}{=} V_n^T \Sigma_V^{-1} V_n + o_P(1)$ for some $V_n \in \mathbb{R}^r$ such that (V_n, S_n, L_n) satisfies ALAN. Furthermore, because T_n is regular, by Theorem 11.12, Σ_{VS} satisfies $\mathrm{span}(\Sigma_{VS}) = \mathrm{span}(H)$. Hence $\Sigma_{VS} = HA$ for some nonsingular $A \in \mathbb{R}^{r\times r}$. Let $U_n = A^{-T} V_n$. Then (U_n, S_n, L_n) satisfies ALAN with $\Sigma_{SU} = \Sigma_{SV} A^{-1} = H$, and

$$T_n \overset{Q_n}{=} U_n^T \Sigma_U^{-1} U_n + o_P(1).$$

The limiting random vector U can be written as $I_{(H)}^{-1} S_{(H)} + R$ where $R = U - I_{(H)}^{-1} S_{(H)}$ and $S_{(H)} = I_{(H)} H^T I^{-1} S$. Note that

$$\begin{aligned} \mathrm{cov}(R, S) &= \mathrm{cov}(U, S) - \mathrm{cov}(I_{(H)}^{-1} S_{(H)}, S) \\ &= \Sigma_{US} - I_{(H)}^{-1} \mathrm{cov}(S_{(H)}, S) \\ &= H^T - I_{(H)}^{-1} I_{(H)} H^T = 0, \end{aligned}$$

where, for the third equality, we have used $\Sigma_{SS} = I$. which implies $R \perp\!\!\!\perp S$ because R and S are jointly Normal by the ALAN assumption. □

By this theorem and its proof, if T_n is any regular QF test for $H_0 : h(\theta) = 0$, then it can be represented by $U_n^T \Sigma_U^{-1} U_n + o_P(1)$, where

$$\Sigma_U = I_{(H)}^{-1} I_{(H)} I_{(H)}^{-1} + \Sigma_R \succeq I_{(H)}^{-1},$$

and $\Sigma_{SU} = H$. Hence the noncentrality parameter of the asymptotic noncentral chi-squared distribution of T_n under $P_n(\delta)$ is

$$\delta^T \Sigma_{SU} \Sigma_U^{-1} \Sigma_{US} \delta = \delta^T H \Sigma_U^{-1} H^T \delta.$$

This implies that the upper bound of the noncentrality parameter is $\delta^T H I_{(H)} H^T \delta$. Furthermore, for any regular QF test that reaches this upper bound, its U_n is asymptotically equivalent to $I_{(H)}^{-1} S_{n,(H)}$. We summarize this result in the next corollary.

Corollary 11.2 *Suppose T_n is any regular QF test. Then the following statements hold.*

1. $T_n \xrightarrow[P_n(\delta)]{\mathcal{D}} \chi_r^2(\delta^T H \Sigma_U^{-1} H^T \delta)$, *where* $\Sigma_U^{-1} \preceq I_{(H)}$.
2. $T_n \xrightarrow[P_n(\delta)]{\mathcal{D}} \chi_r^2(\delta^T H I_{(H)} H^T \delta)$ *if and only if T_n can be written in the form* $T_n \overset{Q_n}{=} U_n^T \Sigma_U^{-1} U_n + o_P(1)$, *where* $U_n \overset{Q_n}{=} I_{(H)}^{-1} S_{n,(H)} + o_P(1)$.

Naturally, we define those QF tests for $H_0 : h(\theta) = 0$ with largest noncentrality parameter $\delta^T H I_{(H)} H^T \delta$ in their asymptotic distribution under $P_n(\delta)$ as the asymptotically efficient tests. All the four tests T_n, W_n, R_n, and N_n are Asymptotically Efficient tests.

11.7 QF tests for estimating equations

In this section we develop QF-tests for estimating equations. To our knowledge, the results presented here have not all been recorded in the statistics literature. The most relevant publications are Rotnisky and Jewel (1990), where a Wald-type statistic was proposed for testing hypothesis for parameters in a Generalized Estimating Equation, and Boos (1992), which developed score tests based estimating equations rather than the likelihood score functions. Of course, the general strategy in Hall and Mathiason (1990) also plays a critical role in the following development, as it has throughout this chapter. We will only consider the case where the parameter of interest and the nuisance parameter are explicitly defined, omitting the general hypothesis $H_0 : h(\theta) = 0$, which can be developed by making analogies to the steps in Section 11.6. Problems 11.24 and 11.25 are devoted to this further generalization.

11.7.1 Wald's, Rao's, and Neyman's tests for estimating equations

Let θ, ψ, and λ be as defined in Section 11.1. Let $X_1, \ldots, X_n$ be an i.i.d. sample from an unspecified distribution P_θ, where $\theta \in \Theta \subseteq \mathbb{R}^p$. Suppose we are interested in testing the hypothesis

$$H_0 : \psi = \psi_0.$$

Let $g : \Theta \times \Omega_X \to \mathbb{R}^p$ be an unbiased and P_θ-square-integrable estimating equation. We estimate the true parameter $\theta_0 = (\psi_0^T, \lambda_0^T)^T$ by solving the equation

$$E_n[g(\theta, X)] = 0. \tag{11.48}$$

Let g_ψ be the first r components of g, and g_λ the last s components of g, and, as in Chapter 9, let

$$J_g(\theta) = E_\theta[\partial g(\theta, X)/\partial \theta^T], \quad K_g(\theta) = E_\theta[g(\theta, X) g^T(\theta, X)].$$

The information contained in g is $I_g(\theta) = J_g^T(\theta) K_g^{-1}(\theta) J_g(\theta)$. Let $J_g(\theta)$ and $K_g(\theta)$ be partitioned, in obvious ways, into the following block matrices

$$J_g(\theta) = \begin{pmatrix} J_{g,\psi\psi}(\theta) & J_{g,\psi\lambda}(\theta) \\ J_{g,\lambda\psi}(\theta) & J_{g,\lambda\lambda}(\theta) \end{pmatrix}, \quad K_g(\theta) = \begin{pmatrix} K_{g,\psi\psi}(\theta) & K_{g,\psi\lambda}(\theta) \\ K_{g,\lambda\psi}(\theta) & K_{g,\lambda\lambda}(\theta) \end{pmatrix}.$$

Here, we do not assume $J_g(\theta)$ to be symmetric: for example $J_{g,\psi\psi}^T(\theta)$ need not be the same as $J_{g,\psi\psi}(\theta)$ and $J_{g,\psi\lambda}^T(\theta)$ need not be the same as $J_{g,\lambda\psi}(\theta)$. Mimicking the efficient score, let

$$g_{\psi\cdot\lambda}(\theta, X) = g_\psi(\theta, X) - J_{g,\psi\lambda}(\theta) J_{g,\lambda\lambda}^{-1}(\theta) g_\lambda(\theta, X).$$

Notice that, in the above definition, the coefficient matrix for $g_\lambda(\theta, X)$ is not $I_{g,\psi\lambda}(\theta)I_{g,\lambda\lambda}(\theta)$, as one might anticipate. We abbreviate $J_g(\theta_0)$, $K_g(\theta_0)$ and $I_g(\theta_0)$ by J_g, K_g, and I_g, respectively, and abbreviate $g_\psi(\theta_0, X)$, $g_\lambda(\theta_0, X)$, and $g_{\psi\cdot\lambda}(\theta_0, X)$ by g_ψ, g_λ, and $g_{\psi\cdot\lambda}$. Let $(J_g^{-1})_{\psi\theta}$ denote the $r \times p$ matrix consisting of the first r rows of J_g^{-1}, $(J_g^{-1})_{\psi\psi}$ the upper-left $r \times r$ block of the matrix J_g^{-1}, and $(J_g^{-1})_{\psi\lambda}$ the upper-right $r \times s$ block of J_g^{-1}. The following lemma gives some properties about $g_{\psi\cdot\lambda}(\theta_0, X)$ that will be useful further on.

Lemma 11.3 *Suppose $g(\theta, X)$ is an unbiased and P_θ-square-integrable estimating equation. If the derivatives, inverses, and moments involved are defined, then*

1. $[J_g^{-1}(\theta)]_{\psi\psi}\, g_{\psi\cdot\lambda}(\theta, X) = [J_g^{-1}(\theta)]_{\psi\theta}\, g(\theta, X)$;
2. $\mathrm{var}[(J_g^{-1})_{\psi\psi}\, g_{\psi\cdot\lambda}] = (I_g^{-1})_{\psi\psi}$;
3. $E[\partial g_{\psi\cdot\lambda}(\theta_0, X)/\partial\psi^T] = (J_g^{-1})^{-1}_{\psi\psi}$;
4. $E[\partial g_{\psi\cdot\lambda}(\theta_0, X)/\partial\lambda^T] = 0$,

where $I_g = I_g(\theta_0), J_g = J_g(\theta_0), g_{\psi,\lambda} = g_{\psi,\lambda}(\theta_0, X)$ in parts 2, 3, and 4.

Proof. 1. By construction,

$$
\begin{aligned}
[J_g^{-1}(\theta)]_{\psi\theta} &= ([J_g^{-1}(\theta)]_{\psi\psi},\ [J_g^{-1}(\theta)]_{\psi\lambda}) \\
&= [J_g^{-1}(\theta)]_{\psi\psi}(I_r,\ [J_g^{-1}(\theta)]^{-1}_{\psi\psi}[J_g^{-1}(\theta)]_{\psi\lambda}),
\end{aligned}
$$

where the second equality is obtained by factoring out the term $[J_g^{-1}(\theta)]_{\psi\psi}$. By Proposition 9.3, $[J_g^{-1}(\theta)]^{-1}_{\psi\psi}[J_g^{-1}(\theta)]_{\psi\lambda} = -J_{g,\psi\lambda}(\theta)J^{-1}_{g,\lambda\lambda}(\theta)$. Hence

$$
\begin{aligned}
[J_g^{-1}(\theta)]_{\psi\theta}\, g(\theta, X) &= [J_g^{-1}(\theta)]_{\psi\psi}(I_r,\ [J_g^{-1}(\theta)]^{-1}_{\psi\psi}[J_g^{-1}(\theta)]_{\psi\lambda}) \\
&= [J_g^{-1}(\theta)]_{\psi\psi}\, [I_r, -J_{g,\psi\lambda}(\theta)J^{-1}_{g,\lambda\lambda}(\theta)]\ g(\theta, X) \\
&= [J_g^{-1}(\theta)]_{\psi\psi}\, [g_\psi(\theta, X) - J_{g,\psi\lambda}(\theta)J^{-1}_{g,\lambda\lambda}(\theta)g_\lambda(\theta, X)] \\
&= [J_g^{-1}(\theta)]_{\psi\psi}\, g_{\psi\cdot\lambda}(\theta, X).
\end{aligned}
$$

2. Because, by part 1, $(J_g^{-1})_{\psi\psi}\, g_{\psi\cdot\lambda} = (J_g^{-1})_{\psi\theta}\, g$, we have

$$
\mathrm{var}[(J_g^{-1})_{\psi\psi}\, g_{\psi\cdot\lambda}] = (J_g^{-1})_{\psi\theta}\, K_g\, [(J_g^{-1})_{\psi\theta}]^T. \tag{11.49}
$$

Note that, for any $p \times p$ matrix A, if $A_{\psi\theta}$ represents the first r rows of A, then $(A_{\psi\theta})^T$ is simply the first r columns of A^T. That is, $(A_{\psi\theta})^T = (A^T)_{\theta\psi}$. Consequently, the right-hand side of (11.49) is

$$
(J_g^{-1})_{\psi\theta}\, K_g\, (J_g^{-T})_{\theta\psi}.
$$

This matrix is simply the upper-left $r \times r$ block of the matrix $J_g^{-1}K_gJ_g^{-T}$; that is, $(J_g^{-1}K_gJ_g^{-T})_{\psi\psi}$. Because $J_g^{-1}K_gJ_g^{-T}$ is the inverse of the information matrix I_g, we have

$$(J_g^{-1})_{\psi\theta}\, K_g\, (J_g^{-T})_{\theta\psi} = (I_g^{-1})_{\psi\psi}.$$

Thus we have the identity in *2.*

3. By definition, for each $i = 1, \ldots r$,

$$\begin{aligned} E[\partial g_{\psi\cdot\lambda}(\theta_0, X)/\partial\psi_i] =& E[\partial g_\psi(\theta_0, X)/\partial\psi_i] \\ & - \partial[J_{g,\psi\lambda}(\theta_0) J_{g,\lambda\lambda}^{-1}(\theta_0)]/\partial\psi_i\, E[g_\lambda(\theta_0, X)] \\ & - J_{g,\psi\lambda} J_{g,\lambda\lambda}^{-1} E[\partial g_\lambda(\theta_0, X)/\partial\psi_i]. \end{aligned}$$

Because $g(\theta, X)$ is unbiased, the second term on the right-hand side is 0, resulting in

$$\begin{aligned} & E[\partial g_{\psi\cdot\lambda}(\theta_0, X)/\partial\psi^T] \\ & = E[\partial g_\psi(\theta_0, X)/\partial\psi^T] - J_{g,\psi\lambda} J_{g,\lambda\lambda}^{-1} E[\partial g_\lambda(\theta_0, X)/\partial\psi^T] \\ & = J_{g,\psi\psi} - J_{g,\psi\lambda} J_{g,\lambda\lambda}^{-1} J_{g,\lambda\psi} = (J_g^{-1})_{\psi\psi}. \end{aligned}$$

4. Similarly, $E[\partial g_{\psi\cdot\lambda}(\theta_0, X)/\partial\lambda^T] = J_{g,\psi\lambda} - J_{g,\psi\lambda} J_{g,\lambda\lambda}^{-1} J_{g,\lambda\lambda} = 0.$ □

Let $\hat\theta = (\hat\psi^T, \hat\lambda^T)^T$ be a solution to the estimating equation (11.48). Let $\tilde\lambda$ be a solution to the estimating equation

$$E_n[g_\lambda(\psi_0, \lambda)] = 0.$$

Let $\tilde\theta = (\psi_0^T, \tilde\lambda^T)^T$. Let $\bar\lambda$ be any $\sqrt{n}$-consistent estimate of λ_0, and $\bar\theta = (\psi_0^T, \bar\lambda^T)^T$. We now give the formal definitions of Wald's, Rao's, and Neyman's test statistics for an estimating equation g.

Definition 11.12 *The Wald's, Rao's, and Neyman's $C(\alpha)$ test statistics for $H_0 : \psi = \psi_0$ based on the estimating equation g are defined, respectively, as*

$$\begin{aligned} W_n(g) =& n(\hat\psi - \psi_0)^T [I_g^{-1}(\hat\theta)]_{\psi\psi}^{-1} (\hat\psi - \psi_0), \\ R_n(g) =& n E_n[g_\psi^T(\tilde\theta, X)][J_g^{-1}(\tilde\theta)]_{\psi\psi} [I_g^{-1}(\tilde\theta)]_{\psi\psi}^{-1} [J_g^{-1}(\tilde\theta)]_{\psi\psi} E_n[g_\psi(\tilde\theta, X)], \\ N_n(g) =& n E_n[g_{\psi\cdot\lambda}^T(\bar\theta, X)][J_g^{-1}(\bar\theta)]_{\psi\psi} [I_g^{-1}(\bar\theta)]_{\psi\psi}^{-1} [J_g^{-1}(\bar\theta)]_{\psi\psi} E_n[g_{\psi\cdot\lambda}(\bar\theta, X)]. \end{aligned}$$

These statistics are generalizations of W_n, R_n and N_n defined in (11.39), (11.40), (11.42): if we take g to be the score function s, then

$$W_n = W_n(s), \quad R_n = R_n(s), \quad N_n = N_n(s).$$

Note that, similar to W_n, R_n, and N_n, $W_n(g)$ requires $\hat\theta$, the solution to the full estimating equation $E_n[g(\theta, X)] = 0$; $R_n(g)$ requires $\tilde\lambda$, the solution to λ-component of the full estimating equation, $E_n[g_\lambda(\psi_0, \lambda, X)] = 0$; whereas $N_n(g)$ just requires any $\sqrt{n}$-consistent estimate $\bar\lambda$ of λ_0.

In the following, We use $J_{g,n}(\theta)$ to denote the matrix

$$E_n[\partial g(\theta, X)/\partial \theta^T].$$

The next theorem shows that $W_n(g)$, $R_n(g)$, and $N_n(g)$ are QF tests with the same asymptotic quadratic form.

Theorem 11.14 *Suppose Assumption 10.3 holds and*

i. $g(\theta, X)$ is an unbiased, P_θ-square-integrable estimating equation and $g(\theta, x) f_\theta(x)$ satisfies $DUI^+(\theta, \mu)$;
ii. $J_g(\theta)$ and $K_g(\theta)$ are invertible and continuous;
iii. the sequence of random matrices $\{J_{g,n}(\theta) : n \in \mathbb{N}\}$ is stochastically equicontinuous in a neighborhood of θ_0;
iv. the true parameter θ_0 is an interior point of Θ.

Then the following assertions hold.

1. If $\hat{\theta}$ is a consistent solution of $E_n[g(\theta, X)] = 0$, then

$$W_n(g) \stackrel{Q_n}{=} U_n^T(g) \Sigma_{U(g)}^{-1} U_n(g) + o_P(1),$$

where $(U_n(g), S_n, L_n)$ satisfies ALAN with

$$U_n(g) = \sqrt{n}(J_g^{-1})_{\psi\psi} E_n(g_{\psi \cdot \lambda}), \quad \Sigma_{U(g)} = (I_g^{-1})_{\psi\psi}, \quad \Sigma_{U(g)S} = -(I_r, 0).$$

2. If $\tilde{\lambda}$ is a consistent solution to $E_n[g_\psi(\psi_0, \lambda, X)] = 0$, then $R_n(g)$ is a QF test of the same form as $W_n(g)$.
3. If $\bar{\lambda}$ is any $\sqrt{n}$-consistent estimate of λ_0, $g_{\psi \cdot \lambda}(\psi_0, \lambda)$ is differentiable with respect to λ, and

$$\{E_n[\partial g_{\psi \cdot \lambda}(\psi_0, \lambda)/\partial \lambda^T] : n \in \mathbb{N}\}$$

is stochastically equicontinuous with respect to λ, then $N_n(g)$ is a QF test of the same form as $W_n(g)$.

Proof. 1. By Theorem 9.5 we have

$$\hat{\theta} = \theta_0 - J_g^{-1}(\theta_0) E_n[g(\theta_0, X)] + o_P(n^{-1/2}).$$

Read off the first r lines of this equation to obtain

$$\sqrt{n}(\hat{\psi} - \psi_0) = -\sqrt{n}(J_g^{-1})_{\psi\theta} E_n[g(\theta_0, X)] + o_P(n^{-1/2}). \tag{11.50}$$

Because $I_g(\theta)$ is continuous and $\hat{\theta}$ is consistent, $I_g(\hat{\theta}) \xrightarrow{Q_n} I_g$, By the invertibility of I_g and Lemma 11.1, this convergence implies $I_g^{-1}(\hat{\theta}) \xrightarrow{Q_n} I_g^{-1}$, which, by the continuous mapping theorem, implies $[I_g^{-1}(\hat{\theta})]_{\psi\psi} \xrightarrow{Q_n} (I_g^{-1})_{\psi\psi}$. Because $(I_g^{-1})_{\psi\psi}$ is invertible, we have, by Lemma 11.1 again,

$$[I_g^{-1}(\hat{\theta})]_{\psi\psi}^{-1} \xrightarrow{Q_n} (I_g^{-1})_{\psi\psi}^{-1}. \tag{11.51}$$

Substituting (11.50) and (11.51) into $W_n(g)$ in Definition 11.12, we have

$$\begin{aligned} W_n(g) &\overset{Q_n}{=} [-\sqrt{n}(J_g^{-1})_{\psi\theta}E_n(g) + o_P(n^{-1/2})]^T[(I_g^{-1})_{\psi\psi}^{-1} + o_P(1)] \\ &\quad [-\sqrt{n}(J_g^{-1})_{\psi\theta}E_n(g) + o_P(n^{-1/2})] \\ &= n[(J_g^{-1})_{\psi\theta}E_n(g)]^T(I_g^{-1})_{\psi\psi}^{-1}[(J_g^{-1})_{\psi\theta}E_n(g)] + o_P(1). \end{aligned}$$

By Lemma 11.3, the right-hand side can be rewritten as

$$n[(J_g^{-1})_{\psi\psi}E_n(g_{\psi\cdot\lambda})]^T(I_g^{-1})_{\psi\psi}^{-1}[(J_g^{-1})_{\psi\psi}E_n(g_{\psi\cdot\lambda})] + o_P(1). \tag{11.52}$$

Let $U_n(g) = \sqrt{n}(J_g^{-1})_{\psi\psi}E_n[g_{\psi\cdot\lambda}(\theta_0, X)]$. By the central limit theorem,

$$\begin{pmatrix} U_n(g) \\ S_n \end{pmatrix} \xrightarrow[Q_n]{\mathcal{D}} \begin{pmatrix} U(g) \\ S \end{pmatrix},$$

where

$$\text{var}\left[\begin{pmatrix} U(g) \\ S \end{pmatrix}\right] = \begin{pmatrix} \text{var}[(J_g^{-1})_{\psi\psi}E(g_{\psi\cdot\lambda})] & (J_g^{-1})_{\psi\psi}E(g_{\psi\cdot\lambda}s^T) \\ [(J_g^{-1})_{\psi\psi}E(g_{\psi\cdot\lambda}s^T)]^T & I \end{pmatrix}.$$

By Lemma 11.3,

$$\text{var}[(J_g^{-1})_{\psi\psi}E(g_{\psi\cdot\lambda})] = (I_g^{-1})_{\psi\psi}.$$

Because $g(\theta, X)f_\theta(X)$ satisfies $\text{DUI}^+(\theta, \mu)$, we have, by Lemma 9.1, $E(gs^T) = -E(\partial g/\partial\theta^T)$. Hence, by Lemma 11.3,

$$(J_g^{-1})_{\psi\psi}E(g_{\psi\cdot\lambda}s^T) = (J_g^{-1})_{\psi\theta}E(gs^T) = -(J_g^{-1})_{\psi\theta}J_g = -(I_r, 0).$$

Thus we have $\Sigma_{U(g)} = (I_g^{-1})_{\psi\psi}$ and $\Sigma_{U(g)S} = -(I_r, 0)$.

2. By Taylor's theorem,

$$E_n[g_\psi(\psi_0, \tilde{\lambda}, X)] = E_n[g_\psi(\theta_0, X)] + [J_{g,n}(\psi_0, \lambda^\dagger)]_{\psi\lambda}(\tilde{\lambda} - \lambda_0)$$

for some $\lambda^\dagger$ between λ_0 and $\tilde{\lambda}$. Because $\{J_{g,n}(\theta) : n \in \mathbb{N}\}$ is stochastically equicontinuous and $\theta^\dagger \xrightarrow{Q_n} \lambda_0$, we have $J_{g,n}(\psi_0, \lambda^\dagger) \overset{Q_n}{=} J_g + o_P(1)$, which implies

$$[J_{g,n}(\psi_0, \lambda^\dagger)]_{\psi\lambda} \overset{Q_n}{=} J_{g,\psi\lambda} + o_P(1).$$

By Theorem 9.5 (as applied to the estimating equation $(\lambda, x) \mapsto g_\lambda(\psi_0, \lambda, x)$),

$$\tilde{\lambda} = \lambda_0 - J_{g,\lambda\lambda}^{-1}E_n(g_\lambda) + o_P(n^{-1/2}).$$

Hence

$$\begin{aligned} E_n[g_\psi(\psi_0, \tilde{\lambda}, X)] &= E_n(g_\psi) + [J_{g,\psi\lambda} + o_P(1)][-J_{g,\lambda\lambda}^{-1}E_n(g_\lambda) + o_P(n^{-1/2})] \\ &= E_n(g_{\psi\cdot\lambda}) + o_P(n^{-1/2}), \end{aligned}$$

where, for the second equality we used $E_n(g_\lambda) = O_P(n^{-1/2})$. Using arguments similar to the proof of part 1, we can show that

$$[I_g^{-1}(\tilde\theta)]_{\psi\psi}^{-1} \overset{Q_n}{=} (I_g^{-1})_{\psi\psi} + o_P(1), \quad [J_g^{-1}(\tilde\theta)]_{\psi\psi} \overset{Q_n}{=} (J_g^{-1})_{\psi\psi} + o_P(1).$$

Hence

$$\begin{aligned} R_n(g) = &[E_n(g_{\psi\cdot\lambda}) + o_P(n^{-1/2})]^T [(J_g^{-1})_{\psi\psi} + o_P(1)][(I_g^{-1})_{\psi\psi} + o_P(1)] \\ &[(J_g^{-1})_{\psi\psi} + o_P(1)]^T [E_n(g_{\psi\cdot\lambda}) + o_P(n^{-1/2})] \\ = &n[(J_g^{-1})_{\psi\psi} E_n(g_{\psi\cdot\lambda})]^T (I_g^{-1})_{\psi\psi}^{-1} [(J_g^{-1})_{\psi\psi} E_n(g_{\psi\cdot\lambda})] + o_P(1), \end{aligned}$$

where the right-hand side is the same as (11.52). The rest of the proof of part 2 is same as that of part 1.

3. By Taylor's theorem,

$$E_n[g_{\psi\cdot\lambda}(\psi_0, \bar\lambda, X)] = E_n[g_{\psi\cdot\lambda}(\theta_0, X)] + E_n[\partial g_{\psi\cdot\lambda}(\psi_0, \lambda^\ddagger, X)/\partial\lambda^T](\tilde\lambda - \lambda_0)$$

for some $\lambda^\ddagger$ between λ_0 and $\bar\lambda$. Because $\{E_n[\partial g_{\psi\cdot\lambda}(\psi_0, \lambda, X)] : n \in \mathbb{N}\}$ is stochastically equicontinuous and $\lambda^\ddagger \xrightarrow{Q_n} \lambda_0$, we have by Corollary 8.2

$$E_n[\partial g_{\psi\cdot\lambda}(\psi_0, \lambda^\ddagger, X)/\partial\lambda^T] \overset{Q_n}{=} E[\partial g_{\psi\cdot\lambda}(\theta_0, X)/\partial\lambda^T] + o_P(1) = o_P(1),$$

where the second equality follows from Lemma 11.3, part 4. Therefore,

$$E_n[g_{\psi\cdot\lambda}(\psi_0, \bar\lambda, X)] \overset{Q_n}{=} E_n[g_{\psi\cdot\lambda}(\theta_0, X)] + o_P(n^{-1/2}).$$

Using arguments similar to those in the proof of part 1, we can show that

$$[I_g^{-1}(\bar\theta)]_{\psi\psi}^{-1} \overset{Q_n}{=} (I_g^{-1})_{\psi\psi} + o_P(1), \quad [J_g^{-1}(\bar\theta)]_{\psi\psi} \overset{Q_n}{=} (J_g^{-1})_{\psi\psi} + o_P(1).$$

Hence

$$N_n(g) \overset{Q_n}{=} n[(J_g^{-1})_{\psi\psi} E_n(g_{\psi\cdot\lambda})]^T (I_g^{-1})_{\psi\psi}^{-1} [(J_g^{-1})_{\psi\psi} E_n(g_{\psi\cdot\lambda})] + o_P(1),$$

where the right-hand side is the same as (11.52). The rest of the proof of this part is the same as that of part 1. □

From Theorem 11.14 we can immediately derive the asymptotic distributions of $W_n(g)$, $R_n(g)$, and $N_n(g)$ under $P_n(\delta)$ for any $\delta \in \mathbb{R}^p$. The proof is straightforward and is omitted.

Corollary 11.3 *Under the conditions in Theorem 11.14, $W_n(g)$, $R_n(g)$ and $N_n(g)$ each converges in distribution to $\chi_r^2(\delta_\psi^T (I_g^{-1})_{\psi\psi}^{-1} \delta_\psi)$ under the local alternative distribution $P_n(\delta)$.*

Note that $(I_g^{-1})_{\psi\psi}^{-1}$ is monotone nondescreasing with respect to I_g in the sense that, if $I_{g^*} \succeq I_g$, then $(I_{g^*}^{-1})_{\psi\psi}^{-1} \succeq (I_g^{-1})_{\psi\psi}^{-1}$. This is because

$$I_{g^*} \succeq I_g \Rightarrow I_{g^*}^{-1} \preceq I_g^{-1} \Rightarrow (I_{g^*}^{-1})_{\psi\psi} \preceq (I_g^{-1})_{\psi\psi} \Rightarrow (I_{g^*}^{-1})_{\psi\psi}^{-1} \succeq (I_g^{-1})_{\psi\psi}^{-1}.$$

Hence, if g^* is the optimal estimating equation among a class of estimating equations $\mathcal{G}$, then the local alternative asymptotic distribution of $W_n(g^*)$, $R_n(g^*)$, and $N_n(g^*)$ under $P_n(\delta)$ have the largest noncentrality parameter among that class, implying that they are asymptotically most powerful compared with any $W_n(g)$, $R_n(g)$, and $N_n(g)$ for $g \in \mathcal{G}$. In other words, an optimal estimating equation leads to an optimal QF test.

11.7.2 QF tests for canonical estimating equations

The QF tests described above are applicable to arbitrary unbiased and P_θ-square-integrable estimations that satisfy some mild additional assumptions. These statistics take simpler forms when the identity $J_g = -K_g$ holds. Recall that this relation does hold for the score function $s(\theta, x)$ under mild conditions; that is,

$$-J(\theta) = K(\theta) = I(\theta)$$

for $J(\theta)$ and $K(\theta)$ defined in (8.15). This relation does not hold for a general estimating equation g, but it is always possible to find an equivalent transformation of g that satisfies this relation.

Specifically, let $g(\theta, X)$ be any unbiased and P_θ-square-integrable estimating equation such that $f_\theta(x)g(\theta, x)$ satisfies $\mathrm{DUI}^+(\theta, \mu)$. All the estimating equations in the class

$$\mathcal{G}_g = \{B(\theta)g(\theta, X) : B(\theta) \in \mathbb{R}^{p \times p}, B(\theta) \text{ is differentiable and invertible}\}$$

are equivalent. That is, $E_n[h(\theta, X)] = 0$ produces the same solution(s) for any $h \in \mathcal{G}_g$. Adopt again the notation

$$E_\theta[g_1(\theta, X)g_2^T(\theta, X)] = [g_1, g_2].$$

In this notation, and in view of Lemma 9.1, we have

$$J_g(\theta) = -[g, s], \quad K_g(\theta) = [g, g],$$

where $s = s(\theta, x)$ is the score function. Let $\tilde{g}(\theta, X) = B(\theta)g(\theta, X)$, and consider the equation $[\tilde{g}, s] = [\tilde{g}, \tilde{g}]$. If this holds then

$$\begin{aligned}
&[s, g]B^T = B[g, g]B^T \\
&\Rightarrow [s, g] = B[g, g] \\
&\Rightarrow B = [s, g][g, g]^{-1} \\
&\Rightarrow B = -J_g^T(\theta)K_g^{-1}(\theta).
\end{aligned}$$

So, if we let

$$\tilde{g}(\theta, X) = -J_g^T(\theta) K_g^{-1}(\theta)\, g(\theta, X),$$

then $\tilde{g}$ satisfies $J_{\tilde{g}} = -K_{\tilde{g}}$, and $\tilde{g}$ is equivalent to g. This motivates the following definition of canonical form of an estimating equation.

Definition 11.13 *Let g be an unbiased, P_θ-square-integrable estimating function such that $g(\theta, x) f_\theta(x)$ satisfies $DUI^+(\theta, \mu)$ with $J_g(\theta)$ and $K_g(\theta)$ invertible. The canonical form of g is $-J_g(\theta)^T K_g^{-1}(\theta) g(\theta, X)$.*

Since the canonical form of an estimating equation is equivalent to the estimating equation, we can assume, without loss of generality, any estimating equation satisfies $J_g = -K_g$. With this in mind, we can redefine $W_n(g)$, $R_n(g)$, and $N_n(g)$ in the canonical form of g.

Definition 11.14 *Suppose g is a canonical estimating equation. The Wald's, Rao's, and Neyman's $C(\alpha)$ tests are defined as*

$$\begin{aligned} W_n(g) &= n(\hat{\psi} - \psi_0)^T [I_g^{-1}(\hat{\theta})]_{\psi\psi}^{-1} (\hat{\psi} - \psi_0), \\ R_n(g) &= n E_n[g_\psi^T(\tilde{\theta}, X)][I_g^{-1}(\tilde{\theta})]_{\psi\psi} E_n[g_\psi(\tilde{\theta}, X)]. \\ N_n(g) &= n E_n[g_{\psi\cdot\lambda}^T(\bar{\theta}, X)][I_g^{-1}(\bar{\theta})]_{\psi\psi} E_n[g_{\psi\cdot\lambda}(\bar{\theta}, X)]. \end{aligned}$$

Note that $W_n(g)$ takes the same form as that in Definition 11.12, but here $I_g = -J_g = K_g$, which is not the case for $W_n(g)$ in Definition 11.12. The forms of $R_n(g)$ and $N_n(g)$ are simplified due to the relation $-J_g = I_g$. The Rao's statistic for a canonical estimating equation takes a particularly simple form. Since $g_\lambda(\psi_0, \tilde{\lambda}) = 0$, $R_n(g)$ reduces to

$$n E_n[g^T(\tilde{\theta}, X)] I_g^{-1}(\tilde{\theta}) E_n[g(\tilde{\theta}, X)],$$

which is of the same form as the score test for likelihood in Definition 11.3 except that s is replaced by g. The quadratic form of $W_n(g)$, $R_n(g)$, and $N_n(g)$ when g is in the canonical form is simplified correspondingly, which is recorded in the next corollary.

Corollary 11.4 *If the conditions in Theorem 11.14 hold and g is in its canonical form, then under Q_n, the statistics $W_n(g)$, $R_n(g)$, and $N_n(g)$ are of the following asymptotic quadratic form*

$$U_n^T(g) \Sigma_{U(g)}^{-1} U_n(g) + o_P(1),$$

where $(U_n(g), S_n, L_n)$ satisfies ALAN with

$$U_n(g) = \sqrt{n} E_n(g_{\psi\cdot\lambda}), \quad \Sigma_{U(g)} = (I_g^{-1})_{\psi\psi}^{-1}, \quad \Sigma_{U(g)S} = -((I_g^{-1})_{\psi\psi}^{-1}, 0).$$

Consequently, they converge in distribution to $\chi_r^2(\delta_\psi (I_g^{-1})_{\psi\psi}^{-1} \delta_\psi)$ under $P_n(\delta)$.

Interestingly, the asymptotic distribution of $W_n(g)$, $R_n(g)$, and $N_n(g)$ are the same whether or not g is in its canonical form, which is not surprising because an estimating equation is equivalent to its canonical form in the sense that they have the same solution(s).

11.7.3 Wilks's test for conservative estimating equations

Up to this point we haven't mentioned the generalization of the likelihood ratio test for estimating equations. This is because an estimating equation $g(\theta, X)$ does not in general correspond to a "likelihood"; that is, there need not be a function $\ell(\theta, X)$ such that

$$\partial \ell(\theta, X)/\partial \theta^T = g(\theta, X).$$

A set of sufficient conditions for $g(\theta, X)$ to possess such a function $\ell(\theta, X)$ are

1. $g(\theta, X)$ is continuously differentiable with $\partial g(\theta, X)/\partial \theta^T$ being a symmetric matrix;
2. Θ is a convex set in $\mathbb{R}^p$.

The convex assumption Θ is not the weakest possible, but is good enough for our purpose. Li and McCullagh (1994) call such estimating equations conservative estimating equations because $\{g(\theta, X) : \theta \in \Theta\}$ forms a conservative vector field. For such estimating equations the line integral

$$\int_C g^T(\theta, X) d\theta,$$

where C is a smooth curve from a fixed point $a \in \Theta$ to an arbitrary point $\theta \in \Theta$, does not depend on the curve C. As such, the integral is a function of θ. We define this integral as the "quasilikelihood function" $\ell(\theta, X)$ for the estimating equation $g(\theta, X)$. McCullagh (1983) introduced such a definition for linear estimating equations. A convenient choice of C is the straight line. Specifically, fix any point $a \in \Theta$, and let $\theta \in \Theta$ be an arbitrary point. Because Θ is assumed convex, the straight line $\{(1-t)a + t\theta : t \in [0, 1]\}$ is contained in Θ, and we can define $\ell(\theta, X)$ as the line integral

$$\ell(\theta, X) = \int_0^1 g^T[(1-t)a + t\theta, X](\theta - a) dt. \tag{11.53}$$

It can be easily checked (Problem 11.20) that $\partial \ell(\theta, X)/\partial \theta = g(\theta, X)$. This motivates the following definition of the Wilks likelihood ratio statistics based on a conservative and canonical estimating equation g.

Definition 11.15 *Suppose that $g(\theta, X)$ is a canonical and conservative estimating equation. Let $\ell(\theta, X)$ be the line integral (11.53). Let $\hat{\theta}$ be a solution to $E_n[g(\theta, X)] = 0$ and $\tilde{\theta}$ be a solution to $E_n[g_\lambda(\psi_0, \lambda, X)] = 0$. The Wilks's statistic for the estimating equation g is*

$$T_n(g) = 2nE_n[\ell(\hat{\theta}, X) - \ell(\tilde{\theta}, X)].$$

The next theorem shows that $T_n(g)$ is a QF-test with of the same asymptotic quadratic form as $W_n(g)$, $R_n(g)$, and $N_n(g)$. Of course, we should keep in mind that $T_n(g)$ requires g to be a conservative estimating equation, whereas the other tests do not.

Theorem 11.15 *Suppose Assumption 10.3 holds and*

1. *$g(\theta, X)$ is an unbiased, P_θ-square-integrable, canonical, and conservative estimating equation;*
2. *$f_\theta(X)$ and $g(\theta, X) f_\theta(X)$ satisfy $DUI^+(\theta, \mu)$ with $J_g(\theta)$ invertible;*
3. *the sequence $\{E_n[\partial g(\theta, X)/\partial \theta^T] : n \in \mathbb{N}\}$ is stochastically equicontinuous;*
4. *$\hat{\theta}$ is a consistent solution to $E_n[g(\theta, X)] = 0$ and $\tilde{\lambda}$ is a consistent solution to $E_n[g(\psi_0, \lambda, X)] = 0$.*

Then $T_n(g)$ is a QF test of the form given in Corollary 11.4. Consequently,

$$T_n(g) \xrightarrow[P_n(\delta)]{\mathcal{D}} \chi^2_r \left(\delta_\psi^T (I_g^{-1})^{-1}_{\psi\psi} \delta_\psi \right).$$

The proof is similar to that of Theorem 11.2 and is left as an exercise (Problem 11.23).

Problems

Many of the following problems require regularity assumptions that are too tedious to be stated completely. We therefore leave it to the readers to impose them as appropriate. In particular, these types of assumptions will be made without mentioning:

1. integrability: certain moments involved, such as means, variances, and third moments, are finite;
2. differentiability and DUI: certain functions of θ are differentiable to a required order, and when needed, the derivatives can be exchanged with integral over a random variable;
3. stochastic continuity: certain sequences of random functions of θ are stochastic equicontinuous.

It is usually obvious where these assumptions should be imposed.

11.1. The notion of a QF test can be extended to the case where Σ_U is singular. Suppose (U_n, S_n, L_n) satisfies ALAN with a singular Σ_U, where $U_n \in \mathbb{R}^r$, $S_n \in \mathbb{R}^p$, and $\Sigma_U \in \mathbb{R}^{p \times p}$. Suppose Σ_U^- is a reflexive generalized inverse of Σ_U; that is,

$$\Sigma_U \Sigma_U^- \Sigma_U = \Sigma_U, \quad \Sigma_U^- \Sigma_U \Sigma_U^- = \Sigma_U^-.$$

See, for example, Kollo and von Rosen (2005). Suppose, furthermore, Σ_U^- is a symmetric matrix. We define a corresponding QF test as any statistic T_n that satisfies

$$T_n \stackrel{Q_n}{=} U_n^T \Sigma_U^- U_n + o_P(1).$$

Show that

$$T_n \xrightarrow[P_n(\delta)]{\mathcal{D}} \chi^2_d(\delta^T \Sigma_{SU} \Sigma_U^- \Sigma_{US} \delta),$$

where d is the rank of Σ_U. (Here, we have used a more general definition of inverse than the Moore-Penrose inverse to accommodate Pearson's test, which is discussed in the next problem).

11.2. This problem concerns Pearson's Chi-square test. Suppose $X_1, \ldots, X_n$ are i.i.d. multinomial variables with k categories and probabilities $(p_1, \ldots, p_k)$. Let $\theta = (p_1, \ldots, p_k)^T$.

1. Show that the MLE of θ is

$$\hat{\theta} = (n_1/n, \ldots, n_k/n)^T,$$

   where $n_i = \sum_{\ell=1}^n I(X_\ell = i)$.
2. Since the MLE is obtained by maximizing $E_n[\ell(\theta, X)]$ subject to $h(\theta) = 1_p^T \theta = 1$, the score function is to be derived from

$$\begin{cases} E_n \ell(\theta, X) - [\partial h(\theta)/\partial \theta] \lambda = 0 \\ h(\theta) = 1 \end{cases}$$

   From this deduce that

$$S_n = \sqrt{n}\, Q \,\mathrm{diag}(\theta)^{-1} E_n(Z),$$

   where $Q = I_p - 1_p 1_p^T$, $Z = (Z_1, \ldots, Z_k)^T$ and $Z_i = I(X = i)$.
3. Let $U_n = \sqrt{n}(\hat{\theta} - \theta_0)$. Show that (U_n, S_n, L_n) satisfies ALAN with

$$\Sigma_U = \mathrm{diag}(\theta) - \theta\theta^T, \quad \Sigma_{US} = Q.$$

4. Show that $\Sigma_U^- = \mathrm{diag}(\theta)^{-1} - 1_k 1_k^T$ is a reflexive and symmetric generalized inverse of Σ_U.
5. In the setting of Problem 11.2, show that Wald's statistic $W_n = n(\hat{\theta} - \theta_0)^T \Sigma_U^-(\hat{\theta})(\hat{\theta} - \theta_0)$ in this case reduces to the Pearson's chi-square test

$$W_n = \sum_{i=1}^k \frac{(n_i - np_i)^2}{np_i}. \tag{11.54}$$

6. Show that $W_n \xrightarrow[P_n(\delta)]{\mathcal{D}} \chi^2_{k-1}(\delta^T Q \mathrm{diag}(\theta)^{-1} Q \delta)$.

11.3. In the setting of Problem 11.2, prove the following statements.

1. The Fisher information is $I(\theta) = Q\mathrm{diag}(\theta)^{-1} Q$.
2. A reflexive generalize inverse of $I(\theta)$ is

$$I^-(\theta) = \mathrm{diag}(\theta) - \theta\theta^T.$$

3. $I^-(\theta) = QI^-(\theta) = I^-(\theta)Q = QI^-(\theta)Q$.
4. Rao's score test, $R_n = nS_n^T I^-(\theta_0)S_n$, also takes the form of Pearson's chi-square test in (11.54).

11.4. In the setting of Problem 11.2, show that the Wilks's likelihood ratio test takes the form

$$T_n = \sum_{i=1}^{k} n_i \log\left(\frac{n_i}{np_i}\right).$$

Derive the asymptotic distribution of T_n under $P_n(\delta)$.

11.5. Suppose $X_1, \ldots, X_n$ are an i.i.d. sample from $N(\theta, \theta)$, where $\theta > 0$. Let $\hat{\theta}$ be the maximum likelihood estimate of the true parameter θ_0.

1. Let T_n be the Wilks's test statistic for testing $H_0 : \theta = 1$. If $n = 100$, find the local asymptotic alternative distribution of T_n at $\theta = 1.1$.
2. Let $U_n = n(\overline{X} - \theta_0)^2/\theta_0$. Derive the asymptotic null distribution (under Q_n) and local alternative distribution of W_n (under $P_n(\delta)$).
3. Derive the Pitman efficiency of U_n and show that it is no greater than 1.
4. Let $M_n = n^{-1}\sum_{i=1}^n (X_i - \overline{X})^2$, and $V_n = n(M_n - \theta_0)^2/(2\theta_0^2)$. Derive the asymptotic null and local alternative distribution of V_n for testing $H_0 : \theta = \theta_0$.
5. Derive the Pitman efficiency of V_n.
6. For which region of θ is V_n more efficient than U_n?

11.6. Let $X_1, \ldots, X_n$ be an i.i.d. sample from probability density function f_θ, which is supported on $(-\infty, \infty)$ and dominated by the Lebesgue measure. For $0 < p < 1$, let $\tau_p(\theta)$ be the pth quantile of X. That is,

$$\int_{-\infty}^{\tau_p(\theta)} f_\theta(x) d\lambda(x) = p.$$

Let T_n be the sample pth quantile (the exact definition of this statistic is not important for our purpose). It is known that T_n has expansion (Bahadur, 1966)

$$T_n \stackrel{\theta}{=} \tau_p(\theta) + cE_n[I(X \le \tau_p(\theta)) - p] + o_P(1/\sqrt{n}).$$

It is also known that T_n is a regular estimate.

1. Use the fact that T_n is regular to derive the value of the constant c.
2. Find the asymptotic null and local alternative distributions of $\sqrt{n}(T_n - \tau_p(\theta))$.
3. Construct a QF test based on T_n for testing $H_0 : \theta = \theta_0$, and derive its asymptotic local alternative distribution.

4. Suppose that the distribution of X_i is $N(\theta, 1)$, and that the p in τ_p is $1/2$. Find the asymptotic local alternative distribution of the QF test constructed in part 2, for testing the hypothesis $H_0 : \theta = 0$, and derive Pitman's efficiency.

11.7. Suppose that $X_1, \ldots, X_n$ are i.i.d. $N(\theta_1, \theta_2)$ where $\theta_1 \in \mathbb{R}$ and $\theta_2 > 0$. Let $\theta = (\theta_1, \theta_2)^T$. We are interested in testing the null hypothesis

$$H_0 : \theta_1 = \theta_2.$$

Let $\hat{\theta}$ be the global MLE and $\tilde{\theta}$ be the MLE under H_0. Let T_n be the Wilks's test statistic:

$$T_n = 2\sum_{i=1}^{n}[\ell(\hat{\theta}, X_i) - \ell(\tilde{\theta}, X_i)].$$

Derive the asymptotic distribution of T_n under $P_n(\delta)$.

11.8. Let $X_1, \ldots, X_n$ be i.i.d. with p.d.f. $f_\theta(x)$, where $\theta \in \Theta \subseteq \mathbb{R}^p$ is a parameter. We are interested in testing the null hypothesis $H_0 : \theta = \theta_0$. Let $\hat{\theta}$ be a consistent maximum likelihood estimate and define

$$R(\theta_0, \theta) = \sum_{i=1}^{n}\left[\frac{f_\theta(X_i)}{f_{\theta_0}(X_i)} - \frac{f_{\theta_0}(X_i)}{f_\theta(X_i)}\right].$$

Show that the statistic $R(\theta_0, \hat{\theta})$ is a QF test, and derive its asymptotic distribution under $P_n(\delta)$. Is this an asymptotically efficient test? (See Li, 1993).

11.9. Let $X_1, \ldots, X_n$ be independent copies of X, where X is a random variable with finite fourth moments. Assume that X has distribution P_θ, where $\theta \in \Theta \subseteq \mathbb{R}^p$ (here, the dimension of θ is irrelevant). Let θ_0 be the true parameter value of θ. Let $\mu_1(\theta), \mu_2(\theta), \mu_3(\theta), \mu_4(\theta)$ denote the first four moments of X. Let $\sigma^2(\theta)$ denote the variance of X. Let $\rho(\theta)$ be the signal-to-noise ratio defined as follows

$$\rho(\theta) = \frac{\mu_1(\theta)}{\sigma(\theta)}.$$

Estimate $\rho(\theta_0)$ by $\hat{\rho} = \hat{\mu}_1/\hat{\sigma}$, where $\hat{\mu}_1$, $\hat{\sigma}$ are the sample mean and sample standard deviation, respectively.

1. Derive the asymptotic distribution of $\sqrt{n}(\hat{\rho} - \rho(\theta_0))$ under $P_n(\delta)$.
2. Based on part 1 construct a QF test T_n, and derive its asymptotic distribution under $P_n(\delta)$.

11.10. Let $X_1, ..., X_n$ be an i.i.d. sample from a distribution whose density is $f_\theta(x)$, with $\theta = (\psi^T, \lambda^T)^T \in \mathbb{R}^p$, $\psi \in \mathbb{R}^r$, $\lambda \in \mathbb{R}^s$. For testing the null hypothesis $H_0 : \psi = \psi_0$, let $\hat{\theta} = (\hat{\psi}^T, \hat{\lambda}^T)^T$ be a consistent solution to $E_n[s(\theta, X)] = 0$, and $\tilde{\lambda}$ a consistent solution to the estimating equation $E_n[s_\lambda(\psi_0, \lambda, X)] = 0$. Let $S_{n,\psi}(\theta) = \sqrt{n}E_n[s_\psi(\theta, X)]$ be the rescaled score for ψ. Let

$$B_n = \sqrt{n}(\hat{\psi} - \psi_0)^T S_{n,\psi}(\psi_0, \tilde{\lambda}, X).$$

This is a hybrid between the Wald's and Rao's score statistic similar to the statistic proposed by Li and Lindsay (1996).

1. Show that B_n is a QF test, derive its asymptotic quadratic form, and its asymptotic distribution under $P_n(\delta)$.
2. Show that B_n can be rewritten as the more compact form

$$B_n = \sqrt{n}(\hat{\theta} - \tilde{\theta})^T S_n(\tilde{\theta}, X),$$

where $\tilde{\theta} = (\psi_0^T, \tilde{\lambda}^T)^T$.

11.11. Under the setting of Problem 11.10, let $S_{n,\psi\cdot\lambda}(\theta, X)$ be the rescaled efficient score $\sqrt{n}E_n[s_{\psi\cdot\lambda}(\theta, X)]$. Let $\tilde{\lambda}$ be any $\sqrt{n}$-consistent estimate of λ_0. Derive the asymptotic distribution of

$$\sqrt{n}(\hat{\psi} - \psi_0)^T S_{n,\psi\cdot\lambda}(\tilde{\theta}; X)$$

under $P_n(\delta)$.

11.12. Under the setting of Problem 11.10, for testing the null hypothesis $H_0 : h(\theta) = 0$, let $\hat{\theta}$ be a consistent solution to $E_n[s(\theta, X)] = 0$ and $\tilde{\theta}$ be a consistent solution to $E_n[s(\theta, X)] = 0$ subject to $h(\theta) = 0$. Show that

$$\sqrt{n}(\hat{\theta} - \tilde{\theta})^T S_n(\tilde{\theta}, X)$$

is a regular QF test. Derive its asymptotic distribution under $P_n(\delta)$. Is this test an Asymptotically Efficient test?

11.13. Suppose $X_1, \ldots, X_n$ are i.i.d. random variables with density $f_\theta(x)$, where $\theta \in \Theta \subseteq \mathbb{R}^p$, $\theta = (\psi^T, \lambda^T)^T$, $\psi \in \mathbb{R}^r$ and $\lambda \in \mathbb{R}^s$, $r + s = p$. We are interested in testing the null hypothesis $H_0 : \psi = \psi_0$. Consider the following procedure for estimating ψ_0. First, estimate λ_0 by $\tilde{\lambda}$, which is the solution to estimating equation $E_n[s_\lambda(\psi_0, \lambda)] = 0$. Second, estimate ψ_0 by $\tilde{\psi}$, which is the solution in ψ to the estimating equation $E_n[s_\psi(\psi, \tilde{\lambda})] = 0$.

1. Show that $\tilde{\psi}$ is not regular at θ_0 unless $I_{\psi\lambda} = O_{r\times s}$.
2. Compare the asymptotic variances of $\sqrt{n}(\tilde{\psi} - \psi)$ and $\sqrt{n}(\hat{\psi} - \psi)$, where $\hat{\psi}$ is the first r components of the MLE $\hat{\theta}$.
3. Construct a QF test based on the asymptotic distribution of $\sqrt{n}(\tilde{\psi} - \psi)$, and show that this test is a regular test despite the fact that $\tilde{\psi}$ in not regular at θ_0. Is this test asymptotically efficient?

11.14. Under the setting of Problem 11.13, suppose $\hat{\theta}$ is the unconstrained MLE, and $\hat{\psi}$ is the first r components of $\hat{\theta}$.

1. Derive the asymptotic linear form of $\sqrt{n}E_n[s_\psi(\psi_0, \hat{\lambda}; X)]$, and construct a QF test based this form.
2. Derive the asymptotic distribution of this QF test under $P_n(\delta)$.
3. Is this QF test regular? Is it asymptotically efficient?

11.15. Under the setting of Problem 11.13, suppose $\hat{\theta}$ is the unconstrained MLE, and $\hat{\lambda}$ is the last s components of $\hat{\theta}$. Let $\tilde{\lambda}$ be the MLE for λ_0 under the null hypothesis $H_0 : \psi = \psi_0$. Assume $r = s$.

1. Derive the asymptotic linear form of $\sqrt{n}(\hat{\lambda} - \tilde{\lambda})$.
2. Assuming $(I^{-1})_{\lambda\lambda} - I_{\lambda\lambda}^{-1}$ is nonsingular, derive a QF test based on the asymptotic linear form obtained in part 1.
3. Show that the QF test in part 2 is regular if and only if $I_{\lambda\psi}$ is nonsingular.
4. Show that, when this QF test is regular, it is asymptotically efficient.

11.16. Under the setting of Problem 11.13, suppose $\hat{\theta}$ is the unconstrained MLE, and $\hat{\psi}$ is the first r components of $\hat{\theta}$. Let $\tilde{\lambda}$ be the MLE for λ_0 under the null hypothesis $H_0 : \psi = \psi_0$.

1. Derive the asymptotic linear form of $\sqrt{n}E_n[s_\psi(\hat{\psi}, \tilde{\lambda}, X)]$, and based on this result construct a QF test.
2. Give a necessary and sufficient condition for this QF test to be regular.
3. Show that, if this QF test is regular and $r = s$, then it is asymptotically efficient.

11.17. Let $X_1, \ldots, X_n$ be an i.i.d. sample from a distribution in $\{P_\theta : \theta \in \Theta\}$, where $\Theta \in \mathbb{R}^p$. Consider testing the implicit hypothesis $H_0 : h(\theta_0) = 0$ versus $H_1 : h(\theta_0) \neq 0$ where h is a mapping from Θ to $\mathbb{R}^r$. Let $\hat{\theta}$ be the global MLE and $\tilde{\theta}$ be the MLE subject to the constraint $h(\theta) = 0$. Suppose the conditions in Theorem 11.8 are satisfied.

1. Derive the asymptotic linear form of $\sqrt{n}(\hat{\theta} - \tilde{\theta})$ under θ_0.
2. Based on part 1 construct a QF test, and derive its asymptotic distribution under $P_n(\delta)$.
3. Is this QF test asymptotically efficient?

11.18. Prove the implication in (11.20).

11.19. In the special case where $\theta = (\psi^T, \lambda^T)^T$ and $h(\theta) = \psi$, prove the following statements:

1. the first equality in (11.19) reduces to

$$E_n[s_\psi(\psi_0, \tilde{\lambda}, X)] = E_n[s_{\psi\cdot\lambda}(\theta_0, X)] + o_P(n^{-1/2}).$$

2. the second equality in (11.19) reduces to

$$\tilde{\lambda} = \lambda_0 + I_{\lambda\lambda}^{-1} E_n[s_\lambda(\theta_0, X)] + o_P(n^{-1/2})$$

3. the efficient information $I_{(H)}$ in Definition 11.9 reduces to $I_{\psi\cdot\lambda}$.

4. the efficient score $s_{(H)}$ in Definition 11.9 reduces to $s_{\psi\cdot\lambda}$.

11.20. Let $\ell(\theta, X)$ be the function defined in (11.53). Show that

$$\partial\ell(\theta, X)/\partial\theta = g(\theta, X).$$

11.21. Suppose that $(X_1, Y_1), \ldots, (X_n, Y_n)$ are i.i.d. random vectors with an unknown p.d.f. $f_\theta(x, y)$, where $\theta \in \Theta \in \mathbb{R}^p$. Suppose the parametric forms of the conditional mean and variance are given:

$$E_\theta(Y|X) = \mu_\theta(X), \quad \text{var}_\theta(Y|X) = V(\mu_\theta(X)),$$

for some known functions $\mu(\cdot)$ and $V(\cdot)$. Consider the class of estimating equations of the form $a_\theta(X)(Y - \mu_\theta(X))$. Recall from Section 9.2 that the optimal estimating function among this class is

$$g^*(\theta, X, Y) = \frac{\partial\mu_\theta(X)}{\partial\theta} \times \frac{Y - \mu_\theta(X)}{V(\mu_\theta(X))}.$$

This is a special case of the optimal estimating equation in Section 9.2 because the function V here is assume to depend on $\theta^T X$ through μ. Suppose $\theta = (\psi^T, \lambda^T)^T$, where $\psi \in \mathbb{R}^r$ is the parameter of interest and $\lambda \in \mathbb{R}^r$ is the nuisance parameter. We are interested in testing $H_0 : \psi = \psi_0$. Assume regularity conditions such as differentiability, integrability and stochastic continuity as appropriate.

1. Let $\hat{\theta}$ is a consistent solution to $E_n[g^*(\theta, X, Y)] = 0$, and $\tilde{\lambda}$ a consistent solution to $E_n[g^*_\lambda(\psi_0, \lambda, X)] = 0$. Find the asymptotic linear forms of $\sqrt{n}(\hat{\theta} - \theta_0)$ and $\sqrt{n}(\tilde{\theta} - \theta_0)$.
2. For a fixed vector $a \in \Theta$, let $\ell(\mu, Y)$ be the function

$$\ell(\mu, Y) = \int_a^\mu \frac{Y - \nu}{V(\nu)} d\nu,$$

and let

$$T_n = 2n\{E_n[\ell(\mu_{\hat{\theta}}(X), Y)] - E_n[\ell(\mu_{\tilde{\theta}}(X), Y)]\},$$

where $\tilde{\theta} = (\psi_0^T, \tilde{\lambda})^T$. Show that T_n is a QF test and derive its asymptotic distribution under $P_n(\delta)$.

11.22. Under the setting of Problem 11.21, assume that

$$\text{var}_\theta(Y|X) = V(\theta^T X).$$

The difference from Problem 11.21 is that, here, we do not assume $V(\cdot)$ depends on $\theta^T X$ through $\mu(\theta^T X)$. In this case, the quasi score function

$$g^*(\theta, X, Y) = \frac{\partial\mu^T(\theta^T X)}{\partial\theta} \frac{Y - \mu(\theta^T X)}{V(\theta^T X)}$$

need not be a conservative estimating equation. For testing $H_0 : \psi = \psi_0$, let $\hat{\theta}$ be a consistent solution to $E_n[g^*(\theta, X, Y)] = 0$, and $\tilde{\lambda}$ a consistent solution to $E_n[g^*(\psi_0, \lambda, X, Y)] = 0$. Let

$$c(\theta, \eta) = \frac{\mu(\eta^T X) - \mu(\theta^T X)}{V(\theta^T X)}[Y - \mu(\theta^T X)].$$

and

$$C(\theta, \eta) = nE_n[c(\theta, \eta) - c(\eta, \theta)].$$

Show that $C(\tilde{\theta}, \hat{\theta})$ is a QF test, and derive its asymptotic distribution under $P_n(\delta)$.

11.23. Prove Theorem 11.15.

11.24. Suppose that $X_1, \ldots, X_n$ are i.i.d. with p.d.f. $f_\theta(x)$, where $\theta \in \Theta \subseteq \mathbb{R}^p$. Let $g : \Theta \times \Omega_X \to \mathbb{R}^p$ be an unbiased and P_θ-square-integrable function. For an integer $1 \leq r < p$, let $h : \Theta \mapsto \mathbb{R}^r$ be a differentiable function. We are interested in testing the hypothesis

$$H_0 : h(\theta) = 0$$

based on the estimating equation $g(\theta, x)$. For convenience, and without loss of generality, assume that g is in its canonical form. Let $H(\theta) = \partial h^T(\theta)/\partial\theta$, and suppose it has full column rank for all $\theta \in \Theta$. Let $I_g(\theta)$ be the information contained in $g(\theta, X)$ and assume that it is nonsingular for all $\theta \in \Theta$. Let

$$\begin{aligned} I_{(H),g} &= [H^T(\theta) I_g^{-1}(\theta) H(\theta)]^{-1} \\ g_{(H)}(\theta, X) &= I_{(H),g}(\theta) H^T(\theta) I^{-1}(\theta) g(\theta, X). \end{aligned}$$

Let $\hat{\theta}$ be a consistent solution to $E_n[g(\theta, X)] = 0$ and $\tilde{\theta}$ a consistent solution to $E_n[g(\theta, X)] = 0$ subject to $h(\theta) = 0$. Let

$$\begin{aligned} W_n(g) &= nh^T(\hat{\theta}) I_{(H),g}(\hat{\theta}) h(\hat{\theta}) \\ R_n(g) &= nE_n[g^T(\tilde{\theta}, X)] I_g^{-1}(\tilde{\theta}) E_n[g(\tilde{\theta}, X)]. \end{aligned}$$

Show that $W_n(g)$ and $R_n(g)$ are QF test, derive their asymptotic quadratic forms, and their asymptotic distributions under $P_n(\delta)$.

11.25. Under the setting of Problem 11.10, suppose we want to test the hypothesis $H_0 : h(\theta) = 0$ based on an estimating equation $g(\theta, X)$. Suppose g is in its canonical form. Let $\hat{\theta}$ be a consistent solution to $E_n[g(\theta, X)] = 0$ and $\tilde{\theta}$ be a consistent solution to $E_n[g(\theta, X)] = 0$ subject to $h(\theta) = 0$. Show that

$$n(\hat{\theta} - \tilde{\theta})^T E_n[g(\tilde{\theta}, X)]$$

is a regular QF test for testing $H_0 : h(\theta) = 0$, and derive its asymptotic distribution under $P_n(\delta)$.

References

Bahadur, R. R. (1966). A note on quantiles in large samples. *The Annals of Mathematical Statistics.* **37**, 37 577–581.

Bera, A. K. and Bilias, Y. (2001). Rao's score, Neyman's $C(\alpha)$ and Silvey's LM tests: an essay on historical developments and some new results. *Journal of Statistical Planning and Inference* **97**, 9–44.

Boos, D. D. (1992). On Generalized Score Tests. *The American Statistician.* **46**, 327–333.

Engle, R. F. (1984). Wald, likelihood ratio, and Lagrange multiplier tests in econometrics. *Handbook of Econometrics*, **2**, 775–826.

Hall, W. J. and Mathiason, D. J. (1990). On large-sample estimation and testing in parametric models. *Int. Statist. Rev.*, **58**, 77–97.

Horn, R. A. and Johnson, C. R. (1985). *Matrix Analysis.* Cambridge University Press.

Kocherlakota, S. and Kocherlakota, K. (1991). Neyman's $C(\alpha)$ test and Rao's efficient score test for composite hypotheses. *Statistics and Probability Letters*, **11**, 491–493.

Kollo, T. and von Rosen, D. (2005). *Advanced Multivariate Statistics with Matrices.* Springer.

Li, B. (1993). A deviance function for the quasi-likelihood method. *Biometrika*, **80**, 741–753.

Li, B. and Lindsay, B. (1996). Chi-square tests for Generalized Estimating Equations with possibly misspecified weights. *Scandinavian Journal of Statistics*, **23**, 489–509.

Li, B. and McCullagh, P. (1994). Potential functions and conservative estimating functions. *The Annals of Statistics*, **22**, 340–356.

Mathew, T. and Nordstrom, K. (1997). Inequalities for the probability content of a rotated ellipse and related stochastic domination results. *The Annals of Applied Probability*, **7**, 1106–1117.

McCullagh, P. (1983). Quasi-likelihood functions. *The Annals of Statistics*, **11**, 59–67.

Neother, G. E. (1950). Asymptotic properties of the Wald-Wolfowitz test of randomness. *Ann. Math. Statist.*, **21**, 231–246.

Neother, G. E. (1955). On a theorem by Pitman. *Ann. Math. Statist.*, **26**, 64–68.

Neyman, J. (1959). Optimal asymptotic test of composite statistical hypothesis. In: *Grenander, U. (Ed.), Probability and Statistics, the Harald Cramer Volume. Almqvist and Wiksell, Uppsala*, 213–234.

Pitman, E. J. G. (1948). Unpublished lecture notes. Columbia Univ.

Rao, C. R. (1948). Large sample tests of statistical hypotheses concerning several parameters with applications to problems of estimation. *Mathematical Proceedings of the Cambridge Philosophical Society*, **44**, 50–57.

Rao, C. R. (2001). *Linear Statistical Inference and Its Applications, Second Edition.* Wiley.

Rotnisky, A. and Jewel, N. P. (1990). Hypothesis testing of regression parameters in semiparametric generalized linear models for cluster correlated data. *Biometrika*, **77**, 485–497.

van Eeden, C. (1963). The relation between Pitman's asymptotic relative efficiency of two tests and the correlation coefficient between their test statistics. *The Annals of Mathematical Statistics.* **34**, 1442–1451.

Wald, A. (1943). Tests of statistical hypotheses concerning several parameters when the number of observations is large. *Trans. Am. Math. Soc.*, **54**, 462–482.

Wilks, S. S. (1938). The large-sample distribution of the likelihood ratio for testing composite hypotheses. *The Annals of Mathematical Statistics*, **9**, 60–62.

Index

B. Li and G. J. Babu, *A Graduate Course on Statistical Inference*,
Springer Texts in Statistics, https://doi.org/10.1007/978-1-4939-9761-9

GPSR Compliance
The European Union's (EU) General Product Safety Regulation (GPSR) is a set of rules that requires consumer products to be safe and our obligations to ensure this.

If you have any concerns about our products, you can contact us on

ProductSafety@springernature.com

In case Publisher is established outside the EU, the EU authorized representative is:

Springer Nature Customer Service Center GmbH
Europaplatz 3
69115 Heidelberg, Germany

www.ingramcontent.com/pod-product-compliance
Lightning Source LLC
Chambersburg PA
CBHW070827300126
38897CB00069B/32
9781493997596